Recent Advances in CROP PHYSIOLOGY

– Volume 1 –

The Editor

Dr. Amrit Lal Singh, Principal Scientist, Plant Physiology at the Directorate of Groundnut Research, Junagadh, Gujarat obtained his M.Sc. (Botany) from Banaras Hindu University, Varanasi and Ph.D. in Botany from Utkal University, Bhubaneswar. He worked on the biological nitrogen fixation by Azolla-Anabaena complex and blue green algal biofertilizers and demonstrated their use in rice. Dr. Singh, joined ICAR as an ARS Scientist in Plant Physiology, in January 1985 and since then working on the various aspects of crop physiology. He is the life member of more than a dozen of scientific societies and published more than 150 research papers, 25 review articles and book chapters, several books manual, and technology bulletins. He is Fellow of the Indian Society for Plant Physiology (ISPP), New Delhi and Indian Society for Oilseeds Research (ISOR), DOR, Hyderabad, and recipient of the J.J. Chinoy Gold medal award 2011 of the ISPP, New Delhi. Dr. Singh served as a Vice-President (2012) of the ISPP New Delhi, visited China, Japan, Tanzania, Turkey and USA and has worked as Referee/Reviewer of the Annals of Applied Biology, UK, Australian J. of Crop Science, Journal of Plant Nutrition, USA and Journal of Plant Nutrition and Soil Science, Germany.

Recent Advances in CROP PHYSIOLOGY

– Volume 1 –

— Editor —

Dr. Amrit Lal Singh

Principal Scientist,
Plant Physiology
Directorate of Groundnut Research,
Junagadh, Gujarat

2014

Daya Publishing House®

A Division of

Astral International Pvt. Ltd.

New Delhi – 110 002

Published by : **Daya Publishing House®**
A Division of
Astral International Pvt. Ltd.
– ISO 9001:2008 Certified Company –
4760-61/23, Ansari Road, Darya Ganj
New Delhi-110 002
Ph. 011-43549197, 23278134
E-mail: info@astralint.com
Website: www.astralint.com

Laser Typesetting : **Classic Computer Services**, Delhi - 110 035

Printed at : **Replika Press Pvt. Ltd.**

PRINTED IN INDIA

प्रो० स्वपन कुमार दत्ता
उपमहानिदेशक (फसल विज्ञान)
Prof. Swapan Kumar Datta
Deputy Director General
(Crop Science)

भारतीय कृषि अनुसंधान परिषद्
कृषि भवन, डा. राजेन्द्र प्रसाद मार्ग, नई दिल्ली - 110001

INDIAN COUNCIL OF AGRICULTURAL RESEARCH
KRISHI BHAWAN, DR. RAJENDRA PRASAD ROAD, NEW DELHI-110001

Foreword

Agriculture plays a pivotal role for food and nutritional security, and in alleviation of poverty. But, agriculture sector has been confronted with numerous challenges linked to food and energy crisis, climate change and natural resources. With beginning of 21st century, India is being recognized as the global power in the key economic sectors with high economic growth, but its slow growth in agriculture sector is major concerns for the future food and nutritional security, as one-third of the country's population lives below poverty line, and about 80% of our land mass is highly vulnerable to drought and floods. Indian agriculture, with only 9% of world's arable land, contribute 8% to global agricultural gross domestic product to support 18% of the world population. Also, India has nearly 8% of the world's biodiversity and many of these are crucial for livelihood security of poor and vulnerable population. Thus, acceleration of agricultural growth along with natural resources conservation is of supreme importance.

As the Global food demand is expected to be doubled by 2050, world must learn to produce more food with less land, less water and less labour by devising more efficient and profitable production systems that are resilient to climate change. Thus, more than ever, we need to produce more food with less land. Also looking to the demand of 2050 all the institutions and agricultural universities need to redesign their research and teaching programmes for harnessing power of science and bringing excellence in agricultural research and education that ensures food, nutrition and livelihood security for all.

The ICAR with the help of SAUs has brought green revolution in agriculture in India through its research and technology development in past and its subsequent efforts have enabled the country to increase the production of food grains by 4-fold, horticultural crops by 6-fold since 1950-51 which made a visible impact on the national food and nutritional security. Using cutting edge technologies, there is tremendous development in agriculture during the last two decades and it is hoped that with ingenuity, determination and innovative partnerships among everyone working in the agricultural sector, we can meet the food needs of 9 billion people by 2050 without irreparably harming our planet. However, all these informations are scattered and need to be compiled and circulated widely.

This series on *"Recent Advances in Crop Physiology"* is a timely effort in this direction, which will act as a reference for directly implementing the available technologies and to help the researchers for planning their future research programme.

Swapan Kumar Datta

Preface

"Food security exists when all people, at all times, have physical, social and economic access to sufficient safe and nutritious food that meets their dietary needs and food preferences for an active and healthy life."

Global food demand is expected to be doubled by 2050, while production environment and natural resources are shrinking and deteriorating. World cereal production has gone 2358 million tonnes (mt) during 2011-12 and is expected to be 2498 mt in 2013. Same time, world cereal utilization in 2013-14 is put at 2418 mt, and 2331 mt in 2012-13. The researchers say that, to feed the world in 2050, yields on maize, rice, wheat, and soybeans will have to rise by 60-110%. But the present projections show an increase of only 40-65%. The researchers also found that most rice and wheat had very low rates of increase in crop yields. In other places, the trajectories of population growth and food production are heading in different directions. About 90% of the world's rice is grown in Asia, on more than 200 million small scale farms (about 1 acre). To keep rice prices affordable as population increase, according to IRRI an additional 8-10 mt of rice will need to be produced worldwide every year. But the International Food Policy Research Institute estimates that by 2050 rice prices will increase some 35% because of yield losses due to climate change.

Malnutrition in form of under nutrition, micronutrient deficiencies and obesity imposes unacceptably high economic costs and improving nutrition requires a multisectoral approach that begins with food and agriculture. A total of 842 million people in 2011–13, or around one in eight people in the world, were estimated to be suffering from chronic hunger, regularly not getting enough food for an active life. The agriculture play its fundamental role in producing food and its processing, storage, transport and consumption contribute to the eradication of malnutrition. Because of better agriculture the total number of undernourished during 2013 has fallen by 17 percent since 1990–92. Agricultural policies and research must continue to support productivity growth for staple foods with greater attention to nutrient-dense foods and more sustainable production systems. Traditional and modern supply chains can enhance the availability of a variety of nutritious foods and reduce nutrient waste and losses. The rice, is the central to existence in many nations, feeds the world, and provides more calories to humans than any other food, and more than a billion people depend on rice cultivation for their livelihoods. Changes in the price and availability of rice have caused social unrest in developing countries and in 2008, when rice prices tripled, as 100 million people were pushed into poverty.

A recent wakeup call report from the Intergovernmental Panel on Climate Change (IPCC) predicts that global food production due to climate change will decline 2% per decade for the remainder of this century compared to food production without climate change even as food demand increases 14% per decade. In 2007, the panel was hopeful that gains in agricultural productivity would more than make up for losses due to climate change. But later research revealed in greater detail the impacts of climate change on sensitive crops and raised questions about how much elevated carbon dioxide levels could increase productivity.

The organic material decays without oxygen, in water-logged rice paddies, soil microbes generate methane, a greenhouse gas with 25 times more warming potential than CO_2. In India, rice methane emission accounts for about 10% of the nation's total greenhouse gas (GHG) emissions. Also, nitrous oxide emissions from rice grown under dryer and aerated conditions, can be as significant as methane emissions which has about 300 times more warming potential than CO_2. It has not yet been estimated what percentage of nitrous oxide emissions in India, or for that matter other rice growing regions in Asia, come from rice cultivation.

If we are unable to double yields on existing cultivated lands, due to food insecurity pressure, we are likely to clear more land for agriculture leaving environmental concerns and efficiency measures a side. This will have a ripple effect, putting additional pressure on already stressed water resources and wildlife habitat, accelerates climate change. This cycle, left unchecked, can only end with farmers competing for increasingly scarce water and arable land in the face of ever more extreme weather – from floods to droughts – brought on by climate change.

These colliding trends indicate that the world must learn to produce more food with less land, less water and less labour by devising more efficient and profitable production systems that are resilient to climate change. Thus, more than ever, we need to produce more food with less land. Farmers must seek out production methods and crop rotations that will be highly productive and have a smaller impact on water quality and quantity, climate and habitat. To do this, we have the tools and technologies that reduce the need for inputs like fertilizer, pesticides and herbicides; innovative irrigation methods that reduce water demand; and methods that reduce greenhouse gas emissions. Using cutting edge technologies, there is tremendous development in agriculture and productivity during the last two decades and it is hoped that with ingenuity, determination and innovative partnerships among everyone working in the agricultural sector, we can meet the food needs of 9 billion people by 2050 without irreparably harming our planet on which we all depend.

However, all these informations are scattered and need to be compiled and circulated widely. This series on Recent Advances in Crop Physiology is an effort in this direction, which will act as a reference for directly implementing the technologies and also to help the researchers for planning their future research.

This book of *'Recent Advances in Crop Physiology'* encompasses 11 chapters written by the experts in the field describing production physiology, abiotic stresses, climatic change and metabolic products. The chapter one on 'Physiology of groundnut under water deficit stress' and chapter three on 'Sugarcane physiology under abiotic stresses', widey covers the physiological behaviour of these crop under abiotic stress and provide a guidelines how to manage these crops under stress conditions. Chapter two covers most of the aspects of 'Climate change and agriculture' and also discuss the pros and cons of the changing climate. Terminal heat stress is a major problem of wheat crop in many countries and there is an exclusive chapter on this aspect elucidating the physiological causes and management strategies to increase the production. As nitrous oxide in plants is an emerging area and major threat as green house gases, the chapter five is devoted on this.

Salicylic acid is emerging as an ameliorative agents for abiotic stresses and the chapter six on 'Salicylic acid: a key to regulate drought stress in chickpea' covers its role in drought management in chickpea. The medicinal plants have importance in daily life and a chapter on 'Production of secondary metabolites from medicinal plants' has been included. As heavy metal pollutants are becoming major hurdle in crop production, a chapter on 'phytoremediation of cadmium through Sorghum' is included to generate interest among researchers and environmental scientists. Mango malformation is a major problem of mango orchard, and chapter nine deals with it highlighting physiological aspects and its management strategies. Though a variety of edible and non-edible oils are available in the market, there is need for oils with certain composition looking to the health benefit and hence a chapter on 'Alteration of fatty acid composition and protein quality of major edible oils' are adding to the quality of the book where all the possible mechanism and methodologies are discussed. Finally, being a C_4 plant sugarcane has got very high yield potential, and a chapter on 'Sugarcane yield potentials and plateaus in tropical and subtropical conditions' are well composed by two renowned sugarcane scientists.

I would like to express my gratitude to all the stalwarts of agriculture and plant biology from various disciplines who has contributed in enhancing agricultural production. Thanks are also due to all the staffs of plant physiology at DGR Junagadh for their help in the various ways. Finally, I would like to express my sincere thanks to Mr. Prateek Mittal for coming forward to take up the responsibility of publishing the series and Mr. Anil mittal and the staff of Astral International (P) Ltd, New Delhi for their care and diligence in producing the book timely.

Dr. Amrit Lal Singh

Contents

Chapter 1

Physiology of Groundnut under Water Deficit Stress

A.L. Singh, Nisha Goswami, R.N. Nakar,
K.A. Kalariya and K. Chakraborty

Directorate of Groundnut Research, P.B. 5, Junagadh – 362 001, Gujarat
E-mail: alsingh@nrcg.res.in, alsingh16@gmail.com

Contents

Abbreviations

ABA: Abscisic Acid
ACP: Acid Phosphatase
ADH: Alcohol Dehydrogenase
AE : PE: Actual Evapotranspiration: Potential Evapotranspiration
AP: Apparent leaf photosynthetic rate
AsA: Ascorbic Acid
ASM: Available Soil Moisture
CaM: Calmodulin
CAT: Catalase
CATD: Canopy temperatures relative to Air
CCC: Chlormequat
CER: Carbon Dioxide Exchange Rate
CGR: Crop Growth Rates
CID: Carbon Isotope Discrimination
CMI: Cell Membrane Integrity
CMS: Cell Membrane Stability
CPE: Cumulative Pan Evaporation
CWSI: Crop Water Stress Index
DAE: Days after Emergence
DAP: Days after Planting
DAS: Days after Sowing
D D-RT-PCR: Differential Display Reverse Transcription-Polymerase Chain Reaction
DM: Dry Matter
DPD: diffusion pressure deficit
DR: Diffusive Resistance

DREB: Dehydration Responsive Element Binding

DS: Drought Stressed

EFV: Extraction Front Velocity

EMS: Early Moisture Stress

ESD: End Season Drought

EST: Esterase

EST: Expressed Sequence Tag

ET/Et: Evapotranspiration

EWL: Epicuticular Wax Load

FC: Field Capacity

FDPP: Fully Developed Pods Plant

FDPR: Fully Developed Pod Ratio

FW: Fresh Weight

FYM: Farm Yard Manure

GCA: General Combining Ability

GM: Genetically Modified

HI: Harvest Index

IW:CPE: Irrigation Water: Cumulative Pan Evaporation Ratios

Kc: Crop Coefficients

ky: Yield Response Factors

LA: Leaf Area

LAI: Leaf Area Index

LE: Latent Heat Flux

IPAR: Intercepted Photosynthetic Photon Flux Density or Photo-synthetically Active Radiation

IUE: Irrigation-use efficiencies

LEA: Late Embryogenesis-Abundant

LER: Land Equivalent Ratio

LIR: Light Interception Rate

LWUER: Land water-use equivalency ratio

LWP: Leaf Water Potential

MC: Moisture Content

MDA: Maloni dialdehyde

MDH: Malate Dehydrogenase

MLT: Multi-Location Trial

MSD: Mid-Season Drought

NDVI: Normalized Difference Vegetation Index

NRA : Nitrate Reductase Activity

O:L ratio : Oleic : Linoleic Ratio

PAR: Photosynthetically Active Radiation

PF : Partitioning Factor

PDB : Pee Dee Beleminite

PEG : Polyethylene Glycol

PGR: Pod Growth Rate

PMA : Phenylmercury Acetate

POD : Peroxidase

PP 333 : Paclobutrazol

PWR : Pod:Shoot Wt. Ratio

QTL : Quantitative Trait Loci

R/S ratio : Root/Shoot Ratio

RDF : Recommended dose of Fertilizer

RILs : Recombinant Inbred Lines

RLD : Root Length Density

RPMP : Relative Plasma Membrane Permeability

RSD : Relative Saturation Deficit

RSW : Reduced Soil Water Supply

Rubisco: Ribulose-1, 5-bisphosphate carboxylase-oxygenase

RUE : Radiation Use Efficiency

RWC : Relative Water Content

SAVI : Soil Adjusted Vegetation Index

SC : Stomatal Conductance

SCA : Specific Combining Ability

SCMR : SPAD Chlorophyll Meter Readings

SD : Saturation Deficit

SDD : Stress Degree Days

SLA : Specific Leaf Area

SMK : Sound Mature Kernels

SMS : Soil Moisture Stress

SMT : Soil Moisture Tensions

SMW : Standard Meteorological Week

SOD : Superoxide Dismutase

SOP : Sulphate of Potash

MOP : Muriate of Potash

SPAD : Soil Plant Analytical Development

SPI : Standardized Precipitation Index
SRWC: Soil Relative Water Content
SSR : Simple Sequence Repeat
T : Transpired Water
T_b: Base Temperature
TCPR: Total Crop Performance Ratio
TE : Transpiration Efficiency
TLWUER: Total Land Water Use Equivalency Ratio
TP : Turgor Potential
TSWV : Tomato Spotted Wilt Virus
TWP : Temporary Wilting Point
V_a: Apparent Sap Velocity
VPD : Vapour Pressure Deficit
VPDla: Leaf to Air Vapour Pressure Deficit
WCU: Water Consumptive Use
W_D: Water Deficit
WRC : Water Retention Capacity
WRSI : Water Requirement Satisfaction Index
WSD : Water Saturation Deficit
WT : Wild Type
WU : Water Use
WUE : Water Use Efficiency
WUEc: Water Use Efficiency Corrected
WW : Well watered

1. Introduction

The groundnut (*Arachis hypogaea* L), though native of South America, due to its wide adaptability the cultivation of this important food legume crop has been spread on almost all soils in the tropical and subtropical countries and now grown in about 128 countries in different agro-climatic zones mainly in semi-arid regions between latitudes 40°S and 40°N. However, its cultivation is settled in southern, eastern and south-eastern part of Asia, western Africa and northern and south America owing to favorable soil and climate and presently it is cultivated on 11.5 m ha land in Asian, 11.5 m ha in African and 1.1 m ha in American countries. Groundnut requires warm growing season with well distributed rainfall in the range of 500-1000 mm and on large scale it is mainly grown in semi-arid and arid regions of India, China, Nigeria, USA, Myanmar, Indonesia, Sudan, Senegal, Argentina and Vietnam, Ghana, Chad, Congo Republic, Mali, Guinea, Niger, Argentina, Brazil, Tanzania, Burkino Faso, and Malawi. More than 85 per cent of the world groundnut production come from food deficit countries with an average productivity of about 1500 kg ha^{-1} and having more than 90 per cent of the groundnut growing areas of the world.

The drought is regarded as the most damaging abiotic stress, in sustainable crop production. Groundnut is grown under a wide range of environments where frequent drought is one of the limiting factors adversely affecting its productivity in rainfed area. Groundnut production, fluctuates considerably as a result of rainfall variability, due to its underground fruiting habit and low moisture availability during drought. There is wide range of groundnut productivity varying from about 500 kg ha^{-1} (poor) in Angola and Mozambique (extremely low about 300 kg ha^{-1}), Madagasker, Namibia, Niger, Uruguay and Zimbabwe, about 3000 kg ha^{-1} (high) in China, Egypt, Syrian Arab Republic, about 4000 in USA, Malaysia, Saudi Arabia, Palestine and Nicaragua, as much as 6400 kg ha^{-1} (very high) in Israel and extremely high (> 12000 kg ha^{-1}) in Cyprus (FAO, 2012). However, the world average yield is around 1600 kg ha^{-1} and about 70 per cent of the world groundnut production occurs in the semi-arid to arid tropics where the average yield is still around 1000 kg ha^{-1}.

Presently, India has the largest groundnut area of about 6 million hectare (24 per cent of the world), producing about 8 million tonne (mt) of pod accounting for only 20 per cent of the world groundnut production, but China with only 18 per cent area contributes 39 per cent of the world production due to better drought and nutrient management practices. Though the average groundnut yield in India is around 1400 kg ha^{-1}, combination of improved genotypes and best agronomic practices recorded more than 6000 kg ha^{-1} pod yield frequently and occasionally upto 8000 kg ha^{-1} indicating that there is tremendous scope to increase the yield through understanding its physiology and water relation (Singh 2004, 2011; Singh *et al.*, 2013).

Drought, considered to be the most complex but least understood of all natural hazards, is an insidious hazard of nature and large historical datasets are required to study these involving complex interrelationships between climatological and meteorological data. Rainfall is an important meteorological parameter, the amount and distribution influence the type of vegetation in a region. Characterization of agricultural drought is essential before undertaking a yield improvement programme in any crop and a simplified model combining ET and water balance concepts with basic data on plant responses to drought is applicable for diagnosing drought types in groundnut. Accordingly the physiological studies and drought tolerance of this crop started in 1980s and by now ample of studies have been conducted. However, due to underground fruiting, indeterminate growth habit and different botanical types still certain aspects of physiology of this crop, especially moisture requirement of pod zone are not very clear. The research is towards the improvement of the performance of the crop and genotypes under varying degrees of stress at various physiological stages of crop growth.

The introduction of an improved genotypes into new region is largely determined by the temperature and phenology and knowledge of crop physiology under water-deficit stress is important for achieving optimal yield under limited water availability and can be used to specify the most appropriate rate and time of specific developmental process to maximize yield. In this chapter an attempt was made to synthesize the impact of water deficit stress on the various component of groundnut physiology and management practices to grow high yielding varieties to increase the productivity.

2. Defining the Drought Stress and various Parameters

There is no universally accepted definition of drought, it is a meteorological term and commonly defined as a period without significant rainfall denoting scarcity of water in a region. The irrigation commission of India defines drought as a situation occurring in any area where the annual rainfall is less than 75 per cent of normal rainfall. Prolonged deficiencies of soil moisture adversely affect crop growth indicating incidence of agricultural drought. It is the result of imbalance between soil moisture and evapo-transpiration needs of an area over a fairly long period so as to cause damage to standing crops and to reduce the yields.

2.1 Soil Type, Climate and Rainfall

Important causes for agricultural drought are inadequate precipitation, erratic distribution, long dry spells in the monsoon, late onset and early withdrawal of monsoon along with lack of proper soil and crop management. Drought is the major abiotic constraint affecting groundnut productivity and quality worldwide. Groundnut plants are drought tolerant because of deep rooting and a water supply-related flexibility in time of flowering and fruiting.

The commercial groundnut cultivation is mainly in Asian (47 per cent of the world groundnut area contributing 62 per cent of the total world production), African (46 per cent area, 28 per cent production) and American (4.8 per cent area and 8 per cent production) countries due to suitable environment and photoperiod matching to the growing season. Though grown in limited area, the productivity of Cyprus (<100 ha area) is highest (>12000 kg ha^{-1}) followed by Israel (2600 ha and 6440 kg ha^{-1}) in the word mainly due to favourable season and high management. On the other hand the productivity of many African countries, are still around 400 kg ha^{-1} mainly due to scanty rainfall (FAO, 2012).

In India generally the groundnut is grown as rainfed crop during rainy season (Kharif) with one or two protective irrigation as it encounter drought and also during Rabi, summer and spring seasons as a irrigated crop. The groundnut, is grown in about 270 districts of India, mostly as rainfed dry lands crop on well drained soils, under vagaries of the weather and only about 25 per cent of its area is irrigated. Presently, Gujarat (30 per cent total area and 36-40 per cent of production), Andhra Pradesh (28 per cent area and 20-28 per cent production), Tamil Nadu (7 per cent area and 11 per cent of production), Karnataka (14.5 per cent area and 10 per cent of production), Rajasthan (6 per cent area and 8.2 per cent of production) and Maharashtra (6.1 per cent area and 5.5 per cent of production) are the main groundnut growing states. The other states growing groundnut are Madhya Pradesh, Orissa, Uttar Pradesh and West Bengal. The occurrence of drought is a major cause of low groundnut productivity.

Table 1.1: Global Scenario of Groundnut Production.

Country	Area (lakh ha)	Production (lakh tonne)	Yield (kg/ha)
China	44.04	149.38	3390
India	55.20	61.06	1105
Nigeria	25.36	28.26	1119
USA	5.17	19.67	3796
Myanmar	8.40	13.36	1590
Senegal	10.30	10.16	975
Indonesia	6.43	7.76	1240
Niger	6.86	3.22	466
World	240.05	375.11	1562

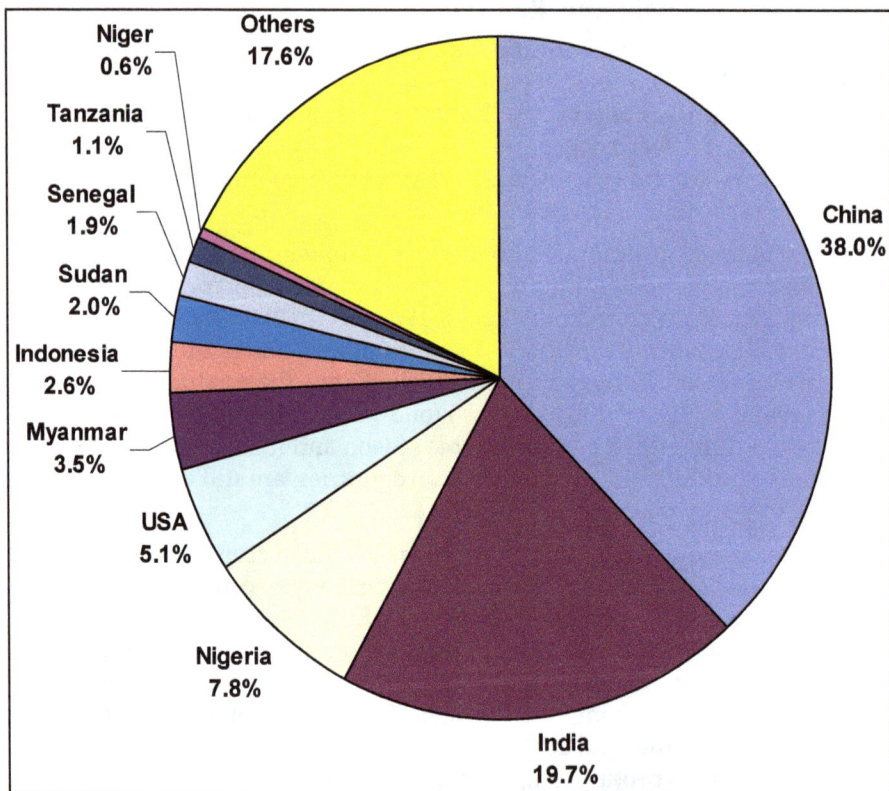

Figure 1.1: Global Scenario of Groundnut Production.

Groundnut is an important crop of the semi-arid tropics where potential yields are frequently reduced by heat and water stress. Studies on occurrence and intensity of the drought during crop growing season revealed the effect of moisture stress on the groundnut yields in the dry lands. The groundnut is relatively drought resistant and important crop of semi-arid regions where evaporation exceeds precipitation for 5-10 months of the year. The productivity (average of both the season) of groundnut during the year 2001 to 2010 though ranged from 700-1460 kg ha^{-1}, in three major groundnut growing states, accounting for about 75 per cent of the total productivity of the country, the productivity was in between 1473-2390 kg ha^{-1} in Gujarat, 1400-2130 kg ha^{-1} in AP and 2100-3730 kg ha^{-1} in Tamil Nadu during rabi-summer season, but fluctuated in between 510-2270, 300-1360, 1150-1880 kg ha^{-1}, respectively in these states during kharif season mainly due to scanty rainfall. Presently, the average productivity of rabi-summer groundnut is about 1850 kg ha^{-1}, much higher than kharif season (1410 kg ha^{-1}) indicating more production potential during this season.

Table 1.2: Area, Production, and Yield of Groundnut in India Since 2001.

Year	Kharif			Rabi/Summer			Total		
	A	P	Y	A	P	Y	A	P	Y
2001-02	54.6	56.2	1030	7.8	14.1	1808	62.4	70.3	1127
2002-03	52.7	30.9	587	6.6	10.3	1548	59.4	41.2	694
2003-04	57.9	52.6	909	8.5	15.1	1771	66.4	67.7	1020
2004-05	52.0	68.6	1320	7.9	12.7	1602	59.9	81.3	1357
2005-06	57.4	63.9	1097	10.0	17.0	1702	67.4	79.9	1187
2006-07	47.8	32.9	689	8.4	15.7	1880	56.2	48.6	866
2007-08	53.0	74.8	1412	11.1	18.8	1691	64.1	93.6	1460
2008-09	52.3	56.4	1077	9.9	17.0	1726	62.2	73.4	1180
2009-10	46.2	38.5	835	8.6	15.8	1830	54.3	54.3	991
2010-11	49.8	66.4	1335	8.8	16.2	1846	58.6	82.7	1411
Average	52.4	54.1	1032	8.8	15.3	1739	61.1	69.3	1134

A: Area (lakh ha); P: Production (lakh tonnes); Y: Yield (kg/ha).

Twenty years of rainfall data at Tirupati for drought classification using aridity index on annual and monthly basis correlated groundnut yields were low due to uneven distribution of rainfall during crop growing season and moisture stress during July and September coincided with moisture critical periods (Sumathi and Subramanyam, 2007).

On deep well-drained sandy soils studies on water use and yield response of groundnuts for three years reveals that, yields were not reduced by droughts of short duration unless the seasonal water use was below 500 mm, however the pod yields were 2.26, 3.00 and 3.82 t/ha with approximately 330, 400 and 460 mm water, respectively (Hammond *et al.*, 1978). The field studies at ICRISAT for 5 years, the groundnuts advanced breeding lines produced greater pod yields on Vertisols (2.02-3.81 t/ha) than on Alfisols (0.61-1.56 t) and there was a strong soil type x genotype interaction. In another study, 4 cultivars water stressed during flowering, pod-set or pod-filling showed that while CGR were greater on Alfisols, these were linearly related to those measured on Vertisols. However, pod growth rates and partitioning of DM to pods showed a strong soil type x genotype interaction and the genotypes developed on the Alfisol maintain relative ranking for total DM on Vertisol, but not necessarily for pod yields (Rao *et al.*, 1992).

For groundnut improvement in India at three regions (Hyderabad, Anantapur and Gujarat), the response relationships between yields and relative available water were estimated and empirical yield distributions were simulated and the alternative risk approaches compared with the traditional stability analysis and recommendations were presented by Bailey and Boisvert (1989). The rainfall data from 1981 to 2003 were categorized into excess, normal, deficit and drought years to know the vegetation cover due to variation in rainfall and identification of the land-use areas facing drought risk, using advanced very high resolution radiometer (AVHRR) sensor's composite dataset for analysing the temporal and interannual behaviour of surface vegetation and land-use classes - crop land (annual, perennial crops), scrub land, barren land, forest land, degraded pasture and grassland were identified using satellite data for excess, normal, deficit and drought years. Normalized difference vegetation indices (NDVIs) derived from satellite data for each land-use class in the drought year, the groundnut crop (0.267) showed the maximum, but the grassland recorded the lowest value of NDVI in all years. The groundnut (0.398), pulses (0.313), sorghum (0.120), tapioca (0.436) and horse gram (0.259), registered higher NDVI values than the perennial crops for the normal year. Among land-use classes, the groundnut witnessed the maximum values of 78.2, 64.5 and 55.2 per cent for normal, deficit and drought years, respectively. The vegetation condition index (VCI) used to estimate vegetation health and monitor drought. Based on the VCI classification, all land-use classes fall into the optimal or normal vegetation category in excess and normal years, whereas in drought years most of the land-use classes fall into the drought category except for sorghum, groundnut, pulses and grasses. These crops sorghum 39.7 per cent, groundnut 55.2 per cent, pulses 38.5 per cent and grassland 38.6 per cent registered maximum VCI values, with sustained under drought conditions suggesting that the existing crop pattern be modified in drought periods by selecting sorghum, groundnut and pulses crops and avoiding onion, rice and tapioca (Muthumanickam *et al.*, 2011).

Drought stress tolerance is seen in almost all plants but its extent varies from species to species and even within species. At whole plant level the effect of stress is usually perceived as a decrease in photosynthesis and growth, and is associated with alteration in carbon and nitrogen metabolism (Mwanamwenge *et al.*, 1999). Drought stress affects the growth, dry matter and harvestable yield, but the tolerance of genotypes to this menace varies remarkably. Quantification of the impact of drought on production of five major kharif crops (rice, groundnut, cotton, bajra, soyabeans), using the standardized precipitation index (SPI) that captures cumulative rainfall deviations at various time scales, computed for 36 meteorological sub-divisions using monthly rainfall data for the period of 1971-2002, identified July as the most drought affected, followed by September, while June and August were near normal. The Correlation between production of major kharif crops (1980-2001) and SPI values reveals that September was the crucial month for defining the crop yield for most of the Kharif crops throughout the country (Chaudhari and Dadhwal, 2004).

2.2 Radiation, Energy and Heat Flux

Estimation of surface sensible and latent heat flux is the most important to appraise energy and mass exchanges among atmosphere, hydrosphere and biosphere. The surface energy fluxes were measured by Kar and Kumar, (2007) over irrigated groundnut during winter (dry) season using Bowen ratio (beta) micrometeorological method in a representative groundnut growing areas of eastern India, at Dhenkanal, Orissa by growing the crop with four irrigations based on phenological stages (branching, pegging, pod development and seed filling) and assessed the crop stress at those times to see if irrigation scheduling could be optimized further. The net radiation R_n varied from 393-437 to 555-612 W m^{-2} during two crop seasons. The soil heat flux (G) was higher (37-68 W m^{-2}) during initial and senescence growth stages as compared to peak crop growth stages (1.3-17.9 W m^{-2}). The latent heat flux (LE) showed apparent correspondence with the growth which varied between 250 and 434 W m^{-2} in different growth stages. The diurnal variation of Bowen ratio (beta) revealed that there was a peak in the morning (9.00-10.00 a.m.) followed by a sharp fall with the mean values varied between 0.24 and 0.28. The intercepted photosynthetic photon flux density or photosynthetically active radiation (IPAR) by the crop was measured and relationship between IPAR and leaf area index (LAI) was established with DAS, which is usedful in developing algorithm of crop simulation model for predicting LAI or IPAR. The stressed and non-stressed base lines were also developed by establishing relationship between canopy temperature and vapour pressure deficit (VPD). With the help of base line equation, $[(T_c-T_a) = -1.32VPD+2.513]$, crop water stress index (CWSI) was derived on canopy-air temperature data collected frequently throughout the growing season. The soil moisture depletion during the crop period when plotted with CWSI at different stages the values of CWSI varied between 0.45 and 0.64 just before the irrigations were applied and at two stages (branching and pegging), CWSI were much lower (0.46-0.49) than that of recommended CWSI (0.60) for irrigation scheduling (Kar and Kumar, 2007).

Studies on radiation and energy budgets over a cropped surface in the Sabarmati river basin, Gujarat, India by recording continuous data on temperature, humidity, wind speed and direction at 1 and 4 m on a 9 m tower, soil heat flux sensible and latent heat fluxes from March to August 1997, a polynomial relationship between residual flux and biomass under different phenological phases was observed, a linear relationship was found between residual flux and plant height under different phenophases and the biomass of crops increased exponentially with increasing AE: PE ratio and a polynomial trend was observed in water deficit, biomass and height (Padmanabhamurty *et al.*, 2001). The groundnut production was directly proportional to light interception and to the ratio between water lost and the vapour pressure deficit from leaf to air. Root growth and development was favoured under limited water supply and high water demand. Leaf conductance to gas exchange were similar at different combinations of soil water content and atmospheric saturation deficit (Goncalves de Abreu, 1988).

2.3 Water-yield Relationship

The evapotranspiration-yield relationships has a strong interaction with timing of drought and drought imposed at (a) emergence to maturity, (b) emergence to peg initiation, (c) start of flowering to the start of seed growth, and (d) from the start of seed growth to maturity during the post-rainy seasons, the amount of water applied during these phases varied in groundnut cv. Robut 33-1 and the greatest reduction in seed yield (28-96 per cent) occurred when stress was imposed during (d), however, decreased irrigation during (b) increased pod yield over fully irrigated control by 13-19 per cent (Rao *et al.*, 1985). The crop coefficient curve facilitates the prediction of groundnut ET in preparation of planting at a new site from estimates of reference crop ET. The crop coefficient (Kc) values of groundnut at different crop-growth subperiods were influenced by evapotranspiration deficits and leaf area development of the crop. On a sandy loam soil of Hyderabad, in fully irrigated crop (W-W-W) the Kc value was low (0.564) during the establishment of plant (0-10 DAS), increased linearly through vegetative period and remained constant at 1.024 from flowering to start of the pod filling period (35-80 DAS), then decreased through pod filling period and reached a lowest value of 0.547 during the final 10 days of the crop period (Devi and Rao, 2003).

A study was conducted to characterize the plant extractable water pattern at four locations in India (Tirupati, ICRISAT, Jalgaon and Junagadh) and one location in Queensland, Australia (Kingaroy) and explore the possibility of clustering the multi-location trial environments based on similar water stress patterns. The APSIM groundnut model was used to compute daily changes in plant extractable soil water (P_{esw}) at each site, by using climate parameters (ambient temperature, radiation, rainfall or irrigation amounts), soil hydraulic parameters and crop parameters (planting and harvest dates). Results from the P_{esw} characterization of experimental sites clearly demonstrated that the crops grown at the multi-location have experienced a wide variation in timing, intensity and duration of crop water deficits during the growing season and that quantification of the P_{esw} during the growing season and clustering of environments based on P_{esw} patterns can assist in understanding the basis of G x E interactions for yield between clusters, and to examine the effect of breeding methods on yield variation within each of the clusters (Rachaputi, 2003).

In Eastern India (Bhubaneswar) during dry season (November-March), the daily moisture use rate, increased gradually and reached the peak value (4.10-4.94 mm) during 55 to 60 days in groundnut cv. AK12-24 where the crop coefficient values followed the same trend as that of crop Et, which were lower at the initial growth stage (0.61-0.80), increased gradually and attained the maximum value of 0.94-1.33 towards the peak period of crop growth and declined thereafter. The crop coefficient value approached unity or slightly exceeded it during the maximum growth stage of the crop. Plants stressed at the early vegetative stage showed lowest crop coefficient value (0.61) at initial stage and the highest value (1.33) at peak crop growth stage, and witholding irrigation at an early stage (14 DAS) resulted in lesser evaporation of water from the soil surface (Kar *et al.*, 2001).

Field data on pod yield and seasonal ET as influenced by irrigation schedules during the summer predicted that the pod initiation and development stage (70 d to harvest) was the most sensitive stage for moisture stress with a yield response factor of 2.10. Water stress during 10 to 40 d was beneficial in enhancing pod yield with a yield response factor of 2.10 (Ramachandrappa and Nanjappa, 1994). During winter at Rajendranagar, Hyderabad, in groundnut cv. ICGS 44 reproductive stage (35-115 DAS) was the most sensitive to a reduction in water supply, whereas water stress in the vegetative stage (10-35 days) had the least effect (Reddy *et al.*, 1996). At Coimbatore, TN, during summer 1994, water stress in groundnuts cv. Co 2 and VRI 2 at flowering, pegging, pod development or pod maturation when compared water stress at pod development was most detrimental on yield (Velu, 1998).

In Western India, a lysimeter experiment on black calcareous vertic Inceptisol at Junagadh, two groundnut cultivars subjected to water stress from the seedling to flowering (24-48 DAE), flowering to pegging (40-60 DAE), pegging to pod development (55-75 DAE) or pod development to maturation (75-95 DAE) decreased pod yields compared with plants given normal irrigations however yield reductions were greatest with stress imposed during the pegging and pod development and lowest with stress imposed from pod development to maturation (Patel and Golakiya, 1988). Further lysimeter trials on Spanish bunch groundnuts cv. J11 and GG 2 revealed that water stress from pegging to pod development gave the lowest pod yields with increased leaf temperature (35°C) markedly lowering photosynthesis. In all stress treatments, GG 2 out yielded J 11 mainly due to lower fluctuations in leaf temperature, stomatal resistance and lower vegetative growth (Patel and Golakiya, 1993). In another lysimeter studies, groundnut subjected to soil moisture tensions (SMT) of 330, 530 or 730 mbar, maximum daily water consumption occurred at 50-80 and 50-65 days in groundnuts grown at the 2 lower and the highest SMT, respectively and increase in SMT decreased total DM yield, but increased unshelled nut yields (Vivekanandan and Gunasena, 1976).

The total dry matter at harvest had positive correlation with TE, leaflet size was negatively correlated with TE under drought stress, the N content in leaves at 80 DAP and the chlorophyll content in leaves during moisture stress (28 days after imposing stress) showed positive relationship with TE. The leaf temperature 28 days after imposition of moisture stress had significant negative relationship with TE under adequately irrigated and simulated drought treatments. The mineral ash content of leaves 80 DAS in Spanish cultivars (ICG 476, ICG 221, ICG 1697, ICGV 86031 and TAG 24) had significant positive correlation with TE in simulated drought treatment (Babitha and Reddy, 2001). Sharma *et al.* (1987) studied the performance of two groundnut cv. under soil moisture stress during rainy season where number of gynophores and pods/plant, 100-seed wt, pod yield and shelling percentage were highest with two irrigations at 50 and 80 DAS and were lowest under rainfed conditions. Irrigation at 80 DAS was more effective than irrigation at 50 DAS. The moisture stress suppressed pod setting more in cv. M 13 than in cv. M 37. Oil content in seeds was not affected by moisture stress in both cultivars.

Studies on water-yield relationship in groundnut showed the yield response factor (k_y) 0.45 and 0.42 under normal irrigation and 1.72 and 1.70 at full deficit irrigation during summer and Rabi seasons, respectively and the pod formation and flowering stages were more sensitive to moisture stress and irrigation during these stages is more important to overcome the yield reduction in groundnut (Thiyagarajan *et al.*, 2010). The groundnut cv. SB 11 grown at 3 levels of water stress applied at 4 growth stages, water stress of 0.8 (ratio of IW: CPE) at any growth stage reduced pod yield, Maximum pod yield obtainable (3.06 t/ha) was predicted to be obtained with 1131 mm irrigation water (Shinde and Pawar, 1982).

The potential (no-water stress) and the lowest (no irrigation) yields for maize, soyabean and groundnut were calculated using three crop growth and water use models - CERES-Maize, SOYGRO, and PNUTGRO where rainfall, temperature and solar radiation were used with these models to identify the 15 most severe drought years in the 53 year record in a 36-county region of Georgia, USA. In the 15 driest years, simulated yield losses averaged 75 per cent for maize, 73 per cent for soyabean, and 64 per cent for groundnut. In irrigated crop acreage of the study area, simulated water withdrawals exceeded 3 million m^3/day, on average, for most of the 130 days between late May and late September (Hook, 1994).

Groundnut drought adaptation mechanisms with a view to developing selection criteria for breeding require survey under (1) drought evasion (ability to complete the development cycle before water deficits occur), (2) drought avoidance (mechanisms such as modified root and leaf morphology which allow the plant to keep its tissues at a high water potential during drought) and (3) tolerance to drought (maintenance of potential turgidity by osmotic adjustments, and tolerance of desiccation due to properties of the cell membrane) (Annerose, 1988). Most breeding programmes in groundnut follow an empirical approach to drought resistance breeding, largely based on kernel yield. Recent advances in the use of easily measurable surrogates for complex physiological traits associated with drought tolerance encouraged breeders to integrate these in their selection schemes. However, there has been no direct comparison of the relative efficiency of a physiological trait-based selection approach (Tr) vis-a-vis an empirical approach (E) to ascertain the benefits of the former.

3. Impact of Water Deficit Stress on Vegetative Growth

3.1. Seed Germination and Seedling Growth

Soil moisture and atmospheric saturation deficit had a large effect on seedling emergence and establishment in groundnuts and best emergence at suboptimum temp was obtained at 40-80 per cent field capacity (FC) (Goncalves de Abreu, 1988). In field groundnut seed germinate well when soil water tension in the surface 30 cm was maintained at <0.6 bar during the growth of parent plants. Soil water tension >15 bar during the growing season reduced the germination of seed by 20, 5 and 5 per cent, of Florigiant, Florunner and Tifspan and yield of sound mature seed by 34, 22 and 7 per cent respectively (Pallas *et al.*, 1977). During germination the groundnut seeds remained in the imbibition phase for up to 25 h followed by a lag phase of 25-55 h and then entered into the germination or growth phase, the diffusivity of seeds increased up to 30 h of imbibition, then remained constant up to 55 h and thereafter again increased up to 75 h. Maximum soluble N and the break of oxygen concentrations constancy occurred at the 45 h of imbibition the time when germination was triggered which became visible 6-8 h later (Golakiya,1989). About 35 per cent moisture is the minimum requirement to initiate the germination process however, germination time was curtailed by increasing seed moisture content and radical emergence in most seeds appeared at 55-60 per cent moisture content (Golakiya, 1989).

Ten crop species evaluated for their relative drought tolerance at the seedling stage by planting seeds in wooden boxes (130 × 65 × 15 cm³) filled with soil by withholding water a week after germination and observing reaction to progressive water stress. Based on percent dead plants and days taken to 100 per cent dead plants, soyabean (*Glycine max*) was the most drought susceptible and cowpea (*Vigna unguiculata*) the most drought tolerant, however, ranking of crops in the increasing order of drought tolerance was: soyabean< black gram (*V. mungo*) < green gram (*V. radiata*) <groundnut < maize (*Zea mays*) < sorghum (*Sorghum bicolor*) < pearl millet (*Pennisetumglaucum*) <bambara nut (*V. subterranea*) < lablab bean (*Lablab purpureus*) < cowpea (Singh *et al.*, 1999).

The relative drought tolerance could be studied at the seedling stage by planting seeds in wooden boxes (130 x 65x 15 cm³) filled with different proportion of sand, loamy sand and sandy loam, irrigating daily till the establishment and withholding water a week after germination and observing reaction to progressive water stress by counting percentage dead plants at various time intervals and days taken to 100 per cent dead plants. Soil with higher sand content induced water stress and with increased clay content and gradual water stress, it may be possible to use this method to detect varietal differences in less drought tolerant crops (Singh *et al.*, 1999). In indoor and field experiments showed that germination potential and germination percentage, growth vigour and vitality index of groundnut seeds increased after soaking with 500 mg/litre of nitrate rare earth element solution (38.7 per cent of rare earth oxide) for 24 h before sowing. Seedlings emerged 1 or 2 days earlier, the emergence rate increased, drought resistance enhanced significantly under moisture stress resulting in strong seedlings and quick root development, leaf area, number of branches, chlorophyll content, proline content and photosynthetic rate also increased (Nie ChengRong *et al.*, 2002).

Water uptake, germination and seedling growth of 12 groundnuts cv. of 4 botanical groups studied under polyethylene glycol 6000 which simulated moisture stress (-1 to -10 bar water potential) water uptake by seeds showed no differences amongst botanical groups in spite of varietal differences (Babu *et al.*, 1985). The simulated water stress by PEG 6000 decreased groundnut germination, germination relative index and vigour index, seeds did not germinate at -10 bar, the activities of acidic and alkaline lipases and protease increased with the progression of germination but decreased with an increase in the stress level. Peroxidase activity was negligible in the cotyledons as well as in the embryonic axis up to day 1 of germination but increased thereafter, increasing level of stress decreased the enzyme activity (Sharma *et al.*, 1987). Seeds of 20 groundnut cultivars watered with a hyperosmotic solution, 20 per cent PEG 6000 at 25°C, and after 8 days, germination percentage, radical root length and dry weight indicated that germination characteristics of groundnuts grown in hyperosmotic solution may be useful as indicators for selection for drought-resistant genotypes (Xue *et al.*, 1997).

Germination studies on four groundnut cultivars (Ex-Dakar, RRB 12, RMP 12, RMP 91) under different osmotic solutions when compared with some agronomic and yield parameters of plants grown under simulated drought conditions in the field, indicated that germination of seeds in polyethylene glycol (PEG), glucose or sodium chloride (NaCl) solutions at 1.2 MPa and 1.8 MPa could be reliably used as a quick and cost effective procedure for screening groundnut cultivars for drought resistance at an early stage of their growth and development and maximum germination percentage, radicle length and dry weight were attained in RMP 91 and Ex-Dakar identified as drought resistant (Mensah and Okpere, 2000). Susceptibility to drought was shown by the relatively greater reduction in yield per plant compared to the resistant cultivars. The PEG-induced water stress imposed by treating the seeds at different water potentials (-0.3, -0.6 and -1.0 MPa) on 10 different groundnut cultivars (TCGS 20, GG 2, TMV 2, JL 24, ICGV 86031, K 134, TAG 24, JL 220, CSMG 84-1 and TCGS 41), showed significant reduction in germination, seedling growth and seedling vigour index with the decreasing water potential from -0.3 to -1.0 MPa. Among the cultivars, ICGV86031 showed the highest resistance to water stress and germination and seedling growth even at the -1.0 MPa where all other cultivars completely failed (Prathap *et al.*, 2006).

In groundnut seedlings of different habit groups for 15 days, the total polyamine level in the root and shoot varied from 2.5-4.9 µ mol/g fresh weight and 2.8-4.6 µ mol/g fw, respectively. In each habit group, roots of GG 2, G 13 and G 20 showed the highest total polyamine content and in the shoots, similar results were obtained except among the spreading cultivars, M13 recorded the highest polyamine content. Artificial stress (PEG 6000) increased the polyamine content in both root and shoots, however, putrescine application just before PEG stress prevented the fall in tissue moisture content in water-deficit seedlings (Vakharia *et al.*, 2003).

Germination of spp. under simulated drought (0.10 bar osmotic pressure) was in the order *D.tortuosum>Crotalaria spectabilis*>soyabean>groundnut. In field trials, the emergence frequency was higher during the sowing dates for groundnut than those for soyabean where drought reduced groundnut seed yields by 83 kg/ha (Hoopper, 1978). In a field trial with 3 groundnut cultivars, the seed hardening with 1 per cent calcium chloride or 2 per cent KH_2PO_4 was most effective, but germination was adversely affected by seed treatment with 1.5 per cent succinic acid (Arjunan and Srinivasan, 1989).The better performance of seedlings of groundnut cv. RS 218 than of cv. MGS7 under soil moisture stress conditions when grown from seeds treated with 0.25 per cent $CaCl_2$ for 8 h was ascribed to a markedly higher accumulation of proline and K in seedlings of RS 218 (Sashidhar *et al.*, 1981).

The bunch groundnut, *Arachis hypogaea* subsp. *fastigiata*, varieties show little seed dormancy and 20-50 per cent of pods germinate in situ due to rains at the pod maturity stage. At Aliyarnagar, 55 high yielding genotypes of subsp. *fastigiata* when grown in the field ICGV 86011, derived for the cross (Dh 3-20 X UAS20) X NcNc 2232, possessed seed dormancy, with pods sprouting 18 days after harvest compared with 2-4 days for the other genotypes. ICGV 86011 has also shown resistance to sucking pests (Varman and Raveendran, 1991).

Seeds of groundnut cultivars soaked in water or in solutions containing 50 ppm Ascorbic acid or 1 per cent calcium chloride for 15 h, dried and germinated in water and polyethylene glycol (at osmotic pressures of -5 and -10 bar) showed that seeds of cv. S 206 and BH 818 in water and cv. NG 268 in calcium chloride gave the highest germination and root length which decreased with increase in osmotic concentration of the solution (Prasad *et al.*, 1974). The germinated 3-day-old seedlings of groundnut cv. Yueyou 551-116, seeds when treated with 5 ppm PP333 resulted in substantial accumulation of ABA in the leaves under both normal and water stress conditions, while the biosynthesis of GA was inhibited by PP333, as ABA and GA could be antagonistic due to competition for a common precursor. The PEG solution (10^{-3} mol PEG/litre), treated seedlings showed little change in plasmalemma permeability and in leaf water content whereas those of the control decreased markedly, suggesting that PP333 could increase drought resistance of groundnut plants (Li and Pan, 1988).

3.2 Crop Growth Rate and Dry Matter Production

In greenhouse the shoot and root growth and root stratification in the soil profile for groundnut, indicated that the leaf number and leaf area decreased and root/shoot ratio increased in response to water stress, greatest deepening of the root system in the soil profile in response to water stress (Pinto *et al.*, 2008). Nageswar Rao *et al.* (1993) discussed various parameters for selection of drought resistant varieties in breeding by taking 10 groundnut genotypes grown with adequate irrigation, and subjected to drought during pod filling (83-113 d after sowing) on a medium deep Alfisol at ICRISAT Centre, Andhra Pradesh, during post-rainy season. Shoot DM accumulation during the drought period was 72-150 g/m² and was closely related to transpiration. The groundnut cv. J 11 subjected to water stress for (a) 6 or (b) 9 days during the vegetative stage (after 30 days growth) or for (c) 6 or (d) 9 days during the reproductive stage (after 45 d), about 80 per cent of plants in (b) died and nodule numbers in (a), (c) and (d) were 47, 57 and 72 per cent, respectively, lower than that of unstressed controls immediately after rewatering and were 55, 28 and 66 per cent 15 days later. 15 days after stress relief, plant N content was only about 50 per cent of that of unstressed plants (Kulkarni *et al.*, 1988).

In field trials at Parbhani groundnuts sown in Feb., with 30 kg N, 60 kg P_2O_5 and 30 kg K_2O/ha and irrigated at different IW: CPE ratios, water stress at an IW:CPE ratio of 0.4 at the seedling stage, flowering, peg formation or maturity reduced number of branches/plant, pod yield and total DM yield (Shinde and Pawar, 1984). The effect of imposed single and double water stress on the growth and yield of three grain legumes, cowpea cv. IT 1627, groundnut cv. Kumawu Red, and Bambara groundnut (*Vigna subterranea* cv. *Jabajaba*) commonly grown in sub-Saharan Africa was studied by providing various treatments: liberal watering until maturity; 7-day dry cycle at 41-47 days after planting (DAP); and dry cycle at 41-47 DAP, followed by liberal watering and another dry cycle from 54-59 DAP. Water stress significantly reduced growth of both cowpea and Bambara groundnut but not groundnut. Groundnut was the most tolerant of post-flowering water stress among the three legumes (Kumaga, 2003).

Groundnut cv. JL 24 was sown on 15 January, 15 February and 15 March at two (50 cm x 6 cm and 25 cm x 12 cm) spacing, and irrigated once during the first phase of flowering (I1), twice during the first phase of flowering and pod initiation (I2) or thrice during the first phase of flowering, pod initiation and development (I3) at Kalyani, West Bengal, India the nodules per plant (78) and nodule dry weight per plant (140) at 90 DAS, as well as pod dry weight (5.88) at 110 DAS, and pod (2336 kg/ha) and haulm yield (6330 kg/ha), were highest in the crop sown on the 15 February. Irrigation had no significant effect on the number of nodules per plant and nodule dry weight at 30 and 90 DAS. However, I3 recorded the highest shoot (21.57 g/plant) and pod dry weight (5.68 g/plant) at 110 DAS, and pod (2420 kg/ha) and haulm yield (6520 kg/ha). Water stress at the pod initiation and development stages reduced pod yield by 13.4 and 44.2 per cent, respectively. Spacing at 25 cm x 12 cm recorded a higher number of nodules per plant at 90 DAS, nodule dry weight per plant, shoot and pod dry weight at 110 DAS and pod yield compared to the 50 cm x 6 cm spacing. Spacing had no significant effect on the haulm yield (Patra *et al.*, 1999). Six groundnut cultivars were grown at Akola, Maharashtra in kharif [rainy] season on deep soil (>120 cm) with 50 per cent water holding capacity or on shallow soil (<20 cm) with 30 per cent water holding capacity. Water stress occurred during the vegetative (-16 bar) and reproductive (-14 bar) growth stages on shallow soil. Water stress decreased root growth, nodule number and DW, pod yield, seed number/plant and shelling percentage (Dhopte *et al.*, 1992).

The rate/temp relation of several developmental processes in groundnut examined in greenhouses at air temp of 19, 22, 25, 28 or 31°C and also the sensitivity of the processes to soil water deficit was examined by applying 30 mm of irrigation immediately after sowing and when tensiometer readings at 0.1 m depth exceeded 20 kPa (wet soil treatment). The relation between rate and temp was linear and the measurements when analysed in terms of thermal time an extrapolated base temp (T_b) was 10°C at which the rate was zero for leaf appearance, branching, flowering, pegging and podding. Leaf appearance and branching were more sensitive to soil water deficit than the other processes examined (Leong and Ong, 1983). Thirty-three groundnut genotypes evaluated under water stress during summer at Bangalore showed wide variability for most of the characters studied. The estimates of PCV and GCV were high for primary branches per plant, biomass per plant and shoot root ratio. Heritability was high for plant height, root-shoot ratio, number of primary branches per plant and number of pods per plant under water stress condition than under non-stress situation. A relatively higher genetic advance as per cent of mean was noticed for plant height, shoot-root ratio, number of secondary branches per plant, number of pods per plant and biomass under water stress conditions under non-stress conditions (Ravi *et al.*, 2008).

The growth and agronomic traits of 18 groundnut cultivars under waterlogging during late vegetative to flowering stage when analysed, waterlogging decreased plant height, number of branches, total pods and full pods, however it promoted pod growth and the ratio of seeds that developed into full pods. Cluster analysis based on agronomic traits was somehow integrated with yield classification after waterlogging (the 18 cultivars classified into 6 types). The most tolerant cultivars (HT type) selectively bred under waterlogging were dwarfed with synchronously increasing number and weight of pods and seeds and promoted the ratio of full pod number and ratio of full pod weight. The findings illustrated that waterlogging tolerant eco-breeding was effective. Because more water was demanded in the late vegetative to flowering stage, the impact of waterlogging was limited and even promoted pod and seed development for most waterlogging tolerant cultivars so long there was water flow and groundnuts not completely submerged. Therefore, groundnut flood impact assessment should focus on flood intensity, growth and development period and varietal tolerance (Wang *et al.*, 2009).

In medium black calcareous soil (Typicustochrepts) groundnut cvGG 2 during summer given cyclic water stress at four stages *i.e.*, flowering (S_f), pegging (S_{pg}), pod formation (S_{pf}) and pod development (S_{pd}) and three levels of potassium (0, 20 per cent K_2O as foliar spray and 60 kg K_2O ha^{-1} as soil application), water stress at pegging or its association with any one or two phenophase(s) reduced the accumulation of dry matter and nutrients, while pod yield was most adversely affected due to water stress during pod development stage. The accumulation of all the nutrients increased with advancement of crop growth except K in which the maximum accumulation was found at 78 DAS. Foliar application of potassium only increased its uptake, while soil application increased N and K uptake by the crop (Sakarvadia *et al.*, 2010).

The crop evapotranspiration (ET) and growth characteristics of groundnut in the transitional humid zone of Nigeria have shown that the total water used (ET) by the crop during 105 days was 303 mm and more amount of water was used between the vegetative and reproductive growth stages of the crop between 20 and 60 DAP, the highest mean leaf area (LAI) obtained was at 75 DAP, dry matter accumulation was highest between 75 and 90 DAP when canopy radiation interception was between 70 and 80 percent and there was a positive correlation (p=0.01) between growth parameters and water use (Idinoba *et al.*, 2008).

3.3 Flowering and Peg Formation

In groundnuts flower production ceased when water stress started, but recovered as soon as plants were rewatered, especially with early drought. A 5-day drought period was most damaging in terms of flower and pod numbers and av. plant wt when imposed 5 weeks after sowing and decreased seed yield from 106-107 (control) to 43-92 g/m^2 (Zaharah, 1986). Several groundnut cv. when grown in field at water deficit during early peg and pod formation (40-82 days), even after re-watering, plants subjected to drought were 3-5 nodes shorter than normal plants and peg and pod numbers at 77 DAS were 51 per cent lower than normal, pod maturity was delayed by 10-11 days (Boote and Hammond, 1981). The flush of late flowers following mid season drought delay maturity. Flowering stopped when soil moisture dropped to wilting point, but fruiting continued (Scandaliaris *et al.*, 1978). The fruiting occurs once the gynophores enter into the soil. The soil physical condition is important and must be wet during the gynophore entering the soil as the gynophore can exert a force equivalent to 3-4 g only.

In a field during summer, groundnut cv. TMV2, moisture stress at various stages of development did not reduce the total number of flowers formed, but reduced the number of gynophores formed from the 1st flush of flowering. However, yield from the 2nd flush compensated so that the total pod yield was not affected by moisture stress treatment. Moisture stress in the early flowering phase (30-45 DAS) was not critical when soil MC was <30 per cent in the top 90 cm (Gowda, 1977). Also in a field trial,

Gowda and Hegde (1986) found that total flowers produced by groundnut plants subjected to soil moisture stress (SMS) between 30-45 DAS were not different from those produced by regularly irrigated plants. The SMS decreased pod yields by 4.7 per cent, as the first flush of flowers produced up to 45 DAS dried up and did not form gynophores. However, a higher percentage of fruit set from the 2nd flush compensated for yield differences.

Flowering pattern and total flowers produced in 6 bunch type groundnut cvs. subjected to intermittent cycles of moisture stress and gynophore length at harvest as an index of pod ontogeny was related to the number and synchrony of flowers produced during any growth period and the proportion of pods produced at different growth stages determined based on gynophore length is suggested as a selection index by Janamatti *et al.* (1986) where the cv. did not differ in the total number of flowers produced/plant in any of the moisture regimes, indicating that the number of flowers was not a constraint in productivity even under moisture stress. Gynophore length of potential pods indicated that under severe stress, only the early formed flowers developed into pods. In stressed plants although the total number of pods decreased, the percentage of total pods with a gynophore length of 0.1 to 3.0 cm increased markedly and in some cv. it was about 55 per cent. A significant burst in flowering on alleviation of stress was the unique feature in the pattern of flowering under moisture stress, particularly when it was imposed just prior to the reproductive phase (Janamatti *et al.*, 1986).

In groundnuts, the leaf area, number of flowers and gynophores, haulm and pod yields, dry wt. of root, RWC of leaf and leaf-water potential decreased continuously with an increase in soil moisture stress from 0-0.3 to 0-14 bar in the greenhouse and from 0-0.5 to 0-20 bar in the field (Patel *et al.*, 1983). A stress of 10 bar at the flowering, pegging and pod-development stages was harmful for haulm and pod yields. When the soil surface was kept dry and the crop was irrigated from the bottom of a pot through soil capillary flow, pod yield was low but haulm yield and number of pegs and flowers were max., indicating that a hard soil hinders gynophore penetration. On re-watering after stress, both leaf area and number of flowers increased simultaneously. Within 40 days from the onset of flowering the crop had put forth 63 per cent of all the flowers produced during the entire period of its growth (Patel *et al.*, 1983).

In field trials in Ceara, Brazil, groundnuts cv. PI 165-317 from the USA, PI 55437 from Senegal and Tatu from Brazil grown on sandy soil and sprinkler irrigated to 100, 88.5, 73 or 61 per cent of the water depth required at soil matric potential of -0.05 MPa reveals that increasing water stress reduced leaf area, number of leaves and flowers, shelling percentage and reproductive efficiency. With 100 per cent irrigation flowering peaked after 5 weeks, but severe water stress shortened this to 2 weeks. Transpiration and yield were correlated with soil water deficit and pod yield was highest for PI 165-3176 under all water regimes (Ferreira *et al.*, 1992).The management practices should aim to optimize the availability of growth resources at the time of pegging in order to ensure that pod initiation is not delayed.

3.4. Leaf Area, SLA, SPAD and Stomatal Studies

Water stress reduces the leaf area, leaf area duration, chlorophyll content and stomatal frequency and this effect varied with stress in the early and late in the growing season. In Bangalore, groundnut irrigated at 0.4, 0.6 or 0.8 CPE during summer the leaf area duration and pod yield increased with increasing irrigation level (Sridhara *et al.*, 1998). Irrigation at 0.8 CPE gave the highest DM accumulation, leaf area and LAI (Sridhara *et al.*, 1995). However, complex nature of physiological traits associated with drought tolerance and the difficulties associated with their measurements in segregating large populations inhibit their use in developing water-use efficient genotypes in breeding programmes. The easily measurable surrogates of transpiration efficiency (TE), a trait associated with drought tolerance - specific leaf area (SLA) and soil plant analytical development (SPAD) chlorophyll meter reading (SCMR), it is now possible to integrate TE through the surrogates in breeding and selection schemes in groundnut. As a noninvasive surrogate of TE, SCMR is easy to operate, reliable, fairly stable and low cost.

Nigam and Aruna (2008) in a study evaluated the drought tolerant characteristic and as to what extent the SCMR measurements can be spread over time by evaluating 18 diverse groundnut genotypes for two physiological traits, SCMR and SLA in post rainy (Nov-Apr) seasons in India by recording observations at different times during and after the release of moisture deficit stress where there was general agreement in genotype and trait performance in both the seasons. The ICGV 99029 and ICR 48, which recorded higher SCMR and lower SLA values in both the seasons, were identified good parents for WUE trait in breeding programmes. Other good parents include ICGS 76, TCGS 647 and TCGP 6. The SCMR recorded at three different times under differing soil moisture deficit showed highly significant correlation with each other. Similarly, SLA at different times also correlated significantly with each other. SCMR and SLA were significantly negatively correlated with each other and the relationship was insensitive to time of observation (Nigam and Aruna, 2008). The SCMR and SLA observations can be recorded at any time after 60 days of crop growth, under moisture deficit conditions and gives groundnut breeders to record these observations in a large number of segregating populations and breeding lines in the field.

The LAI was not affected before 40-45 DAS but was reduced by 20-25 per cent in unirrigated plants between 60 DAS and final harvest (Black *et al.*, 1985). In cv. GAUG 1, water stress increased the leaf proline content and stomatal resistance and decreased nitrate reductase activity (NRA), RWC, transpiration and seed yield and application of 40 kg K/ha increased the proline content, stomatal resistance, NRA, RWC, DM accumulation and yield. Seed oil contents were higher in the water-stressed than in unstressed plants. K increased the oil contents in stressed and unstressed plants (Umar *et al.*, 1991). The RWC decreased with onset of water stress and with increase in plant age. The extent of decrease in RWC was greater in the control than in plants treated with antitranspirants; RWC was higher with Sunguard, alachlor and Rallidhan than other treatments (Amaregouda *et al.*, 1994). Rao *et al.* (1998), in a remote sensing ground truth experiment to monitor a groundnut crop under non-stressed conditions in Brazil (during Sept.–Dec.) reported that, the canopy reflectance in the red and near infrared wave bands of the thematic mapper of two vegetation indices, soil adjusted vegetation index (SAVI) and the normalized difference vegetation index (NDVI) were correlated very well with both the LAI and biomass and can be used to estimate LAI and biomass of a groundnut crop.

The effects of elevated atmospheric CO_2, in combination with water stress in groundnut cv. Kadiri-3 studied by Clifford, (1995) revealed that the effects of future increase in atmospheric CO_2 concentration on stomatal frequency in groundnuts are likely to be small, especially under water stress, but that the combination of associated reductions in leaf conductance and enhanced assimilation at elevated CO_2 will be important in semi-arid regions. The CO_2 exerted significant effects on stomatal frequency only in irrigated plants. The effects of drought on leaf development outweighed the smaller effects of CO_2 concentration, although reductions in stomatal frequency induced by elevated atmospheric CO_2 were still observed. Under irrigated conditions with unrestricted root systems, an increase in atmospheric CO_2 from 375 to 700 ppm. Decreased stomatal frequency on both leaf surfaces by up to 16 per cent; in water stressed plants, stomatal frequency was reduced by 8 per cent on the adaxial leaf surface only. Elevated CO_2 promoted larger reductions in leaf conductance than the changes in stomatal frequency, indicating partial stomatal closure. As a result, the groundnut stands grown at elevated CO_2 utilized the available soil moisture more slowly than those grown under ambient CO_2, thereby extending the growing period. Despite the large variations in cell frequencies induced by drought, there was no effect on either stomatal index or the adaxial/abaxial stomatal frequency ratio (Clifford, 1995).

In central India Azam Ali, (1984) studied the interaction between population and water stress in 4 populations of groundnut by estimating, transpiration, stomatal resistance (rs), boundary layer resistance (ra), vapour concentration difference between leaf and air (deltachi) and LAI and the frequency distributions of rs, ra, deltachi and seasonal changes in LAI when plotted to analyse the dependence of transpiration rate on each variable both per unit area of leaf surface (E1) and per unit land surface (Ee), for estimates of E1, both rs and deltachi were of similar importance, exerting a far greater influence than changes in ra. However, in terms of Ee, changes in LAI were far more important than in any other variable, particularly late in the season when water was scarce. The ability of this technique to describe temporal and spatial variations as well as the dominant environmental and physiological influences on transpiration may outweigh any small loss in accuracy of estimates.

At Tirupati, India, Babitha, (2006) screened 111 Spanish and 110 Virginia groundnut genotypes for moisture stress and high temperature tolerance during post-rainy season, and classified into 3 groups, *i.e.* low, medium and high for SPAD chlorophyll meter reading (SCMR), SLA, chlorophyll fluorescence ratio and membrane injury, where majority of the Spanish genotypes had medium SCMR (45-50) and SLA (125-150), while most of the Virginia cultivars had high SCMR (>50) and medium SLA (125-150), majority of the Spanish and Virginia genotypes had high membrane injury of >60 per cent. Genotypic variation was observed for SCMR, SLA, chlorophyll fluorescence ratio and membrane injury. The Spanish genotype JAL 07 had a low membrane injury (37 per cent). Incidentally, it also maintained high SCMR (52) and low SLA (101.6 cm²/g), indicating that it can tolerate both water deficit and high temperature.

There is equal number of stomata both on the upper and lower epidermis of leaves in groundnut. Unlike other crops, in groundnut the stomata remains open during drought, but are more sensitive to light. The groundnut does not have complete stomatal control over transpiration loss but some control is achieved through folding and orientation of leaves parallel to the incoming radiation. Leaflet area, stomatal frequency, stomata number and stomatal size evaluated in seven groundnut cultivars in the summer and rainy seasons, the cultivars TAG 24 and Somnath, show higher WUE. In TAG 24, Somnath, TG 22 and TKG 19A, reduction of leaflet area was associated with increased stomatal number and frequency on the adaxial surface. However, no differences were observed for stomatal length and breadth. Reduced leaflet area with corresponding increase in stomatal frequency and number of stomata on adaxial surface appeared to be related to WUE in TAG 24 and Somnath (Badigannavar *et al.*, 1999).

In field at Redland Bay, Queensland, groundnut cultivars with reduced soil water supply (RSW) during early reproductive development, total biomass production of two Virginia type cultivars (Virginia Bunch and Q 18801) was greater than that of a Spanish type cultivar (McCubbin), the RUE and transpiration efficiency of Q18801 were significantly greater than those of McCubbin. The RUE of the stressed crops was only about 45 per cent of those that were fully irrigated. Throughout RSW, noon leaf water potential was lowest in McCubbin and under increasing soil water deficit, the leaves of McCubbin tended to wilt, while the Virginia cultivars displayed active leaf folding and the ratio of the fraction of radiation intercepted by the canopy to LAI was always lower in the Virginia type cultivars. For a given LAI, this phenomenon may have allowed these cultivars to decrease the effective atmospheric demand within the canopy, while maintaining radiation interception at saturation for photosynthesis. The consequence of this, given that the supply of water from the roots did not differ, was that Q18801 was able to maintain a higher LAI and greater crop transpiration efficiency (ratio of biomass production to transpiration) than McCubbin. Thus existence of differences among cultivars in transpiration efficiency under drought may prove useful in improving adaptation of groundnut to these environments (Chapman *et al.*, 1993c).

3.5. Leaf Water Potential, Transpiration, RWC and Osmotic Adjustment

The leaf area, which is in rapid growth stage during the vegetative and flowering stages tends to be most affected due to water stress causing reduction in photosynthesing surface and crop growth rate. The most common symptom of water stress in the fields is stunted growth as water stress first affects the cell enlargement rather than cell division, but long exposure of water stress inhibited cell division also. The stem length is reduced more markedly than leaf size and the leaf arrangement becomes more compact. During vegetative and reproductive stages the net assimilation rate is inversely proportional to LAI and it is possible to define optimum LAI for maximum dry matter production in particular area. Turgor potential (ψ_p) and leaf extension rate (R) are reduced at high saturation deficits and R is linearly related to ψ_p between 0900 to 1600 hr., in driest condition. The groundnut maintains high leaf water content even in dry soil and also continues photosynthesising at lower leaf water content than other crops.

The water flow in intact plant under high soil moisture condition is for growth and transpiration and two concepts are expressed about the driving force for transpiration water flow: one consider the water potential differences between the root and leaf as the primary force while the other consider hydrostatic and osmotic pressure differences as the factors determining water flow. In order to sustain plant growth and hydration, water must be continuously supplied to the leaves as it is lost by transpiration. This becomes difficult under low soil moisture condition. The ability of groundnut genotypes to maintain water supply to leaves measured by apparent sap velocity (V_a) was 0.8-1.1 cm min^{-1} and declined with stress in field (Ketring et al., 1990).

The relative water content (RWC) is the water-relation component that seems most directly related to cell-hydration. Other factors such as osmotic adjustment and apoplastic water content contribute to cell turgor through maintaining high RWC and it can be readily measured for larger plant populations. Mild drought induces in plants regulation of water loss and uptake allowing maintenance of their leaf relative water content (RWC) within the limits where photosynthetic capacity and quantum yield show little or no change. The most severe form of water deficit is desiccation - when most of the protoplasmic water is lost and only a very small amount of tightly bound water remains in the cell. Groundnut has an enhanced capacity for maintaining leaf water content against a soil water deficit. However, RWC of the leaf did not recover to its original value on re-watering after severe stress. The leaf water potential and relative water content were negatively correlated with a correlation coefficient of -0.95. The linear regression equation was $\Psi_L = 64.8 - 0.61$ RWC. Perceptibly, stressed plants have lower RWC than non-stressed plants. RWC of non- stressed plants ranged from 85 to 90 per cent, while in drought stressed plants, it may be as low as 30 per cent (Babu and Rao, 1983).

The RWC decreased in groundnut varieties upon induction of drought stress with sharp decrease, but in varieties K 1375 and R 9251 more than 90 per cent relative water content was observed with a decrease of 24 per cent RWC in K 1375 as compared to control, while 44 per cent of decrease was observed in R 9251 as compared to control (Sharada and Naik, 2011). Ramana Rao (1994) reported reduction in RWC of groundnut in both simulated stress and rainfed treatments compared to fully irrigated treatments at 9, 18 and 27 days after imposition of water stress. Increasing moisture stress from 0 to 2, 4 and 6 atm decrease the leaf RWC, increased water saturation deficit and relative saturation deficit, decreased DM accumulation in the shoot and increased it in the root, decreased RGR, increased specific leaf weight, decreased non-reducing sugar and increased reducing sugar in groundnut seedlings (Sharma et al., 1985).

The stomata occupy a central position in the pathways for both the loss of water from plants and the exchange of CO_2 and thus provide the main short-term control of both transpiration and photosynthesis, though the detailed control criteria on which their movements are based are not well understood and are likely to depend on the particular ecological situation.

Using simple models one can investigate the role of stomata in the control of gas exchange in the presence of hydraulic feedbacks and to clarify the nature of causality in such systems. Comparison of a limited number of different mechanistic models of stomatal function is used to investigate likely mechanisms underlying stomatal responses to environment.In greenhouse, a drought resistant groundnut cvs. Nigeria 55437 and IAC Tupa a drought sensitive one when compared, the Nigeria 55437 showed high leaf diffusion resistance (Rs) and low leaf water potential (p_{siw}) and transpiration (E) when subjected to water stress, and had a high proline concentration (P) even under irrigation. There were negative correlations between p_{siw} and Rs, and p_{siw} and P, and positive correlations between leaf temperature (Tf) and Rs, Tf and P, and Rs and P under water stress (Nogueira *et al.*, 1998). A biochemical test using 0.1M EDTA was developed for detection of moisture stress in youngest fully opened leaves, and the cultivars showing 2.77 per cent decline in the pH of the leaves extract in EDTA at -10 bar were categorised as drought tolerant (Dwivedi *et al.*, 1986). Leaves of groundnut have also been shown to accumulate proline under moisture stress (Misra *et al.*, 1992), the level of which was significantly correlated with the level of activity of glutamate-oxaloacetate transaminase in leaves (Yadav *et al.*, 1993).

Most part of India, shortage of water is caused by uneven distribution of rains, gaps between rain events and field water losses rather than from low seasonal or annual rainfall totals. The groundnut grown in the micro catchment during the rainy season, utilized 364-733 mm water in evapotranspiration (ET) and deep percolation (P) (Rathore *et al.*, 1996). In a water balance studies in M.P. in a 1.05 ha field, with a 0.09 ha farm pond (which stored excess water from the wet season) 28-37 per cent of seasonal rainfall was available as surface runoff from a microcatchment (0.66 ha growing groundnut) for collection in the pond and is sufficient to prevent drought stress (Rathore *et al.*, 1996).

Leaf thermocouple hygrometers and specially fabricated stem thermocouple hygrometers were evaluated by Pallas (1978) in groundnut plants under well-watered drought conditions in a growth chamber. When soil-water stress was low and plant-water movement was near steady state, the 2 sensors gave similar water potential values. When soil-water stresses were imposed or when plant process varied cyclically (*e.g.* photosynthesis, transpiration), stem hygrometers sensed dynamic changes in the plant's water potential more consistently than did leaf hygrometers placed in leaves with intact cuticles. Thus, both the stem and leaf hygrometers may be useful for sensing water potential changes of groundnuts under field conditions Rosario and Fajardo (1988) compared 4 high-yielding, 1 low-yielding and 5 intermediate-yielding groundnut cultivars in pot at field water capacity throughout or at 50 per cent field water capacity 14-80 DAS and then watered to FC for 1 week and again subjected to water stress until harvest, significant interactions between cultivar and water stress on stomatal resistance, leaf water potential and root DW were observed and WUE decreased under water stress in all cultivars with a marked decrease in cv. TMV 2 (intermediate yielding) and EG Bunch (low yielding).

At wilting point the transpiration rate in groundnut decreased to 67 per cent. The diffusive resistance increased during drought and after withholding the irrigation for 2-6 days, the RWC was only 30-38 per cent of that at full turgidity. The leaf water potential (ψ_l) is an important parameter, which is measured through psychometric in laboratory; however in field the pressure chamber technique and hydraulic press methods are most common. Saturation deficit (SD) is an important agroclimatic factor controlling the potential evaporation. The groundnut crops are often irrigated or grown on stored moisture during the post-rainy season when SD exceeds 3-4 kPa. This SD has a major effect on the water-use rate and the growth of groundnut as the WUE is inversely proportional to SD. SD more than 2.5 kPa accelerated the depletion of soil moisture reserves and greatly reduces LAI by lowering the turgor potential of expanding leaves. Because expanding leaves are more sensitive to moisture deficit than pods, the partitioning of dry matter is likely to be affected by SD.

The response of leaf area expansion to atmospheric saturation deficit (SD) and soil moisture deficit when examined in terms of leaf water potential and turgor potential by growing groundnut plant at different levels of SD and control with soil kept irrigated to FC, the SD accelerated the depletion of soil moisture reserves in the unirrigated stands and greatly reduced LAI, leaf number/plant and leaf size, but the effect on leaf size was greater than number. SD had less effect than soil water deficit on leaf production. Turgor potential and leaf extension rate were both reduced at high SD and leaf extension rate was linearly related to turgor potential between 0900 and 1600 h. However, leaf extension rate and turgor potential were poorly correlated between 0400 and 0700 h in the driest treatment (Ong *et al.*, 1985).

Mature leaves of well watered field grown groundnut cv. Florunner and Early Bunch were detached and changes in leaf water potential of these leaves and leaves of 10 genotypes subjected to water stress in the field when compared, detached leaves tugor potential (TP) decreased to zero at -1.2 to - 1.3 MPa leaf water potential (ψ_l) and 87 per cent RWC for both cv. Small errors in leaf solute potential (LSP) determined by thermocouple psychrometry, resulted in artefactual negative TP values. In the field tests zero TP occurred at LWP of -1.6 MPa indicating that water relations of groundnut were similar to other crops with no unique drought resistance mechanism (Bennett *et al.*, 1981).In pots, the groundnut cvs. ICGS 1, ICGS 5, ICGS 11, ICGS 44, ICGS 76, ICG (FDRS) 55 and Girnar 1 water stressed at the flowering stage for 4 d, increased stomatal diffusive resistance and proline content and decreased transpiration rate, total chlorophyll, carotenoids and polyphenol contents. The better performance of cv. ICGS 11, ICGS 44 and Girnar 1 under water stress was related to good stomatal conductance and increased proline level for osmotic adjustment (Patil and Patil, 1993).The groundnut leaves entered permanent wilting status at <55 per cent RWC (Nwalozie and Annerose, 1996).

The pressure chamber for measuring leaf water potential in groundnuts when compared with the thermocouple psychrometer in field at Kingaroy on groundnut cv. Q18801, Red Spanish and McCubbin, the pressure chamber over-estimated leaf water potential by an average of 0.4 MPa over the range -0.5 to -5.0 MPa. The study concluded that the pressure chamber technique could be appropriate in comparative studies of groundnut water stress, however, for absolute measurements, as in the calculation of leaf turgor potential, either a correction factor be applied or, preferably, the thermocouple psychrometer technique used (Wright *et al.*, 1988). The RWC of the leaf (97-81 per cent) and water potential of the leaf (-10 to -38 bar) declined with a decrease in soil-water potential from -0.5 to -20 bar. The RWC of the leaf did not recover to its original value on re-watering after severe stress. Cv. TG 17 tolerated water stress better than cv. GAUG 1 (Patel *et al.*, 1983).

The temperature and water stress effect on plant water status, stomatal conductance and water use in groundnut cv. Robut 33-1 studied by Black *et al.* (1985) in controlled environment greenhouses at 25, 28 or 31°C at two treatment (irrigation to half of the stand whenever soil water potential at 10 cm reached -20 kPa and the other half received no further irrigation after sowing, when the soil profile was at FC), the Leaf water potential (LWP), turgor potential (TP) and stomatal conductance (SC) were reduced in unirrigated plants at 29 DAS, when LAI was still below 0.5; SC was more strongly affected than water status. These differences persisted throughout the season as stress increased but SC was poorly correlated with LWP and TP. The decreases in SC and LAI reduced canopy conductance by up to 40 per cent. The conservative influence of decreased SC in unirrigated plants was negated by increases in leaf-to-air vapour pressure difference caused by their higher leaf temp. Transpiration rates were similar in both treatments and the lower total water use of the unirrigated stand resulted entirely from its smaller LAI. Unirrigated plants made less vegetative growth but produced more pegs and pods, although impaired pod-filling reduced pod yields by approx. 35 per cent (Black *et al.*, 1985).

Extensive root system combined with the ability to extract moisture under soil moisture deficits can delay dehydration and prolong the effective production period. Both of these can be evaluated at seedling stages in breeding lines. Matthews *et al.* (1988a) observed that the genotypes with limited irrigation, in central India, transpired similar total amount of water (220-226 mm) over the seasons, but produced different amounts of shoot dry matter varying from 390-490 g m². Joshi *et al.* (1988) identified GG 2 as drought tolerant variety and JL 24 as sensitive cultivar. The lower leaf water potential and diffusive resistance (DR), higher transpiration rate and quick recovery of stomatal activity after relief of the stress and maintenance of low leaf water potential even at high RWC were the main reason for resistance in GG 2. Moderate water deficit at pre-flowering phase (without irrigation from 21 to 50 DAS) showed higher mean stomatal conductance, crop growth rate, pod growth rate, and yield than the control getting regular irrigation alfisol and vertisol.

Nageswara Rao *et al.* (1988) reported that applying two irrigations at 11 and 21 DAS followed by withholding irrigation for 30 days (up to 50 DAS) and again irrigation (at 50 per cent FC) at 10 days intervals showed the mean stomatal conductance 8.3-11.5 mm s⁻¹, CGR 12.2-13.1 g m⁻² day⁻¹, pod growth rate (PGR) 9-9.9 g m⁻² day⁻¹, partitioning factor (PF) 74-76 per cent and pod yield 5.3-5.5 t ha⁻¹ in Robut 33-1 groundnut, however, the crop irrigated to 50 per cent FC at 10 days intervals showed the stomatal conductance 6.9-10.4 mm s⁻¹, CGR 8.8-13.5 g m² day⁻¹, PGR 6.4-10.2 g m² day⁻¹, PF 73-75 per cent and pod yield 4.6-4.7 t ha⁻¹. The responses of crops to drought stress under five farming systems in China reported that the differences in water potentials between the air and crop leaves were >100 times higher than those between crop leaves and soil in the surface 10 cm layer and the differences between the latter were also >100 times higher than those between soils in the 10 cm and 70 cm layers (Zhang *et al.*, 1999). However, the differences in water potentials were lower with minimum tillage and narrow ridge tillage than with conventional tillage and wide ridge tillage respectively, indicating the serious water stress in the former treatments. The diurnal variation of water potentials in the soil-plant-atmosphere continuum indicated that groundnut is more tolerant of drought than soyabean and maize, leaf water potential decreased with the increase in soil water potential and its relationship could be expressed by binomial equations. With ridge tillage, leaf water potential was more related to the water potential of the deeper soil layers than to that of the soil surface under minimum tillage. Seasonal drought in the region is caused by combined water stress and high temperature, water potentials of soils and leaves increased with the increase in soil and air temperatures, respectively and the effect of temperature on water potentials of soils and leaves was influenced by farming practices and crop communities (Zhang *et al.*, 1999).

Effects of 5 drought (Early, middle and late drought periods of short duration of 35 days and extended early and midseason droughts of 70 days) imposed by withholding irrigation studied on Tifton loamy sand measuring the yield, percentage sound mature kernels (SMK), germination, leaf water potential and leaf diffusion resistance; drought progressively decreased yields as duration and lateness of occurrence in the season increased (Pallas *et al.*, 1979). In maize, groundnut and pearl millet, the water stress and aging of leaves within various canopies when studied after withholding irrigation for 7 d, the leaf temp., transpiration and PAR decreased with leaf aging, while diffusive resistance increased with aging, shading and water stress. Leaf temp., diffusive resistance and PAR of the crops were in the order: groundnuts < maize < pearl millet (Golakiya, 1989). In pot, groundnut cvs. Dh 330 and TMV 2 subjected to water stress at flower initiation (30 DAS), peg formation (40 DAS) or pod formation (70 DAS) by withholding water for one week, the leaf diffusive resistance and proline accumulation were highest and transpiration rate lowest with water stress at 70 DAS and cv. Dh330 was more tolerant to water stress as it had higher leaf diffusive resistance, lower transpiration rate and greater proline accumulation than TMV 2 (Koti *et al.*, 1994).

In Perkins, Oklahoma, irrigation at 5.5 cm/week or no irrigation (except rainwater) affected the leaf relative water content (LRWC) of groundnut cvs. Comet and Florunner grown on a Teller loam soil at soil relative water content (SRWC) above 50 per cent, the mean LRWC was about 85 per cent, and was affected more by evaporative demand than by SRWC. Below 50 per cent SRWC, LRWC was highly correlated with SRWC. The predicted SRWC when turgor pressure potentials approached zero was about 45 per cent. This SRWC threshold occurred under rainfed conditions in the 3 years study at 59, 56 and 64 DAP during flowering and pod formation. Thus, the SRWC vs. soil water pressure potential curve of Teller soil may be useful for predicting limiting levels of soil water for groundnut and that limiting levels of soil water may occur well above the classically defined lower limit of soil water availability (Erickson *et al.*, 1991).

3.6. Photosynthesis and Fluorescence

The plant water and soil moisture stress when investigated on photosynthesis, stomatal aperture and transpiration in groundnut, maximum photosynthesis occurred at a soil moisture content (MC) of 50-60 per cent (of the maximum water-holding capacity) and at a leaf DPD (diffusion pressure deficit) of 1-1.5 atm. Photosynthesis of groundnut was adversely affected by a high soil MC. The photosynthesis, stomatal aperture and transpiration were all affected by low soil MC, before the leaves showed visible signs of wilting and hence stomatal aperture may be used as a physiological indicator of necessity of irrigation (Chen and Chang, 1972).Bhagsari *et al.* (1976) reported that in groundnut cvs. Florunner and Tift-8, the relation between net photosynthesis and RWC measured in single leaf during a 5 to 6day period without water was similar, diffusive resistance was 0.5-2.5 s/cm in control plants, 30-35 s/cm in stressed plants and 18-20 s/cm in stressed groundnut cv. Florunner, P.I. 149268 and the wild *Arachis monticola*. The water potential of immature pods was equal to or slightly higher than that of leaves at the end of the outdoor trial. There were no differences among cultivars in response to water stress and it is concluded that net photosynthesis in groundnut is controlled by water stress in a manner quantitatively similar to that in other crop species (Bhagsari *et al.*, 1976).

Wang *et al.* (1995) in a field at Laixi, China, two high-yielding groundnut cultivars (Luhua 11 and Haihua 1) characterized by few branches, medium duration of growth, and large pods studied for the physiological characteristics related to their canopy morphology such as light distribution through the canopy, light interception rate (LIR) at different stages of plant development, diurnal and seasonal variations in photosynthesis, dark respiration, and the effects of water deficiencies and listed Physiological indices for the groundnut canopy that would ensure yield of 7.0-7.5 t ha^{-1}. The major area of photosynthesis was the upper one-third of the canopy which received four-fifths of the solar radiation and contained about half of the total leaf area. The LIR of the canopy remained low (50 per cent of solar radiation reached the ground) until the flowering-pegging stage but rose to a peak which was maintained until the early-mid pod-filling stage. The diurnal variations in photosynthesis on a clear day showed a peak between 1100 and 1300 h. No sign of 'noon rest' or light saturation was evident under natural conditions so the dominant factor was intensity of solar radiation. Dark respiration was maximum at 1100-1500 h and minimum at 0300-0500 h, and was strongly dependent on air temperature. Over the growing season the net photosynthetic rate increased slowly during the seedling stage and accelerated from the flowering-pegging stage until midpod formation when it reached a peak. Photosynthesis was more sensitive to water deficiency than dark respiration. Physiological indices for the groundnut canopy that would ensure yields of 7.0-7.5 t ha^{-1} are listed (Wang *et al.*, 1995).

The effect of water deficit on nodulation, N_2-fixation, photosynthesis, total soluble sugars and leghaemoglobin in nodules was investigated where Nitrogenase activity completely ceased in groundnuts at -1.7 MPa. With increasing water stress, the acetylene reduction activity decreased gradually in groundnuts. Nodule FW declined in groundnuts when leaf water potential decreased by 1.0 MPa, but no nodule shedding was noticed even at a higher stress level in groundnuts. Photosynthesis and stomatal conductance were more stable in groundnuts than in *V. unguiculata* under water stress.There was a sharp increase in total soluble sugar and leghaemoglobin in the nodules of groundnuts with water stress, but no definite trend in V. unguiculata (Venkateswarlu *et al.*, 1989).

Matsunami and Kokubun, (2003) studied the specific differences in the photosynthetic responses of four legume crops to daily fluctuations of evaporative demand and to identify physiological attributes responsible for the differences, soyabean (cv. Enrei), adzuki bean (cv. Dainagon), cowpea (cv. Kuromidori) and groundnut (cv. Chibahandachi) in field of Tohoku University, Japan, and apparent leaf photosynthetic rate (AP) and gas exchange parameters and water potential and transpiration rate were measured at midday where leaf-to-air vapour pressure deficit (VPDla) varied from 1.72 - 3.44 kPa the four species responded differently, the AP of soyabean and that of adzuki bean decreased with increasing VPDla whereas the activities of cowpea and groundnut were greatest at VPDla around 2.53 kPa. The leaf water potentials of soyabean, adzuki bean and cowpea reached their lowest at VPDla around 2.53 kPa, while that of groundnut was fairly constant over the VPDla range of 1.72 to 3.44 kPa. The transpiration rates of soyabean, adzuki bean and cowpea were greatest at VPDla ~2.53 kPa and decreased beyond that range of VPDla, while groundnut transpired actively with increasing VPDla. AP of soyabean was correlated with leaf water potential, whereas that of cowpea and groundnut was correlated with transpiration rate. With respect to water relations, groundnut was the most tolerant of increasing VPDla among the four species tested, presumably because it maintained higher water potential and transpiration rate than the other species under the condition of high VPDla.

Clifford *et al.* (1993) reported that the primary effects of elevated CO_2 on growth and yield of groundnut stands were mediated by an increase in the conversion coefficient for intercepted radiation and the prolonged maintenance of higher leaf water potentials during drought stress. Groundnut cv. Kadiri3 grown in controlled-environment greenhouses at 28 °C (±5°) under 2 levels of atmospheric CO_2 (350 or 700 ppmv) and 2 levels of soil moisture (irrigated weekly or no water from 35 d after sowing) where elevated CO_2 increased the maximum rate of net photosynthesis by up to 40 per cent with an increase in conversion coefficient for intercepted radiation of 30 per cent (1.66 - 2.16 g MJ^{-1}) in well-irrigated conditions, and 94 per cent (0.64 - 1.24 g MJ^{-1}) on a drying soil profile. Elevated CO_2 increased DM accumulation by 16 per cent (from 13.79 to 16.03 t ha^{-1}) and pod yield by 25 per cent (from 2.7 to 3.4 t ha^{-1}), but the HI was not affected. The beneficial effects of elevated CO_2 were enhanced under severe water stress; DM production increased by 112 per cent (from 4.13 to 8.87 t ha^{-1}), and a pod yield of 1.34 t ha^{-1} was obtained in elevated CO_2, whereas comparable plots at 350 ppmv CO_2 only yielded 0.22 t ha^{-1}. There was a corresponding decrease in HI from 0.15 to 0.05. Following the withholding of irrigation, plants growing on a stored soil water profile in elevated CO_2 could maintain significantly less negative leaf water potentials for the remainder of the season than comparable plants grown in ambient CO_2, allowing prolonged plant activity during drought. On a drying soil profile, allocation in plants grown in 350 ppmv CO_2 changed in favour of root development far earlier in the season than plants grown at 700 ppmv CO_2, indicating that severe water stress was reached earlier at 350 ppmv CO_2 (Clifford *et al.*, 1993).

Subramanian and Maheswari, (1990) reported that groundnut plants adapt to water stress by slowing down tissue dehydration. Groundnut plants, subjected to water stress at flowering by withholding irrigation for 4 d and leaf water potential (ψ_l), transpiration rate, stomatal conductance and photosynthetic rate measured daily in the morning until the young leaflets folded vertically where LWP, transpiration rate and photosynthetic rate decreased progressively with increasing duration of water stress, indicating that plants under mild stress postpone tissue dehydration, stomatal conductance decreased almost steadily during the stress period indicating that this was more sensitive than water loss during the initial stress period. Sharma *et al.* (1993) reported that groundnuts cv. M 13 grown in pots by withholding water for 1- 4 d at flowering (30 DAS), reduced photosynthetic rate, leaf water potential and transpiration rate, water stress was mild in the 1st and 2nd day, but stomatal conductance decreased progressively with prolonged water stress. Photosynthetic rate was 0.30 mg CO_2 m^{-2} s^{-1} after 4 d of stress compared with 0.80 mg in control plants. Phosphoenolpyruvate carboxylase activity progressively increased with stress whereas the activities of RuBP carboxylase and NADP-glyceraldehyde-3-phosphate dehydrogenase decreased gradually (Sharma *et al.*, 1993).

During rainy seasons groundnut cv. JL24 (recommended for irrigated areas), TMV 2 and Kadiri 3 (recommended for dryland areas) grown on red sandy loam soil (Alfisol) under irrigated or dryland conditions, seasonal average photosynthetic rate, transpiration rate, stomatal conductance and leaf water potential were lower and canopy temperature higher under dryland than irrigated conditions due to drought. TMV 2 and Kadiri 3 stomatal conductance was lower by 25 per cent and photosynthesis by 22 per cent, however, transpiration rate was lower by only 9 per cent. At peak flowering JL 24 had significantly higher canopy temperature than TMV 2 and Kadiri 3. Thus measurements of integrated physiological processes, rather than of instantaneous rates of assimilation and water loss, are likely to reveal a basis for cultivar differences in water limited environments (Subramanian *et al.*, 1993).

The response of photosynthesis to soil moisture availability is signicantly faster in the tolerant groundnut genotypes compared to susceptible plants, and this appears to be, in part, caused by better stomatal control. Kameshwara Rao *et al.* (2009) suggested that light-saturated photosynthesis (A) was similar in the tolerant (COC 041) and susceptible (COC 166) genotypes of groundnut before water stress was applied (0 d) and at the end of the stress period (7 d), but (A) was higher in the susceptible genotype during stress exposure (3 d) because of higher stomatal conductance. On the second day after re-watering, A and gs were higher in the tolerant genotype reecting faster recovery from stress, but 7 d after re-watering both genotypes had fully recovered. Instantaneous leaf-level WUE (A/gs; WUE) was higher in the tolerant genotype prior to and during the onset of water decit because of lower gs. However, upon return to saturated soil moisture conditions, assimilation rates and stomatal conductance increased rapidly (within 48 h), and WUE was lower in the tolerant genotype compared to the susceptible genotype. Although the susceptible genotype showed a slower recovery following the stress treatment, both genotypes recovered to pre-stress levels of photosynthesis 7 d after re-watering. As a consequence of water stress, the photosynthetic machinery may be reversibly and partially de-activated to reduce detrimental loss of water, but may be rapidly activated upon re-watering. This is an essential trait for production agriculture where plants are continually exposed to intermittent irrigation events in both rain-fed and irrigated conditions. In tolerant groundnut genotype (COC 041), a 30 per cent decrease in irrigation levels in the eld only reduced yield by 15–20 per cent compared with 90 per cent in the susceptible genotype, which is a substantial difference in real-world application. This yield response appears to be correlated with a rapid down regulation of photosynthesis via stomatal closure and decrease in photosynthetic machinery which could minimize water loss from the plant and prevent cellular damage. Additionally, higher epicuticular wax may minimize additional water loss from the tolerant genotype. Upon re-irrigation, the tolerant genotype rapidly regains a signicant percentage of its non-stressed photosynthetic capacity. Interestingly, the tolerant genotypes fail to exhibit up regulation of many proteins known to be responsive to stress, suggesting that under the similar levels of available soil moisture, the tolerant genotype does not experience the same level of stress as experienced by the susceptible genotype. Although both tolerant and susceptible plants recover full photosynthetic capacity within 1 week of re-irrigation, it is perhaps the recovery phase and potential energy costs associated with cellular repair that ultimately have an impact on signicantly reducing yield in the susceptible genotype (Kameswara Rao, 2009).

To study the effect of drought on the mechanisms of energy dissipation, Lauriano (2006) conducted a trial using two-month-old *Arachis hypogaea* cvs. 57-422, 73-30, and GC 8-35submitted to three treatments: control (C), mild water stress (S_1), and severe water stress (S_2). Photosynthetic performance was evaluated as the Hill and Mehler reactions. These activities were correlated with the contents of the low and high potential forms of cytochrome (cyt) b559, plastoquinone, cyt b563, and cyt f. Under mild water stress the regulatory mechanism at the antennae level was effective for 57-422 and GC 8-35, while in the cv. 73-30 an overcharge of photosynthetic apparatus occurred. Relative to this cv. under S1 the stability of carotene and the dissipative cycle around photosystem (PS) II became an important factor for the effective protection of the PSII reaction centres. The cyclic electron flow around PS I was important for energy dissipation under S_1 only for the cvs. 57-422 and 73-30.

Lauriano (2004) measured the photosynthetic response of three *Arachis hypogaea* L. cultivars (57-422, 73-30, and GC 8-35) grown for two months under water available conditions, severe water stress, and 24, 72, and 93 h following re-watering. At the end of the drying cycle, all the cultivars reached dehydration, relative water content (RWC) ranging between 40 and 50 per cent. During dehydration, leaf stomatal conductance (gs), transpiration rate (E), and net photosynthetic rate (PN) decreased more in cvs. 57-422 and GC 8-35 than in 73-30. Instantaneous water use efficiency (WUEi) and photosynthetic capacity (Pmax) decreased mostly in cv. GC 8-35. Except in cv. GC 8-35, the activity of photosystem I (PS I) was only slightly affected. PSII and ribulose - 1, 5-bisphosphate carboxylase/oxygenase (RuBPCO) were the main targets of water stress. After re-watering, cvs. 73-30 and GC 8-35 rapidly regained gs, E, and PN activities. Twenty-four hours after re-watering, the electron transport rates and RuBPCO activity strongly increased. PN and Pmax fully recovered later.

Measurement of chlorophyll (Chl) a uorescence constitutes one of the oldest approaches to investigate photosynthesis, the rst Chl uorescence experiments being reported more than 70 years ago. Monitoring uorescence induction (FI) has become a widespread method for probing photosystem II (PSII), mostly because it is non-invasive, easy, fast, and reliable, and requires relatively inexpensive equipment. When dark-adapted photosynthetic samples are excited with actinic light, FI is characterized by the initial uorescence level (F_0 or O), which represents excitation energy dissipated as photons before it reaches open reaction centres, and a subsequent rise from F_0 to maximal level (Fm or P), related to a series of successive events that lead to the progressive reduction of the quinone molecules located on the acceptor side of PSII.

Kameshwara Rao *et al.* (2009) evaluated 17 genotypes from the US groundnut mini-core collection and three check cultivars using the chlorophyll uorescence screening technique following imposition of water-decit stress. Of these 20 genotypes, ve identied as stress-tolerant (COC 041, COC 384, COC 249, COC 149, TMV 2) and ve stress-susceptible (COC 166, COC 227, COC 068, Tamrun OL 02, ICGS 76) genotypes based on observations of chlorophyll uorescence yield, whole-plant WUE (mg mass produced per g of water used, and specic leaf area. Tolerant genotypes were characterized by smaller percentage changes in chlorophyll uorescence yield during water-decit stress, higher WUE during well-watered and decit stress conditions and slightly higher SLA, compared with susceptible genotypes. The most tolerant genotype (COC 041) exhibited a similar decline in uorescence yield over time in both water-stressed and well-watered plants. In contrast, the most susceptible genotype (COC 166) exhibited a marginal decline in uorescence yield in the water-stressed plants after 72 h of incubation.

In groundnut leaf dehydration enhanced variable fluorescence yield immediately and the model of pattern alteration of fluorescence curves by dehydration included three stages: an increase of variable fluorescence and the steady state of fluorescence, the formation of a fluorescence "plateau" and an eventual decline of fluorescence. The maximum yield of fluorescence was usually reached as the plateau appeared. The water content at which point the plateau was completely formed was critical and may be species/genotype dependent. Leaf rehydration could restore transient fluorescence before this point was reached. The dehydration-rehydration cycle experiments indicated that the block on the photoreducing side of PSII was reversible, whereas the change on the water splitting side was irreversible. Comparing superoxide dismutase (SOD) levels in leaves of different plants, the wild groundnuts had more SOD than a cultivated groundnut based on leaf FW. Both cyanide-sensitive and insensitive SOD isoenzymes were found in leaves of wild groundnuts. In contrast, the isoenzymes in leaves of cultivated groundnuts were cyanide-sensitive and mainly chloroplastic isoenzymes. Paraquat reduced SD activity markedly in groundnut leaves (Wu, 1987).

Clavel *et al.* (2006) obtained genotypic and treatments responses in groundnut on some of the uorescence parameters, in particular the structure-function-index values that traduce the status of photochemical apparatus. The technique of chlorophyll uorescence, as it is rapid, sensitive and non-destructive, could therefore become a useful method for determining variations in tolerance of the photosynthetic apparatus in breeding for resistance to drought. Nevertheless, the role and value of the fluorimetric responses for maintaining yield under drought needs to be clarify due to difference existing in groundnut drought adaptation strategies. Future selection schemes for improvement of drought adaptation in the cultivated groundnut species are therefore possible by using ûuorescence parameters in association with yield-based studies in the eld. Thus Chlorophyll a fluorescence is a highly versatile tool, for researchers studying photosynthesis as well as for those working in broader fields related to physiology of plants.

Kalariya *et al.* (2012b) studied the chlorophyll fluorescence and net photosynthesis under water deficit conditions in six groundnut (*Arachis hypogaea* L.) cultivars by withholding irrigation from 31 to 60 DAS (between beginning bloom to beginning seed) and 60 to 87 DAS (beginning seed to beginning maturity). The soil moisture content was 17 per cent and 18 per cent at 0-15 and 15-30 cm soil depth in well watered plots which decreased to 9 per cent and 10 per cent respectively in water deficit plots during on 60 DAS. A similar trend was followed during 60-87 DAS also. The RWC of groundnut leaves decreased from 92 in control to 88 under water deficit condition at 60 DAS with the least decrease in ICGS 44. The water deficit stress has decreased maximum efficiency of PS II (*Fv/Fm*) and proportion of absorbed energy utilised for photochemistry (ÖPS II), net photosynthesis rate and thereby linear electron transport rate*J* but, increased minimum fluorescence (F_0) and non-photochemical quenching (*NPQ*) 60 and 87 DAS (Kalariya *et al.*, 2012a). Water deficit stress increased minimum fluorescence (F_0) and non-photochemical quenching (*NPQ*) but, decreased maximum efficiency of PS II (F_v/F_m) and proportion of absorbed energy utilised for photochemistry (ÖPS II)at 60 DAS due to water deficit stress imposed 30-60 DAS in groundnut (Kalariya *et al.*, 2012b).

The water deficit condition during 30-60 DAS has significantly increased F_0 and *NPQ* but decreased the maximum quantum yield of PS II (F_v/F_m) from 0.81 in control to 0.77 at 60 DAS which was again resumed to 0.80 after 48 hours of withdrawal of stress. The rate of photosynthesis which was 29 and 36 μmol m^{-2} s^{-1} in well irrigated plots decreased to 26 and 28 μmol m^{-2} s^{-1} with a deduction of 11 and 30 per cent at 60 DAS and 87 DAS, respectively. Variety TAG 24 showed better stress recovering capacity with high photosynthesis under both control as well as water deficit condition whereas, data on chlorophyll fluorescence parameters showed that variety ICGS 44 was least affected to damage via photoinhibitory action (Kalariya *et al.*, 2013).

3.7. Translocation

Soil moisture deficits imposed between sowing and pod initiation or between pod initiation and final harvest of groundnuts by regulating irrigation and $^{14}CO_2$ on 5-6 occasions between 50-97 DAS were studied in a controlled environment by Stirling *et al.* (1989). The study reveals that leaves were the primary sites of $^{14}CO_2$ fixation, though their contribution generally declined late in the season; whereas fixation by stems was initially low but increased sharply when stress was released in the late-irrigated stands. ^{14}C-fixation by stem apices and pegs also rose sharply following irrigation of the late-stressed stands. Leaves were the primary source of assimilates, but translocation tended to decrease as the season progressed, even in the late-irrigated stands. Stems were initially the major sinks, but their sink activity disappeared almost completely when stress was released in the late-irrigated stands. Assimilate import by stem apices declined progressively and pod sink activity was negligible in the late-stressed stand, but both increased markedly when early-season stress was released. Leaf water status showed marked diurnal variation; pegs showed less variation and maintained much higher turgor levels, largely because of their lower solute potentials. Marked osmotic adjustment occurred in expanding but not in mature leaves, allowing them to maintain higher turgor levels during periods of severe stress. This adjustment was rapidly lost when stress was released. The observed changes in assimilate production and partitioning preceded detectable changes in bulk turgor levels (Stirling *et al.*, 1989).

Clavel *et al.* (2005) assessed the field productivity of four Sahelian groundnut cvs during three crop seasons in Bambey (Senegal) and same cvs grown in rhizotrons were subjected to early drought stress and to a desiccation test to assess cell membrane tolerance, where differences were found between cultivars with respect to pod yield, biomass production, WUE, stomatal regulation and cell membrane tolerance and to cope with water deficit two strategies were identified. The first was characterised by rapid water loss, late stomatal closure and low cell membrane damage during drought which were found in the semi-late Virginia cv 57-422 and, into a lesser extent, in the early Spanish cv Fleur 11. The biomass production was boosted in both the cvs under favourable conditions in rhizotrons but the semi-late cv had poor pod yield under end-of-season water deficit conditions. The second strategy involved opposite characters, leading to the maintenance of a highe water status, resulting in lower photosynthesis and yield. This characterised the early Spanish cv 73-30, and also, to some extent, the early Spanish cv 55-437. Earliness associated with high WUE, stomatal conductance and cell membrane tolerance, were the main traits of Fleur 11, a cv derived from a Virginia x Spanish cross, which was able to maintain acceptable yield under varying drought patterns in the field. These traits, as they were detectable at an early stage, could therefore be efficiently integrated in groundnut breeding programmes for drought adaptation (Clavel *et al.*, 2005).

The environmental factors influencing pod yield in groundnuts operated mainly through their effect on the timing of pod initiation and the size of reproductive sink, as defined by the number of pods set (Stirling, 1989). During the post-rainy season, the effect of artificial (bamboo screens) or the onset of rapid pod growth to final harvest and natural (sorghum) shading on growth and development of groundnuts when examined in terms of the associated changes in leaf and soil temperature and plant water status, the artificial shading with bamboo screens providing approxemately 45 per cent shade substantially reduced leaf and soil temperature relative to the crop grown alone, this had little effect on plant morphology and DW at final harvest (Stirling, 1989). Root respiration at 30 d after sowing (DAS) was increased under water-stressed conditions. Yield was negatively correlated with root respiration

30 and 60 DAS. Cv. JL 24 was the most tolerant of water stress, with a yield reduction in stressed conditions of 32.1 per cent compared with an average of 46.7 per cent, and a 66.9 per cent yield reduction in the most susceptible cultivar (TAG 24) (Dhopte *et al.*, 1992).

3.8. Biochemical Parameters

Dwivedi *et al.* (1986) described a biochemical test for prediction of drought resistance in groundnut cultivars based on chelation of the youngest fully opened leaves with 0.1 M EDTA and extraction of organic substances and osmotically-active compounds; the degree of decrease in the pH of EDTA extract of leaves is related to drought resistance and the cultivars showing <2.8 per cent decrease in the pH of EDTA extract of leaves at -10 bar water stress are considered as drought resistance. Drought resistance in 22 cvs. was determined using this test and examined in relation to stomatal resistance, transpiration rate, saturated water deficit, pH water deficit, pH water deficit index and decrease in pod yield under water stress (Dwivedi *et al.*, 1986). Water stress at 2-6 atm decreased protein contents and increased amino acid and proline contents in leaves of 15 d old groundnut seedlings, decreased peroxidase and nitrate reductase activities and increased ribonuclease activity and proline content and ribonuclease activity could be used as indication of water stress in groundnut seedlings (Sharma *et al.*, 1990).

The epicuticular wax load (EWL) on leaves reduces transpiration and improves crop WUE and, genotypic differences were observed in EWL of 12 genotypes grown in the rainy season at 45 DAS by Samdur *et al.* (2003) and the values of EWL ranged from 0.91 mg dm^{-2} in Chico to 1.74 mg dm^{-2} in PBS 11049, with a mean of 1.27 mg dm^{-2}. The mean values were 1.10, 1.58, 2.05 mg dm^{-2} at 45, 75, and 95 DAS, respectively. In both dry seasons, genotypic differences were found in the EWL and effect of various moisture deficit treatments and their interactions with the genotypes were observed with values ranged from 0.653 to 2.878 mg dm^{-2}. The highest EWL was found in PBS 11049 (2.24 mg dm^{-2}). The EWL increased with age of the crop, with greater increase under moisture deficit stress. Thus the genotypic differences exist in EWL of groundnut and this EWL increases with increased crop age and more pronounced under protracted moisture deficit stress (Samdur *et al.*, 2003). The methyl jasmonate (0-125 mg/litre) application in groundnut grown in nutrient solution and treated at the 3- or 4-leaf stage with polyethylene glycol where methyl jasmonate decreased stem height, leaf area and transpiration rate, and increased leaf thickness, the size of water storage cells in leaves, ABA and proline contents of leaves and peroxidase activity in stems. Methyl jasmonate decreases water loss, and may help groundnut seedlings adapt to drought (RuiChiand HuangQing, 1995).

In response to environmental changes in temperature, oxygen or water levels stress proteins occur in groundnut seeds during maturation and curing because these processes are known to be associated with water deficit and anaerobic metabolism in seeds and to test this hypothesis, a polyclonal antibody against dehydrin, a plant stress protein, was used by Chung *et al.* (1998) and the immunoblot analyses showed that a number of dehydrin-related stress proteins were detected in groundnut seeds of different maturity and curing stages, of these, only two were induced during seed curing and maturation. One (protein a) is potentially a groundnut maturity marker because it was shown to occur only in uncured fully mature seeds. Immunoblot analyses of alcohol dehydrogenase (ADH), an enzyme known to be induced in mature groundnut seeds, showed that ADH was not recognized by the antibody. This suggests that ADH is probably not related to protein a or dehydrin (Chung *et al.*, 1998).

Groundnut cultivars grown at different water regimes (75, 50 and 25 per cent FC corresponding to mild, moderate and severe water stress, respectively) decreased leaf dry matter and relative water content with water stress in both cultivars, with K 134 being more tolerant compared to JL 24. The total leaf protein content decreased, whereas the free amino acid and protease activity increased with the increase in water stress. A 2-3 fold increase in proline content was recorded by K 134 and JL 24, respectively, in response to water stress. The better maintenance of RWC, dry matter accumulation, free amino acid and proline and protease activity of K 134 makes it a more drought-tolerant cultivar than JL 24 (Madhusudhan *et al.*, 2002).Generally 400-1000 mg proline m^{-2} accumulated in vegetative parts. Accumulation of K^+ in leaf was identified as one of the good parameter for drought tolerance in groundnut (Arjunan *et al.*, 1988). The PP 333 (Paclobutrazol) caused the accumulation of indigenous ABA and the inhibition of GA biosynthesis in groundnut leaves and increases the drought resistance of groundnut seedlings (Ling and Rui-chi, 1988). Nautiyal, *et al.* (2001) reported the response of groundnut to various aspects of deficit irrigation practices during vegetative phase.

The mild and moderate drought (20 or 30 days) at different crop growth stages in groundnut cv. GG 2 witholding irrigation for 20 days, the total carbohydrate concentration and its fractions did not change except during the pod maturity stage where an increase was observed. After 30 days without irrigation, an increase in the total carbohydrate concentration and its fractions at all crop growth stages was observed. However, a significant reduction in the total carbohydrate and sugar contents (mg/kernel) was observed for both mild and moderate drought at any crop growth stage. Total lipid concentration in the kernel significantly decreased in response to drought for 20 or 30 days at different crop growth stages except for 30 days of drought at the pod maturation stage. The content of total lipid/kernel decreased at all crop growth stages. The proportions of non-polar and polar lipids and individual phospholipids were not influenced by drought (Kandoliya *et al.*, 2000). The benzyladenine soaking of seeds followed by PEG 6000 induced water deficit stress in leaf of 15 day old seedlings and among the groundnut cultivars GG 2, GG 3, GG 4, GG 5, GG 7 and J 11, the GG 2 showed highest RWC, greater accumulation of proline, ascorbic acid and reducing sugars whereas, cv. J 11 had the lowest RWC and showed lower proline and sugar (Dhruve and Vakharia, 2007). Increasing the level of PEG, simulated water stress showed reduction in RWC whereas the level of free amino acids, proline and sugars increased. Soaking of seeds in benzyladenine helped in amelioration of PEG simulated stress by maintaining higher RWC and maintaining adequate level of osmolytes (Dhruve and Vakharia, 2007).

Cell suspensions of the drought-susceptible groundnut cv. JL 24 and drought-resistant Kadiri 3 when inoculated into growth medium containing 2-15 per cent

PEG, both cultivars showed rapid logarithmic growth up to 21 d and then reached a plateau. Cell growth decreased significantly and linearly with increasing PEG concentration. With 15 per cent PEG cell DW after 21 d decreased by 51 per cent in Kadiri 3 and by 44 per cent in JL 24 compared with the control without PEG. When 25 d old whole plants were deprived of water for 6 d in the greenhouse, Kadiri3 maintained higher leaf water potential under water stress than JL 24. Proline accumulation was higher and tissue K content lower in cell culture and whole plants of JL 24 than in Kadiri 3. Since responses to stress in cell culture and in whole plants differ, the need is stressed for care in extrapolation of experimental data to predict field responses (Venkateswarlu *et al.*, 1993).The PEG stress tolerance in TMV 2 and JL 24 when compared at the fifth subculture stage on the stress to stress (selected) and control to stress (non-stress) medium showed further stress tolerance capacity of selected tissues in comparison with non-selected tissues. The JL 24 had a greater capacity to grow at all levels of stress (-0.6 to -0.1 MPa) and its tolerance was associated with higher values of pressure potential, amino acids and proline accumulation as compared to TMV 2 (Purushotham *et al.*, 1998).

3.9. Enzymes Activity

In greenhouse pot trials, 9 groundnut cultivars with different degrees of resistance to water stress, when subjected to water stress for 43 days from 15 DAS and then normal water supply, the water stress resistant cultivars showed similar patterns of peroxidase activity (Santos dos *et al.*, 1997). In 1-week-old groundnut seedlings water stress of -3 to -7 bar for 24 - 168 h, increased activities of both peroxidase and IAA oxidase with increase in the duration of water stress, increases in water stress levels showed smaller increase in the peroxidase activity but greater increase in the IAA oxidase activity, as compared with the control treatments (Mathews, 1988).Two groundnut cultivars (Jun 40 and GG 2) when compared for water deficit tolerance based on the RWC and membrane stability index. The results revealed that both these cultivars were tolerant. Cultivar Jun 40 accumulated superoxide (O_2^-) radicals to a higher level and had higher activity of the scavenging enzyme superoxide dismutase. Increased activity of ascorbate peroxidase [L-ascorbate peroxidase] in the leaves of stressed plants of Jun 40 compared to GG 2 appeared to be responsible for the lower H_2O_2 content. GG 2 showed less lipid peroxidation than Jun 40 under water deficit stress (Mittal *et al.*, 2006).

At USDA-ARS National Peanut Research Station Dawson, Georgia, the groundnuts cv. Florunner grown for 120 days, harvested at maturity, windrow dried for 4 days and sampled from day 4 further dried to 10 per cent moisture content and sorted by pod colour, shelled and the seeds stored at -80°C from which the enzyme extracts were prepared and colour assays developed. The study reveals that the activities of glycolytic enzymes (aldolase [fructose-bisphosphatealdolase], glyceraldehyde-3-phosphate dehydrogenase, pyruvate decarboxylase and alcohol dehydrogenase (ADH)) increased during groundnut maturation and curing, suggesting that these processes are associated with anaerobic conditions, furtete the enzyme activities were higher in cured groundnuts than in non-cured groundnuts, indicating anaerobic conditions were more severe in the former and the increase of

ADH is primarily due to the increased activities of glycolytic enzymes preceding ADH in the alcohol fermentation pathway (Chung *et al.*, 1997).

Common with other abiotic stresses, drought causes increased production of activated oxygen species (ROS) that inactivate enzymes and damage cellular components (Shao *et al.*, 2007, 2008). Oxidative stress occurs when the defence capacity of plants is broken by the formation of free radicals. Since water availability is usually the main factor affecting groundnut productivity in dry regions, strategies aiming at improving sustainable use of water and plant drought tolerance are urgent. Plants scavenge and dispose of these ROS by use of antioxidant defence systems present in several subcellular compartments. The enzymatic antioxidant system includes superoxide dismutase (SOD), peroxidase (PER), catalase (CAT), and the ascorbate and the glutathione cycle enzymes. The primary antioxidant metabolites are ascorbate, α-tocopherol, β-carotene and glutathione. Superoxide dismutase accelerates the conversion of superoxide to hydrogen peroxide; catalase and peroxidase degrade hydrogen peroxide (Santos and Almeida, 2011). SOD constitutes the first line of defense via detoxification of superoxide radicals (Sairam *et al.*, 2000). The lower membrane stability index reflects the extent of lipid peroxidation, which in turn is a consequence of higher oxidative stress due to water stress conditions. Production of activated oxygen species (AOS) is one of the major secondary responses of stress.

Plants respond to various biotic and abiotic stresses threats by an efficient antioxidative system. The exogenous application (spraying on 20 day-old plants) of salicylic acid (0.014 and 0.028 per cent SA), *Acalypha fruticosa* chloroform leaf extract (1.0 per cent) and Neem Oil formulation (0.2, 0.5 and 1.0 per cent NO) on groundnut plants increased oxidative enzyme activities, total phenols and protein contents at 24, 48, 72 and 96 h after treatment with a quick response due to *A. fruticosa* which induced maximum enzyme activities (13.1 IU g-1 FW POD and 0.48 IU g-1 FW PPO, respectively, at 96 h after treatment). The total phenols, H_2O_2 and protein contents were also high in *A. fruticosa* treated plants followed by those treated with NO (1 per cent) (War *et al.*, 2011). The *A. fruticosa* extract and neem oil influenced the metabolic system in plants and induced the oxidative response that could defend plants against a variety of stresses (War *et al.*, 2011).

The effects of drought stress induced from initial flowering of drought-susceptible and -tolerant groundnut cultivars when studied, the SOD activity decreased and the protein content increased during the early stress period. The SOD did not vary between the susceptible and tolerant cultivars, but the protein content in the tolerant cultivar was higher. With further exposure to drought, the SOD activity increased and the protein content decreased and at one stage, the SOD activity in the tolerant cultivar was greater than that in the susceptible cultivar, whereas the protein content did not vary significantly. An additional protein band was detected under drought stress, but no significant variation in protein components was recorded between the cultivars. The drought tolerant characteristics were the number of pods, yield per plant and seed weight in Yuhua 13 and FDRS10; SOD activity and protein content at the early stress period in Nankang Zhisizi; and SOD activity in Mashan Eryang. Under drought stress, water potential was higher in the dragon-type cultivars than in the other types of cultivar (Jiang Hui Fangand Ren Xiao Ping, 2004).

A procedure was developed to screen groundnut plants for water stress tolerance based on in vitro growth and regeneration on a PEG containing medium and evaluated using 6 groundnut cultivars on media containing MS salts + B5 vitamins supplemented with 2 mg 2,4-D and 0.5 mg BAP [benzyladenine]/litre. PEG-4000 (5 per cent w/v) was incorporated into this media. Direct callusing of embryonic axes and regeneration on the PEG-containing medium was followed by rooting in non-stress media and plants regenerated on the stress medium exhibited higher osmolality than the control plants (regenerated on media without PEG-4000). The described procedure is a rapid, taking 3-4 months to complete, compared with protocols based on repeated passage of callus or suspension cultures in stress media which require almost one year (Venkateswarlu *et al.*, 1998).

Leaf discs taken from the seedlings of three groundnut cultivars differing in drought resistance when put in PEG-6000 solutions with osmotic potentials of -0.25, -0.75, -1.25 and -1.75 MPa and treated for 12 h. Water potential in the young leaves was shown to decrease under osmotic stress, while the rate of generation of superoxide free radicals in leaves increased. MDA content and the activity of SOD, POD [peroxidase] and CAT [catalase] tended to change with the variation in oxygen generation. The process of MDA increase was negatively correlated with water potential and positively with RPMP (relative plasma membrane permeability). GSH (glutathione oxidized form) and AsA (ascorbic acid) content dropped under osmotic stress. The cv. Baipi No.1 was highly resistant to drought, which was manifested in the smaller extent of increase in RPMP, oxygen radical generation and MDA content, and in slower decrease in SOD, POD and CAT activity and AsA content (Chen *et al.*, 2000).

Moderate water stress (-0.075 MPa) induced by PEG-6000, and its interaction with 20 mM calcium chloride studied by Usha *et al.*(1999) in groundnut seedlings aged (24-168 h) reveals that protein content in the embryonic axis of water stressed seedlings became progressively lower but increased in stressed seedlings treated with 40 dS/m $CaCl_2$. Addition of $CaCl_2$ to stressed seedlings increased proline oxidase activity and decreased proline accumulation. Accumulation of putrescine in the embryonic axis was observed in water stressed seedlings, associated with increased arginine decarboxylase activity. Addition of $CaCl_2$ to the water stressed seedlings decreased putrescine content and arginine decarboxylase activity. Thus Ca alleviates water stress by modulating the levels of proline and putrescine, which are the principal reserves of nitrogenous compounds.

In another study on water stress induced by PEG 6000 and the ameliorative effect of Ca^{2+} on changes in calmodulin, Ca^{2+}, protein contents and protease [proteinase] activity during seedling growth of two groundnut cvs. TPT 1 and TPT 4, Sulochana and Savithramma, (2002) reported quantitative variations of these in cotyledons and embryonic axis of seedlings and higher protease activity, lower calcium, calmodulin contents were observed in PEG-treated seedlings, however $CaCl_2$-treatment maintained higher levels of calcium, calmodulin contents and lower levels of protease activity. A protein with 76 kDa molecular weight was observed in PEG-treated seedlings, whereas lower molecular weight proteins of 24, 30 and 33 kDa were observed in $CaCl_2$-treated seedlings (Sulochana and Savithramma, 2002). Further studies on these cultivars seedlings treated with PEG, showed an increase in lipid peroxidation, peroxidase activity and decrease in CaM content, superoxide dismutase and catalase activities (Sulochana *et al.*, 2002). The calcium chloride treated seedlings maintained higher levels of CaM content, SOD and CAT activities suggesting amelioration of adverse effects of PEG on membrane deterioration by calcium chloride by modulating the lipid peroxidation and peroxidase activity and maintaining higher levels of CaM and scavenger enzymes such as SOD and CAT and the cv. TPT 4 appears to be more tolerant to water stress than the TPT 1 (Sulochana *et al.*, 2002).

In continuation the moderate water stress (-1 MPa) induced by PEG-6000 and its interaction with 20 mM $CaCl_2$ when studied on calmodulin (CaM) and proline contents and activity of proline oxidase in cotyledons and embryonic axis during seedling growth of groundnut cultivars (TPT 1 and TPT 4), the CaM content in cotyledons of water-stressed seedlings decreased progressively after treatment, however, $CaCl_2$-treated seedlings maintained higher levels of CaM and proline oxidase activity and lower levels of proline content (Sulochana and Savithramma, 2001). Calcium appears to ameliorate the water stress by maintaining higher levels of CaM and lower levels of proline by modulating the activity of proline oxidase (Sulochana and Savithramma, 2001). The acid phosphatase (ACPH) activity was low in the seedlings subjected to water stress, but, higher in $CaCl_2$-treated seedlings and Ca^{2+} maintained higher levels of ACPH activity in the seedlings of TPT 4 than that of TPT 1 indicating that Ca^{2+} modulates the levels of enzyme activity under water stress (Sulochana and Savithramma, 2003).

White grubs larvae damage the root system of groundnut Chitra cv plants. Simulated whitegrub damage was created by cutting the roots of plants at 10, 20 and 30 cm from the top soil surface at 30 DAS in one set and at 60 DAS in the other set under both normal and drought conditions. The root cut reduced RWC, transpiration and the reduction was highest in plants with roots cut at 10 cm and the lowest in plants with roots cut at 30 cm. Proline content and peroxidase activity increased with the increase in the per cent root cut and the magnitude of increase in proline and peroxidase was almost 2-3 times of the control, in plants whose; roots were cut at 30 or 60 DAS. The roots damaged at 60 DAS reduced the seed yield to a greater extent than the roots damaged at 30 DAS (Yadav *et al.*, 2007). Low soil temperature inhibited nodulation and nitrogen fixation of rhizoma groundnut [*Arachis glabrata*] cv. Florigraze. Water stress limited the development of new nodules and stimulated nodule senescence. N fertilizer application stimulated specific and total nitrogenase activity in spring but inhibited nodule and total nitrogenase activity in summer and autumn. During establishment, Florigraze competed strongly with associated *C. dactylon* for available N fertilizer (Valentim, 1989).

Nodulation and N_2 (C_2H_2) fixation were studied in cowpeas and groundnuts during water stress and recovery by withhold water for 5, 6 or 7 d, and leaf water potential and nitrogenase activity measured before and 2, 5, 24, 48, 72 and 120 h after rewatering. In plants relieved from water stress, leaf water-potential recovered much faster than nitrogenase activity in both species, and the recovery ability depended on the intensity of water stress. Nitrogenase activity recovered much faster in groundnuts than in cowpeas, even from a relatively higher level of stress. This was partly due to the lack of nodule shedding in groundnuts, while severe shedding of nodules in cowpeas did not permit the recovery of activity, particularly at higher stress levels. Leghaemoglobin was more stable in groundnut than in cowpea nodules. Water stress led to accumulation of total soluble sugars in groundnut nodules, while there was no change in cowpeas (Venkateswarlu *et al.*, 1990).

Water status and N metabolism of groundnut cv. M-13 and M-145 were examined during a period of water stress and recovery. 10-day-old seedlings grown in controlled environment were exposed to sol. of PEG (mol. wt. 6000, osmotic potential -5 or -8 bar) for 1 wk when the stress was relieved by replacing the PEG sol. with a nutrient sol. M-13, despite its lower relative turgidity during stress, was better able to preserve its protein conc. and nitrate reductase activity and recover to a normal state within 3 days of the relief of stress, but M-145 failed to recover within this period (Saini and Srivastava, 1981).In a growth chamber, groundnut cv. Yue You 551-116 seedlings at the 3-leaf stage treated with 5 ppm PP 333 (paclobutrazol) or 150 ppm. CCC [chlormequat], when one day later, PEG solution at 10-3 mol/litre was used to create water stress in the rhizosphere, both PP 333 and CCC treatments increased peroxidase activity compared with untreated seedlings, though the zymogram patterns of the 2 treatments were different. More bands were observed in the isozymograms of the treated seedlings than in the control (Li and Pan, 1990).

At low water potentials root elongation still continues, while shoot growth and elongation have completely ceased at similar water potentials. This differential response of root and shoot is an adaptation by plants to avoid excessive dehydration while tapping moisture available at lower depths of the dehydrating soil. The stress regulator abscisic acid (ABA) has been implicated in this unique adaptation of plants to water stress and this has been demonstrated convincingly in maize using ABA-deficient mutants. The ABA accumulating capacity of 2 distinct cultivars of finger millet and groundnut, differing in root elongation under water stress, the root ABA content was not significantly different in water-stressed plants of these cultivars, although differences in root growth and root elongation at low psi_w were distinctly different. Tissue sensitivity to ABA and compartmentation under stress could influence these observed differences in root growth in the cultivars rather than the ABA accumulating capacity (Suma *et al.*, 2006).

Groundnut genotypes at normal and water stress conditions for 30 days were studied in the greenhouse in order to investigate the enzymatic behaviour and to determine the best enzyme systems for the production of molecular markers associated with water stress. Normal and stressed leaflets and root tissues were collected at 45 DAS and analysed by via PAGE with a continuous buffer system and studied esterase (EST), peroxidase (POX), malate dehydrogenase (MDH), superoxide dismutase, acid phosphatase (ACP) and leucineaminopeptidase [cytosol aminopeptidase] where POX and ACP in the roots gave the highest contribution to the differentiation process between genotypes and treatments; The EST and MDH contributed towards genotypes differentiation, only under stressed conditions (Santos dos *et al.*, 1998). Under water deficit condition, production of reactive oxygen species in terms of H_2O_2 and superoxide radical (SOR), and lipid peroxidation was more at both flowering and pod development stages compared to vegetative stage in groundnut. The study concluded cultivars ICGS 44 and TAG 24 showed better tolerance capacity by maintaining higher relative water content and antioxidant enzyme activities, and sustaining much less membrane injury due to imposition of water-deficit stress (Chakraborty *et al.*, 2012).

3.10. Molecular Physiology

The eco-physiological responses of groundnut are studied in details, however little is known about the molecular events involved in its adaptive responses to drought. The involvement of membrane phospholipid and protein degrading enzymes as well as protective proteins such as "late embryogenesis-abundant" (LEA) protein in groundnut adaptive responses to drought were studied and Partial cDNAs encoding putative phospholipase Da, cysteine protease, serine protease and a full-length cDNA encoding a LEA protein were cloned and their expression in response to progressive water deficit and rehydration was compared between cultivars differing in their tolerance to drought by Drame *et al.* (2007) and differential gene expression pattern according to either water deficit intensity and cultivar's tolerance to drought and a good correspondence between the molecular responses of the cultivars and their physiological responses was found. Molecular characters, as detectable at an early stage, could therefore be efficiently integrated in groundnut breeding programmes for drought adaptation.

An incomplete half-diallel cross performed on an original population under recurrent selection for drought adaptation analysed, the study confirmed the weak heritability of yields and the best predictor of pod yield was the pod yield itself. In contrast, the study of the genetic correlations showed that a selection for high haulm yield could lead to poor pod maturity under drought constraint. The selection indices were performed and used to estimate genetic gains relative to the main agronomic characters according to selection pressure. The genetic variability of phenological, agronomic and physiological characters was studied in two series of quasi-isogenic early lines where genetic variability was expressed in these lines despite its closeness. The correlations between yield and fluorescence parameters were significant but not stable across lines and environments showing that groundnuts have different drought adaptation strategies according to genetic background and drought pattern. At the molecular level with three reference cultivars involving both recurrent parents of the precedent study, the gene transcript kinetics under drought, obtained using RT-PCR, showed that phospholipase D and cysteine proteinase gene expressions were stimulated by stress in the most susceptible cultivars, while there was higher LEA gene expression in the resistant one (Clave *et al.*, 2007).

Improvement of drought tolerance is an important area of research for crop breeding programmes. Recent advances in the area of crop genomics offer tools to assist in breeding (Varshney *et al.*, 2005, 2006). The identification of genomic regions associated with drought tolerance would enable breeders to develop improved cultivars with increased drought tolerance using marker-assisted selection (Ribaut *et al.,* 1996). There are RFLP (Restricted Fragment Length Polymorphism) maps of wild type x cultivar crosses but the polymorphisms are too low for a cultivated x cultivated species cross; therefore, new markers are needed (Burow *et al.,* 2001). A considerable number of SSR sequences have been identified from groundnut genome by several research groups (Hopkins *et al.,* 1999; He *et al.,* 2003; Ferguson *et al.,* 2004; Moretzsohn *et al.,* 2005; Proite *et al.,* 2007; Cuc *et al.,* 2008) which would enable breeders and molecular biologists to use this wide variability for further improvement of drought tolerance in groundnut. Differential display reverse transcriptase PCR was used to identify genes induced and suppressed in groundnut seed during drought. A total of 1235 differential display products were observed in irrigated samples, compared to 950 differential display products in stressed leaf samples (Jain *et al.,* 2001). Many families of transcription factors including AP2/EREBP (AhWSI 279), bHLH (AhWSI 111, AhWSI 40), bZIP (AhWSI 20), CCAAT box (AhWSI 117), Homeobox (AhWSI6 11) which showed differential expression in groundnut under drought stress.

Twenty two groundnut genotypes evaluated for water stress-regulated proteins by Katam *et al.* (2007) and to determine possible role of these proteins in groundnut acclimatization and adaptation to drought in pots under greenhouse conditions till 120 days and subjected to water stress by withholding irrigation at soil water potential 10, 16, 28 and 38 centibars at 7, 14, 21 and 28-day stress periods, respectively. The study shows that both the drought-tolerant and drought-susceptible genotypes responded similarly during the brief stress for 3 days, but during prolonged stress conditions (>3 days) drought-tolerant genotypes (Vemana, K 1375) are able to maintain expression of certain proteins (molecular weight between 14 kDa and 70 kDa) while in drought-susceptible genotypes (M 13, JL 24), these proteins are suppressed indicating their role in stress tolerance. Evaluation of groundnut genotypes with diverse drought tolerance characteristics for determining differences in their response to drought stress showed varying levels of protein expression, suppression or over-expression of leaf proteins among the genotypes. Polypeptide matches of the selected 2D-resolved proteins were found to be similar with the proteins of ultraviolet-B repressive rubiscoactivase, glyceraldehyde-3-phosphate dehydrogenase, ribulosebisphosphate carboxylase, phosphoribulokinase, cytochrome b6-f complex and oxygen-evolving enhancer protein (Katam *et al.,* 2007).

The NAC transcription factors existed differentially in plant are the new transcription regulatory factors with multiple biological functions. Two NAC-like genes from groundnut were cloned by RT-PCR and RACE methods, named AhNAC2 and AhNAC3 (Gene Bank accession Nos. EU755023 and EU755022), which contained an ORF of 1050 bp and 1008 bp and encoded 349 and 335 amino acids, respectively. Gene sequence analysis showed that the putative protein of both genes contained a conserved NAC domain and highly different C terminal, which were the typical characteristics of NAC transcription factors. The transcription levels of the 2 genes when investigated by semi-quantitative RT-PCR, and the result showed that the expressions of AhNAC2 and AhNAC3 genes were enhanced by ABA, GA$_3$, water stress and cold stress, respectively. Furthermore, the 2 genes expressed constitutively in groundnut tissues and their expression patterns were different in various tissues. In conclusion, AhNAC2 and AhNAC3 genes isolated from groundnut were new members of the NAC transcription factor family and their comparison to RD26 (AT4G27410) revealed a high amino acid homology, they play key roles in ABA signal transduction and drought response in groundnut (Liu and Li, 2009).

A study to test whether DREB1A gene driven by stress inducible rd 29A promoter could have an effect on groundnut root growth under water deficit using lysimetric system and changes in the ET response and in the rooting pattern upon exposure to water deficit in 5 transgenic events and their wild type (WT) parent clearly indicated that DREB1A induced a root response under water deficit conditions (Vadez *et al.*, 2007).This response enhanced root growth under water deficit in particular in the deep soil layers. Consequently, water uptake under water deficit was enhanced, up to 20-30 per cent in some transgenics compared to the WT. This water uptake was well related (r^2=0.91) with the root dry weight below the 40 cm soil depth. Finally, it appeared that the putative effect of DREB1A on root under water stress conditions was due to an effect on the root/shoot ratio, which was dramatically increased under water stress in all transgenic lines (Vadez *et al.*, 2007).

Transgenic plants carrying genes for abiotic stress tolerance are being developed for water stress management. Structural genes (key enzymes for osmolyte biosynthesis, such as proline, glycine-betaine, mannitol and trehalose, redox proteins and detoxifying enzymes, stress-induced LEA proteins) and regulatory genes, including dehydration-responsive, element-binding (DREB) factors, Zinc finger proteins, and NAC transcription factor genes, are being used. Using Agrobacterium [Rhizobium] and particle gun methods, transgenics carrying different genes related to drought tolerance have been developed in rice, wheat, maize, sugarcane, tobacco, Arabidopsis thaliana and groundnut (Gosal *et al.*, 2009). In general, drought stress-tolerant transgenics are either under pot experiments or under contained field evaluation. Molecular markers are being used to identify drought-related quantitative trait loci and their efficient transfer into commercially grown cultivars (Gosal *et al.*, 2009).

Novel stress responsive genes were identified following subtractive hybridization of cDNA synthesized from RNA isolated from stress and unstressed groundnut leaves. One of the cloned genes (Gdi15) exhibited increased expression in stressed leaves. Sequence analysis indicated that it has significant homology with flavonol 3-O-glucosyltransferase, a gene involved in anthocyanin biosynthesis and shows increased expression of flavonol 3-O-glucosyltransferase under desiccation stress (Gopalakrishna, 2001).

Drought and high temperatures are conducive to *Aspergillus flavus* infection and aflatoxin contamination. The molecular tools, proteomics, DD-RT-PCR (differential display reverse transcription-polymerase chain reaction), expressed sequence tag (EST) and gene chip technology (macro/microarray) to study gene expression in response to drought stress, and genetic transformation, were studied by Guo *et al.* (2003) using DD-RT-PCR to display genes expressed in groundnut and maize grown under drought stress vs. irrigated conditions use EST/microarray technology to study the whole genome as influenced by drought stress in maize and groundnut, and *A. flavus* ESTs to better understand the genetic control and regulation of toxin biosynthesis (Guo *et al.*, 2003).

3.11. Leaf Membrane Injury

Two groundnut cultivars grown for 13 weeks under water controlled conditions in pots, the cultivar Falcon (F) showed characteristics of drought tolerance, while cultivar Local (L) showed those of drought susceptibility. Falcon showed an osmotic adjustment mechanism that enables it to withstand short-term drought stress. The membranes of the Falcon were less injured under drought stress and maintained higher RWC (water saturation deficit, WSD) and relatively low relative saturation deficit (RSD) as compared with the cultivar Local. Additionally, proline was substantially more accumulated in this cultivar. Therefore, cultivar Falcon was classified as drought tolerator and cultivar Local as drought avoider. The relative water content (RWC), relative saturation deficit (RSD), cell membrane integrity (CMI) and proline content were effective criteria for detecting drought tolerance strategies taking into account the growth stage and duration of the stress period, while the water retention capacity (WRC) did not show any significant relation with drought tolerance (Quilambo, 2004).

The effect of different physiological indexes on cell injury of detached groundnut leaves under osmotic stress showed significant correlations between water potential, relative plasma membrane permeability (RPMP), MDA, superoxides and superoxide dismutase activity (SOD). The correlations between catalase (CAT), peroxide (POD), glutathione synthase (GSH), ascorbic acid and water potential, RPMP, superoxides, MDA and SOD varied depending on the cultivars. The effects of water potential on the productive elasticities of RPMP, MDA, superoxides and SOD were negative, while the effects of SOD on the productive elasticities of RPMP, MDA and superoxides progressively decreased. Similarly, a progressive decrease was observed in the effect of superoxides on the productive elasticities of RPMP and MDA. The productive elasticities and marginal yields varied with cultivars exhibiting varying drought resistance (Zhu *et al.*, 2001).Groundnut seedlings grown in nutrient solution containing 0-400 mg methyl jasmonate/litre, and transferred at the 3-leaf stage to PEG solution for 3 d, Methyl jasmonate decreased permeability of the plasma membrane, and increased activities of superoxide dismutase and catalase, during drought stress. Methyl jasmonate also induced the formation of new superoxide dismutase isoenzymes, increased ascorbic acid content, and decreased levels of malondialdehyde (Pan *et al.*, 1995).

Cell membrane stability (CMS) in suspension cultures of groundnuts cv. Kadiri 3 and JL 24 studied after incorporation of various doses (0-20 per cent) of PEG in the culture medium showed a negative relationship between PEG concentration and membrane stability measured as electrolyte leakage. The CMS values in the cell cultures correlated well with the whole plant tissue and permitted the differentiation of cultivars based on their known response to drought stress. The cell membrane stability was lower (more electrolyte leakage) in culture as compared with the intact plant tissue. Kadiri 3, the drought tolerant cultivar maintained higher CMS than JL 24, the drought susceptible one. Increasing PEG levels, K concentration in cultured cells declined in both cultivars, however, Kadiri 3 maintained higher K values than JL 24 accompanied with greater cell membrane stability. Total soluble sugars also increased with increasing stress in both cultivars; CMS test can be used under in vitro conditions to differentiate the drought tolerant and susceptible cultivars, and that the cellular K level has a positive relationship with membrane stability (Venkateswarluand Ramesh, 1993).

Maintenance of membrane integrity under stress reflects broadly intrinsic tolerance. Leaf discs of 1 cm diameter from 14 day old plants of groundnut subjected to rapid (open Petri dish; dry filter paper) and slow (closed Petri dish) desiccation in a growth chamber at 35°C, 60 per cent RH and 700 μein/m^2s^{-1} light intensity at water loss about 60 per cent, membrane integrity (percentage leakage), was maintained in groundnut in both rapid and slow desiccation stress (Gopalakrishna, 2001). In variety Florman INTA and 6 pure lines under water stress, 9 physiological variables when measured along with yield and its components, and oleic to linoleic acid ratio, the lines fell into groups contrasting in drought tolerance, with the most precise grouping classing Manfredi 420 as tolerant, Florman INTA as susceptible and the rest as intermediate. The correlations between physiological variables and yield traits show that the most tolerant lines were those which could keep their stomata most open during drought stress, without major alterations in membrane stability (Collino *et al.*, 1994).

3.12. Diseases

Groundnut wilting caused by *S. rolfsii* [*Corticium rolfsii*] often occurs in wet and hot summer, but severe epidemics have also been observed when wet periods follow protracted dry ones in the groundnut producing regions. March *et al.* (1999) studied the influence of drought stress in predisposing groundnut plants to wilting in cv. Florunner by 8 weeks old subjected to different soil-water regimes and inoculated with sclerotia of *C. rolfsii* in greenhouses. On the inoculation day, leaf water potential (ψ_l) of a fully penultimate expanded tetrafoliate leaf when measured using a pressure chamber, leaf PSI of non stressed plants ranged from -0.2 to -0.4 MPa while in wilted plants ranged from -1.2 to -1.4 MPa. Water stress enhances infection by *C. rolfsii* but drench-irrigated plants maintaining soil moisture at water holding capacity did not show infection. Facultative parasites like *C. rolfsii* are usually favoured by conditions that weaken or stress the host (March *et al.*, 1999).

4. Impact on Reproductive Growth and Yield Components

There are three major aspects of drought, duration, intensity and timing relative to crop phenophases which vary independently. Water stress delay pod initiation, and the major cause of variability in pod yield and HI is the delay between peg initiation and onset of rapid -pod growth, because once pods were initiated, the proportion of dry matter allocated to reproductive sink was relatively conservative (Stirling and Black 1991). The period of reproductive growth stages in groundnut occurs over a period of nearly two months and moisture stress has a depressing effect on flowering, stem growth and nodulation. No flowering occurs during the stress, but once the stress is removed, there is a flush of flowering depending on the growth stages and sometimes it results in more flowers than control (Singh 2003). The Virginia groundnuts, due to their longer duration, are more tolerant to drought than Spanish and Valencia however the later due to short duration escape the late season drought.

Experiments conducted on Warin soil series at the Agricultural Development Research Center (ADRC), Khon Kaen Province, indicated that adequate water supply should be maintained in order to get optimum yield and yield decreases due to water stress at different growth stages in order of water stress at seed development > at early pod filling > at early growth > at early pegging (Uthai *et al.*, 1993).

Plant population influenced both the temporal and spatial patterns of water use, with high density crops extracting water from lower depths sooner than low density crop. High water use prior to early pod filling in high density crop was associated with more rapid leaf area development (Wright and Bell 1992b). The more rapid water extraction in a high, compared with a low, population density groundnut crop is associated with greater root production at depth (Nageswara Rao *et al.*, 1989). The seed yields with drought for 3 weeks starting week 5 or 6 were 65.5 and 34.1 g/ m^2, resp., compared with 81.9 g without water stress (Zaharah 1986).

4.1. Seed Growth and Pod Development

The early and continuous availability of water until the start of pod filling results in large canopy and during the period of drought stress the transpirational demand increases. The ratio of pod number: peg number reduced from 0.8 in normal irrigated crop to nearly 0.15 in stressed crop (Harris *et al.*, 1988). Moisture stress at flowering reduced phytobiomass and pod yield by limiting the number of mature pods per unit area as compared to stress at pegging and pod formation stages (Rao *et al.*, 1986). Under water stress there is poor pod filling that reduced kernel size, shelling, SMK per cent and lipid content of kernel. Nageswara Rao *et al.* (1989 b) in a study observed that when water deficit occurred during seed filling phase, genotypic yield potential accounted for approximately 90 per cent of the variation in pod yield sensitivity to water deficit, and further elaborated that it is unlikely that breeders will be able to combine high yield potential with low sensitivity to drought spanning the seed filling phase, therefore other important strategies are necessary.

Stirling (1989) in a controlled-environment greenhouse study a finite quantity of water applied to groundnut cv. Kadiri at different stages of the growing season by imposing two levels of soil moisture deficit by withholding or applying limited amounts of irrigation at regular intervals during two periods; sowing to pod initiation and pod initiation to final harvest where shoot DM yields were hardly affected but pod yields were more than 4-fold lower in early- than in late-irrigated stands. The degree-day requirement for peg initiation was similar in all treatments but late irrigation delayed pod development by about 200 degree days. The effect of timing of irrigation on pod yield operated mainly through its influence on the duration of pod production, which was closely linked to the rate and duration of canopy expansion late in the season. The insensitivity of pod yield to early moisture deficits reflected the extreme plasticity of growth and development in groundnuts, since most processes resumed rates similar to the pre-stress levels in early-irrigated stands once stress was released (Stirling 1989).

Bennett *et al.* (1990) in experimental constructed root tube-pegging pan apparatus to allow physical separation of groundnut rooting and pegging zones with independent control of soil water in both zones where satisfactory shoot growth and pod development of plants occurred. Using the apparatus, the effects of air-dry and moist (7-12 per cent water by wt) pegging zone soil on seed and pod formation when examined, soil water deficits in the pegging zone decreased the number of pods which reached full expansion from 61-48 per cent. Total pod and seed wt, growth rates of pods and seeds and individual pod and seed wt/plant also decreased in air dry treatment. Drought conditions reduced the av. seed wt. of sound mature seed in Florigiant and Florunner but not in Tifspan (Pallas *et al.*, 1977).

Effects of drought on groundnut seed development and quality when studied in Shandong, using cv. Baisha 1016, at soil water-holding capacity controlled at 55-65 per cent, 60-70 per cent and 50-60 per cent at the flowering, fruiting and ripening stages, resp., the most significant effect of drought for 30 days on seeds was at the seed development stage, resulting in 25.1 per cent decrease in 100-seed wt. A drought period of 30 days at the flowering caused a 24.7 per cent decrease of 100-seed wt., while at the the ripening stage it resulted in a 14.6 per cent decrease in 100-seed wt and also reduced seed size and increased the number of shriveled seeds (Yao *et al.*, 1982). The effects of skipping one irrigation during pod initiation, pod development or pod maturity and application of potassium fertilizer (0, 24, 32 and 40 kg K_2O) on the yield and yield components of groundnut cv. Giza 5 were determined in a field in Egypt where water stress, at pod initiation and pod development, reduced the number and weight of pods and seeds per plant, shelling percentage, pod yield, oil and protein percentages and yield. Potassium application increased the number and weight of pods per plant, number of seeds per plant, 100-seed weight, pod yield, oil and protein percentages, and protein yield compared with the control (Ali 2001).

Golakiya and Patel (1992) imposed 10 combinations of water stress at flowering, peg formation, pod development and pod maturity of J 11 and GG 2 cultivars in lysimeters, and reported that the growth, was curtailed by water stress at any of the growth stages with stress at the flowering stage being the most inhibitive, however yield was reduced most by water stress at pod development due to reduced crop reproductive efficiency. Pod yield was reduced by reductions in fertility index, seed formation coefficient and seed formation efficiency and GG 2 was more drought tolerant than J 11. In Akola, groundnut cv. JL 24 recorded higher pod yield when sown in the last week of June, but sowing extended beyond this period, up to September resulted in progressive reduction in yield even though irrigations were applied when necessary (at 80 mm CPE) to avoid moisture stress across delayed sowings. Number of effective pegs, developed pod number and pod dry weight were influenced by variations in atmospheric temperatures, particularly minimum temperature, and the relative humidity indicating that warmer temperatures and higher relative humidity during crop growth period favourably influenced the yield contributing characters and finally the pod yield (Karunakar *et al.*, 2002).

In Andhra Pradesh the effect of mid-season drought (MSD) and end season drought (ESD) on the yield and yield attributes of 20 groundnut genotypes (11 advanced breeding lines and 9 released varieties) studied by withholding irrigations between 50-100 DAS and 100 DAS to the final harvest, respectively, as against crop receiving full irrigation during the whole crop duration, revealed genotype differences, and genotype environment interaction for pod yield, shelling percentage and HI.

Reduction under both stress conditions was observed for the number of mature pods, pod yield, shelling percentage, 100-kernel weight and HI, indicating that these characters were significantly affected by MSD and ESD, however, percentage sound mature kernel and oil content were reduced only under ESD as it coincided with the seed maturation and oil formation, there was maximum reduction in the number of mature pods (47 per cent) under MSD, followed by pod yield (29.7 per cent) while under ESD for pod yield (41 per cent), followed by the number of mature pods (33 per cent), indicating that number of mature pods was most sensitive to MSD and pod yield to ESD (Suvarna *et al.*, 2002).

In southern Telangana on Alfisols, groundnut generally suffers from mid-season drought reducing crop yields where 18 groundnut genotypes along with a local control (TMV 2), crop experienced a 27 day long dry spell from the beginning of pod initiation to full seed development in 1994, and for 16 days from the beginning of peg initiation to the beginning of pod development in 1995, the pod yields ranged from 0.58 (ICGS 88) to 2.41 t/ha and only K 134 and ICGV 86347 were superior genotype with high yield, but, the yield superiority of these two genotypes was not reflected in their ancillary characters (Thatikunta and Durgaprasad, 1996).

4.2. Partitioning and Harvest Index

The pod yield is a function of transpired water (T), transpiration efficiency (TE) and harvest index (HI), and the TE derived from measurements of carbon isotope discrimination in leaves indicated only small variation (Wright *et al.*, 1991). The variation in HI accounted for the large proportion of variation in yield and recommended to make selection for high HI. The reproductive development is sensitive to drought resulting to poor yield. The strategies to combat drought in genotypes are (i) early production of flowers, pegs and pods, with subsequent filling of the pods at a moderate, but essentially at constant rate despite the drought, (ii) faster development of later developed pegs into pods once water become available after drought late in the season, Drought stress effects on groundnut depend primarily on the stress pattern because genotypic variation is usually of secondary significance. Comparison of various groundnut genotypes indicated that the Acc 847, 55-437 and GNP 1157 tended to have the tallest plants, greatest shoot DM and leaf area and highest pod and seed yields under stress conditions. Seed yields under stressed and non-stressed conditions were 5.40 and 12.17, 4.08 and 10.62 and 4.56 and 18.55 g/plant for Acc 847, 55-437 and GNP 1157, respectively (Rosario and Fajardo, 1988).

In Argentina with two different regimes of water (irrigated (IRR) from sowing to maturity, no water between 47-113 DAS) on groundnut cultivars Florman INTA and Manfredi 393 INTA, the fraction of PAR intercepted, (f), leaf area, pod and vegetative above-ground biomass and leaf carbon dioxide exchange rate (CER) measured periodically during the water deficit period, and leaf area index, degree of leaf folding, canopy extinction coefficient, radiation use efficiency (RUE), partitioning factor, (p), and harvest index (HI) calculated from the measurements. Under water stress, f was reduced in both varieties and the reduction was proportionally higher in Florman INTA as a consequence of a higher leaf area reduction and degree of leaf folding. However, f remained higher in Florman INTA than in Manfredi 393 INTA due to the enhanced capacity of the former to generate leaf area under non-limiting water supply. RUE values due to their ability to maintain a higher leaf CER were higher in Manfredi 393 INTA than in Florman INTA, both under irrigation as well as under severe water deficit, where they were obtained using a two-parameter exponential model. Partitioning (p) to pods under irrigation was greater in Manfredi 393 INTA than in Florman INTA, as a result of a longer pod filling period and higher p. Towards the end of pod filling, there was a rapid increase of p in Florman INTA, but too late to improve its HI. Under water stress, the time course of p for both varieties was lower than in the IRR treatments and consequently, HI at harvest was reduced. Low HI values could be attributed to some extent to the mechanical impedance of the upper soil layer, caused by water deficit. Mechanical impedance alters the relation among p and HI values obtained under irrigation and water stress. However, even if it is accounted for, cultivars with high HI under IRR conditions usually have high HI under water deficit (Collino *et al.*, 2001).

A two years of study at the ICRISAT Sahelian Centre, near Niamey, Niger, to select groundnut cultivars tolerant of drought and to investigate selection techniques using 36 cultivars known to differ in yield potential grown under rainfed and irrigated conditions. Crop growth rate (C) and partitioning coefficient (p) were estimated from phenological and final harvest data. The correlation between years was greater for partitioning than for pod yield (implying a higher heritability for p than yield). Tolerance as determined by a drought susceptibility index for pod yield (Sy), crop growth rate (Sc) and partitioning (Sp) to reproductive sinks showed 13 cultivars to be drought tolerant for either C or p or both. The Sahelian cultivars 796, 55-437 and TS 32-1 were the most consistent for drought tolerance. Partitioning was the most important yield component affecting yield differences among cultivars (Ndunguru *et al.*, 1995).

Chapman *et al.* (1993a) in a greenhouse experiment at Queensland, groundnut cultivars grown by withholding water for a period of 3 weeks during 46- 67 DAS, Robut 33, the cultivar with the highest HI, produced greater yield than either Virginia Bunch or McCubbin mainly due to higher number of pods, compared with the other cultivars. When water was withheld from 61-78 DAS, the Virginia type cultivars (Virginia Bunch and Robut-33) produced a greater yield than the Spanish type cultivar (McCubbin). Peg initiation was sensitive even to mild water deficit, but elongation of pegs halted by water deficit could continue after rewatering. This may be an important attribute particularly where intermittent drought occurs. In both water-deficit treatments, peg initiation and elongation in all cultivars halted after about 80 per cent of the extractable soil water had been exhausted. The yield advantage of Robut-33 was mainly in producing a large number of pods prior to water deficit, and in partitioning a greater amount of biomass to pods after rewatering. The Virginia type cultivars were also apparently better able to tolerate the effects of severe water deficit (Chapman *et al.*, 1993a).

In another study Chapman *et al.* (1993b) at Redland Bay in Queensland, groundnut cultivars subjected to a period of reduced water supply during early reproductive development differed in growth responses during and after the period of water deficit and Q 18801, a Virginia type cultivar with a high HI under non water-limiting conditions, yielded higher than Virginia Bunch and McCubbin (a Spanish type). During the period of water deficit, all cultivars produced similar number of pegs and pods, but greater proportions of these were converted to pods in Q 18801 and McCubbin than in Virginia Bunch. Water deficit delayed the start of the period of rapid pod growth by about 15 d and hence extended the time required to reach maturity. After rewatering, the number of pegs and pods and the leaf area index of Virginia Bunch and McCubbin increased rapidly. In contrast, Q 18801 partitioned more assimilate to pods, achieving a higher average growth rate of individual pods, and consequently a higher total yield of pods and seeds. Thus, selecting cultivars with increased HI (via rapid pod growth at the expense of excess canopy growth) under irrigated conditions may also increase yields following a drought during early reproductive development (Chapman *et al.*, 1993b).

Further study in Queensland, water stress from 84 DAS to maturity, pod and seed yield was reduced by 30 per cent in the Virginia cultivars (Virginia Bunch and Robut 33) and by 45 per cent in a Spanish type cultivar, McCubbin (Chapman *et al.*, 1993d). The Virginia type cultivars also extracted water from a greater depth. Robut 33, a cultivar with a high HI, produced the greatest yield under both well-watered and water-stressed conditions. In all cultivars, potential pod number had been almost achieved prior to the start of the period of water deficit. However, low pod number rather than small pod size, was mainly responsible for the decrease in yield. Part of the yield advantage of Robut 33 lay in initiating a large number of pods prior to the period of water deficit (Chapman *et al.*, 1993d). This greater synchrony of development compared with the other Virginia type cultivar, created a greater sink for assimilate prior to the period of water deficit. During the first 3 weeks period of water deficit, Robut-33 had the highest CGR and was thus able to produce the most pod biomass. Thus characteristics of early and rapid pod growth and high HI were more important in determining yield under water deficit than the amount of water extracted from the soil (Chapman *et al.*, 1993d).

Parameters related to drought tolerance were studied by Chavan *et al.* (1992) in 29 genotypes grown during the rainy seasons under natural conditions or deprived of moisture for 20 days after flowering, where among the spreading group DVR 50 gave the highest yield under both natural and moisture-stress conditions (1268 and 885 kg/ha, respectively), and had a high HI under both conditions and among the bunch group, CGC 4018 gave the highest yield under both conditions (1561 and 1067 kg/ha, respectively). But the best drought index was recorded for LG 42 in the spreading group (14.3) by ICGS 35-1 in the bunch group (18.4). Overall, DVR 50 was the most efficient under both conditions for pod yield, HI and plant-water status (leaf-water potential). Greenberg *et al.* (1992) evaluated 36 groundnut genotypes of varied origin for yield, crop growth rates (C) and partitioning to reproductive sinks (p) in 3 trials at Niamey by altering the irrigation and sowing date so as to vary the amount of water available either throughout the crops' life or through the grain filling phase establishing 5 different environments. Although differences in C existed, differences in the stability of p were the dominant attribute of genotypes adapted to the drought prone Sahelian region but these differences were more attributable to tolerance of temperature and/or humidity than water stress. Canopy temperatures relative to air (CATD) were strongly correlated with the value of C, but not with yield (Greenberg *et al.*, 1992).

Drought resistance is an important character for increasing groundnut yields in the sub-humid, dry zone of Sri Lanka and to determine the effect of soil water deficit on vegetative growth and seed yield, and the physiological basis of yield of groundnut under using seven groundnut genotypes (Tissa, ANKG 2, Red Spanish, N 45, ICGV86015, ICGV 86143 and ICGV 86149) de Costa *et al.* (2001) studied under well-watered (90 per cent available water) and water-stressed (30 per cent available water) conditions in pots (12 kg) in a glasshouse at Maha Illuppallama, Sri Lanka, where water stress significantly reduced leaf area, final total dry weight and seed yield. Final total dry weight and yield showed significant genotypic variation under both water regimes but did not show significant genotype x water regime interaction. The highest seed yield under water-stressed conditions was by ICGV 86015, but by ICGV 86149 under well-watered conditions. A greater partitioning of dry matter to seeds (*i.e.* greater HI) was required to achieve high groundnut yields under water stress, but dry matter partitioning was not a yield-determining parameter under well-watered conditions where a greater capacity for total biomass production was required to achieve high yields. Under water-stressed conditions, groundnut yields were positively correlated with pod number per plant, seed weight and the number of primary roots per plant (de Costa *et al.*, 2001).

4.3. Yield and Yield Attributes

The yield is a function of many plant and environmental factors and moisture stress play an important role particularly the stage at which moisture stress occurs. The yield losses (per cent) due to mid season drought are estimated as: Yield loss (per cent) = $100(1 - D_y / W_y)$, where, W_y is the pod yield under adequate irrigation and D_y is the pod yield under drought treatment. The yield reductions have been reported to be 22, 18, 47 and 47 per cent, respectively when drought was imposed from 10-30, 30-50, 50-80 and 80-120 DAS respectively (Billaz and Ochos 1961). The greatest yield reduction in 50-80 days stress treatment corresponds to, peak flowering to early pod filling stage. Meisner and Karnok (1992) observed the pod yield reduction of 49 and 37 per cent by water stresses imposed at 50-80 and 80-120 DAP and suggested that adequate moisture during this period is critical for obtaining maximum yield. The pod yield potential accounted for less of the variation in drought sensitivity (15-64 per cent) in the early and mid-season droughts (Nageswara Rao *et al.*, 1989b). For these circumstances it may be possible to identify genotypes with both high yield potential and relatively low drought sensitivity. Correlation and path coefficients were worked out for nine traits involving 40 hybrids and 14 parents in groundnut where pod yield exhibited significant positive association with pods per plant, dry matter production, kernel weight and harvest index. Path analysis revealed maximum direct effect of pods per plant followed by dry matter production, and kernel weight on pod yield (Vaithiyalingan, 2010).

Ravindra *et al.* (1990) observed that moisture stress at flowering (45-70 DAS) and pod development (60-90 DAS) phases was highly detrimental to leaf area development, dry matter production, pod formation and yield in comparison with stress at the vegetative phase (20-50 DAS) and the reduction in yield was 57 and 66 per cent, respectively due to moisture stress at these stages. The recovery of growth from water stress was better after relief at the vegetative phase than at later growth phases. Nautiyal *et al.* (1999a) reported that transient soil-moisture-deficit stress for 25 days, at the vegetative phase (20-45 DAS) followed by two relief irrigations at an interval of 5 days, resulted in closely synchronized flowering, greater conversion of flowers to pods and higher pod yield and total biomass accumulation indicating that stress in the vegetative phase was beneficial for groundnut growth and pod yields, but was highly detrimental when imposed at flowering (40-65 DAS) and pod development (60-85 DAS). Nageswara Rao *et al.* (1985) while studying with Robut 33-1 (a 140-150 days crop) observed maximum reduction in kernel yield when stress was imposed during seed filling phase *i.e.* 93 DAS onwards however the treatment receiving 12-15 per cent less water than control during the early phase (line source irrigation at 11 and 21 days followed by no irrigation up to 50 DAS) increased the pod yield by 13-19 per cent over fully irrigated (irrigation at 50 per cent FC) treatment (Nageswara Rao *et al.*, 1988). The increase in yield due to pre-flowering drought is mainly due to promotion of root growth during water stress which promoted subsequent growth during pod fills and inhibition of number of vegetative sites (leaves and branches).

On a medium black calcareous soils at Junagadh water stress at flowering, pegging, pod formation and pod development stages of groundnut cv. GG 2 gave pod yields of 1.43, 1.18, 1.07 and 1.00 t/ha compared with 2.25 t/ha with normal irrigation (Sakarvadia and Yadav 1994). The groundnut cv. GAUG 1 grown in field at -0.3 bar (normal), -0.6 bar and -0.9 bar (corresponding to irrigations at intervals of 7-11, 13-16 and 17-21 days, respectively) water stress adversely affected all the growth, yield and quality parameters like oil and protein content, however potassium application (40 kg K_2O ha^{-1}) decreased these negative effects of water stress on yield and quality (Umar *et al.*, 1997). Water stress for 25-30 days during the vegetative, flowering or pod development stages in Spanish-type groundnuts cv. J 11 and GG 2 reduced pod yield. Water stress during the vegetative stage had less of an effect on pod yield, the number of mature pods/plant, number of nodules/plant, plant DM and N uptake than water stress at later developmental stages (Kulkarni *et al.*, 1988).

The moisture stress imposed by withholding irrigation at early vegetative, flowering, peg penetration, pod initiation, or pod development stage in groundnut cv. AK 12-24 in Bhubaneswar, India, during the rabi the highest number of pods and 100-kernel weight were recorded for non-stressed plants and highest shelling percentage obtained with irrigation based on stress day index (six irrigations) and with the control. Plants stressed at the peg penetration stage produced the lowest number of pods per plant and 100-kernel weight, while those stressed at the flowering stage had the lowest shelling percentage, and lowest yield. Stress at the early vegetative stage, which had the least effect on yield and yield attributes, shortened the flowering period and induced effective pod-filling (Kar *et al.*, 2002).

Groundnut cv. TMV 2 exposed to moisture stress at different growth stages by sowing on different dates indicated that moisture stress during the vegetative, flowering, pegging, pod setting and early pod development stages markedly reduced pod yield, but a high pod yield of 2.62 t/ha was obtained even though the crop was exposed to moisture stress during the pod development and maturation stages (Raju *et al.*, 1981). The groundnut cv. SB XI irrigated at irrigation water IW: CPE ratios of 0.75 or 0.5 for groundnuts (40 mm/irrigation) during the periods 0-40, 40-80, 80-120 d after sowing, irrigation at IW: CPE ratio of 0.5 in all 3 periods gave 24 per cent less yields than with irrigation at IW: CPE ratio 0.75 throughout in groundnuts. Water stress (IW: CPE 0.5) only during the early or late period of growth did not reduce yields substantially, on the other hand water stress 40-80 d after sowing was most harmful (Patil and Gangavane, 1990). In an lysimeter experiment during summer at Junagadh water stress at flowering (28-48 d after emergence), pegging (40-60 d), pod development (55-75 d) and pod maturation (75-95 d) stages reduced pod yield by 27, 45, 56 and 6.0 per cent, respectively in J 11 and 13, 15, 38 and 6 per cent respectively in GAUG 10 (Golakiya, 1993).

The moisture stress effect using water requirement satisfaction index (WRSI) was studied on the pod yield of groundnut cultivars ICGS 44 and TMV 2 at Hyderabad, India during kharif season where a proportional reduction in pod yield with decreasing WRSI was observed and a yield prediction model derived from the WRSI and pod yield was prepared (Reddy *et al.*, 2003). In Hyderabad, on Alfisol, during rabi season groundnut grown under control (T_1) and moisture stress between 40-75 DAS (T_2); 30-75 DAS (T_3); 85-105 DAS (T_4); 105-135 DAS (T_5); 20-40 DAS (T_6) reveals that moisture stress at different phenophases showed no significant difference in final biomass production at 120 DAS, but affected the assimilate partition to pod (Kumar and Reddy 2003). Under stress-free environment during crop growth period, last 25-30 days growth period, prior to maturity no dry matter accumulation took place either by giving three irrigations (T_1) or without any irrigation (T_5). Even prolonged moisture stress of 45 days coinciding pre-flowering to seed initiation (30-75 DAS) did not affect the ultimate dry matter production. The pod yield was highest in T6 (2002 kg ha^{-1}) which was on par with T_1 (1971 kg ha^{-1}), while T_5 resulted to a significantly lower pod yield (1770 kg ha^{-1}). Irrigations at sowing, 10, 40, 60, 70, 80, 90, 105 and 115 DAS instead of 14 irrigations at 10 days interval during rabi season had similar effect on pod yield (Kumar and Reddy 2003).

Nigam *et al.* (2005) in a field trial consisting of 192 genotypes (96 each Tr and E selections, using genetic material from three common crosses and one institute-specific cross from four collaborating institutes in India total seven crosses each contributing six genotypes) grown in a 4x48 alpha design in 12 season x location environments in India, the selection efficiency of Tr relative to E, RETr, was estimated using the genetic concept of response to selection. Based on all the 12 environments, the two selection methods performed more or less similarly (RETr=1.045). When the 12 environments were grouped into rainy and post-rainy season, the relative response to selection in Tr method was higher in the rainy than in the post-rainy season (RETr=1.220 vs 0.657) due to a higher genetic variance, lower G x E, and high h2. When the 12 environments were classified into four clusters based on plant extractable soil-water availability, the selection method Tr was superior to E in three of the four clusters (RETr=1.495, 0.612, 1.308, and 1.144) due to an increase in genetic variance and h2 under Tr in clustered environments. Although the crosses exhibited significant differences for kernel yield, the two methods of selection did not interact significantly with crosses. Both methods contributed more or less equally to the 10 highest-yielding selections (six for E and four for Tr). The six E selections had a higher kernel yield, higher transpiration (T), and nearly equal transpiration efficiency (TE) and Harvest Index (HI) relative to four Tr selections. The yield advantage in E selections came largely from greater T, which would likely not be an advantage in water-deficient environments. From the results of these multi-environment studies, it is evident that Tr method did not show a consistent superiority over E method of drought resistance breeding in producing a higher kernel yield in groundnut. Nonetheless, the integration of physiological traits (or their surrogates) in the selection scheme would be advantageous in selecting genotypes which are more efficient water utilisers or partitioners of photosynthates into economic yield. New biotechnological tools are being explored to increase efficiency of physiological trait-based drought resistance breeding in groundnut (Nigam *et al.*, 2005).

Forty six Spanish and Virginia groundnut genotypes were evaluated for stability for pod yield and 4 associated traits over 3 microenvironments where, 100-seed weight was stable across environments, but no genotype was superior over all 3 environments. Genotypes 559, 563 and 551 performed better under irrigated conditions, while genotypes 564, 565 and 571 tolerated moisture stress and 4081-4 and 4069-1B were stable under both irrigation and moisture stress (Reddy and Gupta 1994).

In a trial during Feb-June, 1988 (hot weather conditions) at Niamey, Niger, 9 groundnut lines were irrigated with 20 or 40 mm water/irrigation every 1, 2 or 3 weeks. Crops gave no pod yield with irrigation every 3 weeks. Genotypes ICGV 87123 and 55-437 gave the highest yields over all irrigation treatments, but gave very low haulm yields. There were large differences in soil moisture contents (0-210 cm layer) between irrigation treatments but not between genotypes. There were considerable differences between genotypes for the difference between crop canopy temp and air temp. This value is related to leaf water potential. A strong negative correlation was found between mid season crop canopy - air temp. difference and pod yield under intermediate water stress conditions (20 mm water/irrigation each week), whereas no correlation was found under low water stress conditions (40 mm water/irrigation each week) or high water stress conditions (20 mm water/irrigation every 2 weeks) (Greenberg and Ndunguru, 1989).

In summer on clay soil at Junagadh the transient soil moisture stress at various growth stages significantly reduced the pod per plants, shelling, 100-kernal weight, HI and oil content and yield of six groundnut genotypes. While, moisture stress at flowering and pod development stages did not affect the productivity of the crop and save about 33 per cent of irrigation water, stress at flowering stage (25-47 DAS) and pod development stage (50-72 DAS) gave 18.5 per cent and 30.6 per cent reduction in pod yield over control, respectively (Vaghasia *et al.*, 2010). In another study at Junagadh, groundnut cvs. J 11 and GG 2 grown in field lysimeters, the greatest yield reduction occurred when water stress was imposed during the pod development stage and GG 2 was more tolerant to drought than J 11 (Patel and Golakiya, 1991). Reduction in pod yields of 4 groundnut cvs. by soil moisture stress of -14 bar during pegging or pod development period was greater than stress during the vegetative growth or flowering period and the stress decreased N, P and K uptake, which varied with cv (Polara *et al.*, 1984).

Correlation and path coefficients were worked out for nine traits involving 40 hybrids and 14 parents in groundnut. Pod yield exhibited significant positive association with pods per plant, dry matter production, kernel weight and harvest index. Path analysis revealed maximum direct effect of pods per plant followed by dry matter production, and kernel weight on pod yield (Vaithiyalingan *et al.*, 2010).

4.4 Seed Quality

Drought stress affects seed quality adversely. Drought-stressed plants lose moisture from seed or pod which lead to the reduction in the seeds physiological activity, thereby increasing the susceptibility to fungal pathogens. Besides affecting food quality, drought stress is also known to alter nutritional quality of seed proteins in groundnut (Diwedi *et al.*, 1992). The pod yield and quality of groundnut are reduced when less than 30 cm water was received by the crop. Water deficit during seed production affects C_2H_2 and CO_2 production during subsequent germination (Ketring 1991). The most consistent response of water deficit was reduction in the fraction of rapidly growing seedlings (those with hypocotyl-radical longer than 20 mm at 72 h of germination). Water stress at pod initiation and development phase reduced germinability, vigour, seed membrane integrity, embryo RNA content, and chlorophyll synthesis and dehydrogenase activity in cotyledons during germination, however, moisture stress at early vegetative phase increased 100 kernel wt., embryo wt. and seedling vigour index (Nautiyal *et al.*, 1991). Thus water deficit during seed development affects subsequent growth of seedlings and could pose a problem in establishment for the succeeding crop. Thus a minimum of 500 mm of water was necessary to produce a crop of seeds with high potential for germination and high proportion of vigorous seedlings (Ketring 1991).

At ICRISAT, the seeds of groundnut cv. Robut 33-1 grown under moisture stress (emergence to maturity, emergence to peg initiation, first flush of flowering to last pod set and beginning of seed filling to maturity), seeds from plants with moisture stress from emergence to initiation of pegs gave higher field emergence, better seedling vigour and resulted in increased pod and seed yields over the other treatments. Use of seeds from crops grown with moisture stress from flowering to end of pod set resulted in yield reduction (Sarma and Sivakumar, 1987). The seeds from drought treated plants gave lower germination, though their germination energy was higher than that of the control. Drought treatment at the seed development stage reduced seed oil content, but at the flowering and ripening stages showed no significant effect on oil content. Drought at the flowering reduced protein content of seeds, while at seed development and ripening stages it gave seeds with higher protein content than control (Yao *et al.*, 1982).

The cotyledons and embryonic axis of groundnut seeds germinating under moisture stress created by PEG-6000, total soluble proteins and lipids decreased with the advancement of germination but the decline was less under moisture stress conditions. Total free amino acids, proline and non-reducing sugars increased both in cotyledons and embryonic axes with the advancement of germination stage but decreased with the increase in moisture stress level, the reducing sugar content, however, was higher under moisture stress, indicating that proline cannot be used as an indicator of stress (Sharma *et al.*, 1989). Eight groundnut cultivars grown at Tifton, Georgia by withholding water at 20-50, 50-80, 80-110 or 110-140 d after planting where drought stress early or late in the growing season had little or no effect on seed oil, protein and mineral contents, but stress during mid-season growth affected all components however, only the decreases in oil and copper contents were consistent for all cultivars (Conkerton *et al.*, 1989).

In spanish groundnut cv. Ak 12-24, J 11, GAUG 1 and GG 3 subjected to soil moisture stress at different crop growth stages, moisture stress during the early vegetative phase resulted in an increase of 100-seed weight and seedling vigour index but stress at the pod initiation/development stage reduced germinability, vigour, seed membrane integrity and embryo RNA content. Moisture stress reduced chlorophyll synthesis and dehydrogenase activity in cotyledons during germination. Growth potential was linearly related to chlorophyll content and dehydrogenase activity during seed germination. Stress during pod development was most detrimental to all physiological and biochemical processes studied (Nautiyal *et al.*, 1991). Experiment conducted at Bijapur to study the effect of water stress (-1, -3, -5, -7, -10 and -12 bar prepared using PEG 6000) in four groundnut genotypes *viz.*, ICG-1930, KRG 228, ICGS 11 and S 206 where seed germination, seedling growth and vigour index decreased significantly with increasing intensity of stress irrespective of cultivars tested, with ICGS 11 and ICG 1930 showing higher seed quality parameters indicating drought tolerance (Pawar, 2011).

Misra and Nautiyal (2005) studied kernel quality components (total sugars, phenolics, protein and fatty acid composition) as influenced by the soil moisture-deficit stress imposed during different phenophases, in the summer season in four Spanish cultivars of groundnut, AK 12-24, J 11, GAUG 1 and GG 2 and observed increase in stearic acid due to stress during pod development in all cultivars except GG 2, increase in palmitic acid only in GAUG 1 and oleic acid in AK 12-24. Compared to the control, soil moisture-deficit stress significantly increased the protein content. There was, however, a greater increase in protein content due to stress during flowering and pod development compared to the stress during vegetative phases. Stress during vegetative (short), and flowering phases significantly reduced the sugar content. The interaction between cultivars and treatments were significant only for the changes in fatty acid composition, protein and sugar contents, but was not significant for phenolic compounds. It is concluded that the changes in the composition of fatty acids and contents of sugars and phenolic compounds are governed more by cultivar and its interaction with the environmental conditions rather than by the time or the intensity of imposed soil moisture-deficit stress.

In groundnut total oil content was not affected by early-season drought (Conkerton *et al.*, 1989; Bhalani and Parameswaran 1992), but declined (up to 3 per cent) under mid-season (50-80 days after sowing (DAS) drought (Conkerton *et al.*, 1989). For late-season drought (110-140 DAS), various studies reported no effect (Conkerton *et al.*, 1989; Musingo *et al.*, 1989) but a decline in total oil content by Bhalani and Parameswaran (1992). No consistent effect on protein content has been documented due to drought stress at any particular growth period; nor was protein content in any specific genotype always reduced or increased by drought stress (*Conkerton et al.*1989). However, Musingo *et al.* (1989) reported that late-season (50 days before harvest) drought caused little change in total protein content of groundnut. The effects of drought on oil, protein and fatty acid contents in 12 genotypes differing in seed quality traits by exposing mid-season (40 and 80 DAS) and the end-of-season (80 DAS until harvest) drought at ICRISAT during post-rainy (November-April) seasons reveal that mid-season drought had no effect on the content of oil, protein and fatty acids other than eicosenoic fatty acid. However, end-of-season drought reduced total oil, and linoleic and behenic fatty acid content, and increased total protein and stearic and oleic fatty acid, genotypic interactions. In ICGVs 88369, 88371, 88381, 88382 and 88403, total oil content remained unaffected while oleic fatty acid content increased under end-of-season drought. These were identified as desirable parents for a breeding program to develop cultivars suitable for rainfed cultivation (Dwivedi *et al.*, 1996).

Fatty acid contents in a genotype are affected by drought stress and seed grade. Composition of fatty acid in crop is affected by drought stress. Regardless of intensity, duration and timing of drought, most of the fatty acid and O/L ratio decreased significantly (Hashim *et al.*, 1993). However, some study reported no major changes in the fatty acid composition, except for oleic acid, due to water deficit stress in summer groundnut (Bhalani and Parameswaran 1992). In field trials at Georgia Florunner groundnut (a) irrigated throughout the growing season (140 days) at 0.2 bar matric potential (total 70 cm water) or (b) irrigated at 0.2 bar potential for 30 days and at 15 bar for 40 days with no further irrigation (total 30 cm water), showed leaf water potential -26 and -12 bar in (b) and (a), and seed yield 6.99 and 5.99 t/ha in (a) and (b), respectively, and musty flavour was detected in (b) but not (a) (Pallas, 1983). In another study at Tifton, Georgia, withholding irrigation in groundnut cv. Florunner at the preflowering (20 DAS), pod formation (50 DAS) or maturation (80 DAS) stages altered the fatty acid composition, oleic:linoleic (O:L) ratio, computed iodine value, alpha -tocopherol and gamma -tocopherol contents and grade of groundnut seed. As groundnuts increased in size, regardless of the stage when stressed, long chain saturated fatty acids, eicosenoic acid [gadoleic acid], O:L ratio and alpha-tocopherol decreased significantly. Groundnuts stressed at maturity showed the greatest decrease in the O: L ratio and the greatest increase in the iodine value (Hashim *et al.*, 1993).

Under drought stress complex carbohydrates and proteins are broken down by enzymes into simpler sugars and amino acid residues, respectively and accumulation of compatible solutes in the cell increases the osmotic potential and reduces water loss from the cell. The rafnose family oligosaccharides (RFOs), such as rafnose, stachyose, and verbascose are soluble galactosyl-sucrose carbohydrates. It was also reported that the expression of enzymes related to the biosynthesis of galactinol and RFOs and their intracellular accumulation in plant cells are closely associated with the responses to environmental stresses (Peters *et al.*, 2007).

Water stress at the pod filling stage reduced kernel yield and nitrogen partitioning to reproductive parts. Water stress at all growth stages reduced nitrogen fixation (Venkateswarlu *et al.*, 1991). Huang and Ketring (1985) in a study grew five groundnut cvs. under rainfed and irrigated conditions during first year and 6 cvs. during 2nd year, where Virginia type cvs. had higher ground cover than Spanish types and water relation components differed between rainfed and irrigated treatments at 67 and 81 DAS, but not at 54 DAS and not among cultivars. However, pod yield and percentage sound mature kernels were significantly reduced under rainfed conditions, but no significant differences were observed among cultivars. Yield reductions due to water stress under rainfed conditions were 83-97 per cent.

The 70 day early season drought caused the greatest reduction in SMK, late season 35-day and midseason 70 day droughts reduced subsequent germination by 5 and 9 per cent, respectively, and plant water stress when relieved and leaf diffusion resistances returned to normal (Pallas *et al.*, 1979). According to Sanders *et al.* (1986) agrometeorological studies must include an awareness of the relationship between environment, maturity, and postharvest quality. The post harvest quality of groundnut is the resultant of the particular set of environmental and cultural practices during pod growth and maturation. Seed composition changes dramatically as the crop matures but has relation with environment. A biochemical basis exists for inferior quality in immature groundnut. Drought stress and soil temperature influence maturation rate and thus had an indirect effect on post harvest quality. In a recent study at Junagadh, prolonged water deficit reduces oil and protein content in groundnut kernels leading to loss of nutritional quality. Accumulation of metabolite like glucose, sucrose and raffinose like oligosaccharides plays important role in drought stress tolerance of groundnut. Differential response between different habit groups showed the runner group are more tolerant to water deficit stress compared to bunch type cultivars (Chakraborty *et al.*, 2013).

4.5. Pod and Seed Size

A minimum of 500 mm of water is necessary to produce a crop of seeds with high potential for germination and high proportion of vigorous seedlings, however the large seed size groundnut require more nutrient and hence more water about 700 mm (Singh *et al.*, 2011, 2013). In general the pod yield and quality of groundnut are reduced when less than 300 mm water was received by the crop. Water stress at pod initiation and development phase reduced germinability, vigour, seed membrane integrity and could pose a problem in establishment for the succeeding crop. The pod size tended to be reduced under drought condition (Blakenship *et al.*, 1983). Large seeds of groundnut have greater consumer preference and fetch higher prices in domestic and international markets. In general cultivars belonging to var. *hypogaea* have larger and heavier seeds and those belonging to var. *fastigiata* have smaller seeds. The size of kernel is one of the important factors for export and normally varieties with hundred seed mass of 60 g or more are considered as large seeded groundnut and are preferred for confectionery purpose.

There is very little work reported on the effect of moisture stress on seed size. Drought at flowering reduced the seed size and increased the number of shrivelled kernels (Yao at al., 1983). A drought period of 30 days at the flowering caused a 24.7 per cent decrease of 100-seed wt, while at the ripening stage it resulted in a 14.6 per cent decrease in 100-seed wt and also reduced seed size and increased the number of shriveled seeds (Yao *et al.*, 1982). Seed size of all cultivars decreased when available water was reduced. Under water stress, Tainan 9 and ICGV 98324 had the biggest seed and seed size of Tainan 9 and ICGV 98324 were reducedby 35 per cent and 29 per cent. Regarding seed size, Khon Kaen 60-3 was the most sensitive cultivar under water stress. Seed size of Khon Kaen 60-3 was reduced by 72 per cent at 1/2 AW.

The National Research Center for Groundnut (now DGR), Junagadh, India, has undertaken programme to develop large seed cultivars more suitable for direct consumption and processing and two seasons of studies on a few promising lines (ICGV33101, PBS 29058 and PBS 29030) possessing large seeded and/or confectionery properties indicated water as the main production factor (Hariprasanna *et al.*, 2004). Kale *et al.* (2000) evaluated eight early bold seeded selections over 4 seasons under irrigated and high management condition and found superior yield in selections TGLPS 2, TGLPS 3, TGLPS 4, TGLPS 6, TGLPS 7 and TGLPS 8 over the large kernel checks *viz.*, TKG 19A and BAU 13, with 71.4 to 80.3 g 100 kernel weight and about 49 to 65 per cent kernels having 100- kernel weight >80 g. Manivel *et al.* (2000) evaluated twelve advanced breeding lines along with controls (B95, Somnath and ICGV 89211) where the 100-seed mass of the test genotypes and controls ranged from 53.9 g (PBS29036) to 76.7g (ICGV 89211) and none of the genotypes were superior over the best control (ICGV 89211). Rajgopal *et al.* (2000) evaluated 118 groundnut accessions for two years. The 100 seed mass ranged from 46.8 g for GG11 to 69.9 g for NRCGs 8939, 5850, 7276, 5505, 750 gave significantly higher 100 seed mass than M 13. Thease all are mainly due to water constraint during pod filling phases as groundnut at Junagadh most of the year face mid season or terminal drought and one or two protective irrigation are required for getting proper pod filling (Singh *et al.*, 2011, 2013). Planted seed size affects the rate of emergence and seedling vigour significantly and positively associated with increased seed size (Gorbet, 1977).

In Tifton, Georgia, withholding irrigation iin groundnuts cv Florunner at the pre-flowering (20 days after sowing, DAS), pod formation (50 DAS) or maturation (80 DAS) stages significantly affect fatty acid composition, oleic:linoleic (O:L) ratio and grade of groundnut seed.

In a study on several groundnut varieties Vorasoot *et al.* (2003) found that under reduced available water, the seed size of all varieties decreased. Under adequate moisture, the variety WEST-20 had the biggest seed size while TG-26 had lowest seed size, but under water stress TAG-24 had biggest seed size. The TG-26 was the more sensitive variety under water stress the seed size of which was reduced by 62 per cent whereas TAG-24 reduces by 41 per cent over control as compare to other varieties. Dry pod size of all varieties decreased under water deficit condition. There was significant interaction between soil water regimes and 100 dry pod weight and TG-26 showed maximum reduction in weight (66 per cent) while TAG-24 showed minimum reduction in weight (38 per cent) over control at higher water stress level (Vorasoot *et al.*, 2003).

In a field trial at Redland Bay, Queensland, from 84 d after sowing to maturity, pod and seed yield was reduced by 30 per cent in the Virginia type of cultivars (Virginia Bunch and Robut-33) and by 45 per cent in a Spanish type cultivar,

McCubbin. Robut-33, a cultivar with a high harvest index (HI), produced the greatest yield under both well-watered and water-stressed conditions. In all cultivars, potential pod number had been almost achieved prior to the start of the period of water deficit. However, low pod number rather than small pod size, was mainly responsible for the decrease in yield (Chapman *et al.*, 1993).

4.6. Aflatoxins and Pest Damages

Drought stress during late stages of pod maturation in groundnut crop during the post rainy season, increased the amount of seed infection by *A. flavus*. A significant, positive, linear relationship was found between water deficit (drought intensity) and seed infection in groundnut genotypes. Genotypic differences for seed infection by *A. flavus* were evident at all levels of drought-stress, but, under the more severe stress conditions, the genotypes resistant to *A. flavus* also had seed infection but low levels (Mehan *et al.*, 1988). *Aspergillus flavus* invasion and aflatoxin contamination in groundnuts are related to drought stress, soil temperature and maturity and small, immature seeds are more likely to be contaminated with *A. flavus* than larger, mature seeds (Sanders *et al.*, 1986).The biochemical composition, fungal contamination, as the tendency toward higher moisture content complicate storage of immature seed and each of these factors predisposes immature seed to rapid quality deterioration in storage (Sanders *et al.*, 1986).

A greenhouse study on seven groundnut genotypes and microplot for two consecutive years to determine peg colonization by *A. flavus* and the effect of drought stress on the susceptibility of shells and kernels to *Aspergillus* colonization and aflatoxin contamination reveals that in general, low soil moisture tension enhanced colonization of shells and kernels and shells of most genotypes were highly colonized after harvest from each moisture regime (Azaizeh *et al.*, 1989). Kernels of all genotypes were more susceptible to *A. flavus* and *A. parasiticus* colonization under both long and short drought stress conditions compared with non-stressed and kernels of TX811956 and TX798736 (short stress treatments) contained lower *Aspergillus* infestation and kernels of genotypes PI 337409 and TX 811956 and TX 798736 contained less aflatoxin (Azaizeh *et al.*, 1989). Fifty groundnut genotypes screened for low aflatoxin contamination under field conditions with two main treatments *i.e.*, irrigated and simulated drought conditions reveals that ICGV 86590, 89104, 94350, 99029, IC 48 and ICGS 76 had low aflatoxin levels (<5 ppb) in both the conditions, but no consistent relationship was observed between seed colonization and aflatoxin production (Sudhakar *et al.*, 2007). Also aflatoxin production in groundnut was negatively related with RWC, pod wall integrity and pod wall moisture content at harvest (Sudhakar *et al.*, 2007).

In Georgia, Kisyombe *et al.* (1985) studied 14 groundnut genotypes grown in rain-shaded field microplots under simulated water stress conditions and in plots under normal rainfall conditions, where J 11 and Lampang proved resistant to aflatoxin under both dry and moist field conditions. Although percentage infection of kernels varied with genotype, ranking of genotypes reported to have drought resistance was consistent under both conditions. When 34 genotypes including those tested in microplots were also evaluated for dry seed resistance in the laboratory, J 11 and PI 337409 were highly resistant. Except for J 11 there was no correlation between genotype rankings for resistance to dry seed infection and resistance under field conditions (Kisyombe *et al.*, 1985).

Three groundnut genotypes (ICG 221, 1104 and 1326) drought stressed during the last 58 d before harvest with 8 levels of stress ranging from 1.1 to 25.9 cm of water and the kernels harvested from these were hydrated to 20 per cent moisture and challenged with *A. flavus*, and Fungal colonization, aflatoxin content and phytoalexin accumulation measured the fungal colonization of non-drought-stressed kernels virtually ceased by 3 d after inoculation, when the phytoalexin concentration exceeded 50 μg/g (fresh wt.) of kernels, but in drought-stressed material fungal colonization was inversely related to water supply (r -0.848 to -0.904, according to genotype), as was aflatoxin production (r -0.876 to -0.912); the phytoalexin concentration was correlated with water supply when this exceeded 11 cm (r = 0.696-0.917) (Wotton and Strange 1987).

Early-season moisture stress intensifies groundnut yield and quality losses associated with combined injury from thrips and post emergence herbicides (Funderburk *et al.*, 1998). Whitegrubs (*Scarabaeid* larvae) are major pests of groundnuts in many parts of the tropics and subtropics, attacking both roots and pods and protocols for simulating white grub damage to groundnut roots were developed to predict how feeding by these pests affects plant growth. Pod yield and total biomass were reduced by root-cutting particularly at 51 DAE. Simulated drought reduced pod yield and total biomass, but reductions due to root cutting, were greatest for non-drought treatments. At both 30 and 51 DAE, pod yield reductions were greatest where roots were cut at only 10 cm below the soil surface, and root cut at 30 DAE was totally compensated for by the development of new roots. But cut at 51 DAE initiated very little root regrowth. Both total biomass and pod yield were strongly correlated with plant water use ($R^2>0.9$). The protocols developed measure the plant responses to simulated scarab damage and could easily be adapted to measure plant responses to actual scarab damage (Brier *et al.*, 1997).

5. Conclusions and Future Research Strategies

Drought is the major abiotic constraint affecting groundnut productivity and quality worldwide. There are three major aspects of drought, duration, intensity and timing which vary with crop phenophases. The groundnut is relatively drought tolerant and an important crop of the semi-arid regions. The plant water-status is the result of a balance between water uptake and loss which has been less understood in groundnut. Though different growth stages have different sensitivity to water deficit, none of these can proceed normally below some minimum water. Groundnut plant contains about 80 per cent of water on fresh weight basis and reduction of the plant water status much below this level causes wilting and affects the rate of several plant functions. The water flow in intact plant under high soil moisture condition is for growth and transpiration and two concepts are expressed about the driving force for transpiration water flow, the water potential differences between the root and leaf as the primary force and hydrostatic and osmotic pressure differences, as the factors determining water flow. The ability of groundnut genotypes to maintain water supply to leaves using apparent sap velocity (V_a) was 0.8-1.1 cm min^{-1} and declined with stress in field. Increasing moisture stress from 0 to 6 atm decrease the leaf RWC, increased water saturation deficit and relative saturation deficit, decreased DM accumulation in the shoot and increased it in the root, decreased RGR, increased specific leaf weight, decreased non-reducing sugar and increased reducing sugar in groundnut seedlings.

The timing of drought has a large impact on the variation. The sensitivity of a genotype to drought increases with yield potential, increasing the closer the drought ends to final harvest. Genotypic variation to drought exists in the water-use ratio with some, being able to accumulate up to 30 per cent more shoot DM with the same total transpiration and HI, and large variations in genotypes to midseason are due to recovery differences after the drought is relieved. Drought stress effects on groundnut depend primarily on the stress pattern because genotypic variation is usually of secondary significance. Pod yield is a function of transpired water (T), transpiration efficiency (TE) and harvest index (H). The yield losses (per cent) due to drought are estimated as: Yield loss (per cent) = 100 (1 - D_y/W_y), where, W_y is the pod yield under adequate irrigation and D_y is the pod yield under drought. In most part of India, in a 110-120 days crop water stress at 45-70 DAS (flowering) and pod development (60-90 DAS) phases was highly detrimental to leaf area development, dry matter production, pod formation causing 40-60 and 50-70 per cent yield reductions, respectively and the recovery of growth from water stress was better after relief at the vegetative phase than at later growth phases. However in a 140-150 days crop maximum reduction in kernel yield when stress was imposed during seed filling phase, *i.e.*, 93 DAS onwards.

The water requirement of groundnut varies with the stages and is lowest from germination to flower formation and reaches maximum during pod formation. However, the utilization of available moisture is greatest during flowering and pod formation and the crop receiving adequate water during these stages only can give equal yield to the well watered crop. During these stages if stress is given and later on water supply is resumed only the vegetative growth is benefited not the reproductive growth of crop. Thus the period of maximum sensitivity to drought occurs between 50-80 DAS, the period of maximum flowering and vegetative growth. The yield reductions have been reported to be 10-15, 15-30, 40-50 and 50-70 per cent, respectively when drought was imposed from 10-30, 30-50, 50-80 and 80-120 days after sowing respectively. Yield decreases due to water stress at different growth stages are in order of stress at seed development > at early pod filling > at early growth > at early pegging.

The period of reproductive growth stages occurs over a period of nearly two months and moisture stress has a depressing effect on flowering, stem growth and nodulation. No flowering occurs during the stress, but once the stress is relieved, there is a flush of flowering depending on the growth stages and sometimes it results in more flowers than control. The Virginia type groundnuts, due to their longer duration, are more tolerant to drought than Spanish and Valencia and the Spanish and Valencia due to short duration escape the late season drought. The flush of late flowers following mid season drought delay maturity and hence late harvesting, where late season rain helps. The fruiting occurs once the gynophores enter into the soil and soil physical condition is important and must be wet during the gynophore entering the soil as the gynophore can exert a force equivalent to 3-4 g only. Under water stress there is poor pod filling that reduced kernel size, shelling, SMK per cent and lipid content of kernel.

Plant population influenced both the temporal and spatial patterns of water use, with high density crops extracting water from lower depths sooner than low density crop. High water use prior to early pod filling in high density crop was associated with more rapid leaf area development. The more rapid water extraction in a high, compared with a low, population density groundnut crop is associated with greater root production at depth.

In order to sustain plant growth and hydration, water must be continuously supplied to the leaves as it is lost by transpiration. This becomes difficult under low soil moisture condition. Some of the management practices avoiding minimizing drought are summarized below:

★ In groundnut, water stress delay pod initiation, and the major cause of variability in pod yield and HI is the delay between peg initiation and onset of rapid pod growth. The management practices should aim to optimize the availability of growth resources at the time of pegging in order to ensure that pod initiation is not delayed.

★ The yield is a function of many plant and environmental factors and moisture stress play an important role particularly the stage at which stress occurs. The water stress affects the vegetative, root and reproductive growth and a proper scheduling of irrigation is required.

★ Moisture stress at flowering reduced phytobiomass and pod yield by limiting the number of mature pods per unit area as compared to stress at pegging and pod formation stages. The variation in HI account for the large proportion of variation in yield, and hence recommended to make selection for high HI.

★ As reproductive development is sensitive to drought resulting to poor yield, the strategies to combat drought in groundnut genotypes are (i) early production of flowers, pegs, and pods, with subsequent filling of the pods at a moderate, but essentially at constant rate despite the drought, (ii) faster development of later developed pegs into pods once water become available after drought late in the season.

★ The early and continuous availability of water until the start of pod filling result in large canopy and during the period of drought stress the transpirational demand increases. The transient soil-moisture-deficit stress for 20-25 days as pre-flowering drought during vegetative phase (20-45 DAS) results in synchronized flowering, increases 10-20 per cent pod yield and save 10-15 per cent water mainly due to promotion of root growth during water stress and inhibition of number of vegetative sites (leaves and branches).

☆ About 500 mm of water is necessary to produce a crop of seeds with high potential for germination and seedling vigor and pod yield and quality of groundnut are reduced when less than 300 mm water was received. Groundnut cultivation may be planned accordingly.

☆ Drought stress and soil temperature influence maturation rate and thus had an indirect effect on postharvest quality. The greatest yield reduction corresponds to, peak flowering to early pod filling stage and adequate moisture during this period is critical for maximum yield.

☆ As gypsum increase early pod development, it provides an escape mechanism from drought. Gypsum applied at flowering increased yield of genotype subjected to drought.

Genetic improvement of crop resistance to drought stress is one component and will provide a good perspective on the efficacy of control strategy through genetic improvement. Selection for drought adaptation under rainfed conditions, though commonly practiced, could be misleading, since it may not reflect the ability of the genotype if the stress occurs during the critical stages of plant development. More efficient selection would require simulated drought conditions, and the use of other indirect selection methods that give a good indication of drought adaptation. Water deficit during seed filling phase, genotypic yield potential accounted for approximately 90 per cent of the variation in pod yield sensitivity to water deficit, and it is unlikely that breeders will be able to combine high yield potential with low sensitivity to drought spanning the seed filling phase, therefore other important strategies are necessary. The pod yield potential accounted for less of the variation in drought sensitivity (15-64 per cent) in the early and mid-season droughts. For these circumstances it may be possible to identify genotypes with both high yield potential and relatively low drought sensitivity.

Seed composition changes dramatically as the crop matures but has relation with environment. Agrometeorological relationship between environment, crop phenology, maturity, and postharvest quality is essential as postharvest quality of groundnut is the resultant of the particular set of environmental and cultural practices during pod growth and maturation. A biochemical basis exists for inferior quality in immature groundnut. The pod yield and quality of groundnut are reduced when less than 30 cm water was received by the crop. Water deficit during seed production affected C_2H_2 and CO_2 production during subsequent germination. Water stress at pod initiation and development phase reduced germinability, vigour, seed membrane integrity thus water deficit during seed development affects subsequent growth of seedlings and could pose a problem in establishment for the succeeding crop. A minimum of 500 mm of water is necessary to produce a crop of seeds with high potential for germination and high proportion of vigorous seedlings. Many a time the superiority of genotype with high yield, in water stress is not reflected in their ancillary characters. *Aspergillus flavus* invasion and aflatoxin contamination in groundnuts are related to drought stress, soil temperature and maturity and small, immature seeds are more likely to be contaminated with *A. flavus* than larger, mature seeds.

Most breeding programmes in groundnut follow an empirical approach to drought resistance breeding, largely based on kernel yield and traits of local adaptation, resulting in slow progress. Recent use of easily measurable surrogates traits associated with drought tolerance encouraged breeders to integrate these in their selection schemes. However, there has been no direct comparison of the relative efficiency of a physiological trait-based selection approach (Tr) vis-a-vis an empirical approach (E) to ascertain the benefits of the former. The drought tolerance contributing factors in groundnuts are: an extensive root system established before maximum leaf area to meet the transpirational demand, recurred and synchronized flowering once stress was relieved, water storage cells in the abaxial side of the leaves to provide water when transpiration was greater than the roots extraction of soil moisture; leaf folding during stress to reduce solar incidence; and transpiration regulated by high stomatal resistance during stress.

In groundnuts drought-stress effects depend primarily on the stress pattern because genotypic variation is usually of secondary significance. The different responses of groundnut cv. to drought when assessed relative to the mean response of all genotypes to drought as two major aspects of drought (duration, intensity, and timing relative to crop phenophases) may vary independently. The timing of drought has a large impact on the variation about the mean response. The sensitivity of a genotype to drought increases with yield potential, increasing the closer the drought ends to final harvest. Genotypic variation in response to drought exists in the water-use ratio of genotypes, with some being able to accumulate up to 30 per cent more shoot DM than others with the same total transpiration.

To ensure survival of crops from water deficit and sustainable production stress resistance may involve avoidance mechanisms preventing exposure to stress, tolerance mechanisms permitting the plant to withstand stress through osmotic adjustment, and acclimation by altering their physiology in response to stress. Innovative biotechnological approaches have enhanced our understanding of the processes underlying plant responses to drought at the molecular and whole plant levels. Hundreds of drought stress-induced genes have been identified and some of these have been cloned. Plant genetic engineering and molecular marker approaches allow the development of drought-tolerant germplasm.

6. References

Ali, E.A. (2001). Effect of water stress and potassium fertilization on yield and yield components of peanut (*Arachis hypogaea* L.). Annals of Agricultural Science, 39: 1425-1434.

Amaregouda, A., Chetti, M.B., Salimath, P.M. and Kulkarni, S.S. (1994b). Effect of antitranspirants on stomatal resistance and frequency, relative water content and pod yield in summer groundnut (*Arachis hypogaea* L.). *Annals of Plant Physiology.* 8: 18-23.

Arjunan A, Srinivasan DS, Vindiyavarman P (1988) Physiological aspects of drought tolerance in groundnut. *Madras Agric. J.*, 75: 5-8.

Arjunan, A. and Srinivasan, P.S. (1989). Pre-sowing seed hardening for drought tolerance in groundnut (*Arachis hypogaea* L.). *Madras Agricultural Journal*, 76: 523-526.

Azaizeh, H.A., Pettit, R.E., Smith, O.D., and Taber, R.A. (1989). Reaction of peanut genotypes under drought stress to *Aspergillus flavus* and *A. parasiticus*. *Peanut Science*. 16: 109-113

Azam Ali, S.N. (1984). Environmental and physiological control of transpiration by groundnut crops. *Agricultural and Forest Meteorology*. 33: 129-140.

Babitha, M., Sudhakar, P., Latha, P., Reddy, P.V. and Vasanthi, R.P. (2006). Screening of groundnut genotypes for high water use efficiency and temperature tolerance. *Indian Journal of Plant Physiology*. 11: 63-74.

Babu, V.R., Murthy, P.S.S., Reddy, D.N. and Ramesh Babu, V. (1985). Water uptake, germination and seedling growth of groundnut (*Arachis hypogaea* L.) cultivars under simulated moisture stress. *Seed Research*. 13:141-154.

Babu, VR and Rao, DVM (1983). Water stress adaptations in the groundnut (*Arachishypogaea* L.) - foliar characteristics andadaptations to moisture stress. *Plant Physiol. Biochem*, 10 : 64–80.

Badigannavar, M., Kale, D.M., Bhagwat, S.G. and Murty, G.S.S. (1999). Genotypic and seasonal variation in stomatal characters in Trombay groundnut varieties. *Tropical Agricultural Research and Extension.*, 2: 10-12.

Bennett, J.M., Sexton, P.J. and Boote, K.J. (1990). A root tube-pegging pan apparatus: preliminary observations and effects of soil water in the pegging zone. *Peanut Science*. 17: 68-72.

Bhagsari, A.S., Brown, R.H. and Schepers, J.S. (1976). Effect of moisture stress on photosynthesis and some related physiological characteristics in peanut. *Crop-Science*. 16: 712-715

Bhalani GK and Parameswaran M (1992) Influence of differential irrigation on kernel lipid profile in groundnut. *Plant Physiol. Biochem*. 19, 11-14.

Billaz, R., Ochos, R., 1961. The stage of susceptibility to drought in groundnut. *Oleagineux* 16, 605-609.

Black, C.R., Tang, D.Y., Ong, C.K., Solon, A. and Simmonds, L.P. (1985). Effects of soil moisture stress on the water relations and water use of groundnut stands. *New Phytologist.*, 100: 313-328.

Boote, K.J. and Hammond, L.C. (1981). Effect of drought on vegetative and reproductive development of peanut. Proceedings, *American Peanut Research and Education Society*, 13: 86.

Brier, H.B., Wightman, J.A., Wright, G.C. and Allsopp, P.G. (ed.), Rogers, D.J. (ed.), and Robertson, L.N. (1997). The effect of simulated whitegrub damage on peanut growth and yield. Soil invertebrates. Proceedings of the 3rd Brisbane Workshop., 153-158.

Burow MD, Simpson CE, Starr JL and Paterson AH (2001) Transmission genetics of chromatin from a synthetic amphidiploid to cultivated peanut (*Arachis hypogaea* L.): Broadening the Gene Pool of a Monophyletic Polyploid Species.*Genetics*. 159, 823-837.

Chakroborty K., K.A.Kalariya and A.L.Singh (2011). Increase in production of reactive oxygen species (ROS) as a secondary response to water deficit stress in groundnut(2011). Proc National seminar on "sustainable crop productivity through physiological interventions" organized by ISPP at Ramnarain Ruia College, Mumbai, November, 24-26. Abs p. 62.

Chakraborty, K., Bishi S.K., Singh, A.L., Kalariya, K.A. and Lokesh Kumar (2013). Water deficit stress alters seeds quality and composition in groundnut. *Indian Journal of Plant Physiology*. 18(2): 136-141.

Chakraborty, K., Kalariya, K.A., Goswami, Nisha and Singh A.L. (2012). Stress induced antioxidant enzyme activity contributes to water deficit stress tolerance in groundnut (*Arachis hypogaea* L.). National Symposium on Biotic and Abiotic Stresses under Changing Climate Scenario, UBKV, Cooch Behar, November, 2012.

Chapman, S.C., Ludlow, M.M. and Blamey, F.P.C. (1993a). Effect of drought during early reproductive development on the dynamics of yield development of cultivars of groundnut (*Arachis hypogaea* L.). *Field Crops Research*, 32: 227-242.

Chapman, S.C., Ludlow, M.M., Blamey, F.P.C. and Fischer, K.S. (1993b). Effect of drought during early reproductive development on growth of cultivars of groundnut (*Arachis hypogaea* L.). II. Biomass production, pod development and yield. *Field Crops Research*, 32: 211-225.

Chapman, S.C., Ludlow, M.M., Blamey, F.P.C. and Fischer, K,S. (1993c). Effect of drought during early reproductive development on growth of cultivars of groundnut (*Arachis hypogaea* L.). I. Utilization of radiation and water during drought. *Field Crops Research*, 32: 193-210.

Chapman, S.C., Ludlow, M.M., Blamey, F.P.C. and Fischer, K.S. (1993d). Effect of drought during pod filling on utilization of water and on growth of cultivars of groundnut (*Arachis hypogaea* L.). *Field Crops Research*. 32: 243-255.

Chaudhari, K.N. and Dadhwal, V.K. (2004). Assessment of impact of drought - 2002 on the production of major kharif and rabi crops using standardized precipitation index. *Journal of Agrometeorology*.; 6 (Special Issue): 10-15.

Chaudhari, K.N. and Dadhwal, V.K. (2004). Assessment of impact of drought - 2002 on the production of major kharif and rabi crops using standardized precipitation index. *Journal of Agrometeorology*. 6(Special Issue): 10-15

Chavan, A.A., Dhoble, M.V. and Khating, E.A. (1992). Effect of artificial water stress on different genotypes of groundnut (*Arachis hypogaea*) in dryland. *Indian Journal of Agricultural Sciences*. 62: 376-381.

Chen YouQiang, Ye BingYing, Zhu-JinMao, Zhuang WeiJian, Chen, Y.Q., Ye, B.Y., Zhu, J.M. and Zhuang, W.J. (2000). Effects of osmotic stress on active oxygen damage and membrane peroxidation in young leaves of groundnut (*Arachis hypogaea*). *Chinese Journal of Oil Crop Sciences.* 22: 53-56.

Chen, C.Y. and Chang, H.S. (1972). Studies on the relationship between several physiological functions of plants and the DPD of leaves and soil moisture stress. *Journal of the Agricultural Association of China.* 80: 26-41.

Chung SiYin, Vercellotti, J.R., Sanders, T.H., Chung, S.Y. (1998). Evidence of stress proteins and a potential maturity marker in peanuts. *Journal of Agricultural and Food Chemistry.*, 46: 4712-4716.

Chung, SiYin., Vercellotti, J.R., Sanders, T.H. and Chung, S.Y. (1997). Increase of glycolytic enzymes in peanuts during peanut maturation and curing: evidence of anaerobic metabolism. *Journal of Agricultural and Food Chemistry.* 1997, 45: 4516-4521.

Chuni Lal, Hariprasanna, K., Rathnakumar, A.L., Misra, J.B., Samdur, M.Y., Gor, H.K., Chikani, B.M. and Jain, V.K. (2009). Response of peanut genotypes to mid-season moisture stress: phenological, morpho-physiological, and yield traits. Crop-and-Pasture-Science. 60: 339-347.

Clave, D., Baradat, P., Khalfaoui, J.L., Drame, N.K., Diop, N.D., Diouf, O., Zuily-Fodil, Y. (2007). Drought adaptation and breeding: the case of groundnut cultivated in Sahel area. Part 2: a multidiciplinary approach to breeding. OCL-*Oleagineux, Corps Gras, Lipides.* 14: 293-308.

Clavel, D., Drame, N.K., Roy Macauley, H., Braconnier, S. and Laffray, D. (2005). Analysis of early responses to drought associated with field drought adaptation in four Sahelian groundnut (*Arachis hypogaea* L.) cultivars. *Environmental and Experimental Botany.* 54: 219-230.

Clavel Danie'le, Omar Diouf, Jean L Khalfaoui and Serge Braconnier (2006). Genotypes variations in uorescence parameters among closely related groundnut (*Arachis hypogaea* L.) lines and their potential for drought screening programs. *Field Crops Research,* 96: 296–306.

Clifford, S.C., Black, C.R., Roberts, J.A., Stronach, I.M. Singleton Jones, P.R., Mohamed, A.D. and Azam Ali, S.N. (1995). The effect of elevated atmospheric CO2 and drought on stomatal frequency in groundnut (*Arachis hypogaea* (L.)). *Journal of Experimental Botany.* 46: 847-852.

Clifford, S.C., Stronach, I.M., Mohamed, A.D., Azam Ali, S.N. and Crout, N.M.J. (1993). The effects of elevated atmospheric carbon dioxide and water stress on light interception, dry matter production and yield in stands of groundnut (*Arachis hypogaea* L.). *Journal of Experimental Botany.* 44: 1763-1770.

Collino, D., Biderbost, E. and Racca, R. (1994). Identification and characterization of physiological variables for the selection of drought-resistant material in groundnut (*Arachis hypogaea* L.). *Avances en Investigacion INTA Estacion Experimental Agropecuaria Manfredi*.1: 25-31.

Collino, D.J., Dardanelli, J.L., Sereno, R. and Racca, R.W. (2000). Physiological responses of argentine peanut varieties to water stress. Water uptake and water use efficiency. *Field Crops Research.*, 68:133-142.

Collino, D.J., Dardanelli, J.L., Sereno, R. and Racca, R.W. (2001). Physiological responses of argentine peanut varieties to water stress. Light interception, radiation use efficiency and partitioning of assimilates. *Field Crops Research.*, 70: 177-184.

Conkerton, E.J., Ross, L.F., Daigle, D.J., Kvien, C.S., and McCombs, C. (1989). The effect of drought stress on peanut seed composition II. Oil, protein and minerals. *Oleagineux-Paris.* 44: 593-602.

Costa, W.A.J.M.de. and Nayakarathne, N.M.R.S. (2001). Effect of two different water regimes on growth and yield of different groundnut (*Arachis hypogaea* L.) genotypes in Sri Lanka. *Tropical Agricultural Research and Extension.* 4: 29-35.

Cuc LM, Mace ES, Crouch JH, Quang VD, Long TD and Varshney RK (2008) Isolation And Characterization of Novel Microsatellite Markers and Their Application for Diversity Assessment in Cultivated Groundnut (*Arachis hypogaea*). *BMC Plant Biology.*8, 55.

Craufurd, P.Q., Wheeler, T. R., Ellis, R.H., Summerfield, R.J. and Prasad, P.V.V. (2000). Escape and tolerance to high temperature at flowering in groundnut (*Arachis hypogaea* L.). *Journal of Agricultural Science.* 135: 371-378.

Dhopte, A.M. and Ramteke, S.D. (1994). Relative variation in dry matter partitioning of peanut genotypes under moisture stress conditions. *Annals of Plant Physiology.* 8: 174-178.

Dhopte, A.M., Ramteke, S.D. and Thote, S.G. (1992). Effect of soil moisture deficit on root growth and respiration, nodulation and yield stability of field grown peanut genotypes. *Annals of Plant Physiology.* 6:188-197.

Dhruve, J.J. and Vakharia, D.N. (2007). Amelioration of polyethylene glycol simulated water deficit stress by benzyladenine in groundnut. Indian Journal of Agricultural Biochemistry. 20: 53-58.

Drame, K.N., Clavel, D., Repellin, A., Passaquet, C. and Zuily Fodil, Y. (2007). Water deficit induces variation in expression of stress-responsive genes in two peanut (*Arachis hypogaea* L.) cultivars with different tolerance to drought. *Plant Physiology and Bioch*, 45: 236-243.

Dutta, D. and Mondal, S.S. (2006).Response of summer groundnut (*Arachis hypogaea*) to moisture stress, organic manure and fertilizer with and without gypsum under lateritic soil of West Bengal. *Indian Journal of Agronomy.* 51: 145-148.

Dwivedi, R.S., Joshi, Y.C., Nautiyal,P.C., Singh, A.L., Ravindra, V., Thakkar, A.N., Koradia, V.G., and Dhapwal, G.S. (1986). Developing parameters to predict drought resistance in groundnut genotypes. Transactions of Indian Society of Desert Technology and University Centre of Desert Studies, 11: 59-65.

Dwivedi, R.S., Joshi, Y.C., Saha, S.N., Thakkar, A.N., and Koradia, V.G.(1986). Canopy orientations and energy harvests in groundnut (*Arachis hypogaea* L.) under water stress. *Oleagineux*, 41: 451-455.

Dwivedi, S.L., Nigam, S.N., Rao, R.C.N., Singh, U.and Rao, K.V.S. (1996). Effect of drought on oil, fatty acids and protein contents of groundnut (*Arachis hypogaea* L.) seeds. Field Crops Research., 48:125-133.

Erickson, P.I., Ketring, D.L. and Stone, J.F. (1991). Response of internal tissue water balance of peanut to soil water. *Agronomy Journal*, 83: 248-253.

Ferguson ME, Burow MD, Schulze SR, Bramel PJ, Paterson AH, Kresovich S and Mitchell S (2004) Microsatellite Identification and Characterization in Peanut (*A. hypogaea* L.), *Theoretical And Applied Genetics*. 108(6), 1064-1070.

Ferreira, L.G.R., Santos, I.F. dos, Tavora, F.J.F., Silva, J.V.da., Dos Santos, I.F., Da Silva, J.V. (1992). Effect of water deficit on groundnut (*Arachis hypogaea* L.) cultivars. Physiological reactions and yields. *Oleagineux-Paris*. 47: 523-530.

Funderburk, J.E., Gorbet, D.W., Teare, I.D. and Stavisky, J. (1998). Thrips injury can reduce peanut yield and quality under conditions of multiple stress. *Agronomy Journal.*, 90: 563-566.

Golakiya, B. A. (1989). Drought response of groundnuts. -Physiology of the groundnut seed germination. *Advances in Plant Sciences* 2: 139-142.

Golakiya, B.A. (1989). Changes in biophysical parameters of maize, groundnut and pearl millet due to shading, water stress and aging. *Indian Journal of Experimental Biology*. 27: 749-750.

Golakiya, B.A. (1993).Drought response of groundnut: VII. Identification of the critical growth stages most susceptible to water stress. *Advances in Plant Sciences.*, 6: 20-27.

Golakiya, B.A. and Patel, M.S. (1992). Growth dynamics and reproductive efficiency of groundnut under water stress at different phenophases. *Indian Journal of Agricultural Research*. 26:179-186.

Goncalves de Abreu FMS (1988). Influence of atmospheric saturation deficit on early growth of groundnut. Dissertation-Abstracts-International,-B-*Sciences and Engineering*. 49: 18B.

Gopalakrishna, R., Ganesh Kumar., Krishnaprasad, B.T., Mathew, M.K., Kumar, M.U. and Kumar, G. (2001). A stress-responsive gene from groundnut, Gdi-15, is homologous to flavonol 3-O-glucosyltransferase involved in anthocyanin biosynthesis. *Biochemical and Biophysical Research Communications.*, 284: 574-579.

Gosal, S.S., Wani, S.H. and Kang, M.S. (2009). Biotechnology and drought tolerance. *Journal of Crop Improvement*. 23: 19-54.

Gowda, A. (1977).Moisture stress and hormonal influence on the flowering and yield in groundnut (*Arachis hypogaea* Linn.).

Gowda, A.and Hegde, B.R. (1986). Moisture stress and hormonal influence on the flowering behaviour and yield of groundnut (*Arachis hypogaea* L.). *Madras Agricultural Journal.*, 73: 82-86.

Greenberg, D.C., Williams, J.H. and Ndunguru, B.J. (1992). Differences in yield determining processes of groundnut (*Arachis hypogaea* L.) genotypes in varied drought environments. *Annals of Applied Biology*. 120: 557-566.

Greenberg, D.C. and Ndunguru, B.J. (1989). Groundnut drought-simulation studies at ICRISAT Sahelian Centre. *International Arachis Newsletter*. 6: 10-12.

Gujrathi, B., Hegde, B.A., Patil, T.M. and Sybesma, C. (ed.) (1984). Relative efficiency of photosynthetic carboxylation and enzyme activities in sorghum genotypes and peanut under water stress. *Advances in photosynthesis research* 4: 399-402.

Guo, B.Z., Yu, J., Holbrook, C.C., Lee, R.D. and Lynch, R.E. (2003). Application of differential display RT-PCR and EST/microarray technologies to the analysis of gene expression in response to drought stress and elimination of aflatoxin contamination in corn and peanut. *Journal of Toxicology*, Toxin Reviews., 22: 287-312.

Harris D, Matthews RB, Nageswara Rao RC, Williams JH (1988). The physiological basis for yield differences between four genotypes of groundnut (*Arachis hypogaea*) in response to drought III. Developmental process. *Expl. Agric.* 24: 215-266.

Hashim, I.B., Koehler, P.E., Eitenmiller, R.R. and Kvien, C.K. (1993). Fatty acid composition and tocopherol content of drought stressed Florunner peanuts. *Peanut Science*. 20: 21-24.

He G, Meng R, Newman M, Gao G, Pittman RN and Prakash CS (2003) Microsatellites as DNA Markers In Cultivated Peanut (*Arachis hypogaea* L.). *BMC Plant Biology*, 3: 3.

Hopkins MS, Casa AM, Wang T, Michell SE, Dean RE, Kochert GD and Kresovich S (1999) Discovery and Characterization of Polymorphic Simple Sequence Repeats (SSSR) in Peanuts.*Crop Science*. 39: 1243-1247.

Huang, M.T. and Ketring, D.L. (1985). Studies of water relations of peanut under rainfed and irrigated conditions. Proceedings, American Peanut Research and Education Society, Inc., 17: 65.

Idinoba ME, Idinoba PA, Gbadegesin A, Jagtap SS (2008) Growth and evapotranspiration of groundnut (*Arachis hypogeal* L.) in a transitional humid zone of Nigeria. *African J. Agric. Res.* 3: 384-388.

Jain AK, Basha SM and Holbrook CC (2001). Identification of Drought-Responsive Transcripts in Peanut (*Arachis hypogaea* L.). *Electronic Journal of Biotechnology*, 4(2): 59-67.

Janamatti, V.S., Sashidhar, V.R., Prasad, T.G., and Sastry, K.S.K. (1986). Effect of cycles of moisture stress on flowering pattern, flower production, gynophore length and their relationship to pod yield in bunch types of groundnut. *Narendra Deva Journal of Agricultural Research*., 1: 136-142.

Jiang HuiFang and Ren-XiaoPing (2004). SOD activity and protein content in groundnut leaves as affected by drought stress. *Acta Agronomica Sinica*; 30: 169-174.

Joshi, Y.C., Nautiyal, P.C., Ravindra, V., Dwivedi, R.S. (1988). Water relations in two cultivars of groundnut (Arachis hypogaea L.) under soil water deficit. *Tropical Agriculture*, 65: 182-184.

Kalariya K.A., A.L.Singh, K. Chakraborty, P.V. Zala and C.B. Patel (2013). Photosynthetic characteristics of groundnut (*Arachis hypogaea* L.) under water decit stress. Indian J Plant Physiol. 18 (2):157–163.

Kalariya K.A., A.L.Singh, K. Chakroborty P.V. Zala and C.B. Patel (2012a). Chlorophyll fluorescence in Groundnut (Arachis hypogeal L.) under water-deficit-condition. Paper presented at a zonal seminar on Physiological and molecular interventions on sustainable crop productivity under changing climatic conditions' jointly organised by ISPP, New Delhi and DMAPR, Anand held on 17th January, Abstracts pp. 24.

Kalariya K.A., A.L. Singh and K. Chakraborty (2012b). Influence of Water Deficit Stress on Chlorophyll fluorescence in Groundnut In: Proceedings of U.G.C sponsored national seminar on "New frontiers of plant science for sustainable development" at Khudra, Odhissa 24-26 January, 2012, pp. 89-100.

Kameswara Rao Kottapalli, Rakwal R, Shibato J, Gloria Burow, David Tissue, John Burke, Puppala N, Burow M And Payton P (2009) Physiology and proteomics of the water-decit stress response in three contrasting peanut genotypes. *Plant, Cell and Environment*. 32: 380–407.

Kandoliya, U.K., Parameswaran, M., and Vakharia, D.N. (2000). Effect of drought on kernel carbohydrates and lipids in groundnut (*Arachis hypogaea* L.). *Advances in Plant Sciences*, 13: 573-579.

Kar, D.N., Ray, M. and Garnayak, L.M. (2002). Effect of moisture stress on yield and water use by rabi groundnut. *Annals of Agricultural Research*, 23:159-161.

Karunakar, A.P., Jiotode, D.J. and Nalamwar, R.V.(2002). Basis of variation in pod yield of kharif groundnut under delayed sowings. *Research on Crops*, 3: 546-550

Katam Ramesh, Basha SM, Vasanthaiah HKN and Naik KSS (2007) Identification of drought tolerant groundnut genotypes employing proteomics approach. *Journal of SAT Agricultural Research*. 5: 1-2.

Ketring DL (1991). Physiology of oilseeds, IX Efftects of water deficit on peanut seed quality. *Crop Science* 31: 459-463.

Ketring DL, Erickson PI, Stone JF (1990) Apparent sap velocity in peanut genotypes under control and stress conditions. *Peanut Sci*. 17: 38-44.

Kisyombe, C.T., Beute, M.K. and Payne, G.A. (1985). Field evaluation of peanut genotypes for resistance to infection by *Aspergillus parasiticus*. *Peanut Science* 12: 12-17.

Koti, R.V., Chetti, M.B., Manjunath, T.V.and Amaregowda, A. (1994). Effect of water stress at different growth stages on biophysical characters and yield in groundnut (*Arachis hypogaea* L.) genotypes. *Karnataka Journal of Agricultural Sciences*, 7: 158-162.

Kulkarni, J.H., Joshi, Y.C., and Bhatt, D.M. (1988). Effect of water stress on nodulation and nitrogen content of groundnut (*Arachis hypogaea* L.). *Zentralblatt-fur-Mikrobiologie*, 143: 4, 299-302.

Kumaga, F.K., Adiku, S.G.K. and Ofori, K. (2003). Effect of post-flowering water stress on dry matter and yield of three tropical grain legumes. *International Journal of Agriculture and Biology.*, 5: 405-407.

Kumar, B.V.M. and Reddy, M.M. (2003). Rescheduling irrigation for increased water use efficiency in rabi groundnut (*Arachis hypogaea* L.). *Indian Journal of Dryland Agricultural Research and Development.* 18: 113-117.

Lauriano JA, Ramalho JC, Lidon FC and M do Céu Matos (2006) Mechanisms of energy dissipation in peanut under water stress. *Photosynthetica.* 44 (3): 404-410.

Leong, S.K. and Ong, C.K. (1983). The influence of temperature and soil water deficit on the development and morphology of groundnut (*Arachis hypogaea* L.). *Journal of Experimental Botany.* 34: 1551-1561.

Li, L and Pan, R.Z. (1990). Effects of PP 333 and CCC on the activity of peroxidase and its isoenzymes in the seedling leaves of *Arachis hypogaea* L. under water stress. *Oil Crops of China.*1: 41-43.

Li.L and Pan, R.Z. (1988). Effect of PP333 on endogenous ABA and GA levels in seedling leaves of *Arachis hypogaea* L. *Oil Crops of China.* 2: 36-39.

Liu Xu and Li-Ling (2009). Cloning and Characterization of the NAC-like gene AhNAC2 and AhNAC3 in peanut. *Acta Agronomica Sinica.* 35: 541-545.

Madhusudhan, K.V., Giridarakumar, S., Ranganayakulu, G.S., Reddy, P.C., Sudhakar, C. (2002). Effect of water stress on some physiological responses in two groundnut (*Arachis hypogaea* L.) cultivars with contrasting drought tolerance. *Journal of Plant Biology*, 29: 199-202.

March, G.J., Marinelli, A., Rago, A. and Collino, D. (1999). Effect of water stress caused by drought on the predisposition of groundnut (*Arachis hypogaea* L.) to *Sclerotium rolfsii* infection. *Boletin-de-Sanidad-Vegetal,-Plagas.*, 25: 523-528.

Mathews, S.T., Latha, V.M., Satakopan, V. N. and Srinikasan, R. (1988). Effect of simulated water stress on IAA oxidase in groundnut (*Arachis hypogea*) seedlings. *Current Science.* 57: 485-486.

Matsunami, T. and Kokubun, M. (2003). Photosynthetic responses of four legume crops to fluctuations of evaporative demand following the rainy season. *Tohoku Journal of Agricultural Research;* 54: 1-11.

Matthews RB, Harris D, Nageswara Rao RC, Williams JH, Wadia KDR (1988). The physiological basis for yield differences between four genotypes of groundnut (*Arachis hypogaea*) in response to drought I. Dry matter production and water use. *Exp. Agric.* 24: 191-202.

Mehan, V.K., Rao, R.C.N., McDonald, D., Williams, J.H. (1988). Management of drought stress to improve field screening of peanuts for resistance to *Aspergillus flavus*. Phytopathology. 1988, 78: 6, 659-663.

Meisner, C.A. and Karnok, K.J. (1992). Peanut root response to drought stress. *Agronomy Journal.* 84: 159-165.

Mensah, J.K. and Okpere, V.E. (2000). Screening of four groundnut cultivars from Nigeria for drought resistance. *Legume Research.*, 23: 37-41.

Meyer, J., Germani, G., Dreyfus, B., Saint Macary, H., Boureau, M., Ganry, F. and Dommergues, Y. (1982). Determination of the effect of 2 limiting factors (drought and nematodes) on the nitrogen fixation (C_2H_2) by groundnut and soyabean. *Oleagineux.*, 37:127-134.

Misra, J.B. and Nautiyal, P.C. (2005). Influence of imposition of soil moisture-deficit stress on some quality components of groundnut, *Arachis hypogaea* L. kernel. *Journal of Oilseeds Research.* 22: 119-124.

Misra JB, Nautiyal PC, Chauhan S, Zala PV (1992). Reserve mobilization and starch formation in cotyledons of germinating groundnut seeds. In Nigam SN (ed.), Groundnut-a global perspective, Proc. of an International Workshop, 25-29 Nov. ICRISAT, Patancheru, India, (In,En. Abstracts.) p. 451.

Mittal, G.K., Arunabh Joshi, Rajamani, G., Mathur, P.N.and Sharma (2006). A Water-deficit induced generation of reactive oxygen species and antioxidants in two Spanish groundnut cultivars. *National Journal of Plant Improvement.* 8: 7-10.

Moretzsohn MC, Leoi L, Proite K, Guimara PM, Leal Bertioli SCM, Gimenes MA, Martins WS, Valls JFM, Grattapaglia D and Bertioli DJ (2005) A Micro Satellite–Based, Gene–Rich Linkage Map For The AA Genome of *Arachis* (Fabaceae). *Theoretical and Applied Genetics.* 111, 1060-1071.

Musingo, M.N., Basha, S.M., Sanders, T.H., Cole, R.J., and Blankenship, P.D. (1989) Effect of drought and temperature stress on peanut (*Arachis hypogaea* L.) seed composition. Journal of Plant Physiology 134: 710-715.

Nageshwar Rao Rc, Sardar Singh, Sivakumar MVK, Srivastava KL and Williams JH (1985). Effect of water deficit at different growth phases of peanut. I. Yield responses. *Agronomy Journal* 77: 782-786.

Nageswar Rao RC, Williams JH, Sivakumar MVK, Wadiya KDR (1988). Effect of water deficit at different growth phases of peanut. II. Response to drought during pre- flowering phase. *Agron. J.* 80: 431-438.

Nageswara Rao RC, Simmonds LP, Azan Ali SM, Williams JH (1989a) Population, growth and water use of groundnut maintained on stored water. I. Root and shoot growth. *Exp. Agric.* 25: 51-61.

Nageswara Rao RC, Williams JH, Singh M (1989b) Genotypic sensitivity to drought and yield potential of peanut. *Agron. J.* 81: 887-893.

Nageshwar Rao Rc, Williams JH,Wadia KDR, Hubick KT and Farquhar GD (1993) Crop growth, water-use efficiency and carbon isotope discrimination in groundnut (*Arachis hypogaea* L.) genotypes under end-of-season drought conditions. *Annals of Applied Biology.* 122: 357-367.

Nautiyal, P.C., Ravindra, S. and Joshi, Y.C. (1991). Moisture stress and subsequent seed viability. Physiological and biochemical basis for viability differences in Spanish groundnut in response to soil moisture stress. *Oleagineux*. 46: 153-158.

Nautiyal PC, Bandyopadhyay A, Zala PV (2001). *In situ* sprouting and regulation of fresh- seed dormancy in Spanish type groundnut (*Arachis hypogaea* L.). *Field Crops Res.*70, 233-241.

Nautiyal, P.C., Ravindra, V., Zala, P.V. and Joshi, Y.C. (1999). Enhancement of yield in groundnut following the imposition of transient soil-moisture-deficit stress during the vegetative phase. *Experimental Agriculture.*, 35: 371-385.

Ndunguru, B.J., Ntare, B.R., Williams, J.H. and Greenberg, D.C. (1995). Assessment of groundnut cultivars for end-of-season drought tolerance in a Sahelian environment. *Journal of Agricultural Science.*125: 79-85.

Nie ChengRong, Li HuaShou, Li Mei, Du JianLian, Nie, C.R., Li HS., Li-M, Du, J.L. (2002). Influence of nitrate rare earth elements on the growth of peanut seedlings under moisture stress. Soils and Fertilizers Beijing., No. 3: 14.

Nigam, S.N. and Rupakula Aruna (2008). Stability of soil plant analytical development (SPAD) chlorophyll meter reading (SCMR) and specific leaf area (SLA) and their association across varying soil moisture stress conditions in groundnut (*Arachis hypogaea* L.). *Euphytica*. 160: 111-117.

Nigam, S.N., Chandra, S., Sridevi, K.R., Manohar Bhukta, Reddy, A.G.S., Rachaputi, N.R., Wright, G.C., Reddy, P.V., Deshmukh, M.P., Mathur, R.K., Basu, M.S., Vasundhara, S., Varman, P.V. and Nagda, A.K. (2005). Efficiency of physiological trait-based and empirical selection approaches for drought tolerance in groundnut. *Annals of Applied Biology*. 146: 433-439.

Nogueira, R.J.M.C., Santos, R.C.dos, Bezerra Neto, E., Santos, V.F. dos., dos Santos R.C. dos and Santos, V.F. (1998). Physiological response of two groundnut cultivars subjected to different water regimes. *Pesquisa Agropecuaria Brasileira.*, 33: 1963-1969.

Nwalozie, M.C. and Annerose, D.J.M. (1996). Stomatal behaviour and water status of cowpea and peanut at low soil moisture levels. *Acta Agronomica Hungarica.*, 44:, 229-236.

Ong C.K., Black, C.R., Simmonds, L.P. and Saffell, R.A. (1985). Influence of saturation deficit on leaf production and expansion in stands of groundnut (*Arachis hypogaea* L.) grown without irrigation. *Annals of Botany.*, 56: 523-536.

Pallas JE Jr, Stansell JR and Bruce RR (1977). Peanut seed germination as related to soil water regime during pod development. *Agronomy Journal*. 69: 381-383.

Pallas JE Jr and Michel BE (1978). Comparison of leaf and stem hygrometers for measuring changes in peanut plant water potential. *Peanut Science,* 5: 65-67.

Pallas, J.E. Jr, Stansell, J.R. and Koske, T.J. (1979). Effects of drought on Florunner peanuts. *Agronomy Journal,* 71: 853-858.

Pan RuiChi, Dou ZhiJie and Ye QingSheng (1995). Effect of methyl jasmonate on superoxide dismutase activity and membrane-lipid peroxidation in groundnut seedlings during water stress. *Acta Phytophysiologica Sinica*. 21: 221-228.

Patel, C.L., Padalia, M.R. and Babaria, N.B. (1983). Growth and plant-water relation in groundnut grown under different soil-moisture stresses. *Indian Journal of Agricultural Sciences*. 53: 340-345.

Patel, M.S. and Golakiya, B.A. (1991). Effect of water stress on yield and yield attributes of groundnut (*Arachis hypogaea* L.). *Madras Agricultural Journal*. 78: 178-181.

Patil, N.A. and Patil, T.M. (1993). Physiological responses of groundnut genotypes to water stress. *Legume Research*,16: 23-30.

Patil BP and Gangavane SB (1990) Effects of water stress imposed at various growth stages on yield of groundnut and sunflower. *Journal of Maharashtra Agricultural Universities*. 15, 322-324.

Pawar KN (2011) Influence of water stress treatment on seed quality of groundnut genotypes. *Research on Crops*. 12(2), 402-404.

Peters S, Mundree SG, Thomson JA, Farrant JM and Keller F (2007) Protection mechanisms in the resurrection plant *Xerophyta viscosa* (Baker): both sucrose and raffinose family oligosaccharides (RFOs) accumulate in leaves in response to water deficit. *J Exp Bot*. 58:1947–1956.

Pinto, C.de.M., Tavora, F.J.F.A., Bezerra, M.A. and Correa, M.C.de.M. (2008). Growth and root system distribution in peanut, sesame and castorbean under water deficit cycles. *Revista Ciencia Agronomica*.39: 429-436.

Polara KB, Patel CL and Pathak RS (1984). Pattern of N, P and K accumulation in groundnut varieties as influenced by soil- moisture-stress at different stages of growth Leaflet angle and radiation avoidance by water stressed groundnut (*Arachis hypogaea* L.) plants. *Legume Research* 7: 95-100.

Prasad, T.G., Sindagi, S.S. and Parvatikar, S.R. (1974). Effect of pretreatment of seeds on germination and root growth in seedlings of groundnut varieties. *Current Research*. 3: 46-47.

Prathap, S., Reddy, S.S. and Swamy, G.N. (2006). Effect of water stress on germination and seedling growth of groundnut genotypes. *Crop Research Hisar*. 31: 58-60.

Proite K, LealBertioli SCM, Bertioli DJ, Moretzsohn MC, Silva FR, Martins NF, Guimaraes PM (2007) ESTs from a Wild Arachis Species for Gene Discovery and Marker Development. *BMC Plant Biology*, 7: 7.

Purushotham, M.G., Vajranabhaiah, S.N., Patil, V.S., Reddy, P.C., Prasad, T.G. and Prakash, A.H. (1998). Development of drought tolerant cell lines in groundnut (*Arachis hypogaea* L.) genotypes *in vitro*. *Indian Journal of Plant Physiology*., 3: 283-286.

Purushotham, M.G., Veerangouda Patil, Raddey, P.C., Prasad, T.G., Vajranabhaiah, S.N. and Patil, V. (1998). Development of *in vitro* PEG stress tolerant cell lines in two groundnut (*Arachis hypogaea* L.) genotypes. *Indian Journal of Plant Physiology*. 3:49-51.

Quilambo, O.A. (2004). Proline content, water retention capability and cell membrane integrity as parameters for drought tolerance in two peanut cultivars. South African Journal of Botany, 70: 227-234.

Rajendrudu, G and Williams, J.H. (1987). Effect of gypsum and drought on pod initiation and crop yield in early maturing groundnut (*Arachis hypogaea*) genotypes. *Experimental Agriculture*, 23: 259-271.

Raju, A.P., Reddi, G.H.S. and Reddy, T.B. (1981). Studies on the productivity of rainfed groundnut under moisture stress using neutron moisture probe. *Journal of Nuclear Agriculture and Biology*, 10: 131-133.

Ramana Rao DV (1994) M.Sc. (Ag.) Thesis, APAU, Hyderabad.

Ramesh Babu, V., Murty, P.S.S., Sankara Reddi, G.H. and Reddy, T.Y. (1983). Leaflet angle and radiation avoidance by water stressed groundnut (*Arachis hypogaea* L.) plants. *Environmental and Experimental Botany*. 23:183-188.

Rao PV, Subba Rao IV, Reddy PR (1986) Influence of water stress at different stages on growth and productivity of groundnut (*Arachis hypogea* L.). *The Andhra Agric. J.* 33, 48-52.

Ravi, K., Gangappa, E., Kumar, G. N. V. and Satish, R.G. (2008). Evaluation of selected groundnut genotypes under water stress and non-stress environments. *Mysore Journal of Agricultural Sciences*. 42: 456-459.

Ravindra V, Vasantha S, Joshi YC, Nautiyal PC and Singh AL (1990) To develop ideotype concept and to identify genotypes with high physiological efficiency for high productivity in spanish bunch and virginia runner groundnut. *Under Micromission-I, on Crop Production Technology of Technology Mission on Oilseeds Production (ICAR)*.

Ravindra, V., Nautiyal, P.C. and Joshi, Y.C. (1990). Physiological analysis of drought resistance and yield in groundnut (*Arachis hypogaea* L.). *Tropical Agriculture*. 67: 290-296.

Reddy, K.R. and Gupta, R.V.S. (1994). Stability evaluation in groundnut *Journal of Maharashtra Agricultural Universities*. 19: 80-83.

Reddy, T.Y., Reddy, V.R. and Anbumozhi, V. (2003). Physiological responses of groundnut (*Arachis hypogaea* L.) to drought stress and its amelioration: a review. *Acta Agronomica Hungarica.*, 51: 205-227.

Reddy, T.Y., Reddy, V.R. and Anbumozhi, V. (2003). Physiological responses of groundnut (*Arachis hypogea* L.) to drought stress and its amelioration: a critical review. *Plant Growth Regulation*. 41: 75-88.

Ribaut JM, Hoisington DA, Deutsch JA, Jian C and Gonzalez De Leon (1996) Identification of QTL under Drought Conditions In Tropical Maize. 1. Flowering Parameters and the ASI.*Theoretical and Applied Genetics*. 92(7), 905-914.

Rosario, D.A.del. and Fajardo, F.F.(1988). Morphophysiological responses of ten peanut (*Arachis hypogaea* L.) varieties to drought stress. *Philippine Agriculturist*. 71: 447-459.

Rui Chi Pan and Gu HuangQing (1995). Effect of methyl jasmonate on the growth and drought resistance of groundnut seedlings. *Acta Phytophysiologica Sinica.* 21, 215-220.

Saini HS and Srivastava AK (1981). Osmotic stress and the nitrogen metabolism of two groundnut (*Arachis hypogaea* L.) cultivars. *Irrigation Science.* 2, 185-192.

Sairam RK, Srivastava GC and Saxena DC (2000). Increased antioxidant activity under elevated temperature: a mechanism of heat stress tolerance in wheat genotypes. *Biol. Plant.* 43:245-251.

Sakarvadia HL and Yadav BS (1994). Effect of water stress at different growth phases of groundnut on yield and nutrient absorption. *J Indian Society of Soil Science,* 42: 147-149.

Sakarvadia HL,Parmar KB, Jetpara PI and Kalola AD (2010). Effect of cyclic water stress and potassium on dry matter and nutrients accumulation in summer groundnut. *Advances in Plant Sciences,* 23: 193-196.

Samdur MY, Manivel P, Jain VK, Chikani BM, Gor HK, Desai S and Misra JB (2003). Genotypic differences and water-deficit induced enhancement in epicuticular wax load in peanut. *Crop Science,* 43:1294-1299.

Sanders, T.H., Blankenship, P.D., Cole, R.J., and Smith, J.S. (1986). Role of agrometeorological factors in postharvest quality of groundnut. Agrometeorology of groundnut. Proceedings of an international symposium, ICRISAT Sahelian Center, 21-26 August 1985: 185-192.

Santos, R.C. dos., Moreira, J.de.-A.N., Cabral, E.L., and dos Santos, R.C. (1997). Study on the role of peroxidase in the phenology of groundnuts subjected to water stress. *Revista de Oleaginosas e-Fibrosas.* 1: 117-124.

Santos, R.C.dos, Freitas, N.S., Falcao, T.M.M.de A., Moreira J. de, A.N., Cabral, E.L., dos, Santos, R.C., de A Falcao, T.M.M. de and AN-Moreira, J. (1998). Isoenzymatic behaviour of peanut genotypes submitted to and water stress. *Revista de Oleaginosas e-Fibrosas.,* 2: 1-11.

Sarkar, S. and Kar, S. (1994). Root sink evaluation of groundnut grown under [water] stressed and non-stressed conditions. *Journal of the Indian Society of Soil Science.* 42: 126-128.

Sarma, P.S. and Sivakumar, M.V.K. (1987). Productivity of groundnut as influenced by use of seed from a crop with moisture stress history. *Field Crops Research.* 15: 207-213.

Sarma, P. S. and Sivakumar, M.V.K. (1990). Evaluation of groundnut response to early moisture stress during the rainy and the post-rainy seasons. *Agricultural and Forest Meteorology.* 49: 123-133.

Sashidhar, V.R., Mekhri, A.A. and Sastry, K.S.K. (1981). Potassium content and proline accumulation following seed treatment with calcium chloride in groundnut varieties. *Indian Journal of Plant Physiology.* 24: 89-92.

Scandaliaris, J., Hemsy, V., Rodriguez, M.E., Munoz, H.L.L. and Cajal, J.A. (1978). Flowering cycle in groundnut and factors which influence it. *Revista Agronomica del Noroeste Argentino. Produccion Vegetal.* 15: 3-54.

Shahid Umar, Bansal, S.K. and Umar, S. (1997). Effect of potassium application on yield and quality of water stressed groundnut. *Fertiliser News.*, 42: 27-29.

Shao HB, Chu LY, Lu ZH, Kang CM (2008). Main antioxidants and redox signaling in higher plant cells. *Int. J. Biol. Sci.* 44, 12–18.

Shao HB, Chu LY, Wu G, Zhang JH, Lu ZH, Hu YC (2007) Changes of some anti-oxidative physiological indices under soil water deficits among 10 wheat (*Triticum aestivum* L.) genotypes at tillering stage, Colloids Surf. B. 54,143–149.

Sharada P and Naik GR (2011). Physiological and biochemical responses of groundnut genotypes to drought stress. *World Journal of Science and Technology.* 1(11), 60-66.

Sharma K, Gurbaksh Singh, Sharma R, Sharma HL (1989). Biochemical changes in groundnut seed during germination under water stress conditions. *Annals of Biology* 5(2), 123-127.

Sharma, K., Gurbaksh Singh and Sharma, H.L. (1987).Effect of simulated water stress by polyethylene glycol (PEG) on germination and some enzymes of groundnut (*Arachis hypogaea* L.). *Annals of Biology*, 3: 77-82

Sharma, K., Gurbaksh Singh and Singh, G. (1987). Effect of water stress on yield and quality of groundnut. *Environment and Ecology.* 5: 647-650

Sharma, P., Nisha., Malik, C.P. (1993). Photosynthetic responses of groundnut to moisture stress. *Photosynthetica*, 29:157-160.

Shinde, G.G. and Pawar, K.R. (1984). Effects of water stress at critical growth stages on growth and yield of groundnut in summer season. *J Maharashtra Agricultural Universities*, 9: 26-28.

Singh AL (2003). Phenology of Groundnut. In: *Advances in Plant Physiology* (Ed. A. Hemantranjan) *Scientific Publishers (India), Jodhpur, India.*6: 295-382.

Singh, B.B., Mai Kodomi, Y. and Terao, T. (1999). Relative drought tolerance of major rainfed crops of the semi-arid tropics. *Indian Journal of Genetics and Plant Breeding*, 59: 437-444.

Sridhara, C.J., Thimmegowda, S., and Krishnamurthy, N. (1995). Effect of moisture stress and irrigation levels at different growth stage on leaf area, leaf area index and dry matter accumulation of groundnut. *Indian Agriculturist.*, 39: 187-192.

Sridhara, C.J., Thimmegowda, S. and Krishnamurthy, N. (1998). Effect of moisture stress and irrigation levels on growth and pod yield of groundnut. *International Journal of Tropical Agriculture*, 1998, 14: 219-222.

Stirling, C.M. and Black, C.R. (1991). Stages of reproductive development in groundnut (*Arachis hypogaea* L.) most susceptible to environmental stress. *Tropical Agriculture.* 68: 296-300.

Stirling, C.M., Ong, C.K., and Black, C.R. (1989). The response of groundnut (*Arachis hypogaea* L.) to timing of irrigation. I. Development and growth. *Journal of Experimental Botany*, 40: 1145-1153.

Stirling, C.M., Black, C.R. and Ong, C.K. (1989). The response of groundnut (*Arachis hypogaea* L.) to timing of irrigation. II. 14C-partitioning and plant water status. Journal-of-Experimental-Botany. 40: 1363-1373

Subramanian, V.B. and Maheswari, M. (1990). Physiological responses of groundnut to water stress. *Indian Journal of Plant Physiology*. 33: 130-135.

Subramanian, V.B., Reddy, G.J. and Maheswari, M. (1993). Photosynthesis and plant water status of irrigated and dryland cultivars of groundnut. *Indian J Plant Physiol*, 36: 236-238.

Sudhakar, P., Latha, P., Babitha, M., Reddy, P.V. and Naidu, P.H. (2007). Relationship of drought tolerance traits with aflatoxin contamination in groundnut. *Indian Journal of Plant Physiology*, 12: 261-265.

Sulochana, C. and Savithramma, N. (2001). Effect of calcium in amelioration of PEG (6000) induced water stress in groundnut (*Arachis hypogaea* L.) cultivars during seedling growth. *Journal of Plant Biology.*, 28: 257-263.

Sulochana, C and Savithramma, N. (2002). Interactive effects of PEG-induced water stress and CaCl$_2$ on the changes in CaM, Ca^{2+}, protein and protease activity in seedlings of groundnut, (*Arachis hypogaea* L.) cultivars. *Journal of Plant Biology.*, 29:271-277.

Sulochana, C. and Savithramma, N. (2002). Influence of calcium in amelioration of water stress through calmodulin, Ca^{2+} and peroxidase activity during seedling growth of groundnut (*Arachis hypogaea* L.) cultivars. *Plant Archives.*, 2: 309-315.

Sulochana, C. and Savithramma, N. (2003). Effect of calcium on acid phosphatase activity in groundnut (Arachis hypogaea L.) seedlings during water stress. Indian Journal of Plant Physiology., 8:186-188.

Sulochana, C., Rao, T.J.V.S. and Savithramma, N. (2002). Effect of calcium on water stress amelioration through calmodulin and scavenging enzymes in groundnut. Indian Journal of Plant Physiology. 7:152-158.

Suma, T.C., Srivathsa, C.V., Lokesha,A.N., Prasad, T.G. and Sashidhar, V.R. (2006). Root abscisic acid synthesizing capacity does not influence differences in root elongation under water stress in contrasting cultivars of groundnut (*Arachis hypogaea* L.) and finger millet (*Eleusine coracana* Gaertn). *Journal of Plant Biology*; 33: 231-235.

Suvarna, Kenchanagoudar, P.V., Nigam, S.N. and Chennabyregowda, M.V. (2002). Effect of drought on yield and yield attributes of groundnut. *Karnataka Journal of Agricultural Sciences.*, 15: 364-366.

Thatikunta R and Durgaprasad MMK (1996). Identifying groundnut genotypes for the Southern Telangana zone in India. *Inter Arachis Newsletter*. I 16: 19-20.

Umar, S., Afridi, M.M.R.K. and Dwivedi, R.S. (1991). Influence of added potassium on the drought resistance of groundnut. *Journal of Potassium Research*. 7: 53-61.

Usha, R., Ramamurthy, N., Bhavani, P.N. and Swamy, P.M. (1999). Alleviation of PEG-6000 induced water stress effects by calcium. Indian Journal of Plant Physiology., 4: 95-99

Uthai Arromratana, Tawatchai Na Nakara, Seri Sukakit, Boonlert Boonyoung (1993). Effect of water stress at different growth stages on yield of groundnut. Department of Agricultural Extension, Bangkok (Thailand). Proceedings of the eleventh Thailand national groundnut meeting. Raingan kan sammana thua-lisong haeng chat khrang thi 11 na rongraem Chansomthara, Ranong 17-21 Phrusaphakhom 2536. Bangkok (Thailand). 568: 273-278.

Vadez, V., Rao, S., Sharma, K.K., Bhatnagar Mathur, P., Devi, M.J.(2007). DREB1A allows for more water uptake in groundnut by a large modification in the root/ shoot ratio under water deficit. *Journal-of-SAT-Agricultural-Research*. 5: 1-5.

Vaghasia, P.M., Jadav, K.V. and Nadiyadhara, M.V. (2010). Effect of soil moisture stress at various growth stages on growth and productivity of summer groundnut (*Arachis hypogaea* L.) genotypes. *International Journal of Agricultural-Sciences*. 6: 141-143.

Vaghasia, P.M., Jadav, K.V., Jivani, L.L. and Kachhadiya, V.H. (2010). Impact of water stress at different growth phases of summer groundnut (*Arachis hypogaea* L.) on growth and yield. *Research on Crops*. 11: 693-696.

Vaithiyalingan M, Manoharan V, Ramamoorthi N (2010). Association analysis among the yield and yield attributes of early seasondrought tolerant groundnut (*Arachis hypogaea* L.). *Electronic Journal of Plant Breeding*. 1(5): 1347-1350.

Vakharia, D.N., Kukadia, A.D. and Parameswaran, M. (2003). Polyamines in response to artificial water stress in groundnut seedlings. *Indian Journal of Plant Physiology*. 8: 383-387.

Valentim, J.F. (1989). Effect of environmental factors and management practices on nitrogen fixation of rhizoma peanut and transfer of nitrogen from the legume to an associated grass. Dissertation-Abstracts-International, -B-*Sciences and Engineering*. 49: 4078B-4079B.

Varman, P.V. and Raveendran (1991). TS New source of seed dormancy in bunch groundnut. *Current Research*. 20: 237-238.

Varshney RK, Graner A and Sorrells ME (2005). Genomics assisted Breeding for Crop Improvement. *Trends In Plant Science*. 10: 621–630.

Varshney RK, Hoisington D and Tyagi AK (2006). Advances in Cereals Genomics and Applications in Crop Breeding. *Trends Biotechnology*. 24: 490–499.

Venkateswarlu, B., Maheswari, M. and Reddy, G.S. (1991). Effect of water stress on N_2-fixation and N-partitioning in groundnut in relation to kernel yield. *Indian Journal of Experimental Biology*. 29: 727-275.

Venkateswarlu, B., Saharan, N. and Maheswari, M. (1990). Nodulation and $N_2(C_2H_2)$ fixation in cowpea and groundnut during water stress and recovery. *Field Crops Res.* 25: 223-232.

Venkateswarlu, B., Mareswari, M., Saharan, N. (1989). Effects of water deficit on N_2 (C_2H_2) fixation in cowpea and groundnut. *Plant and Soil.* 114: 69-74.

Venkateswarlu, B., Mukhopadhyay, K. and Ramesh, K. (1993). Effect of PEG induced water stress on cell cultures of groundnut in relation to the response of whole plants. *Oleagineux Paris.*, 48: 463-468.

Venkateswarlu, B. and Ramesh, K. (1993). Cell membrane stability and biochemical response of cultured cells of groundnut under polyethylene glycol-induced water stress *Plant Science Limerick.* 90: 179-185.

Venkateswarlu, B., Mukhopadhyay, K., Jayjayanti Choudhuri., Choudhuri, J.(1998). An in vitro regeneration protocol for screening groundnut plants for water stress tolerance. *Legume Research.*, 21:1-7.

Wang CaiBin, Sun YanHao, Chi Yu Cheng, Cheng Bo, Tao ShouXiang, Chen DianXu, Wang CB, Sun YH, Chi YC, Cheng B, Tao SX, Chen DX, (ed.) Nigam, S.N. *et al.* (eds.), (1995). Studies on light interception, photosynthesis, and respiration in high-yielding groundnut canopies. Achieving high groundnut yields: Proceedings of an International Workshop, Laixi, Shandong, China, 25-29 Aug 1996: 171-180.

Wang Liu Deng, LiLin, Zou Dong Sheng and Liu Fei (2009). Effect of waterlogging on growth and agronomic trait of different peanut varieties. *Zhongguo Shengtai Nongye Xuebao Chinese Journal of Eco Agriculture*, 17: 968-973.

War AR, Shanmugavel Lingathurai, Paulraj MG, War MY, Savarimuthu Ignacimuthu (2011). Oxidative response of groundnut (*Arachis hypogaea*) plants to salicylic acid, neem oil formulation and *Acalypha fruticosa* leaf extract. *American J Plant Physiology*, 6(4): 209-219.

Wotton, H.R., and Strange, R.N. (1987). Increased susceptibility and reduced phytoalexin accumulation in drought-stressed peanut kernels challenged with *Aspergillus flavus. Applied and Environmental Microbiology*, 53: 270-273.

Wright G.C., Bell MJ (1992). Plant population studies on peanut (*Arachis hypogaea* L.) in subtropical Australia. 3. Growth and water use during a terminal drought stress. *Aust. J. Exp. Agric.* 32: 197-203.

Wright GC, Hubick KT, Farquhar GD (1991). Physiological analysis of peanut cultivar response to timing and duration of drought stress. *Aust. J. Agric. Res.* 42: 453-470.

Wright, G.C., Rahmianna, A., and Hatfield, P.M. (1988). A comparison of thermocouple psychrometer and pressure chamber measurements of leaf water potential in peanuts. *Experimental Agriculture*, 24: 355-359.

Xue HuiQin, Gan XinMin, Gu ShuYuan, Sun LanZhen, Xue HQ, Gan XM, Gu SY, and Sun LZ (1997). Hyperosmotic solution used to study the relationship between seed germination characteristics of peanut and drought resistance. *Oil Crops of China.* 19: 30-33.

Yadav SK, Singh AL, Misra JB, Mathur RS (1993). Effect of protected moisture stress on biochemical constituents of groundnut leaves. International conference on biotechnology in agriculture and forestry. Feb. 15-18,1993. New Delhi, India, pp. 65.

Yadav Neelam, Yadav VK and Yadava CPS (2007). Influence of simulated whitegrub damage of roots on water relation parameters in groundnut. *Indian J of Plant Physiology.*, 12: 173-177.

Yao JP, Luo YN and Yang XD (1982). Preliminary report on the effects of drought on seed development and quality of early groundnut. *Chinese Oil Crops Zhongguo Youliao*, 3: 50-52.

Zaharah H (1986). The effect of water deficit at different growth phases on the performance of groundnut. Miscellaneous Crops Res. Div., MARDI, Serdang, Selangor, Malaysia. MARDI Research Bulletin. 14: 126-131.

Zhang Bin, Zhang TaoLing, Zhao QiGuo, Zhang B, Zhang TL and ZhaoQG (1999). Relationship between water potentials of red soil and crop leaves under five farming systems and their responses to drought stress in the dry season. *Acta Pedologica Sinica.* 36: 101-110.

Chapter 2
Climate Change and Agriculture

S.D. Singh and K.S. Muralikrishna

*Centre for Environment Science and Climate Resilient Agriculture,
Indian Agricultural Research Institute, New Delhi – 110 012
E-mail: sdsingh16b@yahoo.in*

CONTENTS

1. Introduction
2. Definition
3. History of climate change
4. Evidence of climate change
   4.1. Physical evidence
        1. Rise in atmospheric temperature
        2. Shifting and shrinking of cooling period
        3. Rise in sea level
        4. Changing pattern of monsoon
        5. Occurrence of natural disaster
   4.2. Biological evidence
        1. Early blossoming of trees
        2. Appearance of grasses in Antarctic
        3. Changing in cropping pattern
        4. Shifting of temperate fruits cultivation towards high latitude
5. Causes of climate change and variability
   5.1. Natural causes
        1. Volcanoes eruption

      2. Super cyclones and earthquake

      3. Low precipitation

      4. Emission of greenhouse gases from natural sources

   5.2. Anthropogenic causes

      1. Deforestation

      2. Over exploitation of fossil fuel

      3. Rapid industrialization

      4. Vehicular emission of gases

      5. Intensive agriculture

      6. Biomass burning

      7. Population explosion

6. Impacts of climate change on agriculture

   6.1. Direct effects of carbon dioxide on crop production

      1. Effect on C_3 and C_4 crop productivity

      2. Quality of crops

   6.2. Indirect effects of CO_2 on crop production

      1. Crop productivity through crop-weed competition

      2. Crop-pest interaction

   6.3. Effect of global warming

      1. Soil water availability

      2. Crop productivity

      3. Water cycle

      4. Crop quality

      5. Crop and pest dynamics

      6. Coastal salanization

      7. Loss of biodiversity

      8. Polar expansion of arable lands

      9. Changes in cropping pattern

      10. Food security

7. Crop adaptation to climate change

   7.1. Natural adaptation

      1. Shifting optimum thermal range

      2. Tolerance to climatic changes (CO_2/temperature)

      3. Escaping mechanism

      4. Compensatory mechanism

   7.2. Anthropogenic adaptation or coping strategies

      1. Agro-physiological manipulation

1. Introduction

As we enter the twenty first century, the world is faced with many difficult problems which possibly human beings had not to confront in the past. Beginning with the Brundtland Commission report, which appeared in 1987 as 'our common future', the question of 'global change' acquired a great significance. It was clearly realized that the pursuit of development based on intensive energy use in the form of fossil fuel and also the increasing rate of deforestation for meeting the food and shelter demands of the increasing population, is perhaps bringing about a change in the atmosphere. There is a continuous increase of gases such as carbon dioxide, methane, nitrous oxide, ozone, chlorofluorocarbons and others which cause greenhouse effect. In simple term, a continuous increase in these gases should lead to a rise in global mean temperature or global warming. This means that the climate of Earth would change, because temperature is an important factor driving the global climate. In fact palaeoclimatic evidence indicates that the climate of Earth, particularly the temperature, has changed in the past, but for the past 10,000 years the global mean temperature has remained at 15°C. It is during the period that the evolution of agriculture has occurred. Thus the agricultural system of today has become adjusted or adapted to the global climate. Any change in this balance can cause effects, which are difficult to define, but some can be anticipated. However, in the recent past it has been argued that present day agriculture also contributes greenhouse gases and hence can be a source of global warming. Therefore, with regards to agriculture there is two major issues; (1) agriculture as a source of greenhouse gases and hence of global warming (2) impact of global warming and hence climate change on agriculture.

2. Definition

Climate change is the variation in either mean state of the climate or in its variables persisting for an extending period, typically decades or longer. It encompasses temperature increase (global warming), sea level rise, changes in precipitation pattern, and increased frequencies of extreme weather events. However, climatic variability refers to sudden and discontinuous seasonal or monthly or periodic changes in climate or its components without showing any specific trend of temporal change. Thus climate change follows a specific pattern of change in climate or its variables over the time.

Global warming and climate change has emerged as an important global concern cutting across geographical and national boundaries. Global warming is defined as the increase in the temperature of globe due to transmission of incoming short-wave radiation from the sun and the absorptivity of outgoing long-wave radiation from the earth. The phenomenon is called Greenhouse effect or more precisely called "Natural greenhouse effect". The term "Greenhouse effect" is derived from the phenomenon of warming effect that take place inside greenhouses (glasshouses) used for off-season cultivation in temperate areas.

Gaseous emissions from human activities are substantially increasing the concentrations of atmospheric greenhouse gases, particularly carbon dioxide, methane, ozone, nitrous oxides and chlorofluorocarbons. Global circulation models predict that these increased concentrations of greenhouse gases will increase average thermal regime of global atmosphere. An increase in the concentration of greenhouse gases such as carbon dioxide in the atmosphere is thought to have been responsible for increasing the air temperature (Hansen *et al.*, 1984). This is called as "Enhanced greenhouse effect", which is the additional effect induced by human activity.

Atmospheric CO_2 concentration has risen by about 37 per cent during the last two centuries to the present level near 370 $\mu mol\, mol^{-1}$ (Keeling and Whorf, 2000; Keeling *et al.*, 2009). The CO_2 concentration is projected to double again during the next century (Alcamo *et al.*, 1996; Carter *et al.*, 2007; Meehl *et al.*, 2007), and to further contribute to a warmer climate. Under the business-as-usual scenario of the Intergovernmental Panel on Climate Change (IPCC), global mean temperatures will rise 0.3°C per decade during the next century with an uncertainty of 0.2 to 0.5 per cent (Houghton *et al.*, 1990). Thus global mean temperatures should be 1°C above the present values by 2025 and 3°C above the present value by 2100. Global mean surface air temperature increased by 0.5°C in the 20th century (IPCC, 1995) There is a general agreement that by the end of this century the atmospheric CO_2 concentration would be between 510 and 760 $\mu L/L$ (ppm) and the temperature would be on an average of 1.4°C to 5.8°C higher than present (IPCC, 2001; Solomon *et al.*, 2007). There may be more frequent occurrences of extreme climate events such as heat waves and drought (IPCC, 2001). Also increasing are atmospheric concentrations of other trace gases (CH_4, N_2O, NO_x, CO) that could intensify global warming. The increase in CO_2 concentration alone is expected to warm Earth by 2 to 4.5°C by the middle of next century, with associated changes in precipitation (Giorgi *et al.*, 1998).

Climate variability is likely to increase under global warming (Katz and Brown, 1992), both in absolute and in relative terms. In the past century, daily minimum nighttime temperature increased at a faster rate than daily maximum temperature in association with a steady increase in atmospheric greenhouse gas concentrations (Karl *et al.*, 1991 and Easterling *et al.*, 1997). Warming is predicted to be greatest at high northern latitudes during autumn and winter (Houghton *et al.*, 1990). Studies by Vinnikov *et al.* (1990) and Jones and Wigley (1990) unanimously point to a rise in average global temperature at a rate of approximately 0.5°C per 100 years. For agriculture purposes, the value averaged across time and space, at best, indicate only general trends. Actual values of temperature regimes experienced by the growing crops are more crucial; averages hardly indicate the real situations. Over the last few decades there has been no change in mean temperature but there has been a clear increasing trend of about 1.5°C in minimum temperature (Sinha *et al.*, 1998).

3. History of Climate Change

If we look back to the climatic change scenario in the extreme, recent and very recent past, there have been several up and down in the magnitude of atmospheric thermal regimes. Climatic change occurs on diverse scale of time and space. The largest changes have occurred on the same time as that of drifting continents. However, large variation such as interglacial period that have marked the climate record during the past 3 million years or so occurred in cycles that lasted tens to hundreds of thousands of years (Schneider *et al.*, 1990). Thus climate change is the normal state of affairs for the earth-atmosphere systems. This suggests that the notion of a stable and stationary climate is an erroneous concept, while that of unceasing climate change may be a more useful mental model to analyze the earth's climate scenarios. Climate of the past billion years have been about 13°C warmer to 5°C cooler than the current climate (Schneider *et al.*, 1990). Prominent in earth's recent history have been the 100,000 year Pleistocene glacial/interglacial cycles when climate was cooler than at present. Global temperature varied by about 5°C through ice age cycles. Some local temperature changes through these cycles were as great as 10-15°C in high latitude regions. During the last major glaciation, ice sheets covered much of North America and northern Europe and sea level averaged 120 cm below current value. Since the last glaciation, there have been relatively small changes of probably less than 2°C as compared to current global mean temperature in global average temperature. During the mid-Holocene epoch between 4000 and 6000 years BP, however, temperature rose significantly, particularly during summer in the northern hemisphere (Folland *et al.*, 1990).

Fluctuations in the distant past are important for analyzing the various causes of climate fluctuations, but substantial fluctuations within recent human history are more compelling for humans. For instance, the so-called medieval optimum occurred from about the 10th to early 12th centuries. There is evidence that Western Europe, Iceland, Greenland were exceptionally warm, with mean summer temperature that were more than 1°C higher than current one. The most notable fluctuation in historical time was that known as little ice age which lasted roughly from 1450 to the end of 19th century, earth's mean temperature at one point was 1°C less than that of today (Lamb,

1982). The effect of little ice age on everyday life is well documented: the freezing of river Thames in London, the freezing of New York harbor, which prompted the people to walk to Staten Island; abandonment of settlement in Iceland and Greenland; and crop failure in Scotland (Parry, 1978). While there has been wide speculation on cause or causes of little ice age, there is no definite explanation. This fluctuation is vital because it ended before the heavy industrialization of the late 19th century began, and is thus believed to be free from anthropogenic activities. Some have argued that increased global temperature in 20th century represents a recovery from the little ice age. However, without a definite cause for global coolness of the little ice age, the concept of recovery from the anomalous cold period remains dubious.

Considerable efforts have been made to analyze the current historical climate record over the past 150 years in order to establish whether there are any trends that could be attributed to greenhouse warming (greenhouse effect). For example, a number of global data sets of near-surface temperature have been developed and analyzed (Nicholls *et al.*, 1996). These data sets have been carefully corrected for errors of bias due to urban heat-island effects, non-homogeneities (such as instrument or location change) and changes in bucket types used to measure sea surface temperature. One of the most carefully constructed data sets (Jones *et al.*, 1994, 1997) indicates there has been a 0.3°C to 0.6°C warming of the earth's surface since the last 19th century. This trend continued through 1998. The global average temperature from January to December 1998 was the warmest on record for the period 1880-1998 (National Climate Data Centre). Specifically, the global average temperature for January until June 1998 was 0.6°C higher than the 1961-90 global mean temperature. However, the distinct warming has not been regionally homogenous, since some regions have experienced cooling during the 20th century.

The diurnal temperature range has primarily decreased in most regions, indicating that minimum thermal level has warmed more than maximum temperature, or that cloudiness has increased in these areas (Karl *et al.*, 1993, 1996). A cooling of the lower stratosphere by 0.6°C has also occurred since 1979. Sea level has increased on an average between 10-25 cm over the past 100 years (Nicholls *et al.*, 1996). This is related to increase in near surface temperature, which has caused thermal expansion of the ocean and melting of glaciers and ice caps. Thermal expansion of the ocean has contributed between 2 and 7 cm to the total increase in sea level. Also there has been a small increase in global average precipitation over land during the 20th century (Dai *et al.*, 1997). Recent investigations indicate that this mean increase has mainly influenced heavy precipitation rates (Groisman *et al.*, 1999). Since the late 1970s, there has been increase in the percentage of the globe experiencing extreme drought or severe moisture surplus (Dai *et al.*, 1998).

Many of the global climate change are analyzed from the point of view of expected combination of change to different variables known as fingerprints. For examples, the combination of cooling of stratosphere, warming of the earth's surface temperature and increased global mean precipitation is expected from increased greenhouse gas-induced climate change. These anticipated combinations of changes and their spatial pattern are based on our understanding of the physics of the earth/ocean/atmosphere system and results from climate models (Santer *et al.*, 1996). In 1990 remotely sensed data of the lower troposphere (about 2.5 km) have been analyzed to compare trend with those from surface observation station. This has led to debate regarding the robustness of results from surface observation compared with remotely sensed data. It has been observed that there was cooling of –0.04°C between 1979 and 1995. According to surface observation, the temperature has increased since 1979 by 0.14°C. This putative discrepancy has been used to question the surface temperature record and to provide evidence that global warming is not occurring.

4. Evidence of Climate Change

There is no doubt that climate is changing. The main reason of climate change is the increase in the concentration of greenhouse gases in the atmosphere due to several natural and anthropogenic activities. The level of GHG in the atmosphere has already increased considerably over the time particularly after the industrial era (1860). There are both physical and biological evidence of climate change, which are discussed below.

4.1. Physical Evidence

4.1.1. Rise in Atmospheric Temperature

There is clear-cut evidence that the atmospheric temperature, which is perhaps the main physical parameter or indicator of greenhouse effect (global warming) is gradually increasing since the period of industrial era (1860). The average temperature recorded in NACL clearly indicate that the average global atmospheric temperature has increased by 0.4-0.7°C by the year 2002 and if the trend is remained like this in the future, the global mean temperature is likely to increase by 4.5°C by the end of 21st century (IPCC, 1995), but, if the Kyoto Protocol is followed properly and the production of GHG especially CO_2 is reduced substantially by the developing and developed nations, increase in temperature can be limited to 2.5°C. Thus continuous increase in thermal regime of the global atmosphere clearly indicated that climate is changing. Similar to the pattern of rise in atmospheric temperature, there is also a great possibility of changing pattern in rainfall over the time, but due to lack of database, this climatic variable is not properly addressed and depicted. But based on our past experience, it is clearly evident that the rainfall intensity and total annual rain fall and number of rainy/cloudy days has decreased substantially. In particular, recent findings by the scientific community suggest that global warming is causing shifts in species spatial distribution that average 6.1 km per decade towards the poles or m per decade upward in the direction predicted by climate change model and that spring is on average, arriving 2-3 days earlier per decade in temperate latitudes. Entire regions are also suffering from the effect of global warming; in particular boreal and polar forest ecosystem. In a recent study published in Nature indicate that the change will occur first in the tropics, for which historical variability is low and biodiversity is highest. This study highlighted the timing of the climate change precisely the climate tipping point (Mora *et al.*, 2013).

4.1.2. Depletion in Rainfall

Definitely the intensity and duration of rainy period have reduced drastically over the time, although the frequency of flood in some areas particularly in

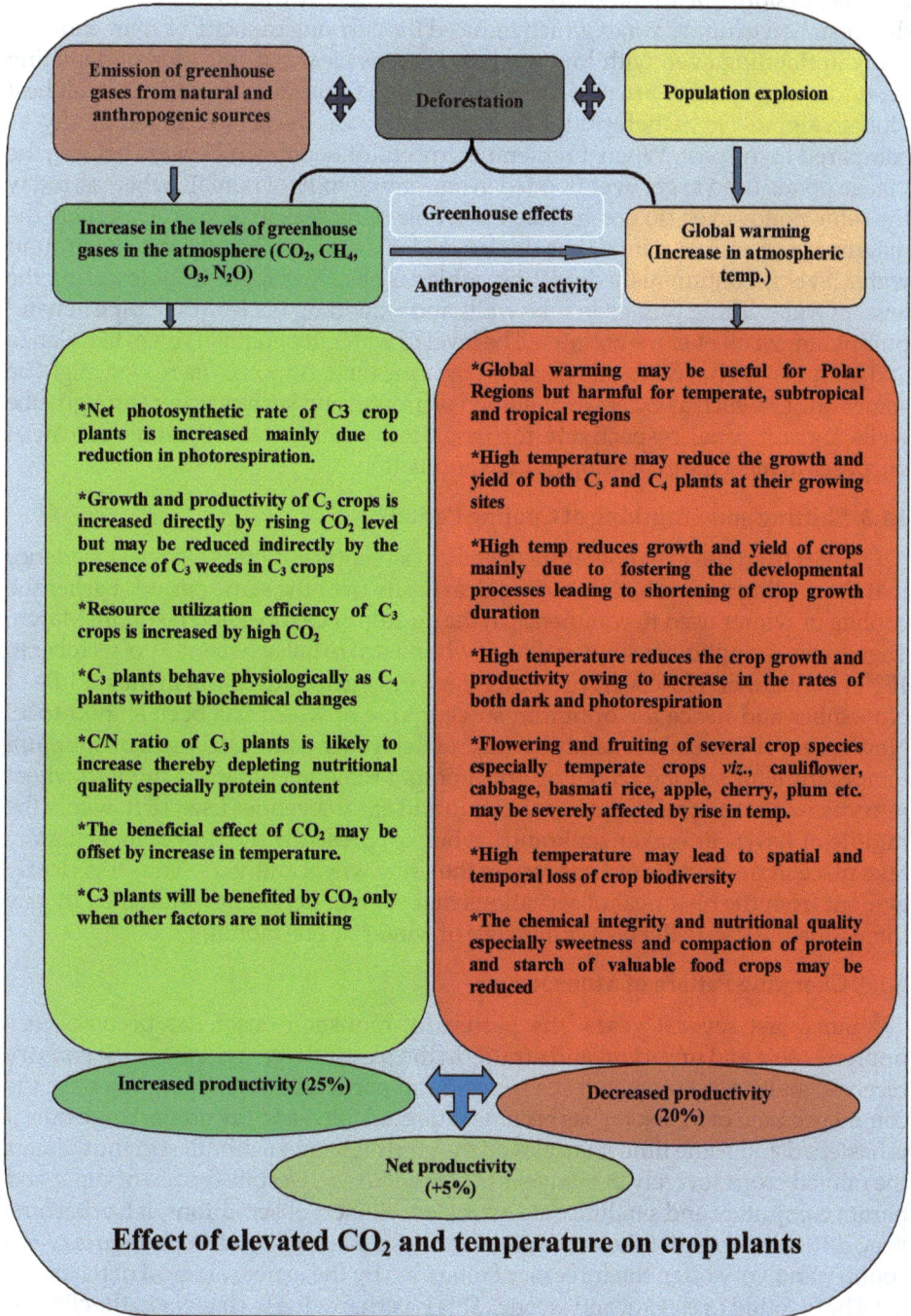

Emission of greenhouse gases from natural and anthropogenic sources	Deforestation

Population explosion

Increase in the levels of greenhouse gases in the atmosphere (CO_2, CH_4, O_3, N_2O)

Greenhouse effects

Anthropogenic activity

Global warming (Increase in atmospheric temp.)

*Net photosynthetic rate of C3 crop plants is increased mainly due to reduction in photorespiration.

*Growth and productivity of C_3 crops is increased directly by rising CO_2 level but may be reduced indirectly by the presence of C_3 weeds in C_3 crops

*Resource utilization efficiency of C_3 crops is increased by high CO_2

*C_3 plants behave physiologically as C_4 plants without biochemical changes

*C/N ratio of C_3 plants is likely to increase thereby depleting nutritional quality especially protein content

*The beneficial effect of CO_2 may be offset by increase in temperature.

*C3 plants will be benefited by CO_2 only when other factors are not limiting

*Global warming may be useful for Polar Regions but harmful for temperate, subtropical and tropical regions

*High temperature may reduce the growth and yield of both C_3 and C_4 plants at their growing sites

*High temp reduces growth and yield of crops mainly due to fostering the developmental processes leading to shortening of crop growth duration

*High temperature reduces the crop growth and productivity owing to increase in the rates of both dark and photorespiration

*Flowering and fruiting of several crop species especially temperate crops *viz.*, cauliflower, cabbage, basmati rice, apple, cherry, plum etc. may be severely affected by rise in temp.

*High temperature may lead to spatial and temporal loss of crop biodiversity

*The chemical integrity and nutritional quality especially sweetness and compaction of protein and starch of valuable food crops may be reduced

Increased productivity (25%)

Decreased productivity (20%)

Net productivity (+5%)

Effect of elevated CO_2 and temperature on crop plants

Brahmaputra basin and north Bihar. The increase in the flood frequency in above river is possibly due to the silting of river basin mainly due to soil erosion led by deforestation in upper zones which reduced the carrying capacity of river and thus leads to flooding even with low rainfall. There is clear cut evidence of depleting ground water in many areas of the country, even the annual rainfall at the highest global rainfall *viz.*, Cherapungi in Meghalaya has decreased substantially as compared to the past. When, I remember my childhood days (40 years before), the village ponds used to get over flooded in one or two spills of rainfall, whereas, today the same ponds even do not get filled in whole rainy season or year. Definitely the magnitude of annual rainfall has depleted/declined significantly and the ground water level is continuously depleting, which is clearly evident from lowering the level of water lifting pumps in tube wells and extending the length of pipe in hand pumps almost all over the country. The level of water in perennial rivers like Ganga and Yamuna etc. has definitely depleted over the time. Although its reason might be something else such as excess exploitation of ground and surface water through tube wells and lift canal respectively for irrigation, drinking and various industrial purposes to meet the exploding human population.

4.1.3. Shifting and Shrinking of Cooling Period

Apart from rise in temperature over the past century, there is also clear evidence that the cooling period has shortened drastically over the same period. Earlier the cooling or winter used to commence in the middle of October and last until March. That is why the optimum sowing time of wheat used to fall between 15th of October to 15th of November. However, presently the winter generally starts in the middle of November and hence the optimum sowing time of wheat has been shifted to 15 November –15 December mainly because of delayed in commencement of optimum temperature for sowing (20°C) by one month. I remember that the modern wheat introduced/developed in the late sixties in and early seventies used to flower in the middle of Jan (near makar sankranti), whereas, presently the flowering generally take place in the middle of February in the same wheat cultivars. Thus it is clearly evident from the historical observations that cooling period has shortened by one month mainly due to delay in commence of winter by one month.

4.1.4. Changing Pattern of Monsoon

Since last several years, the pattern of monsoon onset has become very unpredictable and uncertain and erratic. In the past different regions of the country received south west monsoon at a specific and particular time, but now a days the commencement of monsoon has become unprecedented and uncertain. Some time it is hastened and some time it is delayed by a fortnight and a month and thus affects the rainfed crops severely in many parts of the country. Despite the use of super and param computers and satellite data on meteorological observations, it has become very difficult to predict the exact time of monsoon arrival in different parts of the country and very often the forecast of monsoon by the meteorological department have been found irrelevant and wrong. This is perhaps due to changes in the climatic variables mainly affected by the anthropogenic activities. In the recent past the onset of monsoon in different parts of the country has become very erratic and thus the specific pattern of monsoon which happened to be very stable in the past has deviated from normal pattern of its occurrence. Since several years the monsoon is not following its usual pattern of arrival in various parts of the country, some time it is too early and too late, which poses severe problems in the schedule of rainfed crop cultivation. Last year the pre or early monsoon in the middle of June in north India followed by a long gap in monsoonal rain has led to damage of crop seedlings and finally the crop failure as most of the farmers sown their kharif crop and later majority of the farmers had no seeds to sow second time. Similarly preparation of rice nursery and their transplanting in early monsoon rain had led to spoilage of rice crop in rain fed areas.

4.1.5. Occurrence of Natural Disaster

As compared to the past, the frequency and intensity of natural disaster such as flood, drought, earthquake, super cyclone etc. have increased, which led to the loss of property and lives. This may perhaps be an indication of climatic change and climatic variability. In the last decade the occurrence of super cyclone in Orissa, severe flood in North Bihar, earthquake in Uttaranchal and Gujarat and recently the Tsunami in South East Asia have indicated the dynamics of climatic change and aberration.

4.2. Biological Evidence

4.2.1 Early Blossoming of Trees

Trees are generally a very good indicator of climate change as the flowering in perennial trees take place as a result of completing the crop-specific required thermal unit/thermal period or degree days. The very good examples could be the mango trees, which flower according to the thermal regime/period in different parts of the country. The mango tree generally flower in October in south India, in December in eastern and central India and middle of February in north India. Due to diverse thermal regime in different parts of the country, mango trees flower accordingly. Since the temperature in the month of October and November is higher in southern and eastern parts of India, hence the mango starts flowering in this month in these areas, whereas, in north India the required temperature for flowering generally occurs in the beginning of spring season *i.e.*, in middle of February. That is why the availability of mango in India starts right from February to March in South India to June to July in north India. But there has been some evidence of flowering mango in December in north India especially during 2004 which was probably due to prevailing higher thermal regime in December. Thus the flowering behavior of mango could be used as a very good bio-monitoring indicator for climate change.

4.2.2. Appearance of Grasses in Antarctica

Apart from melting of ice caps and glaciers in the Antarctic region, there has been some evidence of growing grasses in these areas during 2004-05 which indicate the possibility of rising atmospheric temperature in these areas as well. The IPCC assessment report 4 (IPCC AR4, 2007) reported increased abundance and distribution of Antarctic spearwort and the Antarctic hair grass attributed to the increasing summer temperatures.

4.2.3. Changing Cropping Pattern

There has been some indication of spatial changes in cropping pattern particularly in the hilly mountain areas of Himachal Pradesh. The traditional regions of apple and other temperate crops which were earlier found suitable for the cultivation of such crops because of prevalence of optimum temperature required for flowering and fruiting of temperate crops, but these regions are gradually becoming warmer due to global warming and may become unsuitable for the cultivation of such temperate crops which require chilling for flowering. Successful cultivation of maize in Bihar during winter season substituting wheat clearly indicates the possibility of warming and climate change as maize is a C4 plant and performs better under higher temperature.

4.2.4. High Latitudinal Shifting of Temperate Crops

Climate change due to global warming may severely affect the spatial distribution of temperate crops especially chilling requiring crops *viz.*, apple, apricot, cherry, plum, keshar, cauliflower, cabbage, pea, etc. at their traditional growing places during on as well as off seasons. The areas of temperate crops cultivation in India are mainly confined to higher latitude provinces such as Kashmir, Arunanchal Pradesh, Uttaranchal, Sikkim, Himanchal Pradesh (Kullu and Manali, Katrayan). The crops grown in these regions may be severely affected in near future due to rise in thermal regime of these regions, as these temperate crops require very narrow range of optimum thermal regime. Thus any change beyond the optimum thermal range of these crops may lead to the breakdown of their reproductive system thereby failing to flowering and fruiting at increased temperature. Hence the cultivation of these crops especially apple has to shift toward higher latitude to meet their optimum thermal level for flowering and fruiting provided the upper zones is edaphically suitable for their cultivation and fall within the national territory. Some crops like cabbage require a specific thermo-photoperiodic optimum for its bolting and flowering. It require long day length coupled with cooling which is possibly met in northern zones like Kullu, Manali or Katrayan in Himanchal pradesh, that is why the seed of cabbage is produced mainly in north H.P. Similarly keshar, a very precious crop only grown in Kashmir because of its specific thermal requirement. If the temperature increases to the extent beyond the optimum thermal range, this crop may also suffer severely from global warming and climate change in near future.

If these hypo-thermal loving crops do not shift their optimum thermal range over the period of climate change, there may be a great spatial change in these crop species and the cultivation of such crops at their traditional belt may probably be abandoned in future. Thus there is a possibility of either their loss (genetic erosion) from their native or traditional place of cultivation and hence to shift to new and non-traditional zones, where low-temperature limit may become suitable for these crops. Thus the climate change in future may cause greater spatial change or loss of these crops from their native place of cultivation. There has already been some indication of shifting in the apple cultivation towards high latitude because of thermal unsuitability of traditional zones for the cultivation of these crops.

5. Causes of Climate Change and Variability

5.1. Natural Causes

5.1.1. Volcanoes Eruption

Climate change and variability have been resulted from the physico-chemical and biological interaction at regional and global levels. Increasing global atmospheric temperature continuously or intermittently depends upon several natural and anthropogenic activities which disturb or imbalance the normal natural processes. Any change or deviation in normal processes either due to natural process such as volcanoes eruption, super cyclones, drought, floods etc. or man made activities like deforestation, over production of greenhouse gases *viz.*, CO_2, CH_4, N_2O, O_3, CFC 11, 12 causes global warming and thereby climate change. Volcano eruption lead to explosion of sub-surface earth and huge amounts of heat, gases, mud etc. are released into the atmosphere which in turn causes periodic zonal or regional warming which may have its impacts on local and regional atmosphere for certain period and vanish gradually over the time. But for some period the immense amount of liberated heat, dust, gases, etc. have the potential to change the atmospheric thermal regime and can influence the water cycle and finally the climatic variability.

5.1.2. Super Cyclones and Earthquakes

Unprecedented or sudden events such as cyclones and earthquake lead to the loss of biological and physical resources and thereby change in the climatic variables for certain period which in turn causes climatic variability. Loss of vegetation and other forms of biological resources either due to natural super cyclones or earthquakes lead to imbalances in abiotic and biotic components of the environments which finally causes variation in climatic and edaphic components such as rainfall, humidity, wind velocity and temperature. The best examples of such events are the incidence of super cyclone in Orissa, which caused severe damage to the coastal vegetation and possibly could be responsible for rising the atmospheric temperature of Orissa during the winter and summer seasons since its occurrence. Similarly the occurrence of biggest natural disaster *viz.* Tsunami in December 2004 had the impact of climatic variation in 2005 in terms of changes in seasonal thermal regime during winter. The year 2005 has several ups and down in the magnitude of cooling and warming and thereby caused climatic aberration.

5.1.3. Low Precipitation

Precipitation, which has tremendous ability to change the atmospheric thermal regime, is one of the major climatic components responsible for climatic variation. It has greater impact on sudden change in the thermal level of the atmosphere. It is generally found that year with good rainfall also causes better cooling and less warming as the water status both in soil and atmosphere influence the magnitude of thermal level of the atmosphere through evapo-transpiration cooling. If the water regime is poor in soil and atmosphere, it is supposed to be warmer as compared to the moist lithosphere and atmosphere.

5.1.4. Emission of Greenhouse Gases from Natural Resources

The role of natural greenhouse gases such as CO_2, CH_4 and water vapor in raising the global atmospheric temperature from -18°C to 15°C (+33°C) has been very significant, which led to the survival and existence of life on the earth. Apart from these two important natural GHG (CO_2 and CH_4), there are some other natural GHG such as water vapor, N_2O, O_3, which are emitted continuously into the atmosphere as a results of nitrification and de-nitrification processes take place in the nature particularly in tropical rainforest. The release of CO_2 in the atmosphere takes place as a result of direct and indirect respiration of biological materials including animals, plants and microorganism and heterotrophic respiration *i.e.* microbial decomposition of dead organic matter buried in the soil of forest and agro-ecosystems. Huge amount of methane is released into the atmosphere as a result of anaerobic decomposition of organic matter in the soil in the presence of methanogen bacteria particularly under reduced paddy rice fields, marshy/swampy lands and water reservoir and lakes etc. The emission of these GHG from natural resources such as forest, agro-ecosystem, lakes, marshy areas etc. have not been found very harmful to the nature as these are emitted naturally into the atmosphere, rather have been found very useful in maintaining the atmospheric temperature conducive for the existence of lives on this beautiful and bountiful earth planet. The greenhouse effect of naturally emitted GHG is said to be 33°C, but the emission of these gases from human activities has been found harmful to the livings. Any natural process can never be harmful to the nature as this takes place mainly to overcome the natural defects created by human activities. Thus the original or natural greenhouse effect (warming caused by natural emission of GHG such as CO_2, and water vapor) is essential for life. But some times emission of CO_2 from natural disaster such as forest burning, volcanoes eruption may be harmful to the nature and natural resources. Similarly emission of CH_4 from waterlogged marshy lands and rain-fed rice paddy fields or from the ruminants may not be harmful to the nature as these are the natural process imparted by the nature to continue the lives on the earth.

5.2. Anthropogenic Causes

5.2.1. Deforestation

Deforestation is one of the major anthropogenic activities responsible for climatic change and variability. The forest acts as a biological sink of CO_2 and source of O_2 and also responsible for cooling the atmosphere through transpiration and thereby rainfall. About one-third geographical area should be covered with forest to maintain the minimum ecological threshold and environmental security. Forest not only cleans the atmosphere from several noxious gases but also provide the shelter for wild lives. Exploding human population especially in developing countries has mainly been found responsible for massive deforestation to expand the agricultural lands to meet their food, fiber and fuel requirements. Several millions hectares of global forest especially the rain forest in Amazon valley and elsewhere have been denuded for agricultural purposes. Thus deforestation has caused significant depletion in CO_2 sink and thus responsible for increasing CO_2 level in the atmosphere and finally causing greenhouse effect. Deforestation thus not only deplete the natural sink of

CO_2 but also several noxious gaseous air pollutants such as SO_2, HF, NO_x, PAN, O_3 etc. as these pollutants are also absorbed by the vegetation. Forest vegetation also acts as a biological fence or barrier in the transportation of these gaseous and metallic pollutants from their source point to non-source points. Thus forest acts as a biological refinery for the polluted air and finally reduces the load of SPM and gaseous pollutants, which in turn enhance the transmission of sunlight to the earth surface.

5.2.2. Emission of GHG from Non-Agricultural Sources

5.2.2.1. *Industries*

Industries have been the major source of greenhouse gas emission since the beginning of industries in mid 19th century (1860). The overexploitation of natural fossil fuel resources in the form of coal and mineral oil has lead to the emission of greenhouse gases such as CO_2 and several noxious gases (pollutants) and SPM, which have highest contribution to global warming (CO_2 alone contribute about 60 per cent). There has been clear evidence that the level of CO_2 in the atmosphere has increased by 25 per cent @ of 0.5 per cent annually when compared to the level in pre-industrial era (1860). About 3 billion tons of CO_2 per year is emitted into the atmosphere globally, to which industries have made significant contribution. Rapid industrialization to meet the various demands of exploding population world wide as well as for more financial and job security, has finally degraded the environment subsequently, especially in the developing countries where no/or poor measures are applied to control the emission of these gases into the atmosphere.

5.2.2.2. *Thermal Power Plants*

Establishment of thermal power plants/houses to generate electricity has also made significant contribution to the emission of CO_2 and several noxious gaseous and metallic pollutants such as SO_2, HF, NO_x and suspended particulate matter (SPM) which directly and indirectly contribute to global warming and cooling as well. The emission of CO_2 leads to the rise in atmospheric temperature mainly by trapping the terrestrial long wave infrared radiation. The emission of gaseous pollutants such as SO_2 and dust/soot/tar particles into the atmosphere reduce the incoming radiation thereby diminishes the light intensity on the earth and thus causes global dimming and subsequently global cooling.

5.2.2.3. *Vehicles and Transportation*

Rapid industrialization and consequently the increase in people income led to make their life more comfortable through purchasing the power driven vehicles for their easy journey. The improvement in purchasing power of common people in developing countries in recent past enhanced the selling of two and four wheel vehicles tremendously and thereby increased level of air pollution. Thus it is a kind of developmental pollution caused by the varieties of vehicles. The level of vehicular pollution is increasing very fast presently mainly because of increase in purchasing power of fuel and vehicles as well. Vehicles directly does not emit any greenhouse gas, but the exhaust of NO_2 from the vehicles photo-chemically reduced to NO and evolve O, which finally produces O_3 (ozone) following reacting with molecular atmospheric O_2. Vehicles also release a number of gaseous and metallic air pollutants such as SO_2, CO, NO_2 and SPM, which are also responsible for causing climate change through modification in radiation level on the earth. The emission of ozone from vehicles not only contribute directly to global warming as a greenhouse gas, but also act as a potential photo-chemical secondary pollutant and finally affect the crop and animal health considerably. Thus the vehicles, which make our life more comfortable in one hand, pollute our environment on the other hand.

5.2.3. Intensive Agriculture

Agriculture is mainly concerned with the release of carbon dioxide, methane, and nitrous oxide, which are important greenhouse gases. The concentration of these greenhouse gases is increasing in the atmosphere and there is a need to determine the extent to which agriculture contributes to these gases.

5.2.3.1. *Carbon Dioxide*

This gas, a substrate of photosynthesis, is utilized during the dark reaction cycle. However, when plants degrade either directly or through animal or microbial chain, it is released back into the atmosphere. Therefore, in agriculture it becomes a cyclic process, because whatever is fixed by the crops is released back into the atmosphere. Consequently through agricultural practices no net efflux of carbon dioxide into the atmosphere. However, if agricultural practices are followed which lead to erosion of soil or clearing of forest for agriculture, the carbon stored in the soil or the vegetation may become a part of atmosphere. However, it is difficult to attribute any major role of agriculture in increasing the atmospheric CO_2 level.

5.2.3.2. *Methane*

Among the various sources which may be substantially contributing to the increasing level of methane in the atmosphere, rice fields and ruminants animals have been identified which have an agricultural base. Rice is grown in approximately 90 per cent of the area in the developing countries of Asia. The number of ruminant live stock is also large in developing countries, because they are a source of animal product as well as draught power in agriculture. The estimates of methane from rice fields were given as 110 Tg annually with a range of 25-170 Tg annually by the IPCC. These estimates differed substantially and were obtained from Europe and USA under incomparable condition of Asia both climatologically and agriculturally. Hence these were doubted by the scientists in Asia, particularly in India, China and Japan. Studies at large number of sites in India has made it clear that methane production from rice paddy fields is much lower, approximately one tenth of what was stated in the IPCC report. It has been observed that methane emission from rice fields is a function of soil characteristics, microbial population, plant growth, genotype, temperature and other factors. Since all the factors were not considered, and the data were extrapolated from a very small sample size to the whole rice growing area, exaggerated values were obtained. It is, therefore, necessary that more detailed data base obtained from the representative sites of rice growing regions.

5.2.3.3. Nitrous Oxide

N_2O is produced mainly from the aerobic soils during the nitrification process of nitrogen mineralization and de-nitrification process during intermittent irrigation of rice fields. Although the nitrogen use efficiency of many crops in tropical regions is poor, but there are no sufficient data to conclude that fertilizer consumption by India and China would become an important cause for the increase of N_2O in the atmosphere. Now technologies have been improved to enhance the nitrogen use efficiency and reduce emission of N_2O.

5.2.3.4. Ozone

Until recent past the non-destructive tropospheric ozone as a pollutant was considered to be emitted from the vehicles in metros, cities and highways, but the satellite picture have shown the presence of ozone more in rural areas, which might be due to photochemical transformation of N_2O released from nitrogen fertilized arable fields and emission of carbon monoxide from biomass burning in rural areas. Thus nitrogen fertilization in agriculture produces nitrous oxide as well as ozone.

5.2.4. Biomass Burning

Since the advent of mechanization in agriculture especially in north India (Punjab, Haryana, western U.P.), the harvesting of wheat and rice crops is done mechanically by the combine in which about 30 per cent of crop stubbles left in the field, which are generally burnt by the farmers before proceeding for successive crop. Some time this also happens in forest where large area of forest is naturally set to fire, which continues for several days. Thus huge amount of stubble and forest biomass burning lead to the emission of CO_2, CO, SO_2, etc. into the atmosphere, which not only causes global warming through greenhouse effect, but also causes reduction in solar radiation and thereby global cooling. The effect of crop stubble biomass and forest burning is not local but also at non-source points as the gases emitted by the same can travel to any distance and causes climatic variation over there. The burning of forest in India, China, Indonesia and America has been reported to have caused global warming and cooling not only at their source-points but elsewhere as well. Melting of glaciers in Europe has been reported to be caused by the biomass burning in India. The formation of black/brown haze (thick gaseous layer in air) hinders the transmission of light on the earth surface, thereby causes global dimming.

5.2.5. Population Explosion

Indeed, global population explosion mainly in the developing countries has lead to the poverty and subsequently pollution, which in turn has become the root cause of all kinds of social-economic and environmental aberration/imbalance throughout the world. Unprecedented increase in global human population has been said to be responsible for trading off the economic reform took place during the last few decade as large amount of national GDP is diverted for the maintenance of ever increasing human population on the cost of further economic development. Population has invited almost all kinds of biotic and abiotic environmental pollutions *e.g.*, soil, water, air and biological pollution. The population led pollution has occurred mainly through anthropogenic activities such as intensive agriculture, industry, transportation, automobile, thermal power and diverse other small and large scale industries in order to meet the various kinds of domestic and luxurious demands for survival of poor and needy people and for enjoyment of prosperous people. Population increase has been also responsible for massive deforestation, which has not only created environmental problems, but also caused loss of biodiversity, air pollution, global warming and finally climate change and crop productivity. Irrespective of developed and developing countries, efforts are being made to enhance to production of commodities required to human beings on the cost of environmental safety and conservation. In other words just for meeting the demands and enjoyment of a single biological environmental component *i.e.*, human beings all other physical and biological environmental components are on the stake. Thus instead of increasing the productivity of commodities required for meeting the demand of human population, it is better to make efforts for reducing human production.

6. Impacts of Climate Change

Atmospheric CO_2 and climatic factors such as temperature, radiation, precipitation, relative humidity and wind velocity are important for growth of vegetation and hence of crop productivity and production. In addition, however, crop productivity is dependent upon soil characteristics, crop variety, energy, fertilizers, irrigation, pesticides, other farm inputs and their application technologies. Therefore in evaluating the potential effects of climate change on agriculture, we will have to consider the non-climatic factors as well. The major changes which are likely to be predicted and projected due to global warming are as follows. (1) Increase in atmospheric carbon dioxide concentration (2) Differential increase in temperature at different latitudes (3) Increase in changes in precipitation (4) Uncertainty in precipitation and (5) Global dimming due to reduction in radiation owing to cloudiness and aerosol. Since all the above factors influence plants and hence crops, we may consider the effects of some of them individually.

The concern on past, present and future weather aberrations, climate trends, and their effects on agriculture has continued to stimulate research as well as public and policy-level interests on the analysis of climate variability and agricultural productivity (Matthews *et al.*, 1996; IPCC, 1996). It is well recognized that climate variability has a wide range of direct and indirect impacts on crop production. The climatic change would cause rise in temperature, changes in precipitation and other physical factors that influence biological systems including agriculture. For example in Philippines; typhoons, floods, and droughts caused 82.4 per cent of the total Philippine rice (*Oryza sativa* L.) losses from 1970 to 1990 (PhilRice-BAS, 1994). Therefore, the United Nations Formulated Convention on Climatic Change (UNFCCC) expressed concern about food security.

Weather and climate have a direct influence on cropping systems. It affects plant growth and development. The fluctuations and occurrences of climatic extremes particularly at critical crop growth stages may reduce yield significantly (Satake and Yoshida, 1978; Peng *et al.*, 1996). Occurrence of abnormal weather episodes during the growing season or during critical development stages may hamper growth processes resulting in yield reduction. This makes climate variability a threat to food production leading to serious social and economic implications (Geng and Cady, 1991; Hossain, 1997). Rising temperatures - now estimated to be 0.2°C per decade, or 1 °C by 2040 (Mitchell *et al.*, 1995) with smallest increases in the tropics (IPCC, 1992) - would diminish the yields of some crops, especially if night temperatures are increased. The temperature increase since the mid-1940s is mainly due to increasing night-time temperatures, while CO_2-induced warming would result in an almost equally large rise in minimum and maximum temperatures (Kukla and Karl, 1993).

Effect of climate change on agriculture can be of three types: 1. The physiological (direct) effect of elevated levels of atmospheric CO_2 on crop plants and weeds. 2. The effect of changes in parameters of climate such as temperature, precipitation, solar radiation on plants and animals. 3. The effect of climate related changes such as rise in sea level on land use. However, a clear understanding of the vulnerability of food crops as well as the agronomic impacts of climate variability enable one to implement adaptive strategies to mitigate its negative effects (Lansigan *et al.*, 2000).

6.1. Effects of CO_2 Enhancement

6.1.1. Direct Effects on Photosynthesis and Crop Yield

Since carbon dioxide is an essential factor (input) of photosynthesis, its increasing level have received considerable attention in evaluating impacts on crop productivity and hence production. There is evidence that increases in atmospheric CO_2 level would increase the rate of plant growth (Cure, 1985; Cure and Acock, 1986). However, there are vital difference between the photosynthetic mechanism of different crops and hence in their response to increasing CO_2 level. Plant species with C3 photosynthetic pathway (wheat, rice, soybean, barley, mustard, beans etc.) tend to respond more positively to increased CO_2 because it suppresses the rate of photorespiration (destructive photosynthesis). However, C4 plants (maize, sorghum, millets, sugarcane etc.) are said to be less responsive to increased CO_2 level. Since these are largely tropical crops and most widely grown in Africa, thus there is suggestion that CO_2 rising level will benefit temperate and humid tropic agriculture more than that in the semi-arid tropics. Hence the effect of climate change on agriculture in some parts of the semi-arid tropics are negative, these may not be partially compensated by the beneficial effects of CO_2 enhancement as they might in other regions. In addition we should note that, although C4 crop plants accounts for only about 20 per cent of the world's food production, where maize alone contribute to 14 per cent of all production and about 75 per cent of all trade grain. It is the major grain used to make up food deficit in famine prone regions and thus any reduction in its output could affect access to food in these areas (Parry, 1990). In most instances the effect of doubling of CO_2 concentration have been considered to be beneficial for photosynthesis and hence for crop productivity. The reviews of Kimball and Cure and Acock and several others based on publications on the effects of CO_2 have been used to predict an improvement in crop productivity and substantially in C3 plants and to a lesser extent in C4 plants. Therefore, the IPCC report gives the impression that increased concentration of CO_2 may lead to improvement in crop productivity as well as their adaptation to stress environments. Atmospheric concentrations of carbon dioxide increased from 315 ppm (parts per million) in 1959 to a current atmospheric average of approximately 385 ppm (Keeling *et al.*, 2009). The global mean temperature has rose to the extent of 0.7°C in the last century (National Oceanic and Atmospheric Administration, 2007), but there has been no any reports regarding its negative impacts on crop growth and productivity, rather it is showing increasing trend, but it is not clear, whether because of rising atmospheric CO_2 level or improved crop varieties or technologies as these factors have advanced immensely in the recent past.

6.1.1 (a) Photosynthesis, Respiration and Growth

Two groups of plants differ in their fundamental carbon fixation chemistry – the C3 and C4 plant groups. The response of growth to elevated CO_2 in C3 (wheat, rice, barley, root crops, legumes etc.) is larger than that of C4 plants (maize, sorghum, millets, sugarcane) (Cure, 1985; Morison, 1989; Jones, 1986; Ainsworth and Long 2005). The world's most troublesome 17 terrestrial weed species are C4 plants in C3 crops. The difference in response to CO_2 may make such weeds less competitive. But C3 weeds in C4 crops may increase as a problem (Morison, 1989).

Photosynthesis is the net assimilation of carbohydrates formed by the fixation of atmospheric CO_2. Increased atmospheric CO_2 level may result in increased photosynthesis of green plants. A doubling of CO_2 increases the instantaneous photosynthetic rate by 30-100 per cent, depending on other environmental conditions (Pearcy and Bjorkman, 1983; Ainsworth and Rogers 2007). Doubling of CO_2 in addition to increasing net photosynthesis in C3 plants at fixed temperature, also raises the optimum temperature for photosynthesis by about 4 -6°C (Acock and Allen, 1985; Osmond *et al.*, 1980; Jones, 1986; Long, 1991). The balance between the photosynthetic gain and loss of carbohydrate from plants through respiration is the resultant net plant productivity or growth.

The developmental process such as flowering is affected in both positive and negative directions by enhanced CO_2 level (Morison, 1989). This affects the progress of crop through different stages of life cycle, thus potentially shortening or lengthening the required growing period. Increased levels of CO_2 can enhance nitrogen fixation by legumes through increased rate of photosynthesis and assimilate availability. However increased temperature can reduce nitrogen fixation and also increase the energy cost of nitrogen fixation (IPCC, 1990).

6.1.1. (b) Stomatal Aperture and Water Use Efficiency

Doubling of CO_2 causes partial stomatal closure both in C3 and C4 plants (Morison, 1987; Cure and Acock, 1986; Kimball and Idso, 1983). But increased growth results in greater leaf area so the total water use per plant in high CO_2 may be similar to that at present condition (Gifford, 1988). Although stomatal conductance is decreased under elevated CO_2, the ratio of intercellular to ambient CO_2 concentration usually is not modified, and stomata do not appear to limit photosynthesis more in elevated CO_2 compared to ambient CO_2 (Drake *et al.*, 1997). Elevated-CO_2 effects on crop evapotranspiration per unit land area (ET) have been small with cotton (Dugas *et al.*, 1994; Hunsaker *et al.*, 1994; Kimball *et al.*, 1994) and spring wheat (Kimball *et al.*, 1995, 1999) supplied with ample nitrogen fertilizer. With rice, under field-like conditions, CO_2 enrichment reduced seasonal total ET by 15 per cent at 26°C but increased the same by 20 per cent at 29.5°C (Horie *et al.*, 2000). A larger decline (-22 per cent) in the daily ET of a C_4 dominated tall grass prairie was reported by Ham *et al.* (1995), and a strong reduction in water use per plant also was observed for a C_4 plant-maize (Samarakoon and Gifford, 1996; Leakey, 2009). Reduced evapotranspiration can have consequences for the hydrological cycle of entire ecosystems, with soil moisture levels and runoff both increasing under elevated CO_2 (Leakey *et al.*, 2009).

Relative enhancement of growth owing to CO_2 enrichment might be greater under drought conditions than in wet soil because photosynthesis would be operating in a more CO_2-sensitive region of the CO_2 response curve (André and Du Cloux, 1993; Samarakoon and Gifford, 1995). In the absence of water deficit, C_4 photosynthesis is believed to be CO_2 saturated at present atmospheric CO_2 concentration (Bowes, 1993; see also Kirschbaum *et al.*, 1996). However, as a result of stomatal closure, it can become CO_2-limited under drought. Some of the literature examples in which C_4 crop species, such as maize, have responded to elevated CO_2 may have involved (possibly unrecognized) minor water deficits (Samarakoon and Gifford, 1996). Therefore, CO_2-induced growth enhancement in C4 species (Poorter, 1993) may be caused primarily by improved water relations and WUE (Samarakoon and Gifford, 1996) and secondarily by direct photosynthetic enhancement and altered source-sink relationships (Ruget *et al.*, 1996; Meinzer and Zhu, 1998).

6.1.2. Indirect Effect on Crop Growth and Productivity

Apart from direct impact of rising CO_2 level on C3 and C4 crops, it also affect the crop growth and productivity through crop weed competition. C3 crops in temperate and subtropical regions could also benefit from reduced weed infestation. Fourteen of world's 17 most noxious terrestrial weeds are C4 plants in C3 weeds. The difference in response to increased CO_2 may make such weeds less competitive. In contrast C3 weeds in C4 crops, particularly in tropical regions, could become more of a problem which may lead to reduction in growth and productivity of C4 plants, although the final outcome will depend on the relative response of crops and weeds to climate changes as well (Parry, 1990). Many of the pasture and forage grasses of the world are C4 plants including important prairie grasses in North America and central Asia and in the tropics and subtropics. The carrying capacity the world's major rangeland is thus unlikely to benefit substantially from CO_2 enrichment (Parry, 1990). The actual amount of increase in economic yield rather than of biological yield that might occur as a result of increased photosynthesis is also problematic. In controlled environment studies, where temperature, nutrients, and moisture are kept optimal, the yield increase can be substantial, averaging about 36 per cent for C3 cereals such as wheat, rice, barley, and sunflower under a doubling of ambient CO_2 concentration. Few studies have yet been published, however, of the effect of increasing CO_2 in combination with changes of temperature and rainfall.

6.1.3. Effect on Crop Quality

Although the growth and productivity of C3 crops have been reported to increase substantially, but the quantitative gain has been on the cost of crop quality as the nitrogen content of vegetative and reproductive tissues have not increased, rather decreased thereby increased C/N ratio, which is an index of poor crop quality especially in respect of protein content. Now there are sufficient experimental evidence indicating the marked reduction in nitrogen content and increase in carbon content of grain/seeds, forage and other plant parts, employing marked reduction in protein level and nutritional level for livestock and humans. This, however, may also reduce the nutritional value of plants for pests so that they need to consume more to obtain their required protein intake. Thus it is evident from the findings that when C3 plants are subjected to higher CO_2 level, the C3 plants transformed physiologically to C4 plants without changing their metabolic pathways and biochemistry and the effect is transient. The evolution of C4 plants from the C3 ones in tropical climate, the biggest biochemical adaptation resulted in larger leaf surface in order to reduce the loss from insect damage.

Elevated CO_2 reported to reduce grain protein content (Taub *et al.*, 2008), macro-elements (Manderscheid *et al.*, 1995), micro-elements (Hogy and Frangmeier, 2008, Anand and Nagarajan, 2011) and to increase carbon/nitrogen ratio (Hogy *et al.*, 2011) mainly due to higher sugars and starch contents (Bazzaz *et al.*, 1985; Strain, 1985). Studies on wheat shows that with increased CO_2 the same yield may be produced with less nitrogen fertilizer but bread-making quality of grain may be diminished (Austin, 1988). More chalkiness and less protein content of rice kernels produced under elevated CO_2 level has been reported by Kobayashi *et al.* (2005). Changing the carbon to nitrogen ratio and carbon to phosphorus ratio of plant material also has consequences on decomposition of organic matter in soil leading to reduction in decomposition rates and decreased ecosystem productivity (Dahlman *et al.*, 1985).

6.1.4. Effect on Resource Utilization Efficiency

The increased growth and productivity of C3 crops under rising CO_2 level at the same level of water, nitrogen, radiation and other inputs clearly revealed that the resource utilization efficiency of crops *e.g.* water, nitrogen and light utilization efficiency are enhanced markedly under such condition, although the nutritional status of crops is impaired considerably. Thus the crops under high CO_2 level behave like C4 plants and can produce greater biomass and economic yields with same amount of inputs and thus lead to greater economy of water, which may be suitable for cultivation under limited water supply of rain fed condition. Rising CO_2 level may have some effect on closure of stomata. This tends to reduce the water requirement of plants by reducing transpiration rate thus improving water use efficiency (the amount of biomass produced per unit amount of water loss through evapotranspiration). Water use efficiency was nearly doubled under enriched CO_2 in potato and total biomass, yield and water use efficiency increased under elevated CO_2, with the largest percent increases occurring at irrigations that induced the most water stress (Fleisher *et al.*, 2008). A doubling of ambient CO_2 concentration causes about a 40 per cent reduction in stomatal aperture in both C3 and C4 plants which may reduce transpiration by 23-46 per cent (Cure and Cock, 1986; Morison, 1987). This might well help plants in environments where moisture currently limits plant growth, such as in semi-arid regions, but there remain many uncertainties such as to what extent the greater leaf area of plants resulted from increased CO_2 will balance the reduced transpiration per unit leaf surface (Allen *et al.*, 1985; Gifford, 1988). Secondly the CO_2 induced partial closure of stomata of crop plants may cause marked reduction in transpiration with enhanced photosynthesis as the mesophyll CO_2 concentration remains higher under increased CO_2 level. It is expected that a doubling of atmospheric CO_2 level from 350-700 ppmv might cause a 10-50 per cent increase in growth and yield of C3 crop plants (such as rice, wheat, soybean etc.) and a 0-10 per cent increase in growth and yield of C4 crops such as maize, sorghum, millets, sugarcane etc. (Warrik *et al.*, 1986). However, much depends upon the prevailing growing conditions. Our present knowledge is entirely based on experiments conducted in field chambers and phytotron and has not yet included extensive study of crop response to increased CO_2 in fields under suboptimal conditions. Thus, despite beneficial effects of CO_2 enhancement on crop growth and productivity and its partial compensation for negative effects of CO_2 induced climatic changes (hyper thermal), it is still paradoxical and uncertain.

6.1.5. Possible Points to be Considered on the Impact of Increased CO_2

The following points need to be emphasized. (1) Most of the experiments conducted on direct effects of CO_2 used about 600 ppm or an even its higher concentration. At 2 x CO_2 climate level resulting from the increase of CO_2 plus other greenhouse gases to the point where the warming was equivalent to that of the doubling of CO_2 alone, the concentration of CO_2 would be around 450 ppm. (2) Experiments were conducted in controlled environments, maintaining optimal temperature for the growth of experimental plants. (3) Plants were protected from diseases and insect pests. (4) Most of the experiments on climate change have been conducted separately for different climatic factors like CO_2 and temperature and hardly there are any reports on the interaction of CO_2 with other climatic factors such as temperature and other greenhouse gases except little information on CO_2 and ozone interaction. Therefore the positive and negative reports of increased level of CO_2 and temperature on crop productivity may not be realistic. Hence the magnitude of climatic change impacts on agriculture is still an enigma because nowhere in any experiments and studies the adaptability characteristics of crops, which is quite obvious and evident has not been taken into accounts while measuring their quantitative response to climate change. Therefore the conclusion of IPCC report is not justifiable and some more points need to be considered.

The past studies on CO_2 fertilization even at 600 ppm or more do not show any phenological effect on crop plants, however it is strongly influenced by thermal level during the growing season and is a strong determinant of crop yield through altering the crop growth duration. Thus the beneficial effects of CO_2 fertilization are likely to be reduced by altered crop growth duration.

We can learn something about the response of crops to increasing concentration of CO_2 from the past records. Between 1888-1990, the CO_2 concentration has increased from 280 ppm to 350 ppm, a net increase of 25 per cent (NOAA/ESRL 2010). If there had been any effect of this increase, it should be reflected in crop productivity. The productivity of chickpea, rapeseed, and mustard were analyzed for India since 1900.

There was a slight increase in productivity, followed by a decline, and subsequently an insignificant change in yield. These records show that at the country level no changes in productivity of crops occurred until an improvement in variety or management, or both, occurred.

In conclusion, the effects of increased atmospheric level of CO_2 on agricultural productivity can not be simulated adequately by growing crops in glass houses or even Phytotron where the climatic condition of growing crops are changed markedly and plants do not get sufficient time to physiologically adjust themselves to altered climate which never happen during climate change in the nature. Therefore, there is uncertainty with regards to responses of crops to an increased level of CO_2.

6.2. Effect of Changes in other Climatic Parameters

In all cereals, pulses, and oilseed crops, the utilizable parts are the grain or seed. Therefore, it is essential that a plant should enter the reproductive phase if there is to be a harvest of economic yield. However, in tuber and root crops, sugarcane, sugarbeet, jute and other vegetative sink crops, the delay or absence of flowering is important to obtain economic yield. Thus the response of flowering to various factors will play a crucial role in determining the impact of climate change on crop productivity and production of a region. There is no evidence to show that increased level of CO_2 influence the time flowering. This character will be influenced significantly by temperature, radiation, water availability and the presence of other greenhouse gases.

6.2.1. Effect of Temperature

Understanding the responses of crop plants to elevated temperature is of great significance as it would be beneficial to adapt the new varieties to heat stress (Halford, 2009). Day length, temperature and their interaction regulate flowering in plants. In most instances, particularly at low and mid latitude, the increase in temperature results in reducing the total growth duration of crops by inducing early flowering and shortening the grain filling period. Several experiments at the ICARDA and elsewhere have shown that the productivity of wheat, barley and chickpea is reduced by spring planting than winter planting. A broad generalization of the effect of increasing temperature on crop duration is reduction in yield per unit area. A rise in temperature will results in decreasing crop production in low and mid latitude. However, in higher latitude such as Europe, Russia, and North America where crop duration is limited by low temperature, the crop duration could increase, resulting in improved productivity. Therefore, the aspect that requires our attention is the interaction between temperature and CO_2. Assuming that there would be some beneficial effects of CO_2, at what thermal level would these be cancelled? In case of rice an additional factor is radiation. An increase in precipitation would be coupled with reduced radiation and increased temperature which causes spikelet sterility. Thus a decrease in productivity of rice can be expected which would not be compensated by increased CO_2 level as photosynthate availability does not limit its productivity under such condition. Calculations on the effect of the interaction of temperature and 450 ppm CO_2 at low and mid-latitudes shows that the beneficial effects of CO_2 on wheat would be cancelled by a 1°C increase in temperature. In New

Delhi an increase of 3°C could cause a loss of 15-20 per cent in grain yield. IRRI, 2003 reported grain yield reduction @ 10 per cent per °C from 30-40°C. Global warming accelerates the slow sea level rise already in progress. The direct impact on agriculture could be permanent flooding of arable land, some lost by increased marine erosion and some frequently inundated due to disrupted river and tidal regimes together with greater storm and high wave incidence. Saltwater intrusion into surface water and ground water cause salinization of farm lands (UNEP, 1989; IPCC, 1990).

6.2.1. (a) Crop Productivity and Quality

In tropics, possible future global warming may result in substantial yield decrease and increased water requirement. Warming not only influences the productivity but also alters the rate of development and grain quality. Global production of annual crops will be affected by the increases in mean temperatures of 2–4°C expected towards the end of the 21st century (Wheeler *et al.*, 2000). Higher night temperature may increase dark respiration of plants, diminishing net biomass production. Higher temperatures may lead to earlier ripening of annual crops, diminishing yield per crop. Crop scientists have attempted to assess the effects of increasing temperature and high carbon dioxide on the atmosphere on the growth and yield of rice using simulation models (Boote *et al.*, 1994; horie *et al.*, 1996, 1997; Mathews *et al.*, 1997). Simulation studies have shown a decrease in duration and yield of crops as temperature increased (Aggarwal and Sinha 1993). Most information on the effects of high temperature on wheat comes from experiments that stressed the whole plant or the spikes (Ford *et al.*, 1976; Sofield *et al.*, 1977; Al-Khatib and Paulsen 1984; Bhullar and Jenner 1985; Jenner 1991; Hawker and Jenner 1993; Wardlaw and Moncur 1995).

Quality traits of Basmati rice are known to be highly influenced by temperature particularly at the time of flowering, grain filling and maturity. Both in scientists' and farmers' circle it is accepted that the aroma formation and retention in grain is enhanced at lower temperature during grain filling stage. From flowering onwards basmati require bright clear sunny days with a temperature range of 25-32°C and comparatively cooler nights (20-25°C), moderate humidity and gentle wind velocity for proper grain and aroma development (Juliano, 1972; Mann, 1987; Singh *et al.*, 2000). Meng and Zhou (1997) observed that the mean daily temperature of 18°C produced best quality rice. Head rice percentage, chalky rice percentage, alkali spreading value, grain amylose and protein contents are also markedly affected by temperature (Singh *et al.*, 2000).

In general temperature correlates negatively with amylose content (AC) (Lee *et al.*, 1996) and positively with gelatinization temperature (GT) (Resurrencion *et al.*, 1977; Li *et al.*, 1989; He *et al.*, 1990). Asaoka *et al.* (1985) reported that ambient temperature during ripening influenced the fine structure of amylopectin and amylose in rice starch apart from the effect on total AC. Dela Cruz *et al.* (1989) stated that AC decreased with increase in temperature whereas gel consistency (GC) and GT did not show any interaction with temperature. In phytotron study, Dela Cruz (1991) observed that GT of basmati 370 was unaffected at a constant day/night temperature of 33°C/25°C. Decrease in amylose content due to increase in temperature affected grain appearance as it resulted in reduced translucency of grains. Higher alkali disintegration value and grain protein content were observed at constant day/night temperature of 22°C/22°C in Japonica rice cultivar Dongjinbyeo (Lee *et al.*, 1996). He *et al.* (1990) studied the effect of high (32°C/27°C) and low (22°C/17°C) temperature in the early and late ripening stages on some physico-chemical properties of starch in outer and inner layer portion of rice grains of a Japonica rice cultivar, Koshihikari. They found that GT increased under high temperature for the inner portion of starch at early ripening stage and for the outer layer portion at late ripening stage. Similarly the fatty acid composition of the lipids of the outer layer starch responded to temperature differently from that of inner layer starch.

Li *et al.* (1989) reported that environmental factors like temperature, photoperiod and relative humidity had little effect on length (L), Breadth (B) and L/B ratio of rice grains compared with that on chalkiness, GT, AC and GC. However, effects varied between cultivars. Grain elongation is also influenced by temperature at time of ripening (Dela Cruz, 1991). Maximum grain elongation was observed at 25°C/21°C - day/night temperature during ripening. This explains differences in elongation between basmati grown in Punjab, which elongates more than the one grown in Dokri (Sind) due to high temperature (Khush *et al.*, 1979). Late planting, coinciding with flowering and maturity in cooler days has been reported to enhance the grain quality but reduction in grain yield in all aromatic rice tested (Singh *et al.*, 1993, 1995; Chandra *et al.*, 1997; Rao *et al.*, 1996 and Thakur *et al.*, 1996).

6.2.1. (b) C3 and C4 Plants

Because temperature increase enhances photorespiration in C_3 species (Long, 1991), the positive effects of CO_2 enrichment on photosynthetic productivity usually are greater when temperature rises (Bowes *et al.*, 1996; Casella *et al.*, 1996). But the integrated impact of a rise in temperature and CO_2 concentration on the yield of rice and wheat (C3 plants) may be negative (Sinha and Swaminathan, 1991). Under North Indian conditions, every 0.5°C increase in temperature resulted in a drop in the duration of wheat crop by 7 days, which in turn reduced yield by 0.45 tonnes per hectare (t/ha). If mean temperature rises by 1-2°C, the adverse impact on yield will be even greater (Sinha and Swaminathan, 1991; Van Diepen *et al.*, 1987). Similar effects were predicted for maize growing areas in central China. An average 3 per cent decrease in maize (C4 plant) yield per 1°C rise is expected (Terjung *et al.*, 1989).

6.2.1. (c) Photosynthesis

Available evidence indicates that changes in temperature of the magnitude predicted during the next century usually have little influence on the extent to which photosynthetic capacity adjusts to CO_2 (Bunce, 1992; Stirling *et al.*, 1997). A slight increase in temperature could contribute to downward regulation, however, if photosynthetic response to CO_2 is more sensitive than is growth to the rise in temperature (Dijkstra *et al.*, 1999) or if the rise in temperature reduces growth of a carbon sink, like seeds (Lin *et al.*, 1997). In both situations, limitations on plant capacity to utilize photosynthate can lead to loss of photosynthetic capacity.

When temperature is below the optimum for photosynthesis, a small increase in temperature can greatly stimulate crop growth. The converse is true when temperature is near the maximum for yield. A small increase in temperature can dramatically reduce yield (Baker and Allen, 1993, Asher *et al.*, 2008). Crop responses to expected increases in temperature also depend on interactions with CO_2 enrichment. High temperatures reduce net C gain in C_3 species by increasing photorespiration. By reducing photorespiration, CO_2 enrichment is expected to increase photosynthesis more at high than low temperature (Long, 1991), and thereby at least partially to offset negative effects of above-optimal temperatures on yield. Wardlaw (1974) observed that in wheat, photosynthesis had a broad temperature optimum from 20 to 30°C with photosynthesis declining rapidly at temperatures above 30°C.

During the vegetative stage, high day temperatures can cause damage to components of leaf photosynthesis, reducing carbon dioxide assimilation rates compared with environments having more optimal temperatures. Sensitivity of photosynthesis to heat mainly may be due to damage to components of photosystem II located in the thylakoid membranes of the chloroplast and membrane properties (Al-Khatib and Paulsen, 1999). Studies comparing responses to heat of contrasting species indicated that photosystem II of the cool season species, wheat, is more sensitive to heat than photosystem II of rice and pearl millet, which are warm season species adapted to much higher temperatures (Al-Khatib and Paulsen, 1999).

6.2.1. (d) Respiration

A rise in mean global nighttime temperatures (Horton, 1995) could enhance carbon losses from crops by stimulating shoot dark respiration (Amthor, 1997). Despite possible short-term effects of elevated CO_2 on dark respiration (Amthor, 1997; Drake *et al.*, 1997), the long-term ratio of shoot dark respiration to photosynthesis is approximately constant with respect to air temperature and CO_2 concentration (Gifford, 1995; Casella and Soussana, 1997).

6.2.2. Crop Water Use

High day temperatures can have direct damaging effects associated with hot tissue temperatures or indirect effects associated with the plant-water-deficits that can arise due to high evaporative demands. Evaporative demand exhibits near exponential increases with increases in daytime temperatures and can result in high transpiration rates and low plant water potentials (Hall, 2001).

Although increased productivity from increased WUE is the major response to elevated CO_2 in a C_3 or C_4 crop that is exposed frequently to water stress (Idso and Idso, 1994; Ham *et al.*, 1995; Drake *et al.*, 1997), changes in climatic factors (temperature, rainfall) may interact with elevated CO_2 to alter soil water status, which in turn will influence hydrology and nutrient relations. With rice, at the optimal temperature for growth, a doubling of CO_2 increases crop water use efficiency (WUE) by about 50 per cent. However, this increase in WUE declines sharply as temperature increases beyond the optimum (Horie *et al.*, 2000).

6.2.3. Plant Development

When grown near their optimal temperature, plants are more likely to reach maximum yields. However, because of environmental fluctuations, temperatures are often higher than optimum, thus increasing the probability of the grain being exposed to extended periods of supra-optimal temperatures. In maize such temperatures are detrimental for kernel growth (Badu-Apraku *et al.*, 1983; Cheikh and Jones, 1994 and 1995; Commuri and Jones, 1999; Commuri and Jones, 2001; Duke and Doehlert, 1996; Engelen-Eigles *et al.*, 2000; Jones *et al.*, 1985; Singletary *et al.*, 1994). Heat stress applied during the cell division stage of maize kernel development had a detrimental effect on maize kernel growth.

Extreme temperatures can cause premature death of plants. Among the cool-season annuals, pea is very sensitive to high day temperatures with death of the plant occurring when air temperatures exceed about 35°C for sufficient duration, whereas barley is very heat tolerant, especially during grain filling. For warm season annuals, cowpea can produce substantial biomass when growing in one of the hottest crop production environments on earth (maximum day-time air temperatures in a weather station shelter of about 50°C); although it's vegetative development may exhibit abnormalities such as leaf fasciations. For monocotyledons, including cool-season and warm-season annuals, high daytime temperatures can cause leaf firing which involves necrosis of the leaf tips and this symptom also can be caused by drought.

6.2.4. Crop Yield

Temperature effects on yield are complex. Crop responses to a change in temperature depend on the temperature optima of photosynthesis, growth, and yield, all of which may differ (Conroy *et al.*, 1994). Heat stress negatively affected grain yield and its components (Tahir and Nakata, 2005). Significant differences were found among rice genotypes in percentage reduction in grain yield, grain weight, grain filling duration, spikelet sterility and harvest index because of heat stress (Singh 2000; Tahir and Nakata, 2005). Peng *et al.* (2005) reported that reduction in grain yield of rice was mainly due to enhanced night (minimum) temperature rather than mean and day (maximum) temperature.

Grain yield of wheat and other cool season cereals is reduced greatly by stress from high temperatures in many regions (Paulsen 1994; Wardlaw and Wrigley 1994). The major phenological phases in wheat include sowing to seedling emergence, seedling emergence to anthesis and anthesis to maturity, the duration of the phases being controlled by day length and temperature (Robertson, 1968). The development rate correlates positively with temperature between a base and an optimum temperature, which is specific. In the moderately high temperature range, wheat yield declines approximately by 3-4 per cent for each 1°C rise in average temperature above 15°C under controlled as well as field conditions (Wardlaw and Wrigley, 1994). In wheat experiments under controlled conditions from 25 to 35°C, mean grain weight declined by 16 per cent for each 5°C increase in temperature (Asana and Williams, 1965). In pot experiments, grain yield decreased by 17 per cent for each 5°C rise (Wattal, 1965). For every 1°C rise in temperature, there was a depression in grain yield by 8 to 10 per cent, mediated through 5 to 6 per cent fewer grains and 3 to 4 per cent smaller grain weight.

Nearly 85 per cent of the kernel mass of wheat is starch from current assimilation during maturation (Jenner *et al.*, 1991), and reduced yield from high temperature is due mainly to diminished deposition of starch in the grain (Bhullar and Jenner, 1985). The supply of sucrose substrate in the endosperm is altered little or increased by high temperature (Ford *et al.*, 1976; Bhullar and Jenner 1986; Jenner 1991), but differential sensitivity of enzymes in the pathway of starch synthesis suggests that the stress affects growth of grain directly (Hawker and Jenner 1993; Keeling *et al.*, 1993). UDP glucose pyrophosphorylase and ADP glucose pyrophosphorylase in wheat grain characteristically increase up to 40°C (Keeling *et al.*1993). Soluble starch synthase, in contrast, has a low temperature optimum, is susceptible to heat inactivation, and limits starch synthesis (Hawker and Jenner 1993; Keeling *et al.*, 1993). The Q_{10} (change in activity per 10°C change in temperature) of soluble starch synthase in isolated endosperms declines above 20°C, which also suggests that enzyme-regulated starch synthesis and its activity depends on grain temperature and duration of exposure (Keeling *et al.*, 1994; Jenner *et al.*, 1995).

Low light coupled with high temperature (38°C) and humidity (RH 94 per cent) induced complete spikelet sterility, enhanced foliage growth and impaired the biological yield (Singh, 2000). Supra-optimal temperature has been shown to be harmful for rice spikelet fertilization (Satake and Yoshida, 1978; Matsui *et al.*, 1997). A 1°C change in mean temperature could result in a change of 17 to 25 days in effective growing season. Modest changes of 0.5°C and 10 per cent precipitation could alter world rice production by ± 10 per cent (Stansel, 1980). When rice is exposed to air temperatures higher than 35°C, heat injuries occur. Clear varietal differences are observed for high temperature injuries at different growth stages. The rice plant appears to be most sensitive to high temperature at flowering. High temperature during anthesis induces high percentages of spikelet sterility (Yoshida *et al.*, 1981). High temperature reduced days to flowering, days to maturity and panicle weight, while increased tiller production (Horie *et al.*, 2000). In the phytotron, temperatures above 33/25°C reduced panicle and grain weight, increased the shoot dry matter, and shoot : root ratio. The fraction of unfilled spikelets was directly related to temperature hence, high temperature reduced grain yield. Nitrogen and chlorophyll contents of all genotypes increased with increasing temperature above 33/25°C (Egeh, 1991). The yield decline at higher temperatures was primarily due to an increase in the number of sterile spikelets and the spikelet sterility was most closely related to the daily maximum temperature averaged over the flowering period (Kim *et al.*, 1996). In glasshouse pot experiments, transferring rice plants (cv. Calrose) growing at 27/22°C to a range of temperatures from 24/19 to 30/25°C 7 days after heading resulted in little variation in kernel size, but above this temperature range there was a significant drop in kernel dry weight (Tashiro and Wardlaw, 1991).

Root temperature is also important especially in case of cereal crops. The upper soil profile, where roots of most annual species are concentrated, reaches nearly the same temperature as the air (Neilsen 1974; McMichael and Burke 1996). Roots are more sensitive to high temperature than shoots of most species, including barley (Power *et al.*, 1963), maize (Allmaras *et al.*, 1964), and wheat (Kuroyanagi and Paulsen, 1988). High root temperature during maturation accelerates senescence of aerial parts, and decreases the grain yield of wheat regardless of the shoot temperature, and roots appear to control whole plant responses to the stress (Kuroyanagi and Paulsen, 1988). Other studies have found that high shoot temperature affects starch accumulation in maize when root temperature is constant (Wilhelm *et al.*, 1999).

Reproductive development of many crop species is damaged by heat such that they produce no flowers or if they produce flowers they may set no fruit or seeds. The reviews of Hall (1992, 1993) discuss the detrimental effects of heat stress on reproductive development that has been reported for cowpea, common bean, tomato, cotton, rice, wheat, maize and sorghum. Controlled-environment studies in which cowpea plants were subjected to separately controlled root and shoot and day and night temperatures demonstrated that pod set (the proportion of flowers producing pods) was damaged by moderately high night temperature of the shoot (Warrag and Hall, 1984 a, b). It was surprising that night temperature would have this effect since much hotter day temperatures did not damage pod set of cowpea. Reciprocal artificial pollinations between plants grown under high and optimal night temperatures indicated the low pod set was caused by male sterility and that the pistils did not appear to be damaged by high night temperature. The detrimental effects of high night temperature on pod set also were shown to occur in field conditions (Nielsen and Hall, 1985b). These experiments a unique experimental approach was used in which plots of cowpea plants were subjected to different increments of higher night temperatures during early stages of flowering using enclosure systems placed over the plots only during the night-time (Nielsen and Hall 1985).

In diverse crops, grain yield is affected by air temperature during the grain-filling period (GFP) (Yoshida, 1981; Evans, 1996; Egli 1998). In most cases, higher temperatures during GFP increase the rate of grain dry matter increase but shorten the duration of the GFP. This results in reduction of mean grain weight or in the percentage of ripened grains. This effect has been noted in rice (Sasaki, 1935; Nagato and Ebata, 1965; Sato and Inaba, 1976; Yoshida, 1981; Tashiro and Wardlaw, 1989), wheat (Wardlaw *et al.*, 1980; Bhullar and Jenner, 1985; Tashiro and Wardlaw, 1989; Tashiro and Wardlaw, 1990; Guedira and Paulsen, 2002), barley (Wallwork *et al.*, 1998), maize (Jones *et al.*, 1981), and soybean (Egli and Wardlaw, 1980; Gibson and Mullins, 1996). Lower temperature reduces the rate of dry matter increase but extends duration of the grain filling, and delays grain maturation although moderate cool temperatures sometimes benefit grain yield (Yoshida, 1981; Nishiyama, 1985; Egli, 1998). The temperature range that affects the grain dry matter increase rate and grain filling duration differs between plant species (Egli, 1998; Tashiro and Wardlaw, 1989).

When grain growth was restricted by high temperature, the main reason for termination of GIR in rice (Inaba and Sato, 1976), wheat (Hawker and Jenner, 1993), and barley (Wallwork *et al.*, 1998) was thought to result from disappearance of enzyme activity relating to starch synthesis of the grains. These results suggest that potential GIR (PGIR) in grains is controlled by decreased metabolic activity under high-temperature conditions, and hence assimilate supply to the grain is only part of the cause of low grain weight.

6.2.5. Biochemical Composition of Plant

Temperature is one of the most important environmental factors governing plant growth and development. Heat stress is a major factor limiting crop yield in many agricultural regions. Temperature affects virtually all plant processes. High temperature caused marked increase in the content of chlorophyll a and b, and organic nitrogen, but reduced the level of sugars, starch, RNA, and hemicellulose in rice genotypes (Singh 2000). Rice cultivars showed differential biochemical response to high thermal stress. In general, level of cellulose in the shoots changed to varying extant in rice cultivars subjected to high temperature. During the early stage of kernel development, heat stress is particularly detrimental to subsequent dry matter accumulation since it causes disruption of cell division, sugar metabolism, and starch biosynthesis in the endosperm (Monjardino *et al.*, 2005). Heat stress reduced N remobilization and increased total non-structural carbohydrate (TNC) remobilization (Tahir and Nakata, 2005). Changes in protein accumulation in cereals by heat stress were reported by several workers (Bhullar and Jenner, 1985; Stone and Nicolas, 1998a, 1998b; Wilhelm *et al.*, 1999).

6.2.6. Partitioning of Organic Nutrients

High temperature during the grain-filling period of wheat shows many negative effects such as inhibition of photosynthetic process and acceleration of leaf senescence (Al-Khatib and Paulsen 1990; Farooq *et al.*, 2011). As a result, a great reduction occurs in nutrient supply for the developing grain from current assimilation. Stem reserves potentially offer a powerful resource for grain filling under any type of stress that impairs current assimilation, including heat stress (Schnyder, 1993; Blum *et al.*, 1994; Blum, 1998; Fokar *et al.*, 1998) and water deficit (Davidson and Chevalier, 1992; Palta *et al.*, 1994; Yang *et al.*, 2000).

Attempts have been made to explore the potential of remobilized stem reserves in supporting grain filling under heat-stress condition (Blum *et al.*, 1994; Fokar *et al.*, 1998; Yang *et al.*, 2002). Stem reserve remobilization may be considered as a useful criterion in breeding varieties for heat-stress environments.

Tillers are important contributors to final grain yield under heat-stress conditions, but it is not clear whether they respond similarly to the main stem concerning utilization of the stem reserves under heat stress. Heat stress accelerates the loss of chlorophyll content and senescence of the leaves. Stay green character and chlorophyll retention in the leaves under heat-stress conditions were considered as expression of heat tolerance (Fokar *et al.*, 1998). However, it has been suggested that the high potential capacity to utilize stem reserves for grain filling might be linked with accelerated leaf senescence (Blum *et al.*, 1994; Fokar *et al.*, 1998; Yang *et al.*, 2000). In addition, it is not clear whether accelerated leaf senescence and loss of chlorophyll from the leaves are also linked with faster N remobilization from the stem.

Research on the effects of brief periods of ear warming after anthesis on ear metabolism have identified differential responses of starch and nitrogen accumulation in grain of four wheat cultivars (Bhullar and Jenner, 1983; 1985; 1986; Hawker and Jenner, 1993; Jenner, 1991 a, b). Warming increased the rate of dry matter accumulation in grain of all the cultivars. Rate of increase in nitrogen accumulation was, however, higher than the increase in total dry matter accumulation. Under long-term exposure to heat stress, increased grain nitrogen concentration is almost entirely as a result of decreased starch content rather than a change in total grain quality (Bhullar and Jenner, 1985). The conversion of sucrose to starch within the endosperm is decreased by elevated temperatures. Furthermore, heat stress effects on final grain weight were associated with reduced levels of soluble starch synthetase activity (Hawker and Jenner, 1993).

In wheat the rate of assimilate movement out of the flag leaf and phloem loading was optimum around 30°C; the rate of assimilate movement through the stem was independent of temperature from 1 to 50°C. Thus, in wheat, temperature effects on translocation result indirectly from direct temperature effects on source and sink activities (Wardlaw, 1974).

6.2.6. (a) Nitrogen

The role of N stored in the stem prior to anthesis has poorly been investigated under heat stress. Over 80 per cent of the total N content accumulates at anthesis accounts for about 50-100 per cent of the total N content of the wheat grains (Daigger *et al.*, 1976; Simpson *et al.*, 1983; Heitholt *et al.*, 1990; Papakosta and Gagianas, 1991; Gebbing and Schnyder, 1999). However, N mobilization, although under genetic control, depends on the environmental conditions (Halloran, 1981; Simmons, 1987; Van Sanford and Mackown, 1987).

Palta *et al.* (1994) found that despite the reduction in post-anthesis uptake of N, grain N accumulation was not affected because remobilization of N to the grain was increased. However, not many attempts were made to study how heat stress affects N remobilization from the stem. When wheat was grown under heat stress conditions during grain filling, pre anthesis stored nitrogen (N) could serve as an alternative source of assimilates (Tahir and Nakata, 2005). The N concentrations of the stem at five day after anthesis (5 DAA) were significantly different among genotypes. N remobilization efficiency across treatments significantly correlated with grain yield, grain weight, harvest index and grain filling duration. Heat stress significantly reduced the N remobilization efficiency of most of the genotypes. The rate of chlorophyll loss from flag leaf positively correlated with N remobilization efficiencies under heat stress suggesting a link between leaf senescence and remobilization efficiency. Thus heat stress negatively affected N remobilization (Tahir and Nakata, 2005).

6.2.6. (b) Carbohydrates

The extent of damage caused by heat stress depends on the time of exposure in relation to the stage of kernel development (Gibson and Paulsen, 1999). By exposing maize kernels to heat stress during the lag phase starting at 5 days after planting (DAP), cell division and ultimately the number of endosperm cells and starch granules are severely reduced (Commuri and Jones, 1999; Engelen-Eigles *et al.*, 2000), which is also associated with a reduction in cytokinin levels (Cheikh and Jones, 1994). Starch metabolism is particularly repressed, which appears to be due to a reduction of the activity of ADP glucose pyrophosphorylase and soluble starch synthase (Commuri, 1997;Duke and Doehlert, 1996;Keeling *et al.*, 1994; Singletary *et al.*, 1994) and Wilhelm *et al.* (1999) showed that when maize kernels are exposed to heat stress at later developmental stages (*e.g.*, the linear fill period), there is a significant repression in starch biosynthesis because of the reduction in the activity of these two enzymes.

Studies conducted with in vitro-grown maize kernels (Cheikh and Jones, 1994, 1995; Commuri and Jones, 1999) and field-grown kernels of several maize inbred and hybrids (Commuri and Jones, 2001) have shown that 4 day heat stress (DHS) at 35°C from 5 days after pollination (DAP) caused a significant repression in starch biosynthesis, and affected its accumulation (Cheikh and Jones, 1995; Commuri and Jones, 2001).

The embryo and endosperm were particularly sensitive to heat stress (Cheikh and Jones, 1995; Commuri and Jones, 1999; Engelen-Eigles *et al.*, 2000). The 2-DHS kernels were either smaller than control or had a normal phenotype, whereas the 4-DHS kernels had ovary cavities that were only partially filled with endosperm and the embryos were of a reduced size. However, the pericarp and the pedicel (maternal tissue) were less affected by heat stress. An increase in kernel pericarp thickness has been previously reported (Cheikh and Jones, 1995; Commuri and Jones, 1999), which, when combined with the smaller size kernels, resulted in pericarp weights similar to control values at 32 DAP.

Under heat stress conditions during grain filling in wheat, pre anthesis stored total non-structural carbohydrates (TNC) can serve as alternative source of assimilates (Tahir and Nakata, 2005). In wheat, heat stress significantly increased TNC remobilization efficiency and significant variation were observed among genotypes (Tahir and Nakata, 2005). TNC concentrations of the stem at five day after anthesis (5DAA) were significantly different among genotypes. TNC at 5DAA negatively correlated with N at 5DAA and harvest index, but the TNC remobilization efficiency under heat stress positively correlated with main stem grain yield, grain weight and harvest index. The rate of chlorophyll loss from flag leaf positively correlated with TNC remobilization efficiencies under heat stress. Heat stress increased TNC remobilization because of the increasing demand for resources (Tahir and Nakata, 2005)

6.2.6. (c) Proteins

Previous studies have reported protein accumulation in cereals to be slightly affected by heat stress (Bhullar and Jenner, 1985; Stone and Nicolas, 1998a, 1998b; Wilhelm *et al.*, 1999). Maize protein accumulation seems less susceptible to heat stress (Bhullar and Jenner, 1985; Wilhelm *et al.*, 1999) and water stress (Ober *et al.*, 1991) than does starch metabolism. In some cases, it has been reported that protein concentration is positively correlated with high temperature during cereal grain growth (Campbell and Davidson, 1979; Campbell *et al.*, 1981; Kolderup, 1975; Stone and Nicolas, 1998a, 1998b). However, considering that starch constitutes approximately 80 per cent of the endosperm mass and that its accumulation is severely repressed by heat stress, the percentage protein increase on a dry weight basis may be simply a reflection of the proportionately lower starch weight (Bhullar and Jenner,

1985; Stone and Nicolas, 1998a; Wilhelm *et al.*, 1999). Therefore, protein content (mass per kernel) rather than protein concentration may provide a more biologically relevant indication of how protein accumulation is affected by heat stress.

6.2.6. (d) Cell Membrane Thermostability

The plasmalemma and membranes of cell organelles play a vital role in the functioning of cells. Any adverse effect of temperature stress on the membranes leads to disruption of cellular activity or death. Heat stress may be an oxidative stress (Lee *et al.*, 1983). Peroxidation of membrane lipids has been observed at high temperatures (Mishra and Singhal, 1992; Upadhyaya *et al.*, 1990), which is a symptom of cellular injury. Heat injury to the plasmalemma may be measured by ion leakage (Chaisompongpan *et al.*, 1990; Hall, 1993). Membrane thermostability can be assessed using electrolyte leakage from leaf disks subjected to extreme temperatures (Blum, 1988). More stable membranes exhibit slower electrolyte leakage. Injury to membranes from a sudden heat stress event may result from either denaturation of the membrane proteins or from melting of membrane lipids which leads to membrane rupture and loss of cellular contents (Ahrens and Ingram, 1988). Enhanced synthesis of an anti-oxidant by plant tissues may increase cell tolerance to heat (Upadhyaya *et al.*, 1990, 1991) but no such anti-oxidant has been positively identified.

6.3. Effect of Soil Water Availability

Reduced water availability in soil below a threshold reduces productivity of almost all the crops. An excess of water may not necessarily benefit crop production, but it may sometimes cause crop losses. Therefore, the occurrence of unpredictable drought has caused concern around the world. In the general circulation model (GCM) the simulation of precipitation is not satisfactory. Nevertheless, it is clear that any reduction in precipitation will not only influence rainy season crops, but also cause shortage for irrigation. In this respect the runoff and water collection could differ very significantly in low latitude in comparison with the mid latitude and high latitudes. In fact there are hardly ant studies which explained the relationship between rainfall, groundwater storage for irrigation and crop production in low and mid-latitudes. Such studies are urgently needed in low latitude because of the nature of precipitation in these areas. Further an increase in temperature by even 1-3°C would cause enhanced evapotranspiration. Depending upon the changes in precipitation and evapotranspiration of certain crops such as rice, the water table is likely to be lowered. Such events have occurred in Haryana and Punjab during two or more consecutive years. Such problems require detailed studies in low and mid latitude to formulate strategies for adapting to climate change.

6.4. Effect of Drought, Heat Stress and other Extremes

Probably most important for agriculture, but about which least is known, are the possible changes in climatic extremes such as the magnitude and frequency of drought, storms, heat waves, severe frost, hailstorm, and other disasters as evident from super cyclone in Orissa, and the super Tsunami which took lives of several people and loss of property, vegetation and coastal salinization (Ring *et al.*, 1989). Some modeling evidence suggests that hurricane will increase with global warming

(Emanuel, 1987). This has important implications for agriculture in low latitude particularly in coastal regions. Since crop productivity often exhibit a non-linear response to heat or cold stress, changes in the probability of extreme temperature events can be significant in the standard deviation of temperature maxima and minima, we should note that the frequency of hot and cold days can be markedly altered by changes in mean monthly temperature. It is presumed that under $2 \times CO_2$ climate scenario the number of days in which temperature would fall below freezing would decrease from a current average of 39 to 20 in Atalanta, Georgia (USA), while the number of above 32°C would increase from 17 to 53 (EPA, 1989). The frequency and extent of area over which losses of agricultural output could result from heat stress, particularly in tropical regions is therefore likely to increase significantly. Unfortunately, no studies have yet been made on such aspect. However, the apparently small increase in mean annual temperature in tropical regions (1-2°C in double CO_2 level) could sufficiently increase heat stress on temperate crops such as wheat so that these are no longer suited to such areas. Important wheat producing areas such as North India could be affected severely in this way (Aggarwal *et al.*, 1995; Parry and Duinker, 1990), whereas, the central zones which are on the verge of Duram wheat cultivation may be completely unfit for the wheat, unless an efforts are made to develop the high thermal tolerant cultivars or the cultivars which shift their optimal thermal range.

An important additional effect, especially in temperate mid-latitudes, is likely to be the reduction of winter chilling (vernalization). Many temperate crops such as apple, plum, cherry, peach and cabbage require a period of low temperature in winter to either initiate or accelerate the flowering process. Low vernalization results in low flower bud initiation and ultimately reduced yield and their spatial shifting from their traditional zones of cultivation to new and non-traditional areas used to be unfit for cultivation due to prevailing low thermal constraint. A 1°C increase in thermal level would reduce effective winter chilling to the extent of 10-30 per cent, thus contributing to a poleward shift of such temperate crops (Salinger, 1989). There is also distinct possibility that as a result of high rate of evaporation, some regions in the tropics and sub-tropics could be characterized by a higher frequency of drought or a similar frequency of more intense drought than at present. Current uncertainties about how regional pattern of rainfall will alter mean that no useful prediction of this can at present be made. However, it is clear in some regions that relatively small decrease in water availability can readily produce drought conditions. In India for example, lower than average rainfall in 1987 reduced food grain production from 152 to 134 million ton, lowering food buffer stock from 23to 9 million tons. Changes in the risk and intensity of drought, especially in currently drought prone regions, represent potentially the most serious impact of climate change on agriculture at global, national and regional levels.

6.5. Effect of Radiation Change

The relationship between radiation and photosynthetic rate has been studied for several crop plants. Depending upon the nature of plant species and other reactants the saturation limits for radiation varies markedly. However, there are several studies on wheat and other crops, which show linear relationship between radiation and photosynthesis within certain range. The effects of reduced radiation with increasing temperature are severe on spikelet sterility and hence on grain yield of rice. The yield of rice varies considerably in monsoon season even in experimental plots because of variation in total radiation. It is thus possible that the productivity of rice would be influenced adversely by increased temperature and reduced radiation, which however may be compensated by increased CO_2 concentration.

7. Critical Types of Climatic Change

7.1. Climatic Extreme

It is not yet clear whether variability of temperature will occur as a result of climate change. However, even if variability remains consistent, an increase in average temperature would result in increased frequency of temperature above particular threshold. Changes in frequency and distribution of precipitation are less predictable, but the combination of elevated thermal level and drought or flood probably constitutes the greatest risk to agriculture in many regions from global climate change (Parry and Duinker, 1990). Now it is clearly evident that climate during the last decade has become very erratic and unpredictable. Monsoon has delayed to some extent and more often there has been pre-monsoon rain. Earlier the east west monsoon used to come at right time but since last decade it has become unprecedented. The year 2004-05 has become highly variable and erratic in respect of thermal regime and precipitation, which has caused several changes in the biological systems. For the last few years there have been changes in the pattern of chilling and thermal level. The regions of low latitude such as eastern U.P. which used to remain slightly warmer than north India earlier, recorded slightly lower thermal regime during the winter as compared to north parts of India *e.g.* Delhi. Due to extremely low temperature during winter in 2003 in Bihar, the Rabi maize crop suffered severely from hypothermal injury, which led to cob sterility and poor grain setting in the same. Similarly the year 1998 was designated as the hottest year.

7.2. Warming in High Latitudes

There is relatively strong agreement on the basis of GCM prediction that greenhouse gas warming (greenhouse effect) will be greater at higher latitudes (IPCC, 1990; 1995), which will reduce hypo-thermal constraint on high latitude agriculture, increase the competition for land there and result in the northward retreat of the southern margin of the boreal forest and expansion of arable lands in polar regions (Parry and Duinker, 1990). However, warming at low latitudes, although less pronounced, is also likely to have a significant impact on agriculture such as widening of tropical zones of agriculture and cropping pattern and agro-biodiversity.

7.3. Polar Advancement of Monsoon Rain

In a warmer world the inter-tropical convergence zones would be likely to advance further pole ward as a result of an enhanced ocean-continent pressure gradient. If this were to occur then total rainfall could increase in some regions of monsoon Africa, monsoon Asia and Australia, though there is currently little agreement on which regions these might occur (IPCC, 1990; 1995). Rainfall could be more intense and frequent, so there is greater possibility of flooding and erosion as it evident from the occurrence of regular flooding and soil erosion in north parts of Bihar mainly due to heavy rainfall in Nepal during monsoon. The often flooding in Brahmaputra river during monsoon also indicates the possibility of greater rainfall in Himalaya regions.

7.4. Reduced Soil Water Availability

Probably the most severe consequence of climatic change on agriculture would result from higher potential evapotranspiration, particularly due to higher atmospheric and edaphic thermal regimes. Even in the tropical regions, where temperature increase are likely to be smaller than elsewhere and where precipitation might increase, the increased rate of moisture loss from plants and soil through evapotranspiration would be substantial (Rind *et al.*, 1989; Parry, 1990). It may be somewhat reduced by greater air humidity and increased cloudiness during the rainy seasons but could be pronounced during dry seasons. There could be three major ways in which increase in greenhouse gases may be important for agriculture. Firstly, increased atmospheric CO_2 level can have direct effect on crop growth rate by enhanced photosynthetic rate of crops especially C3 crops and indirectly through crop-weed competition. Secondly, GHG induced changes of climate may alter the thermal levels, rainfall and radiation which can affect crop and animal productivity. Finally, rise in sea level may lead to loss of farmland by inundation and to increasing salinity of surface and groundwater in coastal regions.

8. Potential Impacts of Climate Change

8.1. Changes in Thermal Limits to Agriculture

Increased in temperature can be expected to lengthen the growing season in areas where agricultural potential is currently limited by the insufficient warmth, resulting in a pole ward shift of thermal limits of agriculture. The consequent extension of potential will be most pronounced in northern hemisphere because of the greater extent here of temperate agriculture at higher latitudes. There may however, be important regional variation in our ability to exploit this shift. For example, the greater potential for exploitation of northern soils in Siberia than on the Canadian Shield may mean relatively greater increases in potential in northern Asia than in northern N America (Parry, 1990; IPCC, 1995). A number of estimations have been made concerning the northward shift in productive potential in mid latitude northern hemisphere countries. These relate to changes in the climatic limits for specific crops under a variety of climatic scenarios, and are therefore not readily compatible (William and Oakes, 1978; Newman, 1980; Blasing and Solomon, 1983; Rosenzweig, 1985; Parry and Carter, 1988; Parry *et al.*, 1989; IPCC, 1995). They suggest, however, that 1°C increase in mean annual temperature would tend to advance the thermal limit of cereal cropping in the mid latitude northern hemisphere by about 150-200 km, and to raise the altitudinal limit to arable agriculture by about 15-200 m. One such study, a logical development of those considered above, has mapped the shift of growing areas of different cultivars or types of the same crop that might occur under altered climate (Rosenzweig, 1985). Wheat growing regions in North America were characterized according to their present day temperature and rainfall regimes, and then re-mapped for the equilibrium climate based on a 2 x CO_2 requirement with the Goddard Institute of Space Studies GCM. Results indicated a substantial northward extension of winter wheat into Canada from its current location on the US Great Plains, a switch from hard to soft wheat in the Pacific Northwest due to increased precipitation, and an expansion of areas in autumn-sown spring wheat in the southern USA due to higher winter temperature. In Mexico, wheat-growing regions would remain the same but greater high-temperature stress may occur.

A similar magnitude of shift of cropping limits has been estimated for Europe. In this region the major climatic determinant of successful ripening of grain maize is the warmth of the growing season. An effective temperature sum (SET) of 850 degree days above a base temperature of 10°C correspond closely with the actual limit of its cultivation today (Carter *et al.*, 1991). This boundary extends from the southwestern margin of England through northern central Europe and central Russia to just south of Moscow. Much of the fertile north European plain is therefore currently too cool for grain maize to mature in all but the warmest years. However, under the 2 x CO_2 equilibrium climates projected by a number of GCM experiments this limit is displaced 200-350 km further north. The entire northern European plains is estimated to be within the limit of grain maize under 2x CO_2 climate, particularly the western part of northern Europe (UK, northern Germany, Denmark) where maritime influence creates a greater sensitivity to warming because greater CO_2 induced temperature increases are expected in winter than in summer. It is worth noting, however, that there is very little agreement between model estimates of precipitation, which can be a critical factor for many crops in Europe and is also important for maize. Consequently, we are at present only able to draw a very imperfect picture of how potential growing regions may shift. Outside North America and Europe, little study has been conducted of the spatial shift of crop potential. One exception is Japan where an estimate has been made of the extension of area in which rice could safely be cultivated without severe risk of crop loss due to frost. On the island of Hokkaido in the north of Japan, the safety cultivable area of irrigated rice is estimated to more than double under the GISS 2 X CO_2 climate, assuming there remain adequate precipitation and the crop is fully irrigated (Yoshino *et al.*, 1988). Corresponding pole ward shift of crops in southern hemisphere, for example for cereals, fruit and vegetable in New Zealand have also been estimated (Salinger *et al.*, 1990). Once more, however, it must be emphasized that these are estimation only of altered potential.

Since it is important to consider those changes of climate that may occur within next 30 years as well as those within the 21st century, more recent impact assessment have begun to evaluate potential effect of time dependent changes in temperature (Carter *et al.*, 1991). Under the present scenario of climate change due to increasing the rate of greenhouse gas emission @ 1.5 per cent per annum, it is indicated that the rate of northward shift of the grain maize limit could approximate 150 km per decade between 1990-2030, and perhaps 240 km per decade from 2030-60. Broadly similar rate of shift are anticipated for several other crops throughout the middle and high latitudes, and it remain to be seen whether rates of adaptation in agriculture can match them. In the UK for example, the effect of warming suggest a pole ward shift of limits for grain maize and silage maize by about 300 km per °C increase in mean annual temperature (Parry *et al.*, 1989). Under a current scenario of GHG emission rate and global warming, the thermal regime is likely to be increased by 0.5°C by the year 2000-2010, about 1.5°C BY 2020-50 and 4.5°C by the end of 2100 (IPCC,1995). This suggests a rate of shift of about 100-150 km per decade. If the emissions are reduced to the level of one third of global warming, the pole ward shift may be reduced to 50-100 km per decade. Tolerable rates of shift in climate resources can thus be used as a guide to target rates of tolerable climatic change. The shift of crop potential described above are examples of world wide relocation of climatic zones that could occur as a result of CO_2 induced changes of climate, particularly a pole ward shift of thermal zones. The GISS 2 X CO_2 climate model serves to illustrate not only the magnitude of possible changes in agricultural potential, but also the adaptive responses likely to be required to re-tune agriculture to altered climatic resources (Parry and Carter, 1988). For instance, perhaps the combination of barley growing and cattle rearing and their fattening, which are successful enterprises in northern Europe today, would be appropriate for Iceland in the future. Due to difference in latitude, however, there are important differences in day length between such regions and thus the prediction is far from fact. Similar situation may happen regarding the spatial shift in apple, cherry, plum, peach and cabbage production in northern parts of India (H.P., Kashmir and A.P.). If the thermal regime increase by 1-2°C in these areas especially the lower traditional zones, the cultivation of these crops, which require low and narrow optimal thermal range for flowering and fruiting, is likely to be affected because of unsuitability of these traditional zone for their cultivation in future and it is expected that there may be upward shift of these crops in order to provide them optimum thermal level. Similarly the seed production of cabbage and off season crops production in lower zones of these states are likely to be affected in years to come if there is continuous increase in thermal level in these areas.

8.2. Shift of Moisture Limits to Agriculture

There is much less agreement between GCM-based projection concerning GHG induced changes in precipitation than that of temperature not only regarding the extent of changes, but also of spatial pattern and distribution throughout the year. For this reason it is difficult to identify potential shift in the moisture limits to agriculture. This is particularly so because relatively small changes in seasonal distribution of rainfall can have disproportionately large effects on the viability of agriculture in tropics, largely through changes in growing period when moisture is sufficient and thus through the timing of critical activity such as planting etc. However, recent reports of IPCC have made a preliminary identification of those regions where there is some agreement amongst 2 x CO_2 experiments with GCM concerning an overall reduction in crop water availability (Parry, 1990; Parry and Duinker, 1990). It is very difficult to combat the impacts of thermal changes on agriculture as compared to changes in water regimes of soil, as it could be managed by irrigating the crops in its scarcity. Since the impacts of climate change in respect of rainfall could be negated with water management, it is very difficult to assess its impact on agriculture. In India, the western arid zones of Rajasthan, which receive very less level of rainfall (<250 mm annually) and suffer from drought every year have been receiving greater extent of rainfall since few last years and it has become now possible to grow the crops successfully during the monsoon.

8.3. Effect on Crop Yield

Whether crops respond to hyper thermal stress with an increase or decrease in yield depends on whether they are determinate or indeterminate, and whether their yield is currently strongly limited insufficient warming. In cold regions very near the present-day limit to agriculture any thermal increase even as much as the 7-9°C indicated for higher latitude under a doubling of CO_2 can be expected to enhance yield of cereals. For example, near the current northern limit of spring wheat production in the European regions of the USSR yields are estimated to increase about 3 per cent per°C, assuming no concurrent change in rainfall. In Finland, the marketable yield of barley increases 3-5 per cent per °C and in Island hay yield increase about 15 per cent per°C (Kettunen *et al.*, 1988; Bergthorssen *et al.*, 1988). Distant from currently low temperature-constrained regions of farming and in core areas of present day cereal production such as in Corn Belt of North America, the European lowland and northern Asia, increase in temperature would probably lead to decreased cereal yield due to shortened period crop growth and development. In eastern England for example a 3°C rise in mean annual temperature is estimated to reduce winter wheat productivity by about 10 per cent although the direct effect of a doubling of ambient CO_2 might more than compensate this negative impact. At other mid-latitude regions of Asia much would depend on possible changes in rainfall. The increase of 1°C temperature in these areas during the crop growing season may cause small extent of reduction in crop yield, though they could increase or decrease substantially if it is accompanied by an increase or decrease of rainfall.

The yield of temperate root and tuber crops such as sugar beet and potato with an indeterminate growth habit can be expected to see an increase in yield with increasing temperature provided these do not exceed their optimum thermal level for crop development (Squire and Unsworth, 1988). However, increase in thermal limit beyond their optimal thermal level would certainly reduce bulking of root and tubers in these crops.

Changes of temperature would also have an effect on moisture available to crop growth, whether or not level of rainfall remained unchanged. In general and at mid latitude, evaporation increases by about 5 per cent for each °C of mean annual temperature. Thus, if mean temperature were to increase in the east of England by 2°C, potential evaporation would increase by about 9 per cent assuming no change in rainfall. The effect of this would be small in the early phase of crop growth and development, but after mid-July the soil moisture deficit would considerably larger than at present and for some crops, this implies substantially increased demand for irrigation (Rowntree *et al.*, 1989). In most of the tropical and equatorial regions of the world, and across large areas outside the tropics, the yield of agricultural crops is limited more by the amount of water received by and stored in the soil than by air temperature. Even in the high mid-latitudes such as in southern Scandinavia too little rain can restrict growth of cereal crops during the summer when evaporation exceeds rainfall. In all these areas the amount of biomass a crop produces is roughly proportional to the amount of water it transpires (Monteith, 1981). This in turn is affected by the quantity of rainfall but not in a straightforward manner. It also depends on how much of the rainfall is retained in the soil, how much is lost through evaporation from the soil surface and how much remains in the soil which cannot be extracted by the crops.

Relatively few studies have been made of the combined effect of possible changes in temperature and rainfall on crop yield and those of that are based on a variety of different methods. However, a review of results from about ten studies in North America and Europe noted that warming is generally detrimental to yield of wheat and maize in these mid-latitude core cropping regions. With no change in precipitation or radiation slight warming (+1°C) might reduce average yield by about 5±4 per cent, and a 2°C warming might reduce average yield by 10±7 per cent. In addition, reduced precipitation might also decrease yield of wheat and maize in these breadbasket regions. A combination of increased temperature (2°C) and reduced precipitation could lower average yield to the extent of 20 per cent.

The quantity and quality of rice especially the scented rice (Basmati rice) which are mainly confined to their cultivation in north India and eastern Pakistan are likely to suffer severely by climatic changes with regards to their aromatic characteristics as these aromatic compounds are volatile and generally lost under high thermal regimes. The grain yield of even non-aromatic rice at low and mid latitude are supposed to suffer from low productivity probably due to enhanced spikelet sterility, shortening of crop growth duration and greater respiratory losses mainly during night.

However, as compared to the crops in which the sinks are reproductive organs such as seeds/grains, fruits etc. (cereals, pulses, oilseeds, fruit crops), the vegetatively propagated tropical crops such as cassava, yams, rhizomes, diascoria etc. are supposed to be less vulnerable to hyper thermal stress, hence the cropping pattern in the tropics is likely to change from reproductive to vegetative propagated crops under changing climatic condition.

Wheat productivity both grain and biomass yield will increase by 7-11 per cent per 100 ppm increase in CO_2 without other environmental changes under well fertilized and irrigated condition, but less under limited fertilization. All the observed response of wheat to elevated CO_2 can be explained by two direct effects (I) increased photosynthesis and decreased photorespiration and (II) decreased stomatal conductance, which in turn may enhance tillers, leaf area, ears, grains photosynthetic capacity and development. A constant 1°C increase in temperature over the whole growing season would decrease yield by 6-10 per cent mainly due to shorter duration of crop growth.

A doubling of CO_2 concentration increased photosynthesis by 30-40 per cent and reduced photorespiration. Both increasing air temperature and CO_2 accelerated the rice development resulted in reduced number of days to heading. Doubling of CO_2 increased rice biomass production under field by about 24 per cent over a relatively wide range of air temperature which was strongly influenced by soil available nitrogen. A doubling of CO_2 substantially increased the yield capacity of rice through enhanced tillers and spikelet production per unit land area. Air temperature deviate from the optimum temperature of 23-26°C, yield declined to near zero at temperature below 15-18°C or above 36-40°C depending on cultivars. This effect is mainly due to spikelet sterility induced by either low or high temperature. Rice yield increased by CO_2 enrichment at optimum temperature declined with increased temperature, which indicates that CO_2 enhanced the susceptibility to high temperature induced spikelet sterility. Rice genotypes differ in high temperature induced spikelet sterility at flowering. High temperature above optimum mean daily temperature produced fewer viable seeds with decreased quality.

High CO_2 fixed more carbon in cotton plants under all level of water and nutrient deficient condition across a wide range of temperature (Reddy *et al.*, 1995). High CO_2 increased stomatal resistance but enhanced leaf area offset that effect resulting in virtually no difference in canopy water use. However, cotton plants failed to show any phenological response to high CO_2 level. High CO_2 level enhanced the vegetative as well as reproductive growth thereby the higher number of fruits and yield. Developmental processes hence were temperature dependent. Average temperature above 30°C caused drastic abscission of young balls. Doubling of CO_2 did not ameliorate of adverse effect of high temperature on cotton balls retention. The area of the world that are marginally too cool to become productive and areas that is marginally too warm to become less productive. No Varietal difference in high thermal tolerance was observed in upland cotton, however in less tolerant pima cotton species, some cultivars were reported to escape the high temperature effect by transpiration cooling.

8.5. Effects on Disease and Pest Dynamics

Studies suggest that increase in thermal regime may extend the geographical range of some insect and weed pests currently limited by low temperature (EPA, 1989; Hill and Dymock, 1989). As with crops, such effects would probably be greatest at higher latitudes. The number of generation per year produced by multivoltine (multigenerational) pests would increase with earlier establishment of pest populations in growing season and increased abundance during more susceptible stages of growth. An important unknown, however, is the effect that changes precipitation amount and air humidity may have on the insect pest themselves and on their predators, parasites and diseases. Increase in temperature may also cause vigor growth and distribution of several weeds which may compete with crops under normal and sun-normal conditions as the weeds have been found to be more aggressive and adaptive to abiotic stresses especially higher temperature. Climate change may significantly influence interspecific interaction between insect pests and their predators and parasites. Under a warmer climate at mid-latitudes there would be an increase in the overwintering range and population density of a number of important agricultural pests such as the potato leafhopper, which is a serious pest of soybean and other crops in USA (EPA, 1989). Assuming the planting dates did not change, water temperature would lead to greater damage to crops. In the USA Corn Belt increased damage to soybean is also expected due to earlier infestation by the corn earworm. Examination of climatic warming on the distribution of livestock diseases suggests that those at present limited to tropical countries may spread into mid-latitudes.

At cool temperate regions where insect pest and diseases are not generally serious at present, damage is likely to increase under warmer condition. In Iceland for example, potato blight currently does little damage to potato crop, being limited by low summer thermal level. However, under doubling of CO_2, which may be 4°C warmer than at present, crop losses to disease may increase to 15 per cent (Bergthorson *et al.*, 1988). Under warmer and more humid conditions cereals would be more prone to diseases such as septoria. In addition, increases in population level of disease vectors may lead to increased epidemics of diseases they carry. Increases in infestation of the bird cherry aphids or grain aphids could lead to increased incidence of barley yellow dwarf virus in cereals. Increased atmospheric temperature during the last few years indicated the possibility of infestation of wheat crops by the aphids which are a regular severe pest of Brassica family. The prevalence of pests and diseases is among the major constraints to achieve higher yield. This is particularly true in the tropics where more disease and pests occur because of higher temperature and humidity, which provide favourable conditions for infestation. The increase in temperature and precipitation in 2 x CO_2 climates could create more conductive conditions for pest infestation. Pests and diseases that are unimportant and cause little harm today could become devastating under changing conditions. Periodic pest and disease led losses in food production resulted in famine conditions as the following events from history remind us. (1) The Irish famine of 1890s due to potato blight epidemics. (2) The wheat less year of 1917 in USA due to stem rust epidemics. (3) The Bengal famine in India in 1943 due to *Helminthosporium* brown spot of rice epidemics. (4) The southern corn blight epidemics in 1970-71 in USA. (5) The rapid shift of brown plant hopper biotype 1 to biotype 2 in 1974-76 in the Philippines and Indonesia causing massive damage to new rice varieties. The shift in biotypes occurred within two years. These examples are reminders that temperature and humidity together can sometimes cause massive losses of crops. The occurrence of such situation may be further accentuated by increased level of CO_2. The legume and oilseed crops are especially more vulnerable to insect and diseases and often an entire crop can be lost.

8.6. Effect on Cropping Pattern and Crop Distribution

Spatial, seasonal and temporal distribution of different crops are governed to a great extent by their climatic, domestic and marketing requirements, thus any change in climate and its components such as radiation, temperature, rainfall etc. at regional, seasonal and global levels may would definitely affect the crop growth, productivity and quality substantially. There are some places which are at the maximum limit of cultivation of some crops *e.g.*, central India is at the maximum thermal limit of cultivation of wheat, potato, late variety of cauliflower, cabbage cultivation, and lower belt of Himanchal Pradesh, J and K and Arunachal Pradesh are at the maximum limit of cultivation for some temperate crops like apple, cherry, saffron, plum etc., and still suitable for the cultivation these crops, but global warming induced climatic changes in future may be responsible for terminating the cultivation of these crops at these places as these areas may not be suitable for the cultivation of these crops, which in turn may compel the farmers to grow another crops which require slightly higher temperature. Similarly, reproductively propagated crop species at tropics may be replaced by the vegetatively propagated crops like root and tuber crops as reproductive sinks have been reported to be more sensitive than vegetative sinks. Thus climate change due to warming may lead to changes in cropping pattern at different places and in different seasons. This is indispensable for ensuring the production of some crops (alternate crops) under warmer condition in near future.

8.7. Effect of Rise in Sea Level on Agriculture

The global warming driven climatic changes are likely to cause substantial rise in sea water level. The magnitude of the same is expected to about 20 ± 10 cm by 2030, 30 ± 15 cm by 2050 and about 80 cm by the end of twenty first century (Warrick and Oerlemans, 1990; IPCC, 1995). The main reason of rise in sea water level is thermal expansion of ocean and partial melting of ice caps and glaciers in the Arctic and Antarctic regions due to greenhouse gases induced global warming in one hand and formation of brown haze and soot in atmosphere owing to biomass burning and industrial emission of gaseous pollutants on the other hand. Thus rise in sea water level may not only be responsible for affecting the agricultural lands through the inundation of low-lying coastal farm and non-farm lands but also increased salinity of coastal groundwater. Twenty seven countries have been identified as being especially vulnerable to seal level rise on the basis of the extent of the land liable to water logging, the population at risk and the ability to take measure against such situation. On the basis of vulnerability scale (1-10) in ascending order, the following most vulnerable countries are: Bangladesh, Egypt, Thailand, China, Indonesia etc. The most severe impacts of sea level rise are likely to emerge from water logging. Southeast Asia would be most affected because of the extreme vulnerability of several large and heavily populated deltaic regions. For example, with a rise of sea level by 1-1.5 m, about 15 per cent of all land and about 20 per cent of all farm land in Bangladesh would be flooded and further 6 per cent would become more prone to frequent flooding (UNEP, 1989), which may in turn cause a loss of about 21 per cent of agricultural production. In Egypt, by a rise of sea level to the extent of 1-1.5 m would cause 17 per cent loss of agricultural production and 21 per cent loss of most productive farmland. The Maldives Island in Indian Ocean would have one half of its land area inundated with a 2 m rise in sea level (UNEP, 1989). The flooding of coastal areas of affected countries is likely to cause loss of biodiversity, which are sensitive to salinity and may thus change spatial and temporal composition in biological community in affected regions. Thus the large coastal regions, which are presently occupied by different crops, may gradually be converted into mangroves ecosystem. In addition to direct farmland loss from flooding, it is likely that agriculture would experience increased costs from salt water intrusion into surface water and groundwater in coastal regions. Further indirect impacts would be likely as a result of need to relocate both farming population and production in other regions. In Bangladesh about 20 per cent of national population would be displaced as a result of rise in sea level to 1-1.5 m.

8.8. Climate Change and Loss of Biodiversity

Many plants and animal species are unlikely to survive climate change. New analyses suggest that 15-37 per cent of a sample of 1103 land plants and animals would eventually become extinct as a result of climate change expected by 2050. For some of these species there will no longer be anywhere suitable to live. Others will be unable to reach places where the climate is suitable mainly because of several physical and biological barriers in their migration and dispersal. A rapid shift to technologies that do not produce greenhouse gases, combined with carbon sequestration could save 15-20 per cent of species from extinction. There are some rare animals which are endemic to confined and specific regions especially islands because of their specific climatic requirements and origin are supposed to more vulnerable to climatic changes than other species of wider climatic range. Global warming induced climate change is likely to pose a serious threat to biological diversity especially the crop species which require narrow optimal thermal range and are grown in a specific and confined climatic regions such as temperate crops like apple, plum, cabbage, cherry, keshar and also some of the aromatic rice like basmati which is endemic to the specific regions of north India and Pakistan as the aroma/scent in these cultivars are developed perfectly in these areas mainly because of specific climatic requirement of these varieties are met in such specific areas where temperature during the grain filling is slightly go down and thus maintain the aroma in these rice. Some animals such as crocodile, which determine their sex due to changes in their thermal regime during incubation and produce male and female to change the temperature of eggs by putting them at different depths in sand are likely to be affected severely in changing climates. Extinction of dinosaur is also said to be due to changes in temperature in the past. Some reptile like Boud's forest dragon, which is found in Queensland, Australia, and 90 per cent of its distribution, would become climatically unsuitable by 2050, on maximum climate warming scenarios. Similarly the population of Comodo dragon lizard which is endemic to Comodo island of Indonesia is very much vulnerable to climate changes.

8.9. Food Security

Food security is a major problem around the world both in developed and developing countries and with and without a climate change. Can the regions or nations, which have been identified as vulnerable, meet their food requirement? The efforts should not only be made toward increased dependency on trade but towards attaining self-sufficiency by each nation or a group of nations. The large countries like India, Brazil and Mexico where large area of arable lands are still under rainfed agriculture and thus food production in these areas is highly uncertain and risky because of their greater vulnerability to natural calamities such as drought, high temperature etc. Thus any changes in climatic components may lead to the problem of food security both at regional and national levels. Even in irrigated agriculture high temperature stress especially during early vegetative and grain filling stages has been found detrimental in wheat crop and low temperature during the reproductive growth phase has been reported to be detrimental in rice and maize crops. In order to maintain the national and regional food production scenario it is indispensable to make the crop production strategies such as alternative cropping, enhancement in input resources and their efficient utilization etc. to cope with changes in climate in days to come. Cultivation of tuber, root and rhizomatous crops instead of grain/seed crops in tropics could be an alternate source of food and energy in future.

9. Climatic Variability and Crop Productivity

Sudden and discontinuous changes in climate or climatic components at regional, national or intercontinental levels may be referred to climatic variability or climatic aberration. Such type of changes in climatic parameters/components such as rainfall, radiation, humidity, wind velocity etc. spatially or temporally is more detrimental to crop productivity as the crop plants are suddenly exposed to a kind of climatic stress where plants do not get sufficient time to modify their biological system to cope with such adverse condition and finally tend to suffer substantially from the occurrence of climatic stress suddenly. However, effect of climatic changes on crop growth and productivity may not be very severe as magnitude of changes in climatic components is generally smaller and follows a definite trend where plants get sufficient time to alter and establish a new biological setup to cope with gradually changing environment, in other words the crop species may be able to adapt themselves to changing climatic components. Hence climatic vulnerability is said to be more detrimental as compared to climatic changes. Climatic variability is reported to be quite often during any time of a season. Sometime there is sudden drought during the time of transplanting or grain filling in rice, or occurrence of high temperature during vegetative or ripening growth phases of wheat, which cause severe loss of production of these crops through their low productivity. Recently in the year 2002 the temperature of eastern U.P. and Bihar during the winter season dropped substantially which caused which improved the productivity of wheat crop but significantly reduced the yield of winter maize mainly due to hypothermal induced injury in pollen and failure of fertilization and grain setting. Similarly in the year 2003 there was sudden spurt in atmospheric by 4°C for a period of one week in the month of March which generally coincide with grain filling period in wheat in north India, and finally reduced the productivity of wheat to a large extent mainly because of poor grain filling (poor grain size and weight) due to shortening of grain filling duration and higher respiratory losses of biomass. Another example of climatic variability was recorded during the winter season of the year 1997, when there was almost no proper radiation throughout the vegetative and reproductive phases of wheat in north India but the unusually the chilling continued until late March which helped in prolonging the grain filling duration and finally the biological loss occurred during preceding growth phases were compensated by the ripening growth phase through better grain filling and thereby grain size. Thus the magnitude of climatic variability is reported to be higher than that of climate change which can cause havoc in national crop production scenario.

9.1. Response of Crops to Climatic Variability

Wheat and rice are the most important staple foods around the world. Of the global wheat production, 41 per cent is produced in developing countries and 59 per cent in the developed countries, while 95 per cent of global rice is produced in developing countries and 92 per cent is produced in Asia alone. Any climatic variation which influences the production of wheat and rice will have a profound effect on global food security particularly in developing countries owing to greater dependency on nature, and poor in technological and economical competency. However, Africa is mostly dependent on the production of coarse and fine millets (sorghum, pearl millet, finger millet etc.) and root and tuber crops. Rice, millet, sorghum, maize, tubers, cassava, sweet potato and yam are mostly grown as monsoon crops and their yields generally fluctuate with variation in monsoon. The production of rice in India, Bangladesh and Indonesia is severely influenced by drought, but it is less affected in China, Korea and Japan where most of it is grown with irrigation. Among the rice producing regions of India the maximum adverse impact of drought occurs where the irrigation resources are limited (rain fed rice in east and north eastern hill regions of India). However, in northern Japan the productivity of rice is limited by the growing season because of low temperature. Such situation may also occur in in regions where rice is grown at high altitudes as in Nepal. Recently in Bihar the Rabi maize crop was severely affected by low temperature during anthesis, which caused complete-partial cob sterility (no grain formation). If suitable varieties of such crops are identified and developed in future, the productivity of these crops could be enhanced due to global warming in such places. But most of the rice is grown in the humid tropics where high temperature is I limiting factor of grain yield. Wherever high yield of rice is obtained as in Korea, Japan, China and parts of India the temperature is generally moderate (mean daily temperature range between 22-26°C). The productivity in such regions is likely to be adversely affected by arise in temperature. Thus the following possibilities can be presumed. In temperate regions the growing season may increase thereby benefiting crop yields. Increased CO_2 level in the atmosphere may also benefit the crops. Increase in disease in insect pest incidence could occur. The demand for irrigation water may increase. However, in tropical regions the growing season may be reduced, hence adversely affecting the crop yield. Higher temperature and humidity may cause greater spikelet sterility in rice. Increased CO_2 may benefit biomass production, but not grain yield. An increase in disease and insect pests incidence could occur and new pest may emerge. The uncertainty of monsoon may cause greater crop vulnerability to water deficit. Cloudiness or low radiation may adversely influence crop yields. A detailed analysis for coarse cereals and pulses is necessary to predict the effects of climate change on these crops. Therefore, depending upon the location, varieties with appropriate duration and tolerance to biotic and abiotic stresses will have to be developed. In addition a great effort will have to be made towards water management.

10. Vulnerability of Agriculture to Climatic Changes

10.1. Crop Vulnerability

Since agriculture is totally depend upon the nature and natural resources for its successful establishment, any changes in the climatic and its components is directly and indirectly related with crop and animal production through their productivity and quality. As compared to regulated farming system where the crop and animal production take place under controlled management condition (irrigation, fertilization, crop protection etc.), the natural or rainfed agriculture is highly prone to natural calamities such as drought, flood, high temperature, insect and disease infestation etc., because under controlled management conditions the negative impacts of climatic variability could be trade off to a great extent by additional supply of inputs such as irrigation, fertilization, pesticide application and the loss in agricultural production could be minimized substantially. However, rainfed agriculture is severely affected by natural calamities or climatic variability as there are little scope for combating the negative impact of adverse climatic conditions. Similarly, the rich farmers will be least affected than those of poor and marginal farmers to any climatic aberration as the poor farmers have neither financial nor technological support to combat the natural calamities, and if the climate aberration continue for 2-3 consecutive years, then there is no option left except their migration from native place to the new places for job and survival. The prevalence of climate change for longer period may force even the rich and large holding farmers to either migrate or alter their farming system from crop and animal production to plantation based farming system or non-agricultural operations. Thus agriculture is likely to be much more vulnerable than any other enterprises to climate change and variability.

10.2. Socio-economic Vulnerability

In rural India where almost all the families are involved in various agricultural activities such as crop, milk, meat, egg production and largely depend on farm income for their livelihood, education, health and other social activities will have to suffer from climatic changes and vulnerability. As the success of agriculture is totally dependent on environmental conditions prevails during the crop growing seasons, the rural people are certainly going to be affected by the climate change and climatic aberration. The magnitude of socio-economic vulnerability depends upon the severity of climatic changes over the time and space. Rural social structure and function depends on the economic status of the villagers and their economic status is governed by the success of crop and animal production and their marketing. If the climate change causes irregular or erratic behaviour in term of flood, drought, epidemics, storm etc the deficit in farm income will lead to migration of people from rural to urban areas and finally result in social destabilization over there in rural places. If the climate variability affects the agricultural production in rural India, the development of rural infrastructure, living standard, purchasing power, education and health would be affected substantially through depleting the farm income of rural people. Among the rural communities, the rich people may be affected more by climatic changes, while the climatic variability may affect more to the poor people.

10.3. Geographical Vulnerability

The impact of climatic changes depends upon geographical position of the places, which varies with the latitudes, altitudes, surrounding vegetation and distance from sea coast. Thus different crop species grown at different places and time may show differential response to climate changes and variability. As compared to the crops grown at lower latitude (tropics), crops grown at higher latitude and altitude may be more affected and vulnerable to the climate change as the magnitude of climate change has been reported to be greater at higher latitude than at lower latitude. Thus the temperate crops like wheat, potato, chickpea, apple, cherry, cabbage, carrot, peas will be affected more than those of tropical crops *e.g.*, rice, maize, sorghum, babana, greengram, redgram, sugarcane, sesame etc. The productivity of vegetative sink crops such as taro, cocoyam, corm, bulbs, rhizomatous etc. will be affected less as compared to reproductive sink bearing crops under warmer climatic condition. Hence there are possibilities of both spatial and temporal vulnerability of crops to climatic changes and variability. If the climate changes due to global warming and cooling and continue in the future, the biogeographical pattern of different crops will also be changed according to their adaptability to changing climatic condition. Some of the temperate and tropical crops such as apple, cherry, plum, cabbage, basmati rice etc. which require specific and narrow range of optimum thermal limits are likely to suffer greatly than those of crops with wider optimum thermal rage of their adaptation.

11. Crop Adaptation to Climate Change

11.1. Natural Adaptation

11.1.1. Shifting Optimum Thermal Range

By virtue almost all the biological systems and organism have tremendous ability to cope with diversified environments through changing their physiological and biochemical makeup-this process is known as adaptation. During the adaptation process the living always tend to function within the permissible range of environmental variable such as temperature, light, humidity, carbon dioxide concentration etc. This optimum range always fluctuates to certain limit known as biological elasticity. If the changes take place within the optimum range, the effect may not be significant but if the changes cross the limit of optimum range, the organism may suffer heavily if they fail to modify themselves accordingly. Under any abiotic stress the livings have two options, either they can modify themselves to cope with new environments for their survival or die if they fail to modify accordingly. Under global climatic change where the atmospheric temperature is rising gradually, most of the crop species and weed species tend to modify their biological activities for the sake of adaptation to high temperature stress. Under such adverse condition the organism may change their optimum thermal range and can set a new range of optimum temperature. This phenomenon is known as shifting optimum range and through this process the plants can cope with the adverse/changing temperature and can establish to new environment with least or without losing their identity, integrity and productivity. Since environmental changes take place gradually, plant species get lot of time to modify their biological processes for adaptation so as to continue their lives. Thus adaptation is a powerful mechanism of livings, which enhance their sustainability.

11.1.2. Tolerance to Climatic Changes

Living organism has the ability to tolerate the adverse/stress situation on the cost of their growth and productivity. Livings when exposed to any adverse condition, they either resist to the adverse condition so that they do not experience stress situation or undergo some shorts of morphological, physiological, biological or anatomical modification and to do so plants have to exert additional amount of energy in order to cope with adverse situation. Plants under hyper thermal stress generally enhance the respiration rate to meet the additional energy requirement without increasing the rate of photosynthesis, thereby reducing the net productivity. The tolerant plant species have ability to synthesize heat shock protein (HSP), which enable them to endure high thermal stress. Some times plants following their exposure to high temperature synthesize some structural chemical compounds such as waxy cuticle on the cost of stored/non-structural compounds which act as insulator for high temperature and thus protect the plants from hyper thermal injury. Effect of high temperature stress generally causes change changes at cytoplasm levels and if the plants are continuously exposed to high temperature for longer period they reset their cytoplasmic tolerance and adapted to that condition. It has been observed that the plants adapted to a particular environmental stress show greater ability to tolerate that condition as compared to the plants grow under normal condition. That is why the wild crop species or a well adapted cultivar to a specific environment is generally used as a donor for transfer of genes into high yielding varieties.

11.1.3. Escaping Mechanism

Escaping mechanism is an adaptation process by which the tolerant plants have the ability to shortcut their normal life span and complete heir life cycle before experiencing the abiotic stress such as drought, high and low temperature stress etc. Several crop varieties have been identified/notified having showed marked reduction in their life span by hastening the phenological process such as flowering and maturity. In fact these plants have biological potential to experience the onset of adverse climatic situation and faster their ontogenic processes thereby complete life duration. Such stress escaping mechanism is prevalent in several germplasm of different crop species and thus offers an opportunity to exploit such potential in order to ensure the crop production under any climatic stress although on the cost of their productivity. Such type of crop cultivar could be used a donor for breeding crop varieties for environmental stress. These types of cultivars in fact have no ability to tolerate or resist the magnitude of environmental stresses. Although the escaping mechanism may seem to be very simple but physiologically and bio-chemically may be complex as the passage of message from roots or shoots to the biochemical regulation process (transduction) at gene level and thereby instructing the regulatory process in advance to hasten the ontogenic process and finally shorten the life duration before experiencing the adverse situation.

11.1.4. Compensatory Mechanism

Nature has provided to all the livings with compensatory mechanism. The livings organism, who suffer from a biotic stress at any point of time during their life cycle are generally compensated following the resumption of normal condition. The growth and productivity components affected by climatic stress during the vegetative phase are compensated to a large extent by the enhancement of other growth and yield components during reproductive or ripening growth phases following the onset of normal condition and finally the degree of losses in crop growth and productivity are makeup substantially. For instance, a crop which suffered from drought or high temperature stress during vegetative growth phase under rain fed ecosystem condition and subsequently receive rainfall during reproductive or grain filling duration, the economic loss caused by drought during vegetative phase is generally compensated by enhancing the growth and physiological activities following receiving rainfall during reproductive and grain filling stages and eventually the crop productivity is reported to be normal. This mechanism is more successful in longer duration crop varieties especially in photo-thermo sensitive species biennial and perennial herb crops such as sugarcane, perennial cotton and pigeon pea, banana, castor etc. as the long duration crop species get sufficient time to compensate the loss through enhancing their physiological activities following the occurrence of normal condition.

11.2. Anthropogenic Adaptation or Croping Strategies

11.2.1. Agro-Physiological Manipulation

Apart from natural biological process of crop adaptation to a-biotic stresses, there are also some possibilities to reduce the adverse effect of high thermal stress on crop growth and productivity through agro-physiological manipulation such as early/late or dense/sparse sowing, higher dose of seed and fertilizers, increasing the number of irrigation etc. When the crops are grown under high temperature than under normal thermal regime, the growth and productivity are reduced drastically mainly because of hastening the phonological stages such as flower primordial initiation, flowering, anthesis and maturity and finally the crop growth duration. In other words the thermal duration of a crop species, which is almost fixed is commenced earlier under higher thermal regime and finally the crop growth duration is reduced substantially. Secondly the higher temperature also enhances the rates of both dark and photorespiration, which ultimately deplete the net productivity of the crops. Thus to maintain the same level of bio-productivity under high thermal regime where the life duration of crop is reduced, the possible measure is to increase the crop growth rate. This could be achieved mainly through better crop production and protection technologies such as increasing the rate of fertilization, frequency of irrigation and agrochemicals (pesticides, growth stimulators and regulators).

The another non-genetic approach/coping strategies for maintaining higher crop productivity under rising atmospheric temperature is the identification of crop species and varieties with greater level of hyper thermal tolerance. This is possible through screening the available crop genotypes at the active germplasm collection center of different crops or cultivation of traditional cultivars already being grown in the regions of high temperature such as central, western and south India.

11.2.2. Crop Diversification

Crop diversification is another potential agronomical measure for minimizing the detrimental effects of global warming on crop productivity and quality. If the high temperature causes significant loss in the productivity of any specific crop grown in a particular region for several years, the cropping pattern could be changed on the basis of thermal tolerance capacity of different crops e.g., the low temperature loving crops Crop diversification e.g., the low temperature loving crops may be altered with the crops which do not require low temperature for optimum growth and productivity. If the high temperature causes injuries in the reproductive organs of plants and result in poor grain/seed setting, the crops of vegetative sink such as yams, tuber, corm, tapioca, taro etc. may be cultivated or the crops/varieties with greater thermal tolerance level could be practiced. It is also possible to by growing C4 crops rather than C3 crop species to exploit the greater adaptation ability of C4 plants to tropical climatic condition. Among C3 group, there are several crop species, which require tropical climate, which may show comparatively less degree of loss in their productivity under warmer condition. For instance, instead of bread wheat, which requires low temperature for optimum growth, durum wheat, barley, chickpea etc. could be grown. Thus monoculture or specialized cropping system could be replaced with diversified cropping system, which offers an opportunity to reduce the risk of global or regional warming. In our traditional farming system, where there was no facility for irrigation and fertilization, farmers used to practice mixed or diversified cropping system so as to ensure the crop production even under severe adverse climatic condition thereby to reduce the risk of adverse climatic condition. Growing different crop species together in the same field would generally never fail completely under adverse climatic condition as the micro-climatic requirement of different crops is different and out of several crops at least some will thrive and grow even under extreme climatic condition. For example the mixed cropping of pearlmillet, pigeon pea, cotton, castor, blackgram during kharif season, and wheat, barley, chickpea, lentil, mustard, safflower during Rabi season in traditional agriculture ever showed minimum degree of climatic risk such as low rainfall, flood, high and low thermal stresses. Under normal thermal regime all the crop species will germinate and grow successfully, but if the temperature is higher during sowing the survival and establishment of wheat seedlings is recorded to be very poor and other crops like barley, chickpea, lentil, safflower establish and grow properly. Thus crop diversification may be a very good option and strategies to minimize the risk of hyper thermal stress.

11.2.3. Genetic Manipulation

Genetic manipulation of improvement of any crop for a specific trait depends upon the presence of that character in the gene pool of that crop species and the combining ability of the individual possesses the desirable trait to be transferred to the receptor plants. Since different crops are distributed geographically within their climatic extremes (minimum to maximum range), thus possess the trait (genes) responsible for their adaptation to diverse climatic stresses such as high and low temperature, drought, low radiation etc. Hence the crop plants adapted to diverse environments generally posses the genes responsible for their ability to tolerate such adverse climatic condition *e.g.*, the rice cultivars which are being grown continuously under deep water/waterlogged conditions or drought and excess salt stresses possess the genes responsible for their adaptation to such environments and are generally used as donor for transferring the desirable genes from these varieties through gene recombination or back crossing. Now efforts are being made to develop the crop varieties with least sensitivity to varying climatic components such as photoperiod and temperature for their wider adaptability to diverse climatic conditions. Evolution of photoperiod and thermal insensitive cultivars of different crops especially wheat, rice and vegetables have made it possible to grow them anywhere and any time round the year. Such types of cultivars are designated as season proof cultivars and have been able to break the limitation of growing them in a particular season. In south India and west Bengal where irrigation facilities are sufficient people are taking three crops of rice in a year. Now-a-day's most of the crop varieties are insensitive to either photoperiod or temperature while there are limited crop cultivars, which are insensitive to both photoperiod and temperature. In fact the photoperiod insensitive cultivars are thermo-sensitive and photo period sensitive cultivars are generally thermo-insensitive. If any cultivar becomes insensitive to both temperature and photoperiod, it is possible to grow them throughout the year, *e.g.*, some mango plants, guava, sunflower, papaya etc. Earlier (20-25 years ago) most of the crop cultivars were sensitive to photoperiod, hence used to be grown in a particular season and are said to be season bound varieties. Just 10 years before it was not at all possible to grow rice, cauliflower, cabbage, carrot, onion, tomato, gourds etc during the off season, but now all these crops are grown during the off season. Now it is possible to breed crop varieties for high thermal stress which may be helpful in sustainability of crop yield under warmer climatic condition.

11.2.4. Biotechnological Approaches

Apart from conventional Mendelian breeding for genetic manipulation several biotechnological measures are now in practice to transfer the selected genes from donors to the any promising cultivars. Presently the biotechnology has become so advanced that genes could be transferred from any one to another organism, which show biological barrier due to reproductively isolation in the community. Now there is almost no biological limitation of transferring the genes across the kingdom, phylum, division and family, which is not at all possible through conventional breeding. But the success of biotechnology especially in higher and complex organism has not yet been achieved substantially mainly due to higher and complex genomic entities in higher organism in one hand and acceptability of foreign gene by the recipient plant. Thus it is indispensable to find the source of genes for high thermal tolerance in the respective crop gene pool through their screening for desirable traits especially the accessions collected from warmer climatic regions.

12. Mitigation Options to Climatic Changes

Finally, the impact of climate change if found detrimental has to be overcome through socio-political and technological manipulations. The global warming is mainly resulted from an increase in the rate of greenhouse gas emission into the atmosphere owing to overexploitation of reserve fossil fuels and deforestation to meet the energy requirement of exploding human population and industrialization for survival in developing countries and for enjoyment in developed countries. The possible measures to mitigate the global warming led climate change are briefly discussed as follows.

12.1. Reduction in Greenhouse Gas Emission

Since global warming is mainly resulted from an increase in the rate of emission of greenhouse gases (CO_2, CH_4, N_2O, O_3, CFC) from natural and anthropogenic activities in one hand and massive deforestation of tropical rainforest and sub-tropical and temperate forest throughout the world since industrialization (1850 onward), efforts are being made at regional, national and international levels to curtail the production of greenhouse gases and bring them to the level of the year 1990. The UNEP conference held at Kyoto, Japan in 1997 popularly known as Kyoto protocol refers to the international protocol to be signed by all the member countries of United

Nations accepting the request for reducing the production of greenhouse gases through various possible measures to protect the lives on earth from global warming effect in future. But except USA, which is a major contributor of greenhouse gas emission and global warming, almost all the nations including Russia and China have signed the protocol and agreed upon to curtail the production of GHG in future. There are several ways and measures to reduce the emission of GHG into the air, which are discussed as under.

12.2. Judicious Utilization of Reserve Carbon

Thousand and million years ago huge amount of biomass was buried into the soil and transformed to fossil fuel oil over the time (reserve carbon), which is now being used as a major source of energy in the forms of coal, diesel, petrol, crude oil, kerosene oil, charcoal, natural gas etc. to meet the energy requirements of thermal power plants, industries, vehicles, rail and various engines, road construction, domestic purpose etc. Due to rapid industrialization and mechanization huge amount of energy is directly and indirectly generated by using these reserve carbons which in turn have led to excess emission of greenhouse gases especially CO_2, the most global warming contributing greenhouse gas into the atmosphere. Annually about 3 billion tons of CO_2 gas is dumped into the atmosphere through various natural and anthropogenic activities. Although CO_2 is an essential natural greenhouse gas responsible for the existing of life on the earth planet through production of organic foods for all the heterotrophs through photosynthesis and raising the atmospheric temperature optimum for their survival and proper growth and development. If the greenhouse gases would not have present in the atmosphere, the mean atmospheric temperature would have been around minus 18 degree Celsius, but the atmospheric temperature is about 15°C that means the greenhouse gases have been responsible for raising the atmospheric temperature by 33°C. Although presence of greenhouse gases in the atmosphere are imperative, but excess emission of these gases into the atmosphere may further raise the global warming and thus climate change and variability. Since CO_2 has maximum percentage of contribution to global warming (55-60 per cent), reduction in its emission is indispensable through judicious use of fossil fuel in the coming years.

12.3. Restriction on Biomass Burning

Advent of mechanization and abandon of draught farm animals in modern agriculture have led to the burning of huge amount of farm residue instead of converting them as farm yard manure either through animal feeding or their microbial decomposition in soil. Thus the farm residue mainly in the form of straw, which was recycled at farm and applied as manures in the fields during traditional agriculture (still used as manures in several parts of our country) is now being burned in the fields after combine harvesting of the wheat and rice crops especially in north India where farm mechanization is adopted intensively. Thus biomass burning has led to emission of carbon monoxide and carbon dioxide along with black soot and smoke, which not only pollute the atmosphere but also warm it and thereby contribute to global warming substantially. Thus complete to partial ban on biomass burning must be imposed in order to reduce the emission of greenhouse gases, which are responsible for global warming and climate change. Another major source of greenhouse gas emission into the atmosphere is forest burning. There have been several incidents of forest burning in different parts of the world *viz.*, USA, Indonesia, China and India, which might have contributed substantially to global warming through emission of greenhouse gases.

12.4. Improvement in Rice Cultural Practices

Submergence and water logging of rice paddy fields are very common during the cultivation of both rainfed and irrigated rice all over the rice grown regions, which causes methane (greenhouse gas) production from rice fields due to anaerobic decomposition of organic residue in submerged soils. Methane is the next major greenhouse gas after CO_2 and contributes to global warming to the extent of 15-20 per cent although its warming potential is about 21 times greater than CO_2. Global rice fields alone contribute about 20 per cent of total methane production from different sources. Thus in order to reduce methane production from rice paddy fields, the cultural practices especially the water management of rice crops need to be modified with least or without reduction in its production *e.g.*, intermittent irrigation instead of continuous flooding in irrigated rice and by creating proper drainage facility in traditional rainfed rice cultivated areas.

12.5. Balance Doses of N Fertilization

Among the greenhouse gases although the contribution of nitrous oxide (N_2O) to global warming is meager (5 per cent) but its warming potential is about 200 times greater than that of CO_2. The main source of N_2O emission is nitrification and denitrification of nitrogenous organic compounds (organic residue and organic fertilizer) in the soil through microbial transformation/decomposition. Apart from emission of N_2O from nitrogenous fertilizer *viz.*, urea it also produced NO_2 during nitrification process, which becomes one of the major source of ozone (O_3) production through photochemical reaction. Thus nitrogenous compounds produces two greenhouse gases (N_2O and O_3) which together account for about 13 per cent contribution in global warming.

12.6. Afforestation–CO_2 Sink Enhancement

Forest vegetation is the largest sink of carbon dioxide mainly through the process of CO_2 assimilation (photosynthesis) and thus responsible for maintaining its proper level in the atmosphere. But due to ever increasing human population and their unprecedented desire for wealth, large areas of forest ranging from tropical forest of Amazon valley to savanna of Africa and temperate forest have been denuded substantially for food, fodder, feed, fiber, fuel, furniture and farming purposes which in turn has led to depletion in CO_2 sink potential and thus caused an increase in its concentration in the air. Therefore it is indispensable to restrict further deforestation in order to check further rise in global temperature through greater fixation of atmospheric CO_2 by maintaining larger forest vegetation on this beautiful and bountiful earth planet. This is perhaps the best measure to keep our planet cool and safe, as it is an efficient, rational and sustainable approach to control global warming and climatic changes.

12.7. Improvement in Vehicular Gaseous Emission

12.8. Human Population Control

Population leads to poverty and poverty leads to pollution, thus directly and indirectly exploding human population has been responsible for every kind and magnitude of environmental degradation and problem. In other words excess and undesirable human population (human weeds) is perhaps the greatest pollutants on the earth, which nullifies all the economic and social developments made in the past. All the anthropogenic activities including, intensive farming and green and white revolutions have been aimed mainly to cater the food demands of ever increasing population and simultaneously to raise the income of the farmers through agriculture. Thus more food production for more people has resulted in several environmental problems, which could be overcome to a great extent by controlling the human population. I personally feel that instead of increasing food production, we should rather stress more on decreasing the unwanted human production, which is perhaps the main root cause of environmental and social degradation. Exploitation of natural resources has limitation of their availability and carrying capacity, their over exploitation to meet the food and other demands of human population, the only one biological component of environment, would certainly cause severe damage to our ecosystem which supports the life of immense biodiversity existing on the earth. Controlling the human population is perhaps easier than increasing further food production at the cost of our environment. Even environment-friendly agriculture cannot be the ultimate solution unless population explosion is controlled logically. Thus human demographic restriction is the only suitable option left for saving environment and life on the earth, which may otherwise be turned into non-living planet.

12.9. Use of Alternate Sources of Energy

The success of green, white and blue revolution in India and rest parts of the world have resulted from the integrated approaches of high yielding crop and animal varieties/hybrids, use and availability of agro-chemicals, land reforms, ensured irrigation, mechanization for various farm operations and better financial supports through nationalized banks. Tractors, threshers, combines, seed drills, have indeed played a crucial role in achieving the targets of green revolution without which the fruits of success would have never been acquired. Thus to perform various agricultural activities, presently these machines and implements are heavily dependent on fossil fuel conventional energy which have their availability and cost limits and priced heavily towards the cost of crop and animal production. However the fact is that several farm operations could be performed by using various forms of non conventional or renewable energy such as solar, wind, biogas through solar oven, photovoltaic cells, wind mills, wind pump, biogas plant etc. without paying recurrent cost, and environmental pollution. The most significant aspects of such forms of energies are that these are never going to be exhausted or finished, and also the establishment of such power plants costs initially only, which last for several years without much maintenance charges. As we know that the conventional source of energy *e.g.*, coal, fossil fuel and radioisotopes are not endless and would be finished after some period, it is better to develop the technologies and establish the power plants which can transform the non conventional energy into usable forms such as mechanical and electrical energy to run the various farm operations. Although these renewable source of energy have not yet been exploited for heavy farm activities such as ploughing, threshing, lifting sufficient water for irrigation or transportation etc., but heavy dependency on the conventional source of energy could certainly be reduced through the integrated approaches of using both forms of energy for several small farm operations and domestic works such as lifting ground water by wind mill pump, solar cells electric motor, biogas engine, cooking by solar oven and biogas, and solar electricity for domestic uses. Thus use of non- conventional source of energy has great potential of power generation, which could be managed at every village level through cooperative and collective approaches.

13. References

Acock, B. and Allen, L.H. 1995. Crop responses to elevated carbon dioxide concentrations. 33-97 in Strain, B.R. and Cure, J.D. (eds.). Direct effects of increasing carbon dioxide on vegetation, US Department of Energy.

Aggarwal, P. K. and Sinha, S. K. 1993. Effect of probable increase in carbon dioxide and temperature on productivity of wheat in *India. J. Agric. Meteorol*, 48: 811—814.

Ahrens, M.J. and Ingram, D.L. 1988. Heat tolerance of citrus leaves. Hort Sci, 23: 747-748.

Ainsworth, E. A. and Long, S. P. 2005. What have we learned from 15 years of free-air CO_2 enrichment (FACE)? A meta-analytic review of the responses of photosynthesis, canopy properties and plant production to rising CO_2. *New Phytologist* 165: 351-372.

Ainsworth, E. A. and Rogers, A. 2007. The response of photosynthesis and stomatal conductance to rising (CO_2): mechanisms and environmental interactions. *Plant, Cell and Environment* 30: 258-270.

Alcamo, J., Kreileman, G.J.J., Bollen, J.C., van den Born, G.J., Gerlagh, R., Krol, M.S., Toet, A.M.C. and de Vries, H.J.M. 1996. Baseline scenarios of global environmental change. *Global Environmental Change*, 6: 261–303.

Al-Khatib, K. and Paulsen, G. M. 1999. High-temperature effects on photosynthetic processes in temperate and tropical cereals. *Crop Sci.*, 39: 119-125.

Al-Khatib, K. and Paulsen, G.M. 1984. Mode of high temperature injury to wheat during grain development. *Physiologia Plantarum*, 61: 363–368.

Al-Khatib, K., and G. M. Paulsen, 1990. Photosynthesis and productivity during high temperature stress of wheat genotypes from major world regions. *Crop Sci.*, 30: 1127-1132.

Allen, L.H. Jr. and Boote, K.J. 2000. Crop Ecosystem to Climate Change. Soybean. In: Climate Change and Global Crop Productivity (Ed. by K.R. Reddy and H.F. Hodges) pp. 133-160, CAB International Publishing, Wallingford, Oxon UK.

Allen, L.H., Jr. Jones, P. and Jones, J.W. 1985. Rising atmospheric CO_2 and evapotranspiration. In : Advances in Evapotranspiration. Proceeding of the National Conference on Advances in Evapotranspiration, pp.13-27. American Society of Agricultural Engineers. St. Joseph, Michigan.

Allmaras, R.R., Burrows, W.C., Larson, W.E. 1964. Early growth of corn as affected by soil temperature. Proceedings of the Soil Science Society of America, 28: 271–275.

Amthor, J. S. 2000. Ann. Bot. (London) 86: 1–20.

Anand, A. and Nagarajan, S. 2011. Impact of climate change on quality of cereals and oilseeds crops. In: Climate change : Impacts and adaptations in crop plants. M.P. Singh *et al.* (Eds.), *Today and Tomorrow,s Printers and Publishers, Daryaganj, New Delhi*, pp. 33-53.

André, M. and Du Cloux, H. 1993. Interaction of CO_2 enrichment and water limitations on photosynthesis and water efficiency in wheat. *Plant Physiology and Biochemistry*, 31: 103-112.

Austin, P. B. 1988. Breeding for increased yield in wheat. Prerequisites achievements and prospects. *Aspects of applied biology.* 17: 289-295.

Azeez, M.A. and Shafi, M. 1966. Quality in rice. Department of Agriculture West Pakistan Technical Bulletin 13: 50

Badu-Apraku, B., Hunter, R.B. and Tollenaar, M. 1983. Effect of temperature during grain filling on whole plant and grain yield in maize. *Can. J. Plant Sci.* 63: 357–363.

Baker, J.T., and L.H. Allen, Jr. 1993. Contrasting crop species responses to CO_2 and temperature: rice, soybean and citrus. *Vegetatio* 104/105: 239–260.

Bazzez, F.A., Garbett, K. and Williams, W.E. 1985. Effects of increasing atmospheric carbon dioxide on plant communities. In Strain, B. R. and Cure, J. D. (eds.). Direct effect of increased carbon dioxide on vegetation, US Department of Energy.

Bergthorsson, P., Bjornsson, H., Dyrmundsson, O., Gudmundsson, B., Helgadottir, A and Jonmundsson, J.V. 1988. The effect of climatic variation on agriculture in Iceland. In: The impact of climatic variation on Agriculture, Vol. I Cool Temperature and Cold Region(Ed. By M.L. Parry T.R. Corter and N.T. Konijn), pp. 232-509. Kluwer, Dordrecht.

Bhullar, S.S. and Jenner, C.F. 1983. Responses to brief periods of elevated temperatures in ears and grain of wheat. *Aust. J. Plant Physiol.* 10: 549-560.

Bhullar, S.S. and Jenner, C.F. 1985. Differential responses to high temperature of starch and nitrogen accumulation in the grain of four cultivars of wheat. *Aust. J. Plant Physiol.* 12: 313-325.

Bhullar, S.S. and Jenner, C.F. 1986. Effect of a brief episode of elevated temperature on grain filling in wheat ears cultured on solutions of sucrose. *Australian Journal of Plant Physiology* 13: 617–626.

Blum, A. 1988. Plant Breeding for Stress Environments. CRC Press, Inc., Boca Raton, Florida, 223.

Blum, A. 1998: Improving wheat grain filling under stress by stem reserve mobilization. *Euphytica.* 100: 77-83.

Blum, A., Sinmena, B., Mayer, J., Golan, G. and Shipiler, L. 1994: Stem reserve mobilization supports wheat grain filling under heat stress. *Aust. J. Plant Physiol.* 21: 771–781.

Boote, K.J., Pickering, N.B., Baker, J.T., Allen, L.H. Jr. 1994. Modelling leaf and canopy photosynthesis of rice in response to carbon dioxide and temperature. *International Rice Research Notes* 19: 47-48.

Bowes, G. 1993. Growth at elevated CO_2: photosynthetic responses mediated through Rubisco. *Plant Cell Environ.* 14: 795–806.

Bowes, G., Vu, J.C.V., Hussain, M.W., Pennanen, A.H. and Allen Jr., L.H. 1996: An overview of how rubisco and carbohydrate metabolism may be regulated at elevated atmospheric (CO_2) and temperature. *Agricultural and Food Science in Finland*, 5: 261-270.

Bunce, J.A. 1992. Light, temperature and nutrients as factors in photosynthetic adjustment to an elevated concentration of carbon dioxide. *Physiol. Plant.* 86: 73–179.

Campbell, C.A. and Davidson, H.R. 1979. Effect of temperature, nitrogen fertilizer and moisture stress on yield, yield components, protein content and moisture use efficiency on Manitou spring wheat. *Can. J. Plant Sci.* 59: 963–974.

Carter, T.R., Jones, R.N., Lu, X., Bhadwal, S., Conde, C., Mearns, L.O., O'Neill, B.C., Rounsevell, M.D.A., Zurek, M.B., 2007. New assessment methods and the characterisation of future conditions. Contribution of Working Group II to the Fourth Assessment Report of the Intergovernmental Panel on Climate Change, 2007. Cambridge University Press, Cambridge, UK, pp. 133–171.

Casella, E. and Soussana, J.F. 1997. Long-term effects of CO_2 enrichment and temperature increase on the carbon balance of a temperate grass sward. *Journal of Experimental Botany*, 48: 1309-1321.

Casella, E., Soussana, J.F. and Loiseau, P. 1996: Long term effects of CO_2 enrichment and temperature increase on a temperate grass sward, In: productivity and water use. *Plant and Soil*, 182: 83-99.

Cassman, K. G. 1999. *Proc. Natl. Acad. Sci.* USA 96: 5952–5959.

Chaisompongopan, N., Li, P.H., Davis, D.W. and Mackhart, A.H. 1990. Photosynthetic responses to heat stress in common bean genotypes differing in heat acclimation potential. *Crop Sci.* 30: 100-104.

Chandra, D., Lodh, S.B., Sahu, S.A., Sahoo, K.M. and Nanda, B.B. 1997. Effect of date of planting and spacing on grain yield and quality of scented rice (*Oryza sativa*) varieties in wet season in coastal Orissa. *Indian Journal of Agricultural Science.* 67: 93-97.

Cheikh, N. and Jones, R.J. 1994. Disruption of maize kernel growth and development by heat stress. *Plant Physiol.* 106: 45–51.

Cheikh, N., and Jones, R.J. 1995. Heat stress effects on sink activity of developing maize kernels grown *in vitro*. Physiol. Plant. 95: 59–66.

Commuri, P.D. 1997. Mechanisms by which high temperature during endosperm cell division causes kernel abortion in maize. Ph.D. diss. (Diss. Abstr. 002509983). Univ. of Minnesota, St. Paul, MN.

Commuri, P.D. and Jones, R.J. 1999. Ultrastructural characterization of maize (*Zea mays* L.) exposed to high temperature during endosperm cell division. *Plant Cell Environ.* 22: 375–385.

Commuri, P.D. and Jones, R.J. 2001. High temperatures during endosperm cell division in maize: A genotypic comparison under in vitro and field conditions. *Crop Sci.* 41: 1122–1130.

Conroy, J.P., Seneweera, S., Basra, A.S., Rogers, G. and Nissel-Wooller, B. 1994. Influence of rising atmospheric CO_2 concentrations and temperature on growth, yield and grain quality of cereal crops. *Australian Journal of Plant Physiology* (Australia). 21(6): 741.

Cure, J. D. and Acock, B. 1986. Crop responses to carbon dioxide doubling: a literature survey. *Agriculture and forest meteorology.* 38: 127-145.

Cure, J.D. 1988. Carbon dioxide doubling response: a crop survey. In: Direct effect of increasing carbon dioxide on vegetation, (Ed. By B.R. Strain and J.D. Cure), pp. 110-116 US DOE/ER-0238, Washington, USA.

Cure, J.D. and Acock, B. 1986. Crop response to carbon dioxide doubling: a literature survey. *Agric. Forest Meteorol.* 38: 127-145.

Dahlman, R.C., Strain, B.R. and Rogers, H.H. 1985. Research on the responses of vegetation to elevated atmospheric carbon dioxide. *Journal of environmental quality.* 14: 1-8.

Dai, A., Fung, I.Y. and Genio, A.D. 1997. Surface observed global land precipitation variation during 1900-1988, *Journal of Climate* 10: 2943-2962.

Daigger, L. A., Sander, D. H. and Peterson, G. A. 1976. Nitrogen content of winter wheat during growth and maturation. *Agron. J.* 68: 15—818.

Davidson, D. J. and Chevalier, P. M. 1992: Storage and remobilization of water-soluble carbohydrates in stems of spring wheat. *Crop Sci.* 32: 186—190.

Dela Cruz, N. 1991. Effect of temperature during grain development on the performance and stability of cooking quality of long grain rice. Research series- Arkansas agricultural Experimental Station, 456: 152-157.

Dela Cruz, N., Kumar, I., Kaushik, R.P. and Khush, G.S. 1989. Effect of temperature during grain development on the performance and stability of cooking quality components of rice. *Japanese Journal of Breeding.* 39: 299-306.

Dijkstra, P., Schapendonk, A.H.C.M., Groenwold, K., Jansen, M. and Van de Geijn, S.C. 1999. Seasonal changes in the response of winter wheat to elevated atmospheric CO_2 concentration grown in open-top chambers and field tracking enclosures. *Global Change Biol.* 5: 563–576.

Drake, B.G., Gonzàlez-Meler, M.A. and Long, S.P. 1997. More efficient plants: a consequence of rising atmospheric CO_2? *Annu. Rev. Plant Physiol. Plant Mol. Biol.* 48: 609–639.

Dugas, W.A., Heuer, M.L., Hunsaker, D., Kimball, B.A., Lewin, K.F., Nagy, J. and Johnson, M. 1994. Sap flow measurements of transpiration from cotton grown under ambient and enriched CO_2 concentrations. *Agricultural and Forest Meteorology*, 70: 231-245.

Easterling, D. R., Horton, B., Jones, P. D., Peterson, T. C., Karl, T. R., Parker, D. E., Salinger, M. J., Razuvayev, V., Plummer, N., Jamason, P. 1997. *Science* 277: 364–367.

Egeh, A.O., Ingram, K.T. and Zamora, O.B. 1994. High temperature effects on leaf exchange. *Phil. J. Crop Sci.* 17: 21-26.

Egli, D.B. 1998. Seed biology and the yield of grain crops. CAB Int., Oxford, UK.

Engelen-Eigles, G., Jones, R.J. and Phillips R.L. 2000. DNA endoreduplication in maize endosperm cells. I: The effect of exposure to short-term high temperature. *Plant Cell Environ*, 23: 657–663.

Evans, L.T. 1996. Crop evolution, adaptation and yield. Cambridge Univ. Press, Cambridge, UK.

Farooq, M., Bramley, H., Palta, J.A. and Siddiqu, K.H.M. 2011. Heat Stress in Wheat during Reproductive and Grain-Filling Phases. *Critical reviews in plant science*, 30:1–17.

Fleisher, D.H., Timlin, D.J. and Reddy, V.R. 2008. Elevated carbon dioxide and water stress effects on potato canopy gas exchange, water use, and productivity. *Agricultural and Forest Meteorology* 148: 1109-1122.

Fokar, M., Blum, A. and Nguyen, H. T. 1998. Heat tolerance in spring wheat. II. Grain filling. *Euphytica* 104: 9-15.

Folland, C.K., Karl, T. and Vinnikov, K.Ya. 1990. Observed climatic variation and change. In : Climate Change: The IPCC Scientific Assessment (Ed.by J.T. Houghton, G.J. Jenkins and J.J. Ephraums). PP. 195-238, Cambridge University Press, Cambridge, UK.

Ford, M.A., Pearman, I. and Thorne, G.N. 1978. Effects of variation in ear temperature on growth and yield of spring wheat. *Aust. J. Plant Physiol.* 3: 337-347.

Ford, M.A., Pearman, I., Thorne, G.N. 1976. Effects of variation in ear temperature on growth and yield of spring wheat. *Annals of Applied Biology.* 82: 317–333.

Gebbing, T. and Schnyder, H. 1999. Pre-anthesis reserve utilization for protein and carbohydrate synthesis in grains of wheat. *Plant Physiol.* 121: 871–878.

Geng, S., and Cady, C.W. (eds.), 1991. Climatic Variation and Change: Implications for Agriculture in the Pacific Rim. Proceedings, U.C. Davis, California.

Gibson, L.R. and Mullins, R.E. 1996. Influence of day and night temperature on soybean seed yield. *Crop Sci.* 36: 1636–1642.

Gibson, L.R. and Paulsen, G.M. 1999. Yield components of wheat grown under high temperature stress during reproductive growth. *Crop Science* 39: 1841-1846.

Gifford, R.M. 1988. Direct effect of carbon dioxide levels concentration on vegetation. In: Greenhouse: Planning for climate change (Ed. By G.I. Pearman), pp.506-519, CSIRO, Melbourne, Australia.

Giorgi, R., Meehl, G.A., Kattenberg, A., Grassl, H., Mitchell, J.F.B., Stouffer, R.J., Tokioka, T., Weaver, A.J. and Wigley, T.M.L. 1998. Simulation of regional climate change with global coupled climate models and regional modeling techniques. p. 427.

Groisman, Y.P., Karl, T,R., Easterling, D.R., Knight, R.W., Jamason, P.F., Hennessy, K.J., Suppiah, R., Page, C.M., Wibig, J., Fortuniak, K., Razuvaev, V.N., Douglas, A., Foreland, E. and Zhai, P. 1999. Changes in probability of heavy precipitation. Important indicator of climate change. 42: 246-283.

Guedira, M. and Paulsen, G.M. 2002. Accumulation of starch in wheat grain under different shoot/root temperatures during maturation. *Funct. Plant Biol.* 29: 495–503.

Halford, N. G. 2009. New insights on the effects of heat stress on crops. *J. Exper.Bot.* 60: 4215–4216.

Hall, A.E. 1992. Breeding for heat tolerance. *Plant Breed Res.* 10: 129-168.

Halloran, G. M., 1981: Cultivar differences in nitrogen translocation in wheat. *Aust. J. Agric. Res.* 32: 535-544.

Ham, J.M., Owensby, C.E., Coyne, P.I. and Bremer, D.J. 1995. Fluxes of CO_2 and water vapor from a prairie ecosystem exposed to ambient and elevated atmospheric CO_2. *Agricultural and Forest Meteorology*, 77: 73-93.

Hansen, J., Lacis, A., Rind, D., Russel, G., Stone, P., Fung, I., Ruedy, R., Lerner, J. 1984. climate sensitivity: analysis of feedback mechanisms. In: Hansen, J., Takahashi, T., eds. Climate process and climate sensitivity. Washington DC: American Geophysical Union, 130-163.

Hawker, J.S. and Jenner, C.F. 1993. High temperature effects on the activity of enzymes in the committed pathway of starch synthesis in developing wheat endosperm. *Aust. J. Plant Physiol.* 20: 197-200.

He, G.C., Kogure, K. and Suzuki, H. 1990. Development of endosperm and synthesis of starch in rice grain. III. Starch property affected by temperature during grain development. *Japanese Journal of Crop Science*, 59: 340-345.

Heitholt, J. J., Croy, L. I., Maness, N. O. and Nguyen, H. T. 1990. Nitrogen partitioning in genotypes of winter wheat differing in grain N concentration. *Field Crops Res.* 23: 133-144.

Hill, M. G. and Dymock, J.J. 1989. Impact of climate change: Agricultural/Horticultural systems. DSIR Entomology Division, submission to New Zealand climate change programme, Department of Scientific and Industrial Research, New Zealand, pp. 16.

Högy, P. and Fangmeier, A., 2008. Effects of elevated atmospheric CO_2 on grain quality of wheat. *J. Cereal Sci.* 48: 580-591.

Högy, P., Brunnbauer, M., Koehler, P., Schwadorf, K., Breuer, J., Franzaring, J., Zhunusbayeva, D. and Fangmeier, A. 2012: Grain quality characteristics of spring wheat (*Triticum aestivum*) as affected by free-air CO_2 enrichment. *Environ. Exp. Bot.* (in press).

Horie, T., Baker, J. T., Nakagawa, H., Matsui, T. and Kim, H. Y. 2000. in Climate Change and Global Crop Productivity, eds. Reddy, K. R. and Hodges,H. F. (CAB International, Wallingford, U.K.), 81–106.

Horie, T., Baker, J.H., Nakagawa, H. Mitsui, T. and Kim, H.Y. 2000. Crop Ecosystem to Climate Change. Rice. In: Climate Change and Global Crop Productivity (Ed. by K.R. Reddy and H.F. Hodges) pp. 81-106, CAB International Publishing, Wallingford, Oxon UK.

Horie, T., Centeno, H.G.S., Nakagawa, H., Matsui, T. 1997. Effect of elevated carbon dioxide and climate change on rice production in east and Southeast Asia. In: Oshima Y, ed. Proceedings of the international scientific symposium on Asian paddy fields. Canada, Saskatchewan: College of Agriculture, University of Saskatchewan, 49-58.

Horie, T., Matsui, T., Nakagawa, H., Omasa, K. 1996. Effect of elevated CO_2 and global climate change on rice yield in japan. In: Omasa, K., Kai, K., Toda, H., Uchijima, Z., Yoshino, M., eds. Climate change and plants in East Asia. Tokyo: Springer-Verlag, 39-56.

Horton, B., 1995. Geographical distribution of changes in maximum and minimum temperatures. *Atmospheric Research,* 37: 101-117.

Hossain, M., 1997. Rice supply and demand in Asia: a socio- economic and biophysical analysis. In: Teng, P.S., *et al.* (Eds.), Applications of Systems Approaches of the Farm and Regional Levels. Kluwer Academic Publishers, Dordrecht, the Netherlands, pp. 263 279.

Houghton, J.T., Collander, B.A. and Ephraums, J.J. (eds.). 1990. Climate Change - The IPCC Scientific Assessment. Cambridge University Press, Cambridge. p. 135.

Hunsaker, D.J., Hendrey, G.R. Kimball, B.A. Lewin, K.F. Mauney, J.R. and Nagy, J. 1994: Cotton evapotranspiration under field conditions with CO_2 enrichment and variable soil moisture regimes. *Agricultural and Forest Meteorology,* 70: 247-258.

Idso, K.E. and Idso, S.B. 1994. Plant responses to atmospheric CO_2 enrichment in the face of environmental constraints: a review of the past 10 years' research. *Agric. For. Meteorol.* 69: 153–203.

Idso, S.B., 1983. The long-term response of trees to atmospheric CO_2 enrichment. *Global Change Biology*, 5: 593-595.

Inaba, K., and K. Sato. 1976. High temperature injury of ripening in rice plant: VI. Enzyme activities of kernel as influenced by high temperature. *Proc. Crop Sci. Soc. Jpn.* 45:162–167.

IPCC, 1996. Houghton, J.T., Meira Filho, L.G., Callander, B.A., Harris, N., Kattenberg, A., Maskell, K. (Eds.), Climate Change 1995: The Science of Climate Change. Cambridge University Press, Cambridge, MA, p. 572.

IPCC, 2001. Climate Change. The Scientific Basis. Contribution of Working Group I to the Third Assessment Report of the Intergovernmental Panel on Climate Change (Eds. Houghton J.T. *et al.*). Cambridge University Press, UK, p. 881.

IPCC. 1990a. *Climate Change: The IPCC Scientific Assessment.* J.T. Houghton, G.J. Jenkins and J.J. Ephraums (eds.). Cambridge University Press, Cambridge.

Jenner, C.F. 1991a. Effects of exposure of wheat ears to high temperature on dry matter accumulation and carbohydrate metabolism in the grain of two cultivars. I. Immediate response. *Aust. J. Plant Physiol.* 18: 165-177.

Jenner, C.F., Denyer, K., Guerin, J. 1995 Thermal characteristics of soluble starch synthase from wheat endosperm. *Australian Journal of Plant Physiology* 22, 703–709.

Johns, T.C., Cammel, R.E., Grossley, J.F., Gregory, J.M., Mitchell, J.F.B., Senior, C.A., Ten, S.F.B. and Wood, R.A. 1997. The second Hadley center coupled ocean-atmosphere GCM: Model description spinup and validation. *Climate Dynamics*. 13:103-134.

Jones, C.A. and Kiniry, J.R., 1986. CERES-maize: A Simulation Model of Maize Growth and Development. Texas A and M University Press, College Station, TX.

Jones, P. D. and Wigley, T. M. L. 1990. Global warming trends. Sci. Am. 263: 4—91.

Jones, P.D., Wigley, T.M.I. and Briffa, K.R. 1994. Global and hemispheric anomalies: land and marine instrumental record. In: Trends 93: a Compendium of data on clobal change, carbon dioxide analysis center (Ed. by T.A. Boden, D.P. Kaiser, R.J. Sepanski and F.W. Stass) pp. 603-608, Oak Ridge National Laboratort, Oak Ridge Tennnessee.

Jones, R.J., Gengenbach, B.G. and Cardwell, V.B. 1981. Temperature effect on in vitro kernel development in maize. *Crop Sci.* 21: 761–766.

Juliano, B. O., 1972. Physio-chemical properties of starch and proteins in relation to grain quality and nutrition value of rice 389-404., In: *Rice Breeding.* IRRI, Manila, Phillippines.

Karl, T. R., Kukla, G. and Razuvayev, V. N. 1991 *Geophys. Res. Lett.* 18, 2253–2256.

Karl, T.R., Jones, P.D., Knight, R.W., Kukla, G., Plummer, N., Razuvayev, V. Gallo, K.P., Lindseay, J., Charlson, R.J. and Pererson, T.C. 1993. Asymetric trend of daily maximum and minimum temperature. *Bulletin of the American Meteorological Society* 74: 1007-1023.

Karl, T.R., Knight, R.W.,Esterling, D.R. and Quayle, R.G. 1996. Indices of climate change for USA. *Bulletin of the American Meteorological Society* 77: 279-292.

Katz, R.W. and Brown, B.G. 1992. Extreme events in a changing climate: variability is more important than averages. *Clim. Change* 21: 289-302.

Keeling, C.D. and Whorf, T.P. 2000. Atmospheric CO_2 records from sites in the SIO air sampling network. *In* Trends: A compendium of data on global change. Carbon Dioxide Information Analysis Center, Oak Ridge Nat. Lab., Oak Ridge, TN.

Keeling, P.L., Bacon, P.J. and Holt, D.C. 1993. Elevated temperature reduces starch deposition in wheat endosperm by reducing the activity of soluble starch synthase. *Planta* 191: 342–348.

Keeling, P.L., Banisadr, R., Barone, L., Wasserman, B.P. and Singletary, G.W. 1994. Effect of temperature on enzymes in the pathway of starch biosynthesis in developing wheat and maize grain. *Australian Journal of Plant Physiology* 21, 807–827.

Khush, G. S., Paule, C. M. and Dela Cruz, N. M. 1979. Rice grain quality evaluation and improvement at IRRI. In: Chemical Aspects of Rice Grain Quality IRRI, Los Bonos, Philippines. pp. 21-31.

Kimball, B. A. and Idso, S. B. 1983. Increasing atmospheric CO_2 effects on crop yield water use and climate. *Agricultural water management*, 7: 55-72.

Kimball, B.A., LaMorte, R.L., Seay, R.S., Pinter, Jr., P.J., Rokey, R.R., Hunsaker, D.J., Dugas, W.A., Heuer, M.L., Mauney, J.R., Hendrey, G.R., Lewin, K.F. and Nagy, J. 1994. Effects of free-air CO_2 enrichment on energy balance and evapotranspiration of cotton. *Agric. for. Meteorol.* 70: 259–278.

Kimball, B.A., Pinter, Jr., P.J., Garcia, R.L., LaMorte, R.L., Wall, G.W., Hunsaker, D.J., Wechsung, G., Wechsung, F. and Kartschall, T. 1995. Productivity and water use of wheat under free-air CO_2 enrichment. *Global Change Biol.* 1: 429–442.

Kirschbaum, M.U.F., Fischlin, A. Cannell, M.G.R. Cruz, R.V.O. Cramer, W. Alvarez, A. Austin, M.P. Bugmann, H.K.M. Booth, T.H. Chipompha, N.W.S. Cisela, W.M. Eamus, D. Goldammer, J.G. Henderson-Sellers, A. Huntley, B. Innes, J.L. Kaufmann, M.R. Kräuchi, N. Kile, G.A. Kokorin, A.O. Körner, C. Landsberg, J. Linder, S. Leemans, R. Luxmoore, R.J. Markham, A. McMurtrie, R.E Neilson, R.P. Norby, R.J. Odera, J.A. Prentice, I.C. Pitelka, L.F. Rastetter, E.B. Solomon, A.M. Stewart, R. van Minnen, J. Weber, M. and Xu, D. 1996: Climate change impacts on forests. In: Climate Change 1995: Impacts, Adaptations and Mitigation of Climate Change: Scientific-Technical Analyses. Contribution of Working Group II to the Second Assessment Report of the Intergovernmental Panel on Climate Change [Watson, R.T., M.C. Zinyowera, and R.H. Moss (eds.)]. Cambridge University Press, Cambridge, United Kingdom and New York, NY, USA, pp. 95-129.

Kolderup, F. 1975. Effect of temperature, photoperiod, and light quantity on protein production in wheat grain. *J. Sci. Food Agric.* 26: 583–592.

Kukla, G. and Karl, T.R. 1993. Nighttime warming and the greenhouse effect. *Envir. Sci. Technol.* 27 (8): 1468-1474.

Kuroyanagi, T, Paulsen, G.M. 1988. Mediation of high-temperature injury by roots and shoots during reproductive growth of wheat. Plant, Cell and Environment 11: 517–523.

Lamb,H.H. 1982. Climate, History and the Modern World. Cambridge University Press, Cambridge, UK, pp. 384.

Lansigan, F.P. and de los Santos, W.L. 2000. Impacts of climate variability on rice production in the Philippines Agriculture, Ecosystems and Environment. *Coladilla J.O.Agronomic,* 82(1): 129-137.

Lawlor, D.W. and Mitchell, R.A.C. 2000. Crop Ecosystem Response to Climatic Change: Wheat. In: Climate Change and Global Crop Productivity (Ed. by K.R. Reddy and H.F. Hodges) pp. 57-80, CAB International Publishing, Wallingford, Oxon UK.

Leakey, A. D. B., Ainsworth, E. A. *et al.,* Elevated CO_2 effects on plant carbon, nitrogen, and water relations; six important lessons from FACE. *Journal of Experimental Botany* **60**, 2859-2876 (2009).

Lee, P.C., Bochner, B.R. and Ames, B.N. 1983. A heat shock stress and cell oxidation. *Proc. Natl. Acad. Sci., USA* 80: 7496-7500.

Li, X., Gu, M.H. and Pan, X.B. 1989. Studies on rice grain quality. II. Effect of environmental factors at filling stage on rice grain quality. *Journal of Jiangsu Agricultural College.* 10: 7-12.

Long, S.P. 1991. Modification of the response of photosynthetic productivity to rising temperature by atmospheric CO_2 concentrations: Has its importance been underestimated? *Plant Cell Environ.* 14: 729-739.

Long, S.P. 1991. Modification of the response of photosynthetic productivity to rising temperature by atmospheric CO2concentrations: Has its importance been underestimated? *Plant, Cell and Environment* 14: 729-739.

Manderscheid, R., Bender, J., Jager, H.J. and Weigel, H.J. 1995. Effects of season long CO_2 enrichment on cereals. II. Nutrient concentrations and grain quality. *Agr. Ecosyst. Environ.* 54: 175 -185

Matthews, R. B., Kropff, M. J., Horie, T. and Bachelet, D. 1997. Agric. Syst. 54: 399–425.

Matthews, R.B., Kropff, M.J. and Bachelet, D. 1997. Simulating the impact of climate change on rice production in Asia and evaluating options for adaptation. *Agricultural Systems,* 54: 399-425.

McMichael, B.L. and Burke, J.J. 1996. Temperature effects on root growth. In Plant roots: the hidden half (2nd edn). Y Waisel, A Eshel and U Kafkafi (Eds.), 383–396. (Marcel Dekker: New York, NY).

Meehl, G.A., Stocker, T.F., Collins, W.D., Friedlingstein, P., Gaye, A.T., Gregory, J.M., Kitoh, A., Knutti, R., Murphy, J.M., Noda, A., Raper, S.C.P., Watterson, I.G., Weaver, A.J., Zhao, Z.C., 2007. Global climate projections. In: Solomon, S., Qin, D., Manning,

M., Chen, Z., Marquis, M., Averyt, K.B., Tignor, M., Miller, H.L. (Eds.), Climate Change 2007: The Physical Science Basis. Contribution of Working Group I to the Fourth Assessment Report of the Intergovernmental Panel on Climate Change. Cambridge University Press, New York, USA.

Meinzer, F.C. and Zhu, J. 1998. Nitrogen stress reduces the efficiency of the C4-CO_2 concentrating system, and therefore quantum yield, in *Saccharum* (sugarcane) species. *Journal of Experimental Botany*, 49:1227-1234.

Merca, F. E. and Juliano, B. O. 1981. Physico-chemical properties of starch of intermediate-amylose and starch/starke, 33: 253- 260.

Miglietta, F., Bindi, M., Vaccari, F. P., Schapendonik, A.H.C.M., Wolf, J. and Butterfield, R.E. 2000. Crop Ecosystem Response to Climatic Change: Root and Tuberous crop. In: Climate Change and Global Crop Productivity (Ed. by K.R. Reddy and H.F. Hodges) pp. 189-212, CAB International Publishing, Wallingford, Oxon UK.

Mishra, R.K. and Singhal, G.S. 1992. Function of photosynthetic apparatus of intact wheat leaves under high light and heat stress and its relationship with thylakoid lipids. *Plant Physiol.* 98: 1-6.

Mitchell, J.F.B., Johns, T.C., Gregory, J.M. and Tett, F.B. 1995. Climate response to increasing levels of greenhouse gases and sulphate aerosols. *Nature*. 376: 501- 504.

Mora, C., A.G. Frazier, Longman, R.L., Dacks, R.S., Walton, M.M., Tong, E.J., Sanchez, J.J., Kaiser, R.L., Stender, Y.O., Anderson, J.M., Ambrosino, C.M., Iria Fernandez-Silva, Giuseffi, L.M. andGiambelluca, T.W. 2013. The projected timing of climate departure from recent variability. *Nature*, 502:183–187.

Morison, J.I.L. 1987. Intercellular CO_2 concentration and stomatal response to CO_2. In: : Stomatal Function(Ed, by E. Zeiger, I.R. Crown and G.D. Farquhar), pp. 229-251. Standford University Press, Standford.

Nagato, K., and Ebata, M. 1965. Effects of high temperature during ripening period on the development and the quality of rice kernels. *Proc. Crop Sci. Soc. Jpn.* 34:59–66.

Neilsen, I.F. 1974. Roots and root temperatures. In 'The plant root and its environment'. (Ed. EW Carson) pp. 293–333. (University Press of Virginia: Charlottesville, VA)

Nicholls, M., Gruza, G.V., Jouzel, J., Karl, T.R., Ogallo, L.A. and Parker, D.E. 1996. Observed climatic variability and change. In:Climate change 1995 (Ed. by J.T. Houghton, L.G. Meira Filtho, B.A. Callender, N. Harrie, A. Kattenberg and K. Maskel) pp. 133-192, The Science of Climate Change. Cambridge University Press, Cambridge, UK.

Nielsen, C. L. and A. E. Hall. 1985a. Responses of cowpea (*Vigna unguiculata* [L.] Walp.) in the field to high night temperature during flowering. I. Thermal regimes of production regions and field experimental system. *Field Crops Res.* 10: 167-179.

Nielsen, C. L. and A. E. Hall. 1985b. Responses of cowpea (*Vigna unguiculata* [L.] Walp.) in the field to high night temperatures during flowering. II. Plant responses. *Field Crop Res*.10: 181-196.

Nishiyama, I. 1985. Physiology of cool weather damage in rice. Hokkaido Univ., Sapporo, Japan.

Ober, E.S., Setter, T.L. Madison, J.T. Thompson, J.F. and Shapiro, P.S. 1991. Influence of water deficit on maize endosperm development. *Plant Physiol*. 97:154–164.

Osmond, C. B., Bjorkman, O. and Anderson, D. J.1980. Physiological process of plant ecology: towards a synthesis and atriplex. In Ecological studies, 36: 1-5.

Palta, J. A., Kobata, T., Turner, N. C. and Filley, I. R., 1994. Remobilization of carbon and nitrogen in wheat as influenced by postanthesis water deficits. *Crop Sci*. 34: 118-124.

Papakosta, D. K. and Gagianas, A.A. 1991. Nitrogen and dry matter accumulation, remobilization and losses from Mediterranean wheat during grain filling. *Agron. J*. 83: 864-870.

Parry, M. 1990. The potential effects of climate change on agriculture and land use. *Advances in Ecological Research* 22: 63-91.

Parry, M.L. 1978. Climate, Agriculture ans Settlement. Dawson and Sons, Folkestone, UK, pp. 214.

Parry, M.L. 1990. Climate Change and World Agriculture. Earthscan, London.

Parry, T.R. Carter and N.T. Konijn pp. 511-614. Kluwer, Dordrecht. Monteith, J.L.1981. Climatic variation and the growth of crops. *Q.J.R. Meteorol. Soc*. 107:749-774.

Paulsen, G.M. 1994. High temperature response in crop plants. In 'Physiology and determination of crop yield'. (eds. AJ Boots, JM Bennett, TR Sinclair and GM Paulsen) pp. 365–394. (American Society of Agronomy: Madison, WI).

Pearcy, R.W., and Bjorkman, J. 1983. Comparative ecophysiology of C_3 and C_4 plants. *Plant Cell Environ*. 7: 1–13.

Peng, S., Ingram, K.T., Neue, H.U., Ziska, L.H. (Eds.), 1996. Climate Change and Rice. Springer/IRRI, Berlin/Los Banos, pp. 374.

Peng, S; Haung, J; Sheehy, J.E; Laza, R.C; Visperas, R.M; Zhong, X; Centeno, G.S; Khush, G.S. and Cassman, K.G. 2006. Rice yield decline with higher night temperature from global warming. Proceeding of the National Academy of Science of United State of America Vol. 1.

Poorter, H., 1993. Interspecific variation in the growth response of plants to an elevated ambient CO_2 concentration. *Vegetatio*, 104: 77-97.

Power, J.J., Grues, D.L., Willis, W.O., Reichmann, G.A. 1963. Soil temperature and phosphorus effect upon barley growth. *Agronomy Journal* 55: 389–392.

Reddy, K.R., Hodge, H.F. and Kimbell. B.A. 2000. Crop Ecosystem Response to Climatic Change: Cotton. In: Climate Change and Global Crop Productivity (Ed. by K.R. Reddy and H.F. Hodges) pp. 168-188, CAB International Publishing, Wallingford, Oxon UK.

Reddy, V.R., Reddy, K.R. and Hodges, H.F. 1995. Carbon dioxide enrichment and temperature effects on cotton canopy photosynthesis, transpiration, and water-use efficiency. *Field Crops Research*, 41: 13-23.

Rind, D., Goldberg, R. and Reddy, R. 1989. Change in climate variability in the 21st century. *Climatic Change*, 14: 5-37.

Robertson, G. W., 1968. A biometeorological time scale for a cereal crop involving day and night temperature and photoperiod. *Int. J. Biometeorol.* 12: 191-223.

Rosenzweig, C. and Parry, M. L. 1994. *Nature*, 367: 133-138.

Rowntree, P.R., Callander, B.A., and Cochrane, J. 1989. Modelling climate change and some potential effect on agriculture in UK. *Journal of the Royal Society of England.* 149: 120-126.

Ruget, F., Bethenod, O. and L. Combe, 1996. Repercussions of increased atmospheric CO_2 on maize morphogenesis and growth for various temperature and radiation levels. *Maydica*, 41: 181-191.

Samarakoon, A.B., and Gifford R.M. 1995. Soil water content under plants at high CO_2 concentration and interactions with the direct CO_2 effects: a species comparison. *Journal of Biogeography* 22:193–202.

Samarakoon, A.B., and Gifford R.M. 1996. Elevated CO_2 effects on water use and growth of maize in wet and drying soil. *Aust. J. Plant Physiol.* 23: 53–62.

Sasaki, T. 1935. Effects of low temperature on grain ripening of rice plants. *Rep. Sci. Assoc.* 10:449–453.

Satake, T., Yoshida, S. 1978. High temperature induced sterility in Indica rice at flowering. *Japanese Journal of Crop Science* 47: 6-17.

Sato, K., and Inaba K. 1976. High temperature injury of ripening in rice plant: V. On the early decline of assimilate storing ability of grains at high temperature. *Proc. Crop Sci. Soc. Jpn.* 45:156–161.

Schneider, S.H. Gleick, P.H. and Meams, L.O. 1990. Prospects for climate change. In: Ammerican Association for Advancement of Science.Climate and water: Climate change, Climatic variability and the Planning and Management of US Water resources. John Viley and Sons, New York, pp. 41-73.

Schnyder, H., 1993: The role of carbohydrate storage and redistribution in the source-sink relations of wheat and barley during grain filling. A review. *New Phytol.* 123: 233—245.

Simmons, S.R., 1987. Growth, development, and physiology. In: E.G. Heyne, ed. Wheat and Wheat Improvement, 2nd edn, pp. 77-113. ASA/CSSA/SSSA, Madison, WI, USA.

Simpson, R. J., Lambers, H. and Dalling, M. J. 1983. Nitrogen redistribution during grain growth in wheat (*Triticum aestivum* L.) IV. Development of a quantitative model of the translocation of nitrogen to the grain. *Plant Physiol.* 71, 4—14.

Singh, R.K, Singh, U.S., Khush, G.S. (Eds.) Aromatic rices, Oxford and IBH Publishing Co. Pvt. Ltd. New Delhi., 2000.

Singh, S. 2000. Growth, yield and biochemical response of rice cultivars to low light and high temperature humidity stress. *Oryza* 37(1): 35-38.

Singh, S.P., Pillai, K.G., Pati, D. and Shobha Rani, N. 1993. Influence of time of planting on grain yield and quality of dwarf scented rice varieties. *Oryza*. 30: 285-288.

Singletary, G.W., Banisadr, R. and Keeling, P.L. 1994. Heat stress during grain filling in maize: Effects on carbohydrate storage and metabolism. *Aust. J. Plant Physiol.* 21: 829–841.

Sinha, S. K., Singh, G. B. and Rai, M. 1998. Decline in Crop Productivity in Haryana and Punjab: Myth or Reality? Report of Fact Finding Committee, pp. 89. ICAR, New Delhi.

Sinha, S.K. 1992. Climate Change and Agriculture. *Indian Farming*, May 1992, pp.13-18

Sofield, I., Wardlaw, I.F., Evans, L.T., Zee, S.U. 1977. Nitrogen, phosphorus, and water contents during grain development and maturation in wheat. *Australian Journal of Plant Physiology* 4: 799–810.

Solomon, S., Qin, D., Manning, M., Marquis, M., Averyt, K., Tignor, M.M.B., Miller, H.L., Chen, Z., 2007. Climate change 2007: the physical science basis. Contribution of Working Group I to the Fourth Assessment Report of the Intergovernmental Panel on Climate Change. Cambridge University Press, New York.

Squire, G.R. and Unswarth, M.H. 1988. Effect of CO_2 and climatic change on agriculture.Contract Report to the Department of the Environment, Department of Physiology and Environmental Science, University of Nottingham, Sutton Bonnington, UK.

Stirling, C.M., Davey, P.A., Williams, T.G. and Long, S.P. 1997. Acclimation of photosynthesis to elevated CO_2 and temperature in five British native species of contrasting functional type. *Global Change Biol.* 3: 237–246.

Stone, P.J. and Nicolas M.E. 1998a. The effect of duration of heat stress during grain filling on two wheat varieties differing in heat tolerance: Grain growth and fractional protein accumulation. *Aust. J. Plant Physiol.* 25:13–20.

Strain, B. R. 1985. Physiological and ecological control on carbon sequestering in terrestrial ecosystems. *Biogeochemistry*. 1:293-332.

Tahir, S. A. and Nakata, N. 2005. Remobilization of Nitrogen and Carbohydrate from Stems of Bread Wheat in Response to Heat Stress during Grain Filling. *J. Agronomy and Crop Science* 191: 106-115.

Tashiro, T., and Wardlaw, I.F. 1989. A comparison of the effect of high temperature on grain development in wheat and rice. *Ann. Bot.* 64:59–65.

Tashiro, T., and Wardlaw, I.F. 1990. The effect of high temperature at different stages of ripening on grain set, grain weight and grain dimensions in the semi-dwarf wheat 'Bakns'. *Ann. Bot.* 65:51–61.

Taub, D.R., Miller, B. and Allen, H. 2008. Effects of elevated CO_2 on the protein concentration of food crops: a meta-analysis. *Global Change Biol.* 14, 565–575.

Terjung, W.H., Ji, H.Y., Hayes, J.T., O'Rourke, P.A. and Todhunter, P.E. 1989. Actual and potential yield for rainfed and irrigated maize in China. *International Journal of Biometeorology*, 28, 115-135.

Thakur, R.B., Pandeya, S.B. and Dwivedi, P.K. 1996. Effect of time of planting on performance of scented rice. *Oryza* 33:107-109.

UNEP. 1989. Criteria for assessing vulnerability of sea level rise a global inventory to high risk areas. UNEP and government of Netherlands. Report. pp. 57.

Upadhyaya, A., Davis, T.D. and Sankhla, M. 1991. Heat shock tolerance and anti-oxidant activity in moth bean seedlings treated with tetayclasis. *Plant Growth Regulation* 10: 215-222.

Upadhyaya, A., Davis, T.D., Larsen, M.H., Walsen, R.H. and Sankhla, M. 1990. Uniconazole-induced thermotolerance in soybean seedling root tissue. *Physiol. Plant.* 79: 78-84.

Van deepen, C.A., Van Keulen, H., Penning de Vries, F.W.T., Noy, I.G.A.M. and Goudriaan, J. 1987. Simulated variability of wheat and rice yields in current weather conditions and in future weather when ambient CO_2 has doubled. In: Simulation Reports CABO-TT, 14., Wageningen, The Netherlands: University of Wageningen.

Van Sanford, D. A., and Mackown, C. T. 1987: Cultivars differences in nitrogen remobilization during grain.ll in soft red winter wheat. *Crop Sci.* 27: 295—300.

Vinnikov, K.Ya, Groisman, P.Ya and Lugina, K. M. 1990: Emprical data on contemporary global climate change: temperature and precipitation. *J. Clim.* 3: 662—677.

Wallwork, M.A.B., Logue, S.J. MacLeod, L.C. and Jenner C.F. 1998. Effect of high temperature during grain filling on starch synthesis in the developing barley grain. *Aust. J. Plant Physiol.* 25:173–181.

Wardlaw, I. F., and Wrigley, C.W. 1994. Heat tolerance in temperate cereals: an overview. *Australian Journal of Plant Physiology* 21, 695–703.

Wardlaw, I.F. 1974. Temperature control of translocation. In: Mechanism of Regulation of Plant Growth. R.L. Bielske, A.R. Ferguson and M.M. Cresswell (eds.). Bull. Royal Soc. New Zealand, Wellington. pp. 533-538.

Wardlaw, I.F., Moncur, L. 1995. The response of wheat to high temperature following anthesis. I. The rate and duration of kernel filling. *Australian Journal of Plant Physiology* 22: 391–397.

Warrag, M. O. A. and Hall, A. E. 1984a. Reproductive responses of cowpea [*Vigna unguiculata* (L.) Walp.] to heat stress. I. Responses to soil and day air temperatures. *Field Crops Res.* 8: 3-16.

Warric, R.A., Gifford, R. and Parry, M.L. 1986. CO_2, climatic change and agriculture. In: The Greenhouse Effect, Climatic change and Ecosystem (Ed. By B. Bolin, B.R. Doos, J. Jager and R.A. Warrick) pp. 393-473. SCOPE 29. John Wiley, Chichester.

Wattal, P.N. 1965. Effect of temperature on the development of the wheat grain. *Indian J. Plant Physiol.* 8: 145-159.

Wilhelm, E.P., Mullen, R.E., Keeling, P.L. and Singletary G.W. 1999. Heat stress during grain filling in maize: effects on kernel growth and metabolism. *Crop Science* 39, 1733–1741.

Yang, J., Zhang J., Huang Z., Zhu Q., and Wang L. 2000. Remobilization of carbon reserves is improved by controlled soil-drying during grain filling of wheat. *Crop Sci.* 40, 1645-1655.

Yong, K.J. and Long, S.P. 2000. Crop Ecosystem response to climate change: Maize and Sorghum. In: Climate Change and Global Crop Productivity (Ed. by K.R. Reddy and H.F. Hodges) pp. 107-132, CAB International Publishing, Wallingford, Oxon UK.

Yoshida, S. 1981. Fundamentals of rice crop science. International Rice Research Institute, Los Banos, Philippines.

Zheng, K.L. and Mackill, D.T. 1982. Effect of high temperature on anther dehiscence and pollination in rice. *Sabrao J.* 14: 61-66.

Chapter 3
Sugarcane Physiology under Abiotic Stress Environments

S. Vasantha, R. Gomathi, S. Venkataramana and P.N. Gururaja Rao

Crop Production Division, Sugarcane Breeding Institute, Coimbatore – 641 007, T.N.

CONTENTS

1. Introduction

Sugarcane is grown in both tropical and sub-tropical areas of the country from 7°to 32°N latitudes. The tropical regions south of 23°N latitude are best suited for cane cultivation and yields are consistently high compared to regions further north. The average yield realized in Tamil Nadu is the highest at about 110 t/ha. The sub-tropical regions experience pronounced winter and consequently have reduced growth period. Besides, the sub-tropical regions also experience poor sprouting and growth during winter affecting the winter ratooning. Ratoon yields are generally low in the country and poor ratoons are one of the major reasons for low productivity. The warm days and cool nights prevailing in Maharashtra and Northern Karnataka during the crop maturity phase provide ideal conditions for sucrose accumulation and this region records the highest sucrose recovery in the country. The cane production in the country is dependent on rainfall and drought spells appearing in regular intervals leading to wide fluctuations in cane area and production.

Sugarcane productivity is mainly dependent on growth, sucrose accumulation and yield, and the environment in which it is cultivated. Abiotic stresses are the most important limiting factors for cane productivity. These stresses include drought, salinity, temperature extremes, heavy metals and radiation which cause detrimental effects on plant growth and yield. These negative factors affect the root function, growth rates, metabolism and in extreme cases lead to dehydration and death. Also, the expected rise in global temperatures indicates that there is an urgent need to understand and improve plant tolerance to these stresses. In India, the productivity losses due to various abiotic stresses vary from 20 to 50 per cent (Dwivedi, 2000). In Maharashtra, a high recovery zone, large areas have gone out of cultivation due to salinity, alkalinity and waterlogging (Zende and Hapse, 1986, Zende, 2002).

Irrigated or dependable rainfall areas offered high yields; however, the average yields remained low in constraint environments. Drought is the primary abiotic stress causing not only differences between the mean yield and the potential yield but also causing yield instability. Drought stress associated with high day temperature causes poor growth and high tiller mortality particularly during primary growth stage which normally coincides with summer months in tropics. High temperatures have deleterious effects on plant photosynthesis, respiration and reproduction. A small increase in temperature results in conspicuous effect on growth and survival. Elevated temperatures cause rapid loss of water resulting in dehydration. In addition, drought coupled with water logging *i.e.* early drought and subsequent water logging in Bihar, U.P. and Orissa is becoming a serious productivity constraint affecting considerable area under sugarcane cultivation. Sugarcane is moderately tolerant to flooding and water logging. However duration of water logging and the physiological stage at which the problem occurs determines the final yield and quality. Salinity is another major constraint in sugarcane agriculture. It is primarily due to irrigation with poor quality water (mostly saline). Continuous irrigation with saline water, improper drainage and inadequate reclamation of saline soils lead to considerable yield losses. It is therefore obvious that as Boyer (1982) pointed out, the crop plants attain only about 25 per cent of their potential yield because of these detrimental effects imposed by environmental stresses. These abiotic stresses (Table 3.1) are location specific, exhibiting variation in frequency, intensity and duration and might occur at any stage of plant growth and development.

Table 3.1: Environmental Stresses Affecting Cane Productivity in different Agro Climatic Zones of India.

Stresses	Peninsular		East Coast		North Central	North Western
	Upper	Lower	I	II		
Drought	+	+	+	+	+	+
Water logging			+	+	+	
Salinity		+			+	+
Alkalinity		+			+	+
Low temperature						+

1.1. Growth Phases in Sugarcane

Sugarcane crop passes through four distinct physiological growth phases *i.e.* germination (0 to 60 days), formative (60 to 150 days), grand growth (150 to 240 days) and maturity (240 to 360 days). Each phase requires a set of specific light, temperature and water availability. Water requirement of an annual sugarcane crop is 300mm, 600mm, 1000mm and 600mm during germination (0-60 days), formative (60 to 150 days) grand growth (150 to 240 days) and maturity (600 mm) phase, respectively. The optimum temperature for growth is around 30°C. The germination process is dependent on temperature. An aerial temperature of 26-33 °C with soil temperature of 23-28°C is favourably suited for initial sprouting and germination of buds. Tillering and establishment of canopy characterize the formative phase. The optimum temperature for tillering ranges from 26-33°C, while higher day temperature in the range of 32-37°C has inhibitory effect. Tillering process is highly photosensitive and mutual shading of leaves and higher interplant competition reduces the tillering. A threshold level of 400-900 hrs sunshine was found optimum for good tiller production in tropics. Since tillering coincides with the summer months, adequate water availability should be ensured to meet the evaporation demands of the crop and to minimize the tiller mortality. The grand growth phase is characterized by cane elongation, canopy closure and completion of vegetative growth. A temperature range of 30-35°C with a relative humidity of about 75 per cent is most suitable for grand growth. Rainfall during this growth phase is essential for higher yields of good quality cane. Sucrose accumulation and maturity follow grand growth phase. A clear day and rain free nights coupled with moderately low temperature is helpful for inducing increased storage of sucrose, lower nitrogen and better quality juice. Rainfall during the maturity period induces a resumption of growth and thus harmful for sucrose formation and accumulation. Limited water supply, moderate relative humidity, 7-9 hrs of sunshine per day, and a temperature of 10-14°C favour ripening process.

2. Sugarcane under Moisture Stress

Water stress remains an ever growing problem and it is the major limiting factor in crop production worldwide (Jones and Corlett, 1992). In India, nearly 60 per cent of the total sugarcane agriculture suffers from lack of adequate water supply mainly because of limited availability of water for irrigation in lift irrigated areas, canal closure during summer in many of canal irrigated tracts, and drought which occur in a cyclic manner (Sundara, 1998). Therefore water stress of varying degrees is experienced at one stage or the other of the crop growth in all most all the sugarcane growing regions of the country.

2.1. Water Requirement and Evapotranspiration

Total water requirement of annual sugarcane crop varies from 1850 mm to 2500 mm. It is estimated that 250 tonnes of water is required for production of a tonne of sugarcane. Daily evaporation in sugarcane fields varies from 8-10 mm. Solar energy, wind velocity, temperature and humidity affect the evapotranspiration. Earlier trials on response of sugarcane to irrigation suggested that maximum tonnage was obtained at Et/Ep of 0.8. Sheath moisture and moisture content of immature nodes also served as useful indices for determining the water requirement of sugarcane crop. For high yield, sheath moisture index at 5th month stage should be high enough (83 -85 per cent), and for higher CCS per cent, proper drying off with sheath moisture index of about 72 per cent at 12 th month was found to be desirable.

2.2. Critical Water Demand Period

Formative growth stage (60-150 days) has been identified as the critical water demand period and stress at this early growth phase had a direct influence on the cane yield and juice quality. Yield reduction up to 60 per cent has been recorded in a typical drought year. Water stress especially during summer months coincides with the formative phase of the crop which affects the final yield through reduction in tiller productivity, number of millable canes, individual cane weight, and finally the cane yield and juice quality (Naidu, 1987).

2.3. Plant Responses

2.3.1. Root System

Extensive root investigations revealed that the sett roots emerge from the root band (present at nodal region of sugarcane sett), and start growing within 24 hr of planting. At the third day, some roots extend at a rate of 10 mm/day and by day 5, the elongation reaches to 20 mm/day. These thin and branched sett roots are replaced by thick; fleshier and less branched shoot roots by 90 days age. Rooting depth, distribution and activity are generally affected by soil water relationships (Naidu and Venkataramana 1993). Generally more root mass occur at less than 50 cm depth in normally irrigated condition while under stress, roots penetrate vertically downwards in the form of a rope. The root system also shows penetrating type roots which reach out for water source and hence longer and thicker roots are seen under drought (Venkataramana and Naidu, 1989). The varieties selected for greater rooting depth suffered the least water deficits as compared to the normally irrigated plants. However, reports of diminished root development under moisture stress has been reported by Sheu and Kong, (1988,1989), Mongelrad (1968), Rao,(2000). Differences in root growth were related to differences in growth of susceptible and tolerant varieties (Mongelrad, 1968a).

2.3.2. Shoot System

Leaf area development is necessary to absorb light for the photosynthetic process. The maximum LAI is generally achieved by about 6 months from planting and then slowly declines. A high LAI produces large structural apparatus for the production of photosynthate and a higher yield. Leaf expansion is very sensitive to stress. Large differences occur in the density of stomata of crop plants. The activity of stomata is greatly affected by external factors such as light, temperature, and humidity. Direct sunlight makes stomata to open, while weak and diffusive light result in closure. This explains the beneficial effect of early morning sunshine on sugarcane. Since drought is common in many sugarcane growing areas, it is important to consider reducing the transpiration and thereby reducing consumptive water use. Transpiration occurs predominantly (>90 per cent) through the leaves while nodal region, which is free from wax deposition. Significant reduction in water loss (10 to 20 per cent) was demonstrated due to passive curling of leaves, which reduce the radiation receipt by leaves thereby reducing water loss and increasing water use efficiency to a greater extent. Cell growth is retarded under mild stress which in turn results in reduced leaf area, followed by reduced sink growth and reduced stem elongation. The major attribute is the drying off of older leaves and stunted growth of stem resulting in a dwarf canopy (Plates 3.1 and 3.2). The young leaves however remain green, but when the stress intensity becomes severe, the entire crop loses its turgidity and drying will be hastened. Characters like leaf thickness, leaf dry weight and leaf area ratio are highly sensitive to drought. Deposition of wax, which is a protective mechanism, is also seen on the upper surfaces of the sugarcane leaves and stem.

2.3.3. Light Interception and Photosynthesis

Sugarcane is one of the most efficient crops capable of converting a maximum of 2-3 per cent of solar energy into organic matter through an efficient photosynthetic

Plate 3.1: Sugarcane Crop at Early Growth Phase (Control).

system (Bull and Glasziou, 1975). It has been estimated that one hectare of sugarcane can produce 100 tonnes of green matter which is more than twice the yield of most other commercial crops (Almazan *et al.*, 2001). Majority of the clones intercepted 60-80 per cent of the radiation at the completion of formative phase. The light falling on the crop surface varied from 1275 to 1950 μ mol m² s⁻¹. In the initial stage of the stress, stomatal closure occurs which reduces transpiration rates and a decrease in leaf water potential which collectively influence the photosynthesis and productivity. The chlorophyll content also decreases resulting in low CO_2 fixation. Drought during the vegetative period tends to slow down the leaf development and canopy expansion. Chlorophyll fluorescence kinetics changed significantly during moisture stress indicating that photosynthetic electron transfer system (PETS), especially PSII and carbon assimilation were inhibited (Luo *et al.*, 2000).The decrease in chlorophyll fluorescence was related to drought tolerance of varieties (Luo *et al.*, 1999). Leaf water potential and stomatal diffusive resistance are the measure of stress intensity and were found to be related to the yield of a variety. These two parameters which were identified as water stress indicators were found useful for screening varieties for drought resistance (Naidu *et al.*, 1983; Venkataramana *et al.*, 1986). The carbon isotope discrimination at 240 days was negatively and significantly associated with leaf area and total dry matter, but with photosynthesis and transpiration, the relationship was not significant (Gururaja Rao, *et al.*, 2008).

Plate 3.2: Sugarcane under Drought at Formative Phase.

2.3.4. Dry Mass Accumulation and Distribution

Sugarcane has the capability of producing 65 MT of above ground dry mass per year. The dry mass production rate ranged from 20 to 35 g/day during active growth phase and the energy conversion efficiency was estimated to reach a maximum of about 1.8 per cent (Ramanujam and Venkataramana,1999). The increase in dry matter was low during periods of incomplete canopy development. The average dry matter produced was either 16.83, 41.23, or 49.41 tonnes/ha or 4.81, 22.41 and 47.48 tonnes/ha under drought at the completion of formative (150days), grand growth (240days) and maturity (360 days), respectively (Venkataramana and Naidu, 1989). The growth analysis studies indicated that net assimilation rate (NAR) and relative growth rate (RGR) were high during early growth phase, but declined with the age of the crop. Leaf area ratio (LAR) and leaf area index (LAI) increased with crop growth under normal irrigation while drought caused 34.62 per cent reduction in LAI(Venkataramana *et al.*, 1984). Harvest index was significantly associated with cane yield, sugar yield and CCS per cent (Naidu and Venkataramana, 1989).

2.3.5. Biochemical Responses

Sugarcane plant responds to the stresses at the biochemical level. The cellular water deficits results in the concentration of solutes, loss of turgor, change in cell volume, disruption of water potential gradients, change in membrane integrity, denaturation of proteins and several other physiological and molecular components. The concentration of malondialdehyde, a lipid peroxidation product doubled as the leaf water potential declined (Venkataramana *et al.*, 1987). Epicuticular wax content was significantly high in drought resistant varieties when compared to drought susceptible types. Cellular membrane thermo stability and electrolyte leakage decreased due to water stress thereby increasing the membrane injury to as high as 85 per cent in susceptible types. Drought tolerant varieties recovered effectively during rehydration (>60 per cent).The capacity to maintain high membrane themostability is an important feature of tolerance to water stress (Venkataramana *et al.*, 1983).

2.3.5.1. Osmoprotection

Recently, interest has been generated on osmotic adjustment, turgor maintenance and growth. Turgor can be maintained by increasing various osmolytes. Accumulation of osmolytes (proline, glycine-betaine, polyamines, sugars etc.) which maintain the turgor and reduce the osmotic potential, help the plant to cope with the drought effect, the phenomenon called as osmoregulation. Osmotic adjustment occurs through increased accumulation of various osmolytes such as soluble sugars, soluble carbohydrates, proline, potassium, sugar alcohols and organic acids. Concomitant with 70 per cent reduction in leaf water potential, the osmotic potential increased in many varieties suggesting an increased accumulation of osmolytes. Under water deficit conditions, the proline accumulation increased several folds in sugarcane and a significant varietal variation was noticed by Rao and Asokan (1978). Drought stress leads to the generation of reactive oxygen species (ROS) which include superoxide anion radicals (O_2), hydroxyl radicals (OH), hydrogen peroxide (H_2O_2) and singlet oxygen ($O·$) which cause damage to the cellular system. Drought enhanced activities of peroxidase and polyphenol oxidase have been reported in popular cultivars of sugarcane (Vasantha and Gururaja rao, 2003). The process of osmoprotection prevents protein denaturation, helps preserve enzyme structures and protects membranes from damage by ROS. Increase in solute concentration or accumulation of solutes causes osmotic adjustment and the compounds that are accumulated during stress are soluble sugars, soluble carbohydrates, proline, potassium, sugar alcohols and organic acids. Osmotic adjustment has a few advantages such as maintenance of cell turgor, continued cell elongation, maintenance of stomatal opening, and photosynthesis and survival under dehydration.

Among higher plants, proline and glycine betaine are the most commonly reported osmolytes and accumulation of these compounds sustains the viability of plants and their yields under moderate and severe drought. Osmotic adjustment helps plants to maintain turgor by increasing various osmolytes thus resulting in good growth. Free proline accumulates in water stressed leaf tissue. Oxidation of proline (to glutamate) in turgid tissues generally prevents accumulation while in stressed tissue proline accumulates only to serve as buffer of nitrogenous substances. The progressive accumulation has been accompanied by a fall in leaf water potential. In several studies proline accumulation was used as a screening test for drought resistance. Another metabolically inert compound called betaine also accumulates under stress. Carlin and Santos (2009) evaluated the sugarcane variety IAC91-5155 under water stress and observed a trehalose accumulation of 25.9 per cent (increase of 0.54 μmol g^{-1} fresh mass weight) at the 60th day under stress, reaching concentrations of trehalose of 2.54 μmol g^{-1} of the fresh weight. Queiroz1 *et al.*, 2011, reported the increase in levels of trehalose and free proline found to confirm what many others have reported: the importance of the osmotic adjustment of plant species, genotypes and cultivars to water deficiency in the soil. These authors declare that this mechanism is related to the accumulation of compatible osmolytes to maintain cell turgidity and facilitate physiological and biochemical processes under drought conditions.

2.3.5.2. Nutrients

Drought imposed during formative phase significantly reduced P content while N and K did not decrease (Ramesh, 2000) contrary to the earlier report of decreasing N and K content by Samuels (1971).

2.3.5.3 Abscisic Acid

Abscisic acid (ABA) accumulates in drought-affected leaves. ABA content enhances the leaf water potential by 1 to 2 bars and thus helps in dehydration postponement. The ABA was also found to possess a direct and stabilizing effect on protoplasm, and drought induced senescence of leaves. Dry matter production by ABA treated plants was greater than that of control. This was due to a greater development of shoot at the expense of roots. External application of abscisic acid (1×10^{-5} M) exerted a regulatory role on stomatal diffusive resistance and helped in maintaining relatively high water potential (Venkataramana and Naidu, 1993). ABA content enhanced the leaf water potential by 1 to 2 bars and thus helped in dehydration postponement and drought induced senescence of leaves.

2.3.5.4. Enzymes

Enzymes such as nitrate reductase, sucrose phosphate synthase, invertase etc. have been found to be regulated by the tissue water status. Nitrate reductase activity is reversible and the extent of loss under stress is to an extent of 30 per cent and the regulation of nitrogen metabolism and the constituent end products are affected in the rate limiting way. Moisture stress induced reduction in the activity of sucrose phosphate synthase and sucrose synthase was reported in popular cultivars of sugarcane, which on rehydration resumed to normal level (Vasantha *et al.*, 2003).

2.3.6. Cane Elongation

Cane elongation is positively correlated with amount of irrigation water (Chang *et al.*, 1968). The rapid cane elongation (60 to 70 per cent) takes place during grand growth during which the seasonal available water will be utilized. Large reduction in stalk number, height, cane yield and sucrose yield were noticed due to drought. Shih and Gascho (1980) reported that stalk elongation was positively and strongly correlated with water content of the elongating and meristematic tissues and cumulative soil water depletion.

2.3.7. Sucrose Accumulation

The sucrose content in cane will be high during maturity period in the normal crop as compared to the stressed cane. Sucrose accumulation increases by about 100 per cent while cane tonnage increases by only 20 per cent during the maturity phase (Gascho and Shih, 1983). Sucrose accumulation begins at the bottom of the stem and progresses upward to the top internodes. After about 11 months age, the sucrose per cent in the normal crop remains constant while the percentage in the stressed crop continue to increase until 14 months.

3. Salinity and Sugarcane

Soil salinity threatens agricultural productivity in 77 mha of agricultural land, including 45 mha (20 per cent of irrigated area) in irrigated and 32 mha (2.1 per cent of dry land) in unirrigated (Munns, 2002). Sugarcane is grown in India in about 5.02 million hectares, and about one fourth of the acreage is affected by salinity, alkalinity and (saline) irrigation water. The salts that largely contribute to salinity include the chlorides and sulphates of sodium, calcium, magnesium and potassium. The electrical conductivity of these soils is more than 4 dS/m, while alkalinity is imparted mainly by sodium carbonate. In such soil the plants are unable to absorb the water and nutrients in adequate quantities due to high osmotic pressure of the soil water.

The salinity effects are aggravated when irrigation water becomes scanty and EC of irrigation water is high (>3.0). With heavy rainfall a temporary relief of salt stress can be observed due to leaching of the salts from root zone. Sugarcane is ranked moderately sensitive to salinity with a threshold value of 1.4 dS m^{-1} (Maas, 1986). Soil root zone EC below 2 dSm^{-1} have no effect on growth and yield: 5-7.0, the yield decreases by 50 per cent and at EC of 8.0, stools of some cultivars are killed and do not survive. A yield reduction of up to 60 per cent has been recorded due to salinity. The symptoms of salt damage are pale green or yellow leaves, scorched tips and margins, reduced leaf area and stunted canopy.

3.1. Relative Salt Tolerance of Sugarcane at Various Growth Stages

Various experiments conducted over the years showed sett germination (bud sprouting) to be the most resistant phase whereas shoot growth following germination being the most sensitive phase to salinity in the life of a sugarcane plant. The severe sensitivity of sugarcane to salinity at various growth stages is manifested by a considerable reduction in growth rate (Plaut *et al.*, 2000). Plants with two or more fully expanded leaves are more resistant than those with one fully expanded leaf. Shoot growth inhibition in sand culture starts, even with a level of 30 meq salts/l whereas root growth inhibition starts by 100 meq salts/l during settling growth indicating root growth to be more resistant than shoot growth. Salinity reduced tillering and other growth parameters, leaf/shoot elongation being the most sensitive and leaf/internode number being the least sensitive parameter. Sugar accumulation in the canes, even though invariably reduced may not show its effect upon juice analysis of the harvested canes in terms of sucrose per cent juice because reduced internode growth at moderate levels of salinity may compensate for reduced accumulation of sugars.

3.2. Sett Germination

Germination is delayed progressively with increasing salinity and reduction in final germination percent observed at higher salinity levels (EC >5 dSm^{-1}). Varietal response to salinity varied with the temperature change. Temperatures below 25°C were more damaging under saline conditions. In sand culture, levels up to 200 meq salts/l did not affect emergence but further growth of the shoot is inhibited at levels 30 meq salts/l and above and the settlings did not survive at levels of 200 meq salts/l beyond 2 leaves stage. Growth of leaf blades showed a maximum reduction compared to stem and sheath whereas that of sett roots was the least affected during sett germination phase (Kumar *et al.*, 1994). Reduced germination with biomass variation for root and shoot has been well documented in several works (Liu, 1967, Chowdhury *et al.*, 2001 and Kumar and Naidu, 1993).

Soil salinity has a profound impact on the crop growth specially so with the process of germination. Germination was delayed under salt treatment and reduction in final germination percent was observed at higher salinity level (EC >5 dSm^{-1}). Higher concentration of NaCl *i.e.*, 0.5N, completely inhibited germination while at moderate levels, the germination was reduced (Rizk and Normand, 1969). Higher reduction in germination of setts with increasing salinity levels were reported for sugarcane. Varieties showed significant difference in germination. Kumar and Naidu, (1993) observed that soil salinity as more damaging for germination of setts at low temperature (below 25°C). Varietal response is a critical factor in determining the final germinant. For instance genotypes like Co 97010, Co 95007 etc. recorded a reduction in germination over 50 per cent indicating their sensitiveness. During germination and early growth of the crop, nitrogen requirement is more, as sink production is at its peak. Nitrate reductase, an important enzyme in N- metabolism showed significant reduction in salinity treatment (Jasmine Rani, *et al.*, 2004).

3.3. Tillering and Early Growth

Tiller production per main shoot decreases under saline as well as sodic conditions. In a study with 10 popular varieties, the reduction in tiller production due to salt treatment was from as low as 16.3 per cent in Co 6304 to as much as 49.8 per cent in Co 86010. Consequently, shoot population was also reduced resulting in poor and patchy field stand. Shoot height, number of internodes, number of leaves and leaf area per plant were significantly lesser in saline soil. Decreased or nil expansion growth of leaves and young internodes results in stunted canopy and poor tillering results in poor crop. Apart from tillering, cane formation was inhibited and the internodes were very narrow suggesting the sensitivity of expansion growth. The reduction in tillering due to soil salinity as well as high salt concentration in irrigation water has been reported in sugarcane (Robinson and Worker, 1965: Syed and El-Swarfy,1972). Restricted growth in terms of reduced shoot height and less green leaf production for photosynthesis was reported (Joshi and Naik, 1977, 1980; Naik and Joshi, 1981). A reduction in the elongation and expansion of sugarcane leaves under salinity has been attributed to a lowered efficiency of growing tissues to utilize sugars for growth (Kumar *et al.*, 1994). Shoot growth rate reduced even under mild salinity (EC of 2dSm^{-1}) in different cultivars of sugarcane (Meinzer *et al.*, 1994).

Ion-toxicity was the main determinant of salt tolerance at the grand growth stage while the osmotic component of NaCl mainly appeared to affect the transport of sucrose to stalks, followed by stimulated sucrolytic activity in the internodes, resulting in reduced final cane yield (Wahid, 2004).

3.4. Yield and Quality Characters as Influenced by Salinity

The cane maturity is delayed by salinity. In some genotypes the juice quality is severely affected so also the sugar yield (Sharma *et al.*, 1997). Reduction in number of millable canes was up to 37 per cent in popular genotypes with tolerant genotypes recording lesser reduction. Cane length, girth, number of internodes showed reduction due to salt treatment, which ultimately reduced the cane weight and yield. Cane yield recorded significant reduction of upto 64 per cent in sensitive genotypes while it was marginal (27 per cent) in tolerant types (Figure 3.1). A decrease in cane yields of the order of 5.45 t/ha for every 1 dSm^{-1}/ha is experienced due to soil salinity. Yield reduction from 20 per cent (Co 86011) to 45 per cent (Co 7219) has been recorded in popular genotypes at $8dSm^{-1}$ (Vasantha, 2003).

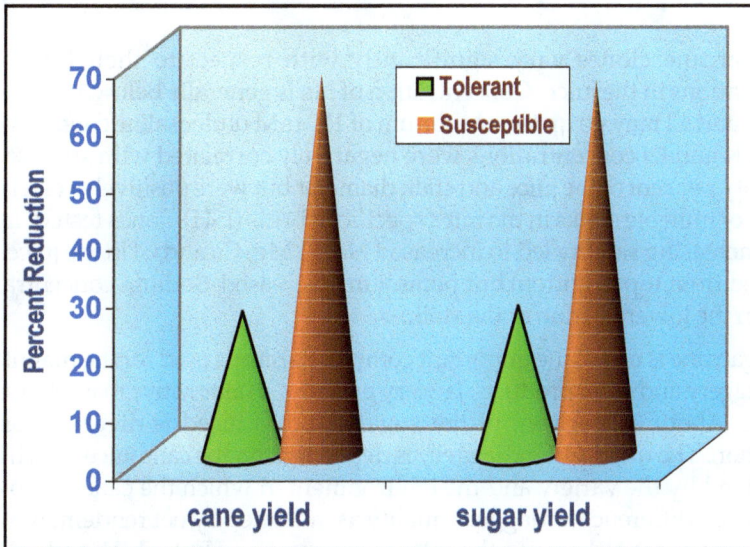

Figure 3.1: Cane and Sugar Yield as Affected by Salinity.

3.5. Juice and Jaggery Quality Characters as Affected by Salinity

Sucrose per cent juice, brix and purity are reduced by salinity. Increased non-sugar solids and salts reduce the purity. The salt content of cane juice ranges from 900-1900 ppm in non-saline soils whereas it ranges from 4000-4500 ppm in saline soils. The electrical conductivity of the juice at harvest increased in all the genotypes under saline conditions due to irrigation with saline water; increased accumulation of Na, K and Cl ions caused a reduction in sucrose per cent juice due to salinity (Figure 3.2). In varieties Co 94012, Co 99004 and Co 97001, the increase in Na and EC was marginal and consequently there was not much reduction in sucrose per cent juice due to salinity.

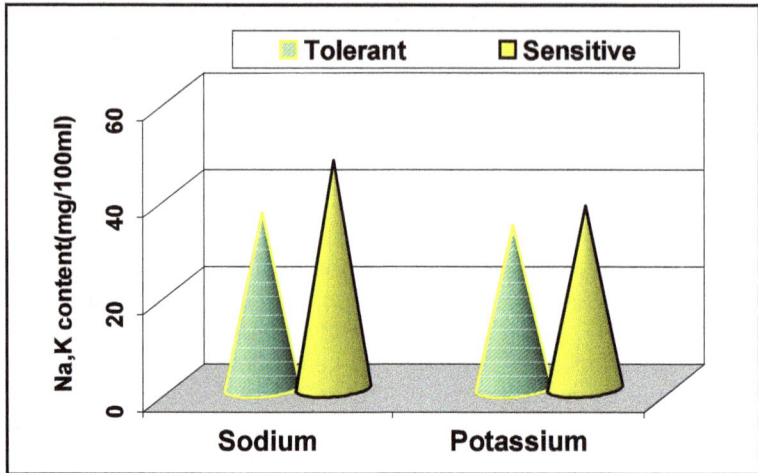

Figure 3.2: Sodium and Potassium Content of the Juice under Salinity.

Sugarcane clones vary significantly with respect to their Na, K and Cl concentrations in the juice. Concentration of Na is generally below 10 mM whereas that of K and Cl may go up to a maximum of 150 mM under saline conditions (EC 7.5 dSm-1). K and Cl concentrations were negatively correlated with sucrose per cent and purity per cent of the juice and stalk diameter but were positively correlated with number of millable stalks in the inter specific hybrids (ISH) clones tested. In another study, increasing salinity led to increased Na, K, Mg, Ca and Cl in the juice. Sucrose increased from top to bottom but potassium decreased. Sodium concentration was higher in the lower portion of the stalk.

Sugarcane is one of the important commercial crops used for the manufacture of sugar, jaggery and other products. Jaggery provides an alternative market to sugarcane growers. About 26 per cent of the sugarcane produced is diverted for jaggery production. The quality of the jaggery is dependent on the cane juice which in turn is determined by the variety and the environment in which the cane is grown. Well pronounced differences in jaggery quality as indicated by net rendement value and colour were observed among the tolerant genotypes. Under high soil salinity, the tolerant genotypes Co 85019, Co 94008 and Co 97008 produced jaggery with poor grade, colour and taste while, the genotypes Co 94012 and Co 99004 produced good quality jaggery even under salinity as sodium and chloride content increased only marginally and cane yield and juice quality were not affected.

Among the tolerant genotypes variations for quality parameters were recorded, which renders them either as better variety for jaggery making or unsuitable for jaggery purpose. In genotypes Co 94012, Co 99004 and Co 97001 the increase in Na and EC was marginal and consequently there was not much reduction in juice sucrose content under salinity. Juice sodium content influences the jaggery quality to a greater extent. The colour and taste of the jaggery decide the market value of the jaggery. In the context of sizeable area of sugarcane being grown under saline soils, there is a need for identification of genotypes like Co 94012 and Co 99004 able to produce good quality jaggery under saline conditions. Content of Na in juice is to be considered essential new criteria than salinity tolerance per se in identifying genotypes' suitability exclusively for jaggery making purpose (Vasantha *et al.*, 2009).

3.6. Root Mass under Salinity

Root growth and mass was studied in a pot experiment. Tolerant genotypes had better root penetration, mass and density as compared to sensitive genotypes (Plates 3.3 and 3.4). Live and rope types roots were noticed in tolerant types under salinity while in sensitive genotypes the roots were less dense with very few fresh roots. The improved root mass perhaps supported more number of canes in tolerant types under salinity.

Plate 3.3: Root Structure in a Sensitive Variety.

Plate 3.4: Root Structure in a Tolerant Variety.

3.7. Physiological and Metabolic Behaviour under Salinity

The crop in an attempt to equilibrate with the osmotic potential of the soil solution under saline conditions, in order to increase water absorption, the plants not only accumulate salts but also stress specific osmolytes such as proline, betaine, etc. Osmotic potential of the leaf tissues increased by 50 to 200 m mole kg^{-1} in salt treatment as compared to normal plants. Proline is one among the widely studied osmolyte that accumulate in response to stress conditions. In popular varieties several fold increase in proline content was observed (Vasantha, 2003).

Salinization leads to a decrease in rates of transpiration, stomatal conductance and CO_2-assimilation of all the leaves present on the plant. The damaging effect increases with time after salinization. Gas-exchange measurements suggested that variation in carbon isotope discrimination (delta) was attributable largely to variation in bundle sheath leakiness to CO_2. Salinity-induced increases in (phi) appeared to be caused by a reduction in C3 pathway activity relative to C4 pathway activity rather than by physical changes in the permeability of the bundle sheath to CO_2 (Meinzer *et al.*, 1994). The rates of transpiration continue to decrease probably due to its effect on stomatal conductance whereas it was not the case with rates of assimilation. The effect appears to be due to their effects on its efficiency to fix CO_2 present in the leaf rather than its deficiency. Accumulation of sugars in the leaves upon salinization appeared to result from reduced rates of their translocation, which in turn appeared to be related with their reduced utilization in the sink tissues. Thus reduced rates of photosynthesis were not directly responsible for reduced growth under saline conditions. Results of another experiment at grand growth phase, the tolerant clones maintained more or less uniform rates of photosynthesis, while the sensitive types showed sharp decline due to salt treatment. Net photosynthetic rates were reduced when the leaf water potential was <-0.9 MPa on diurnal basis, suggesting the sensitivity of the photosynthesis process to water potential gradients. During grand growth phase, the tolerant clones maintained more or less uniform rates of photosynthesis, while the sensitive types showed sharp decline due to salt treatment The reasons for reduced photosynthesis include stomatal closure, and feedback inhibition due to reduced sink activity. A reduction in stomatal conductance may result from the osmotic effects of salinity. Net photosynthetic rate reduced when the leaf water potential was <-0.9 MPa on diurnal basis, suggesting the sensitivity of the photosynthesis process to water potential gradients. Fluctuations in photosynthetic rate during hours of day in stress free environment and stressful environment would account for the variation in net photosynthetic rate and photosynthate production.

Long term salinity effects were studied in tolerant and sensitive genotypes of sugarcane in order to understand the mechanism of salinity tolerance. Water and osmotic potentials were distinctly different between tolerant and sensitive genotypes during grand growth phase (150-240 DAP). Failure of osmotic adjustment (with only marginal increase in osmotic potential in saline condition) coupled with reduced photosynthetic rate resulted in poor dry matter production in sensitive genotypes *viz.*, Co 97010, Co 95007 and Co 97009. On the contrary, the tolerant genotypes exhibited better osmotic adjustment, minimal reduction of photosynthesis and less reduction in biomass production. The long term maintenance of water status, osmotic adjustment, maintenance of high photosynthetic rate and biomass production are essential features for a sugarcane genotype to perform as tolerant type (Vasantha *e al.*, 2010).

Total biomass on an average was reduced by 41 per cent under salt treatment with tolerant clones showing only a moderate reduction of 28 and 17 per cent during formative and grand growth phases respectively, whereas, the sensitive clones showed reductions of 60 and 71 per cent at formative and grand growth phases respectively. In sugarcane, the major share of biomass is diverted towards stem after the completion of grand growth phase. In tolerant types, the per cent of stem dry mass remained more or less uniform both in control as well as under salt treatment whereas in sensitive types, the dry matter allocation to stem reduced sharply. Biomass accumulation varied in accordance with the potential of a genotype during grand growth phase. The photosynthetic and transpiration rates and leaf water potential showed little difference during the early stage of crop, the clones tolerance to salinity was determined by their performance later during the grand growth stage (Vasantha *et al.*, 2010).

3.8. Lipid Peroxidation and Cell Membrane Injury

Malondialdehyde (MDA), a lipid per oxidation product, varied from 0.85 µg g^{-1} to 1.667 µg g^{-1} in control while it varied from 1.28 to 2.51µg g^{-1} under saline conditions. Tolerant genotypes recorded lesser average increase of ~28 per cent in MDA while sensitive genotypes recorded nearly double the increase of ~57 per cent thereby indicating for a greater damage to the membrane system. Cell membrane injury test conducted with popular varieties showed significant variation in their tolerance capacity. Cell membrane stability is a measure to test the membranes biophysical/ biochemical properties. Under stress situations, the cell membrane loses the selectivity of ions and macromolecules resulting in heavy influx/efflux of essential ions from the cells. A resistant genotype maintains the cell membranes selectivity thereby support the maintenance of growth and metabolism. Cell membrane injury test conducted with popular varieties showed significant variation indicating their tolerance capacity. In tolerant genotypes the MDA content increased by 36 per cent while in sensitive genotypes the increase was 57 per cent (Vasantha *et al.*, 2008).

3.9. Oxidative Enzymes under Salinity Stress

3.9.1. Peroxidase Activity

Peroxidase activity increased from1.6 to 5.1 folds in response to salinity. The increase was highest in tolerant genotype (Co 85019) and less in sensitive genotype (Co 95016). Pox isoforms (cytosolic) were recorded in leaf sample. It is interesting to note that two low molecular forms with faster mobility were induced under higher salinity level only in tolerant genotypes, suggestive of its role in tolerance behaviour. Isolated chloroplast lysate also showed induced isoforms under high salt condition (Vasantha *et al.*, 2008).

3.9.2. Super Oxide Dismutase (SOD) Activity

SOD activity increased marginally in response to high salt condition in varieties Co 85019 and Co 95003 and in other varieties activity was on par with control plants.

SOD isoforms (five in all) were similar in both control and salt treatment. A single isoforms each was located in chloroplast and mitochondrial lysate. Either treatment or genotypic influence could not be detected in isoforms of SOD or in its activity (Vasantha *et al.*, 2008).

3.9.3. Ascorbate Peroxidase

Ascorbate peroxidase activity increased by two fold in tolerant varieties while in sensitive types the increase was only marginal. However, APX isoforms failed to show any variation due to high salt treatment (Vasantha *et al.*, 2008).

3.9.4. Enzymes of Sucrose Metabolism

The enzymes of sucrose metabolism *viz.*, sucrose synthase, sucrose P.synthetase activity declined due to salinity. The tolerant genotypes showed relatively less reduction (Gomathi and Thandapani, 2004).

Progressive stress responses enlighten us about the metabolic changes during stress adaptation in tolerant types and any flaw that reflect on the metabolic failures resulting in sensitive behaviour. The salinity (NaCl) effect was noticed in the sensitive variety Co 95007, on day two with poor growth. The visual symptoms *i.e.*, yellowing of leaves and salt injury in leaves were noticed on day seven in the sensitive variety. Progressive stress responses were studied in contrasting sugarcane genotypes to elucidate the stress adaptative features with regard to physiological and biochemical characters. Varieties showed differences with respect to parameters studied from the day four. The tolerant variety Co 85019 maintained stability of plastid pigments (Chlorophyll and Carotenoids), higher proline level and increased activity of oxidative enzymes *viz.*, POX, SOD). Sensitive genotype suffered heavy loss with regard to these characters. Lipid peroxidation, a measure of damage to the membrane system was high in sensitive variety and difference between genotypes became significant from day four, indicating the progressive nature of adaptation in tolerant and its failure in sensitive variety (Vasantha and Rajalakshmi, 2009).

3.10. Strategies Adopted by Plants to become Salt Tolerant

The studies indicated that salinity tolerance in sugarcane involves improved expansion growth of leaves, shoot, better internode and stalk length at harvest, osmoregulation, ROS enzyme activities as well as less reduction in photosynthesis, biomass production, and cane yield and sucrose per cent juice. However, jaggery characteristics differ widely even among the tolerant types. Wherever juice Na and Cl content is less the jaggery qualities are maintained, while higher Na and Cl content produced poor quality jaggery.

3.11. *In vitro* Studies

Patade *et al.* (2006) studied the effects of salt and drought stresses on irradiated cells of sugarcane and obtained plants tolerant to higher salt stress. Gandonou *et al.* (2006) studied the effects of salt stress by exposing the callus to a single level of 68 mM NaCl, and observed that higher levels of leached-out and retained Na^+ in the NaCl-treated than in the control calli, implies that sugarcane can be considered as a Na^+-excluder. On the contrary, the retained K^+ content was significantly higher in the control than in the NaCl-treated calli. It is noteworthy that growth retardation and reduced cell viability were associated with a conspicuous increase in Na^+ but a corresponding decline in K^+ concentrations, demonstrating a typical glycophytic nature of sugarcane. The lower growth rate and reduced cell viability could be related to the failure in the maintenance of higher K^+/Na^+ ratio under salt stress. Errabii *et al.* (2007) reported the accumulation of Na^+ and Cl^- but a decrease in K^+ and Ca^{+2} under salt stress in sugarcane calli. Taken together, the results suggest that the accumulation of salt ions (Na^+ and K^+) and osmolytes (proline and glycine betaine) may have an important role in osmotic adjustment in sugarcane cells under salt stress.

4. Sugarcane under Water Logging

Water logging drastically reduces the growth and survival of sugarcane worldwide and cane yield reduction is estimated between 15-45 per cent. A considerable area under sugarcane crop in several parts of India (Assam, Bihar, and West Bengal, and eastern Uttar Pradesh, coastal region of Andhra Pradesh, Tamil Nadu, Kerala and Karnataka) are exposed to stagnant water for two to three months during monsoon season. Sugarcane is fairly tolerant to flooding and water logging. It was observed that sugarcane crop was susceptible to water logging in the first 3-4 months, somewhat tolerant at 4-9 months age and helped by it in maturity beyond that age. Higher water table during active growth phase adversely affects stalk weight and plant population resulting yield loss at the rate of about one ton per acre for one inch increase in excess water (Carter and Floyed, 1974; Carter, 1976).

Some physiological effects of cane are found due to waterlogging are (*i*) transpiration rates are reduced due to stomatal closer, (*ii*) rate of photosynthesis is considerably reduced presumably that cause the reduction of effective leaf area, (*iii*) growth rates are drastically reduced during water-logging, (*iv*) higher respiration rate of submerged organs compared to leaves. A shift in respiratory metabolism from aerobic to anaerobic pathways is one of the main effects of oxygen deficiency causing from waterlogging. The effects of water logging on respiration rate depend on the varieties and its physiological age. It is also reported that under waterlogging condition some morphological, anatomical, physiological and biochemical changes take place in plant for sack of adaptation/survival (Barcly and Crawford, 1982; Gomathi and Chandran 2009).

4.1. Germination

Studies conducted in Australia indicated that waterlogging decreased germination if soil was saturated for more than 3 days (McMohan *et al.*, 1993). Flooding at planting affects emergence. A sugarcane variety Cp 89-2376, less than 6 days of flooding has more emergences and CP 72-2086 had externally low emergence (Glaz, 2003).

4.2. Root System

In the absence of oxygen, root hairs die and eventually the roots blacken and rot with the results entire underground root system gets choked and root respiration is also impaired. Because of the insufficient and inadequate root system absorption of nutrients and water is seriously affected. Nutrient absorption is further affected by their unavailability.

During natural waterlogging of the soil, roots exposed to hypoxia condition under such situation roots able survive either by inducing biochemical acclimation or by anatomical acclimation. Following sensing of partial oxygen deficiency, genes coding for so called anaerobic proteins (HIPs protein) are up –regulated at transcriptional and post- transcriptional levels and the HIPs are necessary for the acclimation (Jackson and Richard, 2003).

4.2.1. Anatomical Acclimation through Aerenchyma Formation

Sugarcane is supported by adventitious roots in waterlogging conditions, which develop possibly as result of hormonal imbalance induced by hypoxia and decreased supply of oxygen to the submerged tissue. These roots remain in the upper layers of the water which are presumably richer in oxygen content. These are adapted to water logging conditions than the original roots because they have much larger intercellular spaces (Van der Heyden *et al.*, 1998). Study examining genetic correlation of sugarcane traits under flood, found that selection for adventitious root development may not increase sugarcane yield under flooding (Sukchain and Dhaliwal, 2005).

All these supportive roots developed during flooded situations, helped in maintenance of root activity by supplying necessary oxygen (Drew, 1997).

4.2.2. Role of Ethylene

Ethylene involvement in adventitious root formation and aerenchyma formation in several crops was summarized by Jackson and Richard (2003). Ethylene synthesis increases under flooded conditions when ACC synthase concentrations increase, which stimulates ACC synthesis. ACC then diffuses to aerated parts of the root and is converted into ethylene by ACC oxidase. Ethylene is far less soluble in water than in air. Therefore, more ethylene is retained inside plant tissues when flooding occurs and ethylene concentrations increase.

4.3. Shoot Growth

It is inevitable that, because of the close functional inter dependence between roots and shoots, stress on roots from water logging also threatens the shoot system. One example of this is the arrest of nitrate uptake that arises from microbial de nitrification and damage to uptake mechanism from an absence of oxygen. Young leaves remobilize the nutrients from older leaves leading to premature senescence of the later (Drew and Sisworo, 1979). The impact of water logging on shoot growth can be observed on changes in growth habit, visual health, internal anatomy, water relations, hormonal and nutritional composition. Water logging can inhibit leaf and stem expansion and tiller production and cause epinastic curvature of petioles, orientation of shoot extension (Jackson, 1990b). Sarkar *et al.* (1999) was observed that plant height after 12 days of submergence showed significant positive association with survival percentage. In sugarcane, varieties which maintained better shoot height and internodal length yielded better under flooding condition (Gomathi, 2009; Gomathi and Chandran, 2009).

4.4. Tiller Production

Flooding during tillering, resulted in greater tiller mortality and reduced stalk population. Flooding at any stage reduced production of new tillers and rate of elongation of the established tillers (Webster and Eavis, 1971) and decrease was more with longer duration of flooding. Varieties also differed in the regard (Rahman *et al.*1989). Studies conducted at SBI, Coimbatore, indicated that the waterlogging stress during formative phase of the (90-170 DAP) caused 13.00, 21.63 and 26.52 per cent reductions in plant height, tiller production and leaf area respectively. However, the reduction was less in the resistant clones (Gomathi and Chandran, 2008).

4.5. Leaf Development and Growth Parameters

The expression of yellowing symptom in leaf, higher stalk mortality, faster drying of lower leaves, reduction in leaf number and size, are the morphological changes due to waterlogging stress. Further, waterlogging resulted in 26.5 per cent, 25.2 per cent and 24.0 per cent mean reductions in, leaf area index (LAI), leaf nitrogen content and total chlorophyll content, respectively (Gomathi and Chandran, 2009; Gomathi, *et al.*, 2010, Pandey *et al.*, 2001).

4.6. Yield and Quality

Cane yield losses depend upon the duration of water logging, stage of crop growth and management practices before, during and after water logging. Yield loss occurs due to stalk mortality, reduced crop growth due to lack of nutrition and water uptake, lodging, cane breakage, etc. About 5-30 per cent loss in yield was reported for 15-60 days of water logging condition created artificially during the late grand growth phase (7.5-9.50 months). A study conducted in tropical India (Tamil Nadu) indicated that in a water stagnation period of 2 months the reduction in cane yield was to the tune of 26-36 per cent in various varieties (Monoharan *et al.*, 1990). A study conducted at SBI, Coimbatore (Plate 3.5) showed that the waterlogging caused 22.4 per cent, reduction in NMC, 45.6 per cent reduction in single cane weight, 30.0 per cent reduction in cane length, 15.9 per cent reduction in internodal length, 17.8 per cent reduction in cane thickness and 40.1 per cent reduction in cane yield (Gomathi and Chandran, 2009; Gomathi *et al.*, 2010).

4.7. Photosynthesis and Partitioning of Assimilates

Under anaerobic condition, photosynthesis declined due to slow diffusion of CO_2 in water and reduced availability of light as result flow rate of assimilates to the roots also decreased. In sugarcane, chlorophyll content reduced under submergence and the reduction was more pronounced in susceptible varieties results in reduction photosynthetic rate and leaf dry matter accumulation (Gomathi *et al.*, 2010). Reports available in sugarcane that the reduced photosynthesis after lowering of the water table to end saturation caused a reduction in biomass by harvest date in the range 40 to 50 per cent in saturated treatments as compared with the control (Giaz *et al.*, 2004; Giaz and Gilbert, 2006).

4.8. Anaerobic Proteins (ANP's) in Response to Flooding Stress

Plants also respond to anoxia by altering the pattern of root protein synthesis.

Plate 3.5: Natural Waterlogging and Clonal Response.
a: Natural waterlogging condition; b: Tolerant clones; c: Susceptible clones.

The proteins which are synthesized as a specific response to anaerobiosis are called the anaerobic polypeptides (ANPs) (Sachs *et al.*, 1980). Flooding stimulated the synthesis of a small group of proteins known as anaerobic polypeptides (ANP's) appear to play an essential role for anoxia survival. All the characterized polypeptides are glycolytic enzymes (Mujer *et al.*, 1993). Among the ANPs, ADH is predominating one and has been extensively studied (Sachs *et al.*, 1980). New synthesized ADH isozymes emerge during flooding in many plants (Lin and Lin, 1992; Liao and Lin, 1995) and with different biochemical properties. Both in leaf and root, specific expression of ANP's *viz.*, 66 kDa, 98 kDa and 132 kDa proteins in response to short term flooding stress was recently reported in sugarcane especially in tolerant genotypes (Co 99006 and Co 8371), indicating their possible role in tolerant behaviour (Gomathi *et al.*, 2010).

4.9. Nitrate Reductase Activity

Reduction of NRase in leaves of waterlogged plants results in the rapid depletion of the nitrate and oxygen is consumed by soil biota and then anaerobic conditions develop. Gomathi and Chandran (2009) found a positive association between nitrogen content of index leaf and nitrate reductase activity (NRase) with flooding tolerance of sugarcane clones exposed to long term flooding stress.

4.10. Alcohol Dehydrogenase (ADH)

Alcohol Dehydrogenase (ADH) is responsible for the synthesis of alcohol and regeneration of NAD in alcoholic fermentation (Russel *et al.*, 1990; Kennedy *et al.*, 1992). This regenerated NAD enables glycolysis to continue under anoxia, thus producing a net 2 moles of ATP per mole of glucose relative to the 38 moles of ATP produced under aerobic conditions through respiration (Davies, 1980). A significant increase in ADH activity was recently reported in sugarcane due to short term flooding (Gomathi *et al.*, 2010).

4.11. Antioxidant System in Response to Waterlogging

Tolerance to wide varieties of environmental stress conditions has been correlated with increased activity of antioxidant enzymes and levels of antioxidant metabolites (Davies 1987). A short-term waterlogging treatment led to an increase in the activities of anti-oxidant enzymes *viz.*, APX,CAT and SOD in sugarcane (Gomathi *et al.*, 2012). Weijun Zhou and Xianqing Lin (1995) concluded that Leaf chlorophyll content and SOD and CAT activities were markedly reduced after plants were waterlogged for 30 days at various stages of growth and results indicated that waterlogging could promote the degradation of chlorophyll, reduce the activities of SOD and CAT, and therefore accelerate leaf senescence.

5. Temperature Extremes

5.1. High Temperature

Temperature is a major environmental attribute influencing the crop yields. While optimum temperature is necessary for growth, metabolism and final productivity, high temperature induces a wide variety of changes in cellular structures and metabolic process. When the magnitude and duration of the heat stress exceeds a threshold, cells are irreversibly damaged and die. Different living systems respond differently to increased temperature. In addition to speeding up of phonological events, high temperatures have deleterious effects on photosynthesis, respiration and reproduction including seedling survival. A small increase in temperature can have pronounced effect on growth and survival. A 10° rise in temperature may be sufficient to kill the plant. Temperature stress in plants is dependent on many factors including the thermal adaptation of the species or genotypes, the duration and the growth stage of the exposed tissue.

High temperatures caused significant declines in shoot dry mass, relative growth rate and net assimilation rate in maize, pearl millet and sugar- cane, though leaf expansion was minimally affected (Ashraf and Hafeez, 2004; Wahid, 2007) Major impact of high temperatures on shoot growth is a severe reduction in the first internode length resulting in premature death of plants (Hall, 1992). For example, sugarcane plants grown under high temperatures exhibited smaller internodes, increased tillering, early senescence, and reduced total biomass (Ebrahim *et al.*, 1998).

5.2. Heat Resistance

Plants normally try to avoid high temperature by adjusting the canopy architecture through insulation, lowered respiratory rates, decreased absorption of radiant energy transpirational cooling etc. Avoidance or tolerance of direct heat injury can be achieved by different means such as presence of substances that inhibit coagulation, and increased alkalinity of the protoplasm, the decrease in free water, greater thermostability of proteins, enhanced protein bond strength, increased concentration of sugars etc.

5.3. Heat Stress Acclimation and Adaptation

Plants manifest different mechanisms for surviving under elevated temperatures, including long-term evolutionary phenological and morphological adaptations and short-term avoidance or acclimation mechanisms such as changing leaf orientation, transpirational cooling, or alteration of membrane lipid compositions. In many crop plants, early maturation is closely correlated with smaller yield losses under high temperatures, which may be attributed to the engagement of an escape mechanism (Adams *et al.*, 2001). Inadequate responses at one or more steps in the signaling and gene activation processes might ultimately result in irreversible damages in cellular homeostasis and destruction of functional and structural proteins and membranes, leading to cell death (Vinocur and Altman, 2005; Bohnert *et al.*, 2006).

One of the most closely studied mechanisms of thermotolerance is the induction of hsps, which, as described in above, comprise several evolutionarily conserved protein families. However, each major HSP family has a unique mechanism of action with chaperonic activity. The protective effects of hsps can be attributed to the network of the chaperone machinery, in which many chaperones act in concert. An increasing number of studies suggest that the hsps/chaperones interact with other stress-response mechanisms (Wang *et al.*, 2004).

5.4. Photosynthesis and Water Relations

Several experiments have shown that all genotypes are sensitive to temperature at one stage or another and phenological stages differ in sensitivity to temperature. Among all plant processes, photosynthesis is essentially an important event in crop growth. The temperature optimum for photosynthesis is broad, presumably because crop plants have adapted to a relatively wide range of thermal environments. Photosynthesis of germplasm adapted to higher temperature environments was less sensitive to high temperature than was germplasm from cooler environments. Genotypes most tolerant to high temperatures had the most stable leaf photosynthetic rates across temperature regimes or they had the longest duration of leaf photosynthetic activity after anthesis and high grain weights. Despite observed negative effects of high temperature on leaf photosynthesis, the temperature optimum for net photosynthesis is likely to increase with elevated levels of atmospheric carbon dioxide. Several studies have concluded that CO_2-induced increases in crop yields are much more probable in warm than in cool environments. Thus, global warming may not greatly affect overall net photosynthesis.

In sugarcane an increased chlorophyll *a:b* ratio and a decreased chlorophyll: carotenoids ratio were observed in the tolerant genotypes under high temperatures, indicating that these changes were related to thermotolerance (Wahid and Ghazanfar, 2006).

Under field conditions, high temperature stress is frequently associated with reduced water availability (Simoes-Araujo *et al.*, 2003). In sugarcane, leaf water potential and its components were changed upon exposure to heat stress even though the soil water supply and relative humidity conditions were optimal, implying an effect of heat stress on root hydraulic conductance (Wahid and Close, 2007).

5.5. Enzymes

High temperature causes denaturation of the protein, coinciding with drop of the function of the enzymes due to the loss of tertiary structure of the protein at high temperature. As a consequence of breaking up the weak molecular bonds that keep polypeptide chain folded appreciably. Those proteins that have disulfide bonds in tertiary structure are relatively resistant to denaturation. Thermal stability of the enzymes can vary to some degree amongst species that grow on different environments. Differential thermal stability of isoenzymes has been reported as one of the parameters associated with high temperature tolerance. Hydroxypyruvate reductase and glutathione reductase are two thermostable enzymes, which can play important role in protecting plants from heat stress.

5.6. Osmoprotectants and Metabolites

Desiccation and dehydration due to heat stress inflict injury to the plants through destabilization of the plasma membrane. Natural osmolytes, especially soluble sugars are widely used as stabilizers of plant oligomeric proteins, protecting them against a variety of adverse conditions. Soluble sugars cause acclimation of photosynthesis to high temperature in desert plants. Non reducing disaccharide, trehalose has been used as stabilizing agent on membrane phospholipids, proteins and cell organelles. Higher concentration of trehalose in desiccation tolerant resurrection plants has suggested that this molecule may be involved in the desiccation tolerance of plants. Proline accumulation in plant tissues under dehydration and desiccation has indicated that it is the most potent osmoprotectant that plays a role in contracting the effect of osmotic and heat stresses. It is suggested that genetically engineered crop plants that overproduce proline under temperature stress might thus, acquire osmo tolerance that is the ability to tolerate environmental stresses such as drought and heat stress.

In assessing the functional significance of accumulation of compatible solutes, it is suggested that proline or GB synthesis may buffer cellular redox potential under heat and other environmental stresses (Wahid and Close, 2007). Similarly, accumulation of soluble sugars under heat stress has been reported in sugarcane, which entails great implications for heat tolerance (Wahid and Close, 2007). Phenolics, including flavonoids, anthocyanins, lignins, etc., are the most important class of secondary metabolites in plants and play a variety of roles including tolerance to abiotic stresses (Chalker-Scott, 2002; Wahid and Ghazanfar, 2006; Wahid, 2007).

The temperature tolerance is a polygenic character and many genes have been characterized, imparting respective high temperature tolerance in transgenic plants. Strategies are to be evolved for breeding high temperature tolerant crops either by hybridization or genetic engineering techniques.

5.7. Heat Shock Proteins (HSPS)

Synthesis and accumulation of proteins designated as 'Heat Shock Proteins' (hsps) were identified due to heat stress. Subsequently it was shown that increased production of these proteins also occurs when plants experience a gradual increase in temperature more typical of that experienced in a natural environment.

Three classes of proteins were distinguished based upon the molecular weight of most HSPs, namely HSP90, HSP70, and low molecular weight proteins of 15 to 30 kDa (LMW HSP). The proportions of the three classes differ among species. In general, heat shock proteins are induced by heat stress at any stage of development. Under maximum heat stress conditions, HSP70 and HSP90 mRNAs can increase ten-fold and LMW HSP increase as much as 200-fold. Three other proteins, though less important, are also considered to be heat shock proteins *viz.* 110 kDa polypeptides, ubiquitin, and GroEL proteins.

In arid and semi-arid regions, dryland crops may synthesize and accumulate substantial levels of heat shock proteins in response to elevated leaf temperatures. The induction temperature for synthesis and accumulation of heat shock proteins in laboratory-grown cotton ranged from 38 to 41°C. Soil water deficits resulting in midday canopy temperature of 40°C or greater were used to study heat shock proteins in field-grown cotton. The polypeptides that accumulated in the dryland leaves but not irrigated cotton leaves had molecular weights of 100, 94, 89, 75, 60, 58 and 21 kDa. In a similar experiment with field-grown soybean, several heat shock proteins were observed in both irrigated and dryland treatments, although levels were greater in the non-irrigated treatments.

The mechanism by which heat shock proteins contribute to heat tolerance is still not certain. One hypothesis is that HSP70 participates in ATP-dependent protein unfolding or assembly/disassembly reactions and that they prevent protein denaturation during stress. If this mechanism is true, then heat shock proteins may provide a significant basis for increasing heat tolerance of crop plants in a global warming situation. The HSPs may play a structural role in maintaining cell membrane integrity during stress. Other heat shock proteins have been associated with particular organelles such as chloroplasts, ribosomes and mitochondria. HSPs thus provide a significant opportunity to increase heat tolerance of crops.

5.8. Extreme Temperature Effects on Crops

There are two major forms of extreme temperature stress on crops - heat and cold. An increase in global temperatures may have either or both of these two acute effects. Increase in temperature will lengthen the effective growing season in areas where agricultural potential is currently limited by temperature stress. Thus, increased temperature will cause a shift of the thermal limits to agriculture. This shift will be especially important for crops such as rice that have tropical centers of origin and adaptation but are also grown in temperate latitudes during warm seasons.

5.9. Long-term Effects of High Temperatures on Crops

More important than acute effects of extreme temperature stress are the chronic effects of continuously warmer temperatures on crop growth and development. Record crop yields clearly reflect the importance of season-long effects on crop yields: crops generally yield the most where temperatures are cool during growth of the harvested component. This chronic effect of high temperature differs significantly from the acute effect of short-term temperature events, because seasonal temperature effects are mostly a result of effects on crop development.

5.10. Effects of Low Temperature

In most cases, plants do not suffer chilling injury until temperature drops below 10°C. Plants can be characterized into three categories with respect to their responses to low temperature *i.e.* chilling sensitive, freezing sensitive and freezing tolerant. Freezing sensitive plants are damaged by exposure to temperatures below 0°C. Cold and freezing temperatures have different effects on sugarcane seed pieces, young cane, and mature cane. Cold stress (but not freezing) seed pieces for three weeks will result in reduced stalk numbers, stalk height, stalk weight and sugar yield. The results of stressing the seed cane will carry over to ratooning. Studies have shown young cane can withstand a few repeated minimal (30 °F) 2 to 4 hour freezes without resulting in reductions in stalk count and vigor. However, repeated exposure to 25 °F for 4 hours reduced stalk counts and vigor. Increasing the number of these freeze events shows additive harmful effects by the reduction of both cane and sugar yields.

Poor sprouting of stubble buds at low temperatures is associated with a lower level of reducing sugars, reduced activity of acid invertase and higher accumulation of IAA and total phenols (Radha Jain *et al.*, 2007). Studies on cold tolerance were reported as early as 1960's (Hartt, 1965), since then many studies have shown that temperature affects the growth and development of the plant in various ways. Being a tropical/subtropical crop sugarcane responds to chilling temperatures with significant alterations in photosynthesis. Severe reduction in photosynthesis has been reported (Inman-Bamber *et al.*, 2010), mainly due to reduced activities of photosynthetic enzymes and sucrose phosphate synthase (Du and Nose, 2002).

For young plant cane and young ratoon cane, most or all of the aboveground primary and secondary shoots that have growing points above ground will be killed by a freeze event. Post-freeze regrowth will come from secondary shoots whose growing points were below ground and protected from the cold. The freeze killing of the young cane shoots may result in poor stands. Early freezes on mature sugarcane slow or stop the production and storage of sucrose by the plant which may lower sugar recovery at harvest. A severe freeze causes changes in crusher juice quality which can be monitored by measuring crusher juice brix, polarity, pH and titratable acidity. With this data, sucrose concentration, purity, and theoretical sugar yield can be calculated. After freeze damage, crusher juice sucrose, purity, and sugar yield decline. At the same time, titratable acidity and gums (*e.g.*, dextran) may increase several fold. These increases are highly variable depending on variety, maturity and growing conditions at the time of and after the freeze. Freezes can also cause a loss of sugarcane stalk weight, with time, after the freeze. Freezing temperatures cause leaf damage first. A very light freeze, between 32 and 29° F for a few hours, may only cause a banded chlorosis or burning of the leaf tips. There will be no significant changes in juice quality unless the freeze lasts much longer than a few hours. A slightly harder freeze for a few hours will cause extensive leaf browning and terminal bud death. Responses of different varieties will differ. A few lateral buds will be killed and the freeze may damage tissue in the top few internodes. Changes in juice quality may become apparent in a week or two and will be variety dependent. A more severe freeze, 23 to 24° F for a few hours, will completely brown leaves and partially or entirely freeze stalks. The terminal bud and all or nearly all the lateral buds will be killed. Deterioration of juice quality may start within a few days. If the low temperature duration is short, some of the heartier varieties may not experience juice quality deterioration for two to three weeks. A freeze below 22° F will almost invariably kill all leaves, buds, and internal stalk tissue to ground level. Stalk splits will be seen, although some may be hard to detect because of closure after the freeze. Juice quality deterioration may become evident a few days after the freeze. There are many variables involved in sugarcane freeze damage. Besides the severity of the freeze, the duration of the freeze is the next most important variable. A steady 30° F freeze for 48 hours could presumably do as much damage as a short 22° F freeze. The importance of early and continuous field damage surveys and juice quality monitoring post-freeze cannot be overemphasized. Completely frozen cane may become unacceptable to the mill within one to two weeks. Warm and humid weather following a freeze increases the rate of deterioration. Stalk damage can be determined by splitting the stalk and looking for frozen or thawed water-soaked tissue. Although standing cane routinely freezes from the top down, temperatures are normally lower near ground level and, therefore, lodged cane is more prone to damage.

Thus it is evident that temperature is one of the most important abiotic factors controlling the growth, development and acclimation of natural and cultivated plant species. Plants have evolved various morphological, anatomical, biochemical and physiological means to cope with the undesirable temperature regimes. Crop cultivation is limited to temperatures operating in a narrow range of 10-30°C. Strategies have to be evolved for breeding low temperature and high temperature tolerant crops either by hybridization or genetic engineering techniques so as to evolve the crops possessing tolerance to high and low temperature limits which are suitable for cultivation beyond traditional agriculture.

6. Molecular Interventions for Mitigation

A number of genes have been reported to be induced by drought, high salinity and low temperature stresses and their products are thought to function in stress tolerance and response (Bray 1997, Thomashow, 1999, Shinozaki *et al.*, 2003). The direct introduction of small number of genes by genetic engineering seems to be a more attractive and rapid approach for improving stress tolerance (Cushman and Bohnert, 2000). Present day biotechnological strategies rely on the transfer of one or several genes that encodes either biochemical pathways or endpoints of signaling pathways. These genes/products protect either directly or indirectly against environmental stresses.

6.1. Drought Stress

Studies were conducted to detect genes expressed under drought by microarray, identified 165 genes responded to drought. Important stress related pathways were repressed in sensitive cultivar. A great number of genes with unknown function were reported which may provide new insight into the tolerance mechanism (Rodrigues *et al.*, 2009). Upregulation of genes encoding for polyamine oxidase, cytochrome-c-oxidase, s-adenosyl methionine (SAM) decarboxylase and thioredoxins, which directly or indirectly involved in the regulation of redox status was reported in sugarcane under drought stress(Prabhu *et al.*, 2011). Rodrigues *et al.* (2009) demonstrated increased expression of a gene encoding a peroxidase in a drought tolerant sugarcane cultivar. This enzyme is responsible for the reduction of H_2O_2 to H_2O and O_2 and a decline in peroxidase activity is considered a limiting step to ROS neutralization in sugarcane (Chagas *et al.*, 2008). The accumulation of the osmolytes trehalose and proline also contributes to the reduction in the damage caused by the accumulation of ROS and is associated with drought tolerance in sugarcane (Zhang *et al.*, 2006; Molinari *et al.*, 2007; Guimaraes *et al.*, 2008).

Area that deserves attention is the response mediated by the phytohormone absisic acid (ABA). ABA is the major plant hormone related to water stress signaling and regulates plant water balance (Riera *et al.*, 2005; Wilkinson and Davies 2010). Rocha *et al.* (2007) found drouight responses in sugarcane analogous to those induced by exogenous ABA application. Both drought and ABA induced the expression of genes encoding a PP2C- like protein phosphatase, a S-adenosylmethionine decarboxylase and two delta – 12 oleate desaturases. Trujillo *et al.* (2008) showed that SodERF3, a sugarcane ethylene responsive factor (ERF), is induced by ABA and drought stress and may be involved in salt and drought tolerance. However, plant response to drought is a complex phenomenon, especially with a polyploid genome like sugarcane (Givet and Arruda 2001). Besides the fact that drought stress involves biochemical networks which are still being elucidated. For example, phosphorus supply improved sugarcane acclimation capacity by affecting plant characteristics related to water status and photosynthetic performance and causing network modulation under water deficit (Sato *et al.*, 2010).

6.2. Salinity

SodERF3 is a new member of the FT-ERF family in sugar cane, closely related to salinity and drought tolerance. The C-terminal repression (EAR) motive in *SodERF3* is different from that described to date for other plants transcription factors (Trujillo *et al.*, 2009).Salts interfere with sugar production in two ways: firstly, by affecting the growth rate and cane yield and secondly, by affecting the sucrose content of the stalk (Rozeff 1995). Patade *et al.* (2006) studied the effects of salt and drought stresses on irradiated cells of sugarcane and obtained plants tolerant to higher salt stress. Gandonou *et al.* (2006) studied the effects of salt stress by exposing the callus to a single level of 68 mM NaCl, and observed that physiological and biochemical indicators could play a crucial role in salt tolerance.

6.3. Temperature Extremes

RNA expression profile of sugarcane subjected to low temperature generated reported about twenty novel transcripts/genes (Nogueira *et al.*, 2003). Expression profiles of the cold inducible genes revealed proteins that are directly involved in chilling tolerance. One sugarcane EST encoding a putative xanthine dehydrogenase (XDH) was induced after cold exposure and might be involved in the protection against oxidative stress due to cold exposure (Nogueira *et al.*, 2003). In sugarcane the heat stress effects are reversible through small heat shock proteins (sHsp), which belong to chaperone family (Tiroli-cepeda and Ramos, 2010).

7. Conclusions and Future Research Needs

Phenomenal progress has been made in recent years in diverse areas ranging from meteorological analysis, water balance models which have provided new avenues to achieve improved sugarcane yields under water limited environments. Sugarcane productivity is mainly dependent on growth, sucrose accumulation and yield, and the environment in which it is cultivated. Abiotic stresses are the most important limiting factors for cane productivity. These stresses include drought, salinity, temperature extremes, heavy metals and radiation which cause detrimental effects on plant growth and yield. These factors affect the root function, growth rates, metabolism and in extreme cases lead to dehydration and death. Using selection methods, it has been difficult to manipulate precisely the individual components of stress tolerance, and thus development of stress tolerant cultivars has not made much headway. Also, the expected rise in global temperatures indicates that there is an urgent need to understand and improve plant tolerance to these stresses.

The sugarcane plant manifests the drought in several ways and hence several morphological, physiological and biochemical traits were given weightage for identifying drought tolerant types. The major attribute is the drying of older leaves and stunted growth of stem, resulting in a dwarf canopy. The inward rolling of top canopy, which is seen in many tolerant varieties, transmits back the irradiance load thus absorbing less quantum of direct sunlight. Wax coating on the leaf surface help prevent water loss from leaf as well as nodal regions of the cane. In addition, various plant responses such as tissue hydration, stomatal behaviour, water potential changes, stability of enzyme systems was found to be helpful in enhancing water use.

A considerable area under sugarcane crop in several parts of India are exposed to stagnant water for two to three months during monsoon season. Sugarcane is fairly tolerant to flooding and water logging. During water-logging i) transpiration rates are reduced due to stomatal closer, ii) rate of photosynthesis is considerably reduced presumably that cause the reduction of effective leaf area, iii) growth rates are drastically reduced during, iv) higher respiration rate of submerged organs compared to leaves.The studies conducted over decades indicated that salinity tolerance in sugarcane involves improved Osmoregulation, ROS enzyme activities as well as less reduction in photosynthesis, biomass production, and cane yield and sucrose per cent juice. However, jaggery characteristics differ widely even among the tolerant types. Wherever juice Na and Cl content is less the jaggery qualities are maintained, as high Na and Cl content produce poor quality jaggery.

Atmospheric temperature is one of the most important abiotic factors controlling the growth, development and acclimation of natural and cultivated plant species. Though sugarcane plants have evolved various morphological, anatomical, biochemical and physiological means to cope with the undesirable temperature regimes, its cultivation is limited to temperatures operating in a narrow range of 10-30°C. Strategies have to be evolved for breeding low temperature and high temperature tolerant crops either by hybridization or genetic engineering techniques so as to evolve the crops possessing tolerance to high and low temperature conditions.

A number of genes have been reported to be induced by drought, high salinity and low temperature stresses and their products are thought to function in stress tolerance and response. The direct introduction of small number of genes by genetic engineering seems to be a more attractive and rapid approach for improving stress tolerance. Present day biotechnological strategies rely on the transfer of one or several genes that encodes either biochemical pathways or endpoints of signaling pathways. These genes/products protect either directly or indirectly against environmental stresses.

Improvement of qualitative traits such as photosynthesis, transpiration and water use efficiency is difficult through conventional breeding because of the complexity of quantifying these traits in a large number of germplasm lines. Therefore, the immediate task is to look for new options to exploit the variability present in our rich germplasm through new techniques. Several stress related genes that alter the expression of important stress traits for improving the plant adaptation have to be fully characterized. The donor sources for various stress resistant features have to be identified and the genes responsible for these traits need to be transferred through molecular breeding.

In recent years, many genes and gene products have been identified which get induced upon exposure to various abiotic stresses. Genes encoding enzymes of the biosynthetic pathways of different osmolytes such as proline, glycine betaine, sorbitol, pinitol, have been cloned and exploited in improving abiotic stress tolerance. Response of crop plants to abiotic stresses involves several gene alterations. Due to stress, more than 100 transcripts are affected and a large number of proteins show up/down regulation. Proteins which are up regulated are synthesized in response to high temperature, drought, salinity and several other abiotic stresses. Studies in these areas hold promise for sustained cane production in all agro-climatic zones of India.

8. References

Adams SR, Cockshull, KE, Cave, CRJ (2001). Effect of temperature on the growth and development of tomato fruits. Ann. Bot. 88, 869–877.

Almazan O, Gonzalez L, Galvez L (2001). Sugar Cane International, 7, 3-8. Journal of Agricultural Science, Cambridge (1992), pp. 119, 291-296. 1992 Cambridge University Press 291.

Ashraf M, Hafeez M (2004). Thermotolerance of pearl millet and maize at early growth stages: growth and nutrient relations. Biol. Plant. 48, 81–86.

Barcly AM, Crawford RMM (1982). Plant growth and survival under strict anaerobiosis. J.Expt.Bot., 33: 541-549.

Bohnert HJ, Gong Q, Li P, Ma S (2006). Unraveling abiotic stress tolerance Mechanisms-getting genomics going. Curr.Opin. Plant Biol. 9, 180–188.

Bray EA (1997). Plant responses to water deficit Trends in Plant Science, 2: 48-54.

Bull TA, Glasziou KT (1975). Sugarcane. In Crop Physiology, case histories, Ed. LT Evans, Cambridge Univ. Press, Cambridge, pp. 51-72.

Carlin SD, Santos DMM (2009). Physiological indicators of the interaction between water deficit and soil acidity in sugarcane. Pesquisa Agropecuária Brasileira 44: 1106-1113.

Carter CE (1976). Excess water decreases cane and sugar yields. proc. Am. Soc.Sug.Cane technol., 6: 44-51.

Carter CE, Floyed JM (1974). Inhibition of sugarcane yields by high water level during dormant season. Proc.Soc.Sug.Cane technol., 4: 14-18.

Chagas RM, Silveira JAG, Ribeiro RV, Vitorello VA, Carrer H (2008). Photochemical damage and comparative performance of super oxide dismutase and ascorbate peroxidase in sugarcane leaves exposed to paraquat-induced oxidative stress. Pestic.Biochem. Physiol. 90: 181-188.

Chalker-Scott L (2002). Do anthocyanins function as osmoregulators in leaf tissues? Adv. Bot. Res. 37: 103–106.

Chang H, Wang JS, Ho FW (1968). The effect of different pan ratio for controlling irrigation of sugarcane in Taiwan. Proc.Int. Soc. Sugarcane Technol. 13: 652-663.

Chaves MM, Maroco JP, Pereira JS (2003). Understanding plant responses to drought – from genes to the whole plant. Functional Plant Biology 2003; 30: 239-264.

Chowdhury MKA, Miah MAS, Ali S, Hossain MA, Alam Z (2001). Influence of Sodium chloride salinity on germination and growth of sugarcane (*Saccharum officinarum*L). Sugarcane International : 15-16.

Cushman JC, Bohnert HJ (2000). Genomic approaches to plant stress tolerance. Curr. Opin. Plant Biol, 3117-124.

Davies DD (1980). Anaerobic metabolism and the production of organic acids. In: The biochemistry of plants. Davies, D.D. (ed.). Vol.2 Acad. Press, New York, USA, pp. 581-611.

Drew MC, Sisworo ET (1979). The development of waterlogging damage in young barley plants in relation to plant nutrient status and changes in soil properties. New Phytologist 82: 301–314.

Drew MC(1997). Oxygen deficiency and root metabolism: Injury and acclimation under hypoxia and anoxia. Annual Review of Plant Physiology Plant Molecular Biology 48: 223-250.

Du YC, Nose A (2002). Effects of chilling temperature on the activity of enzymes of sucrose synthesis and the accumulation of saccharides in leaves of three sugarcane cultivars differing in cold sensitivity. Photosynthetica 40: 389-395.

Dwivedi RS (2000)Adaptability mechanisms of sugarcane cultivars to abiotic stresses. In Sugarcane Production: Strategies and Technology (Shahi HN, Lal M, Sinha OK, Srivastava TK, eds), Technical Bulletin No. 40, Indian Institute of Sugarcane Research, Lucknow, pp. 42-45.

Ebrahim MK, Zingsheim O, El-Shourbagy MN, Moore PH, Komor, E (1998). Growth and sugar storage in sugarcane grown at temperature below and above optimum. J. Plant Physiol. 153: 593–602.

Errabii T, Benard JG, Essalmani CH, Idaomar M, Skali-Senhaji N (2007). Growth, proline and ion accumulation in sugarcane callus cultures under drought-induced osmotic stress and its subsequent relief. Afr. J. Biotechnol., 5(16): 1488-1493.

Ferraudo, Samira Domingues Carlin3, Marcelo de Almeida Silva (2011). Biochemical and physiological responses of sugarcane cultivars to soil water deficiencies Sci. Agric. (Piracicaba, Braz.),68(4). 469-476.

Gandonou CB, Errabii T, Abrini J Idaomar M Senhaji NS (2006). Selection of callus cultures of sugarcane (Saccharum Sp.). tolerant to NaCl and their response to salt stress. Plant Cell Tissue Organ Cult. 87: 9-16.

Gascho GJ, Shih SF (1983). Sugarcane. In Crop-Water Relations, Ed Teare ID and MM Peet, Wiley Interscience, New York, 445-479.

Giaz B, Gilbert RA (2006). Sugarcane response to water table, periodic flood, and foliar nitrogen on organic soil. Agronomy Journal 98: 616 – 621.

Giaz B,Morris DR, Daroub SH (2004). Sugarcane photosynthesis, transpiration and stomatal conductance due to flooding and water table. Crop Sci. 44: 1633 – 1641.

Gilbert RA, Curtis R, Rainbolt, Dolen R, Morris, Andrew C, Bennett (2007). Morphological Responses of Sugarcane to Long-Term Flooding.Agron. J. 99(6): 1622-1628.

Glaz B (2003). Sugarcane genotypes emergence after rain-caused flood. Sugarcane International. Issue 35.

Gomathi R, Chandran K (2008). Annual Report, SBI, Coimbatore 39-40.

Gomathi R Chandran K (2008). Annual Report, SBI, Coimbatore 39-40.

Gomathi R, Chandran K, Gururaja Rao PN, Rakkiyappan P (2010). Effect of Waterlogging in Sugarcane and its Management" Published by The Director, Sugarcane Breeding Institute (SBI-ICAR), Coimbatore. Extension Pub.No. 185.

Gomathi R, Chandran K (2009). Effect of waterlogging on growth and yield of sugarcane clones. Sugarcane Breeding Institute (SBI-ICAR), Quarterly News letter Vol. 29 (No. 4).

Gomathi R, Thandapani P (2004). Influence of salt stress on yield and quality of sugarcane genotypes (*Saccharum officinarum* L).Indian Sugar: 117-124.

Gomathi R Gowri, Manohari Rakkiyappan P (2012). Antioxidant enzymes on cell membrane integrity of sugarcane varieties differing in flooding Tolerance. Sugar Tech14: 261-265.

Grivet L, Arruda P (2001). Sugarcane genomes: depicting the complex genome of an important tropical crop. Curr Opin Plant Biol 5: 122-127.

Guimaraes ER, Mutton MA, Mutton MJR, Ferro MIT, Ravaneli GC, Silva JA (2008). Free proline accumulation in sugarcane under water restriction and spittlebug infestation. Sci Agric 65: 628 – 633.

Hall AE (1992). Breeding for heat tolerance. PlantBreed. Rev. 10, 129– 168.

Hartt CE (1965). Effect of temperature upon translocation of C^{14} in sugarcane. Plant physiol 40: 74-81.

Inman-Bamber NG, Bonnett GD, Spillman MF, Hewitt MH, Glassop D (2010). Sucrose accumulation in sugarcane is influenced by temperature and genotype through the carbon source-sink balance. Crop Pasture Sci 61: 111-121.

Jackson MB, Richard B (2003). Physiology, biochemistry and molecular biology of plant root systems subjected to flooding of the soil. In H. de Kroon and E. J. W. Visser/9eds.), Root ecology. Springer – Verlag, Berlin, Heidelberg.

Jackson MB (1990b). Communication between the roots and shoots of flooded plants. In: Importance of root to shoot communication in the responses to environmental stress (Davies, W.S. and Geffocat, B., Eds.), British Society for Plant Growth Regulation Bristol. U.K. pp. 115-133.

Jasmine Rani GM, Vasantha S, Kanimozhi K (2004). Influence of Salinity on Growth and Biochemical Characters during Germination Phase in Sugarcane. Plant Archives 4(2): 447-452.

Jones HG, Corlett JE (1992). Current topics in drought physiology. Journal of Agricultural Science;119: 291-296.

Joshi GV, Naik GR (1977). Salinity effect on growth and photosynthetic productivity in sugarcane variety Co 740. Indian Sugar 27(6): 329-332.

Joshi GV, Naik GR (1980). Response of sugarcane to different types of salt stress. Plant and Soil 56(2): 255-263.

Kennedy RA, Rumpho ME, Fox TC (1992). Anaerobic metabolism in plants. Plant Physiology100: 1-6.

Kumar S, Naidu KM (1993). Germination of sugarcane setts under saline conditions. *Sugarcane* 4: 2-5.

Kumar S, Naidu KM, Sehtiya HL (1994). Causes of growth reduction in elongating and expanding leaf tissues of sugarcane under saline conditions. Australian Journal of Plant Physiology 21: 79-83.

Liao CT, Lin CH (1995). Effect of flood stress on morphology and anaerobic metabolism of Momordicacharantia.Environmental Experimental Botany 35: 105-113.

Lin CH, Lin CH (1992). Physiological adaptation of wax apple to waterlogging. Plant Cell Environment 15: 321-328.

Liu LJ (1967). Salinity effects on sugarcane germination, growth and root development. J.Agric. Univ. P.R. 51: 201-209.

Luo Jun, Zhang MuQuing, Lu JianLin, Lin YanQuan (1999). Chloroplast fluorescence parameters, MDA content and plasma membrane permeability in sugarcane and their relation to drought tolerance.J. J. Fujian Agricultural University 28 (3): 257-262.

Luo Jun, Zhang MuQuing, Lu JianLin, Lin YanQuan (2000). Effects of water stress on the chlorophyll a fluorescence induction kinetics of sugarcane genotypes. J. Fujian Agricultural University 29 (1): 18-22.

Maas EV (1986). Salt tolerance of plants. Applied Agricultural Research 1: 12-26.

Manoharan ML, Duraisamy K, Krishnamurthy SV, Vijayaraghan H, Muthkrishnan K (1990). Performance sugarcane varieties in waterlogged condition. Maharastra Sugar15(11): 39-45.

Mc Mahon GG, Chapple PA, Ham GJ, Saunders M, Brandon R (1993). Planting sugarcane on heavy clay soils in the Burdekin.Proc.15th conference of the Australian Society of Sugarcane technologist, Cairns, Queensland, 27-30.

Meinzer FC, Palut Z, Saliendra NZ (1994). Carbon isotope discrimination, -exchange and growth of sugarcane cultivars under salinity. Plant Physiology 104: 521-526.

Molinari HBC, Marur CJ, Daros E, Campos MKF, Carvalho JFRP, Bespalhok Filho JC, Pereira LFP, Vieira LGE (2007). Evaluation of the stress-inducible production of proline in transgenic sugarcane (*Saccharum* spp.): osmotic adjustment, chlorophyll fluorescence and oxidative stress. Physiol Plant 130: 218-229.

Mongelrad JC (1968a). Effects of planting methods on yield of sugarcane under subsurface irrigation. Hawaiian Planter's Rec. 58(22): 315-322.

Mongelrad JC (1968). Further notes on the use of Sin bar for the selection of drought resistant sugarcane varieties. Annl.Rept.M S R I Mauritius: 87-89.

Mujer CV, Rumpho ME, Lin JJ, Kennedy RA (1993). Constitute and inducible aerobic and anaerobic stress proteins in the *Echinochloa* complex and rice. Plant Physiology101: 217-226.

Munns R (2002). Comparative physiology of salt and water stress. Plant Cell Environ. 25, 239-250.

Naidu KM, Venkataramana (1993). Sugarcane In "Rooting pattern of tropical crops" Ed. M.A. Salam, Tata McGraw Hill (India). New Delhi pp. 169-187.

Naidu KM (1987). Potential yield in sugarcane and its realization through varietal improvement- present status and future thrusts. Sreenivasan, T.V. and Premachandran, M.N. (eds.)., Sugarcane Breeding Institute, pp. 19-55.

Naidu KM, Venkataramana S (1989). Harvest Index in water stressed sugarcane varieties Sugarcane (London). No. 6: 5-7.

Naidu KM, Ramana Rao TC, Shunmugasundaram S (1983a). Varietal stability in sugarcane under conditions of drought.Proc. Int. Soc. Sugarcane Technol., 18: 682-690.

Naik GR, Joshi GV (1981). Effect of pre planting treatment with growth substances on sugarcane productivity under saline conditions. Deccan Sugar Technologists Association.31st Annual Convention. A 77-86.

Nogueira FTS, De Roza Jr VE, Menossi M, Uhan EC, Arruda P (2003). RNA Expression profiles and data mining of sugarcane response to low temperature. Plant Physiol 132: 1811-1824.

Pandey DM, Pant RC, Roy DK (2001). Changes in chlorophyll content of three sugarcane (*Saccharum officinarum* L.). cultivars in relation to waterlogging and planting methods. Crop Research (Hisar). 21 (3): 360-363.

Pandey OP, AshaKumari, Haque H (2000). NAD-Alcohol dehydrogenase and Superoxide Dismutase activity in *Zea mays* under Hypoxia and post-Hypoxia stress Regime. Journal of Plant Biology27(1): 71-73.

Patade VY, Suprasanna P, Bapat VA (2008). Effects of salt stress in relation to osmotic adjustment on sugarcane (*Saccharum Officinarum* L.). callus cultures Plant Growth Regul. 55: 169-173.

Patade VY, Suprasanna P, Kulkarni UG, Bapat VA (2006). Molecular profiling using RAPD technique of abiotic stress (salt and drought). tolerant regenerants of sugarcane Cv. Coc-671. Sugar Tech 8: 63-68.

Plaut Z, Meinzer F, Federman E (2000). Leaf development, transpiration and ion uptake and distribution in sugarcane cultivars grown under salinity. Plant Soil 218: 59-69.

Prabhu G, Kawar PG, Pagariya MC, Prasad DT (2011). Identification of water deficit stress upregulated genes in sugarcane. Plant Mol Biol Rep 29 doi.

Radha Jain AK, Shrivastava S, Solomon RL, Yadav (2007). Low temperature stress-induced biochemical changes affect stubble bud sprouting in sugarcane (Saccharum spp. hybrid). Plant Growth Regul 53: 17–23.

Rahman ABMM, Martein FA, Terry ME (1989). Physiological response of sugarcane to flooding stress. Proceedings of International Society Sugarcane Technology 20(2): 668-676.

Ramanujam T, Venkataramana S (1999). Radiation interception and utilization at different growth stages of sugarcane and their influence on yield. Indian J Plant Physiol. 4: 85-89.

Ramesh P (2000). Effect of drought on nutrient utilization, yield and quality of sugarcane *(Saccharrum officinaram).* varieties. Indian J.Agron. 45(2): 401-406.

Rao KC, Asokan S (1978). Studies on free proline association to drought resistance in sugarcane. Sugar J., 40: 23-24.

Rao PN (2000). Cane management under drought conditions or water stress. Proc. Annl.Conv. S.T.A.I. 61: 3-8.

Riera M, Valon C, Fenzi F, Giraudat J, Leung J (2005). The genetics of adaptive responses to drought stress: abscisic acid-dependent and abscisic acid-independent signaling components. Physiol Plant 123: 111-119.

Rizk TY, Normand WC (1969). Effect of salinity and Louisiana Sugarcane. International Sugar Journal 71: 221-230.

Robinson FE, Worker GF (1965). Growth of sugarcane in areas irrigated with Colarado river water. California Agriculture 19(8): 2-3.

Rocha FR, Papini-Terzi FS, Nishiyama MY Jr, Vencio RZN, Vicentim R, Duarte RDC, Rosa VE Jr. Vinagre F, Barsalobres C, Medeiros AH, Rodrigues FA, Ulian EC, Zingaretti SM, Galbiatti JA, Almeida RS, Figueira AVO, Hemerly AS, Silva-Filho MC, Menossi M, Souza GM (2007). Signal transduction-related responses to phytohormones and environmental challenges in sugarcane. BMC Genomics 8: 71.

Rodrigues FA, Laia ML, Zingaretti SM (2009). Analysis of gene expression profiles under water stress in tolerant and sensitive sugarcane plants. Plant Sci 176: 286-302.

Rozeff N (1995). Sugarcane and salinity-a review paper. Sugarcane 5: 8-19.

Russell DA, Wong DML, Sachs MM (1990). The anaerobic response of soybean. Plant Physiology 92: 401-407.

Sachs MM, Freelins M, Okimoto R (1980). The anaerobic proteins of maize. Cell20: 761-767.

Samuels G (1971). Influence of water deficiency and excess on growth and leaf nutrient element content of sugarcane. Proc. International Soc. Sugar Cane Technol. 14: 653-656.

Sarkar RK, Sahu RK, Suriya Rao AV, De RN (1999). Correlation and path analysis of certain morpho-physiological characters with submergence tolerance in Rain - fed lowland rice. Indian Journal Plant Physiology4(4): 346-348.

Sato AM, Catuchi TA, Ribeiro RV, Souza GM (2010). The use of network analysis to uncover homeostatic responses of a drought-tolerant sugarcane cultivar under severe water deficit and phosphorous supply. Acta Physiol Plant 32: 1145-1151.

Sharma SK, Sharma P, Upal SK (1997). Influence of salt stress on growth and quality of sugarcane. Indian Journal of Plant Physiology 2: 179-180.

Shih SF, Gascho GJ (1980). Relationship among stalk elongation, leaf area and dry biomass of sugarcane. Agron. J., 72: 309-313.

Shinozaki K, Yamaguchi-Shinozaki K, Seki M (2003). Regulatory network of gene expression in the drought and cold stress responses. Curr Opin Plant Biol 6: 410–417.

Simoes-Araujo JL, Rumjanek NG, Margis-Pinheiro M (2003). Small heat shock proteins genes are differentially expressed in distinct varieties of common bean. Braz. J. Plant Physiol. 15, 33–41.

Sukchain DS, Dhaliwal LS (2005). Correlations and path coefficients analysis for aerial roots and various other traits in sugarcane under flooding. Annuals of Biology21: 43 – 46.

Sundara B (1998). Sugarcane Cultivation, Vikash Pub. House, New Delhi, pp. 292.

Syed MM, El- Swaify SA (1972). Effect of saline water irrigation on NCO 310 and H 50-209 cultivars of sugarcane. Tropical Agriculture 49(4): 337-346.

Thomashow MF (1999). Plant cold acclimation: freezing tolerance genes and regulatory mechanisms. Annu Rev Plant Physiol Plant Mol Biol 50: 571–599.

Tiroli-Cepeda AO, Ramos CHI (2010). Heat causes oligomeric disassembly and increases the chaperone activity of smaller heat shock proteins from sugarcane. Plant Physiol Biochem 48: 108-116.

Trujillo LE, Carmen Menéndez, María E Ochogavía, Ingrid Hernández, Orlando Borrás, Raisa Rodríguez, Yamilet Coll, Juan G Arrieta, Alexander Banguela, Ricardo Ramírez, Lázaro Hernández (2009). Engineering drought and salt tolerance in plants using SodERF3, a novel sugarcane ethylene responsive factor Biotecnología Aplicada 26(2): 168-171.

Trujillo LE, Sotolongo M, Menendez C, Ochogavia ME, Coll Y, Hernandez I, Borras-Hidalgo O, Thomma BPHJ, Vera P, Hernandez L (2008). SodERF3, a novel sugarcane ethylene responsive factor (ERF), enhances salt and drought tolerance when overexpressed in tobacco plants. Plant Cell Physiol 49: 512-525.

Van der Heyden, Ray CJ, Noble R (1998). Effects of waterlogging on young sugarcane plants. Australian Sugarcane 2: 28 – 30.

Vasantha, Gururaja Rao PN, Ramanujam T (2003). Effect of moisture stress on sucrose synthesizing enzymes in promising genotypes of sugarcane J. Plant Biol.30: 15-18.

Vasantha S, Gururaja Rao (2003). Influence of moisture stress on the activity of oxidative enzymes in sugarcane India Journal Plant Physiol. 8: 405-407.

Vasantha S (2003). Varieties for salt affected soils. 35th meeting of Sugarcane Research and Development workers of Tamil Nadu. Sep 8-9, 2003. Agenda notes pp. 25-29.

Vasantha S, Rajalakshmi S (2009). Progressive changes in biochemical characters in sugarcane genotypes subjected to NaCl treatment. Indian Journal of Plant Physiology 14(1): 34-38.

Vasantha S, Gomathi R, Rakkiyappan P (2009). Sodium content juice and jaggery quality of sugarcane genotypes under salinity. Journal of Biological Sciences 1(1). : 33-38.

Vasantha S, Gururaja Rao PN, Venkataramana S, Gomathi R (2008). Salinity induced changes in the antioxidant response of sugarcane genotypes. Journal of Plant Biology 35 (2): 115-117.

Vasantha S, Venkataramana S, Gururaja Rao PN, Gomathi R (2010). Long term salinity effect on growth, photosynthesis and osmotic characteristics in sugarcane. Sugar Tech. 12(1). 5-8.

Venkataramana S, Gururaja Rao PN, Naidu KM (1983). Evaluation of cellular membrane thermostability for screening drought resistant sugarcane varieties, Sugarcane (London). No. 4: 13-15.

Venkataramana S, Gururaja Rao PN, Naidu KM (1986). The effects of water stress during the formative phase on stomatal resistance and leaf water potential and its relationship with yield in ten sugarcane varieties. Field Crops Res. 13: 345-353.

Venkataramana S, Naidu KM (1993). Abscisic acid effect on water stress indicators in sugarcane Plant Physiology and Biochem 20: 1-4.

Venkataramana S, Naidu KM (1989). Root growth during formative phase in irrigated and water stressed sugarcane and its relationship with shoot development and yield. Indian J. Plant Physiology 32: 43-50.

Venkataramana S, Naidu KM, Singh (1987). Membrane thermostability and nitrate reductase activity in relation to water stress tolerance of young sugarcane plants. New phytologist 107 (2): 335-340.

Venkataramana S, Mohan Naidu K, Sudama Singh (1987). Membrane thermostability and nitrate reductase activity in relation to water stress tolerance of young sugarcane plants. New Phytol. 107: 335-340.

Venkataramana S, Shunmugasundaram S, Naidu, KM (1984). Growth behaviour of field grown sugarcane varieties in relation to environmental parameters and soil moisture stress.Agric. and For. Meteorol. 31: 251-260.

Vinocur B, Altman A (2005). Recent advances in engineering plant tolerance to abiotic stress: achievements and limitations. Curr.Opin.Biotechnol. 16, 123–132.

Wahid A (2004). Analysis of toxic and osmotic effects of sodium chloride on leaf growth and economic yield of sugarcane. Bot. Bull. Acad. Sin. 45: 133–41.

Wahid A (2007). Physiological implications of metabolites biosynthesis in net Assimilation and heat stress tolerance of sugarcane sprouts. J. Plant Res. 120: 219–228.

Wahid A, Close TJ (2007). Expression of dehydrins under heat stress and their relationship with water relations of sugarcane leaves. Biol. Plant. 51: 104–109.

Wahid A, Ghazanfar A (2006). Possible involvement of some secondary metabolites in salt tolerance of sugarcane. J. Plant Physiol. 163: 723–730.

Wang W, Vinocur B, Shoseyov O, Altman A (2004). Role of plant heat-shock Proteins and molecular chaperones in the abiotic stress response. Trends Plant Sci. 9: 244–252.

Webster PWD, Eavis BW (1971). Effect of flooding on sugarcane growth 1.Stage of growth duration of flooding. Proceeding of International Society Sugarcane Technology 14: 708-714.

Weijun Zhou, Xianqing Lin (1995). Effects of waterlogging at different growth stages on physiological characteristics and seed yield of winter rape *(Brassica napus L.)*. Field Crops Research 44: 103-110.

Wilkinson S, Davies WJ (2010). Drought, ozone, ABA and ethylene: new insights from cell to plant community. Plant Cell Environ 33: 510-525.

Zende GK (2002). Soil water management in waterlogged soils. Bharatiya Sugar 27(5): 7-8, 10,12,14,16,18,20.

Zende NA Hapase DG (1986). Development of saline and alkaline soils in Maharashtra under sugarcane cultivation, their reclamation and mamgement. Bahratiya Sugar 11(11): 41-48.

Zhang SZ, Yang BP, Feng CL, Chen RK, Luo JP, Cai WW, Liu FH (2006). Expression of the *Grifola frondosa* trehalose synthetase gene and improvement of drought-tolerance in sugarcane (*Saccharum officinarum* L.). J Integr Plant Biol 48: 453-459.

Zhang HX, Hodson JN, Williams JP, Blumwald E (2001). Engineering salt-tolerant *Brassica* plants: characterization of yield and seed oil quality in transgenic plants with increased vacuolar sodium accumulation. Proc. Natl. Acad. Sci. USA 98, 12832–12836.

Chapter 4

Terminal Heat and Wheat Production in India

T.P. Singh

National Bureau of Plant Genetic Resources, New Delhi – 110 012
E-mail: tpsy60@gmail.com

CONTENTS

1. Introduction

Common wheat (*Triticum aestivum* L.) belongs to Poaceae family. It is the most important staple food crop in the world and second important crop in India after rice. It is cultivated in about 28.0 million ha area with a production of 93.90 million tonnes and productivity of 3140 kg/ha (2011-12). Global reports indicate a loss of 10-40 per cent in crop production by 21st century because of impact of climate change (Dutta *et al.*, 2013). Its straw is used as fodder for large population of cattle in the world. Wheat is now grown extensively in countries with tropical and subtropical climates and the importance of high temperature has been recognized by CIMMYT in Mexico which has selection programme for high temperature tolerance (Reynolds *et al.*, 1994). The productivity and yield of wheat is significantly influenced by selection of suitable varieties, soil and environmental conditions as well as the management factors. The environmental stresses like water stress (water logging and drought), temperature (heat and chilling) and salinity are the major problems of wheat growing areas which substantially reduce the yield and quality of wheat (Singh and Purohit, 1995). Drought and heat are major abiotic streeses that decrease wheat quality and productivity to a considerable extent. Continual heat stress is a problem in about 7 m ha while teminal heat stress is a problem in about 40 per cent of the irrigated wheat growing areas of the world. High temperature during grain filling stage in wheat is a major constraint and reduces the number of grain per ear, grain weight and subsequently the harvest index, resulting in reduced grain yields (Bansal *et al.*, 2013). Wheat is grown in almost all cropping environments of the world, except in the humid lowland tropics. Winter wheat under rain-fed conditions dominates in Europe, USA, Ukraine and southern Russia, followed by spring-sown. Spring wheat in semi-arid conditions in the developed world. In India, wheat is mainly grown during the winter season, planted during October-November and harvested in March-April. The Northwestern Plains Zone (NWPZ) comprising Punjab, Haryana, Delhi, parts of Rajasthan, and western Uttar Pradesh and the Northeastern Plains Zone (NEPZ) including eastern UP, Bihar, Jharkhand, Orissa, West Bengal and Assam respectively, account for 40 and 33 per cent of the total wheat area grown (26 m ha). Nagarajan (2005) in an in-depth analysis arrived at the causes for declining wheat productivity in the country: (i) decrease in the use of fertilizers (N, P and K) in the NWPZ and NEPZ; (ii) micronutrient imbalance in soils; (iii) fatigue in genetic gain in varietal development and terminal heat stress in the NWPZ; and (iv) low minimum support price fixed by the Government for the procurement of wheat. The requirement of wheat for 2020 is estimated at 109 mt. Though India is the second largest producer of wheat in the world, the average productivity is 2907 kg/ha against 4738 kg/ha in China and 7926 kg/ha in UK. Productivity in India is only 27.3 per cent of that in UK.The wheat crop in UK and NW European countries grows at much lower temperatures all through the crop duration with no water stress, and the grain-filling takes place over an extended period of 60 days at temperatures below 20°C. In India, the lower productivity is due to shorter crop duration and period of grain-filling and higher temperatures during crop growth, particularly during grain-filling. Crop duration varies from 150 days in the north and goes down to 100 days in Maharashtra and further south, with corresponding decline in yield. The pre-anthesis phase that determines the yield potential of the crop includes germination to seedling emergence; (ii) canopy development and conversion of the vegetative meristem to reproductive meristem and spike initiation; (iii) spikelet development – first spikelet initiation to the formation of the terminal spikelet; and (iv) ear emergence and anthesis. Post-anthesis phase includes the period of grain development. It has already been established that heat stress can be a significant factor in reducing the yield and quality of wheat (Stone and Nicolas 1995). In Pakistan terminal heat stress is a major reason of yield decline in wheat due to delayed planting. Similarly heat stress is a major challenge to wheat productivity in India (Joshi *et al.*, 2007). Late planted wheat suffers drastic yield losses which may exceed to 40-50 per cent. In the period leading up to 2020, demand for wheat for human consumption in developing countries it expected to grow at 1.6 per cent per annum, and feed at 2.6 per cent per annum. The global average wheat yield will have to increase during the coming 25 years from 2.6 to 3.5 tonnes ha^{-1}. Late planting of wheat in India is common due to the intensive cropping system, often delays the sowing of the crop up to the middle of January, particularly in North West India where it is generally sown after harvest of paddy, sugarcane, pigeonpea and potato. As a result, a portion of maturity period of the crop pushed forward and thus has to face higher temperature of the summer with hot spells often occurring at the time of maturity (Abrol *et al.*, 1991) the late sown wheat is more affected by high temperature stress leading to reduced yield and quality (Wardlaw and Wringley, 1994). Morphologically, similar wheat varieties have showed different degree of tolerance to post anthesis high temperature stress. Hence breeding for heat stress tolerance can open new insights significant in Indian agriculture. There are several physiological traits that are associated with heat tolerance. Photoassimilation, chlorophyll retention, chlorophyll a:b ratio, canopy temperature depression, stomatal conductance, membrane stability are some of the examples (Shukla *et al.*, 1997). Photosynthesis, respiration, conversion of sugar into starch in developing grain has been found to be affected by high temperature (Rawson, 1992). It has been reported that high temperature causes membrane damage resulting in electrolyte leakage and this leakage has been shown to be related to temperature and drought tolerance of the wheat genotypes (Deshmukh *et al.*, 1996). Reynolds *et al.* (1994) showed significant correlation among yield, photosynthetic rate and stomatal conductance and premature loss of chlorophyll associated with heat sensitivity. Wheat is a sink limited crop and high temperature during grain filling causes the production of shriveled grains due to forced maturity (Savin and Slafer, 1991). Late sown crop gets exposed to mean maximum temperature of about 35°C during grain growth and causes yield reduction of 270 kg ha^{-1} per degree rise in temperature (Rane *et al.*, 2000). Temperature adversely alters the growth and development of wheat during the early phase of panicle emergence, grain set and grain development. High temperature reduces the yield drastically due to its deterimental effect on metabolism and duration of phenological phases (Jenner, 1991). Temperature is the key factor that influences the phenological development and grain yield of wheat crop (Wardlaw and Wringley, 1994, Stone and Nicolas, 1994). Wild relatives of crops which have survived under strong natural selection pressure can be particularly useful as a source of genes for specific adaptive traits (Braun *et al.*, 1996). So far genetic variability for heat tolerance was searched only *T. aestivum* and *T. durum*. Modern wheat varieties are well adapted to controlled cultural practices, but they are generally not highly tolerant to extreme environmental stresses such as high temperature (Morgunov, 1994).

The physiological approaches of breeding for heat tolerance can provide a tool by identifying the traits that can be easily measured and used as a selection criterion for identifying heat tolerant genotypes. In course of evaluation, the organisms have acquired a whole set of distinct overlapping stress response systems, helping them to survive and propagate under multitude of unfavorable environmental conditions. Therefore there is a wide range in the temperature tolerance both between species and amongst genotypes within a species. This will be essentially confined to the latest understanding of the physiological, biochemical and molecular responses of organisms to heat stress and the effect of heat stress on molecular function, cellular processes, plant growth and development and crop production. Abiotic stress adversely affect almost all major field-grown plants belonging to varied ecosystems. The severity of abiotic stresses is on the rise due to the practice of intensive cultivation in farming areas as well as due to environmental deterioration caused by the greenhouse effect. In marginal or arid lands, environmental factors such as drought, high salinity, high temperature and flooding are serious problems. These stresses cause a great amount of loss that lead to instability in crop production. To address these problems, identification of traits leading to development of stress tolerant plants is of paramount importance.

2. Botany and Distribution

Triticum turgidum durum, the main modern variety of tetraploid durum or macaroni wheat, is widely grown in drier areas of the world such as those found in India, the Mediterraneoan Basin, the farmer Soviet Union, Argentina and the rain-fed areas of the North American great plains. Wheat belongs to the genus *Triticum*, which originated about 10,000 years ago. Polyploid *Triticum* arose when two diploid wild grasses crossed naturally to produce tetraploid wheat, which today includes cultivated durum wheat (*Triticum turgidum* L. var. group *durum* Desf. 2n=4x=28). Tetraploid wheat later outcrossed to goat grass (*T. tauschii*, considered a troublesome weed in many wheat-growing areas) and gave rise to hexaploid bread wheat (*T. aestivum* L. em Thel. 2n=6x=42).*Triticum aestivum* is the most widely grown wheat in the world today. Generally called bread or common wheat its flour is best suited for making bread. Bread wheat encompasses several thousand cultivars that are widely adapted over a range of environments and grown worldwide. Chromosomal configuration at meiosis have unequivocally demonstrated that diploids, tetraploids and hexaploids have either one (AA), two (AABB) or three (AABBDD) sets of seven chromosomes each and that the A genome is common to durum and bread wheats. This led to the inference that bread wheat is an allohexaploid. CIMMYT's wheat genebank contains almost 122 000 accessions representing; more than 50 years of breeding, collecting and acquisition. Hexaploid wheat and triticale germplasm recently have been stored in both base and active collections. Durum wheats and wild relatives are maintained as active collections, in accordance with the ICARDA-CIMMYT agreement. International Wheat Information System has two major components; The Wheat Pedigree Management System which assigns and maintains unique wheat identifiers and genealogies, and the Wheat Data Management System, which manages performance information and data on known genes. Several accessions of diploid wild relatives that have the A, B or D genomes are potential candidates for use in interspecific corsses. The different genepools with the annual and perennial grasses of the Triticeae tribe also provide tremendous genetic variability for improving wheat (Dewey 1984). However, in contrast with the *Triticum/Aegilops* spp., the perennial genera we use in our intergeneric crosses are genomically quite diverse and rather difficult to cross with wheat. Hence, accomplishing beneficial alien transfers through intergeneric hybridization is quite time consuming.

2.1. Pre-breeding

Wheat wide crossing at CIMMYT was established during the late 1970s. At that time, the main goals were intergeneric hybridixation and subsequent introgression of agronomically important traits into modern high-yielding strains, and *Aegilops Hordeum, secale* and *Thinopyrum* spp. received the most attention. Major achievements to date include the registration of Karnal bunt and *Helminthosporium sativum* resistant stocks and the relaease in Mexico and Pakistan of Karnal bunt-resistant stocks and the release in Mexico and Pakistan of Karnal bunt-resistant and salt-tolerant cultivars, respectively (Mujeeb-Kazi and Hettel 1995). During the late 1980s, the wide crosses section expanded to include interspecific hybridization and started exploiting the variability locked in the three genome donors of modern wheat. As a result, more than 500 synthetic wheats having exotic A, B or D genomes were produced and are being maintained at CIMMYT. These synthetics are proving to be extremely valuable sources of resistance to various biotic and abiotic stresses, as well as of yield-related traits (Mujeeb-Kaizi and Hettel 1995).

3. Growth Parameters

3.1. Plant Height

Plant height showed significant positive associations with aerial biomass LAI, LAD, ears per plant and dry weight (Sharma *et al.*, 1991). Rawson *et al.* (1996) observed that increasing mean temperature from 12°C to 20°C resulted in reduction in plant height by 16 per cent).

3.2. Leaf Area Index

Leaf area index was positively correlated with chlorophyll content and yield and negatively correlated with plant height (Kler and Bains, 1989). Leaf area index showed significant positive association with leaf area ratio (LAR) and leaf area duration (LAD) but negative association with ear number per plant (Sharma *et al.*, 1991). Das *et al.* (1993) found peak value of LAI, 73-85 days after sowing irrespective of sowing date and nitrogen rate. Ostapenko (1993) concluded that final yield was highly correlated with leaf area index and photosynthetic potential. Sharma *et al.* (1994) observed that LAI had high degree of association with grain yield. Frederick and Camberato (1995) found that rate of decline in LAI during grain fill were more rapid in 1993 compared to 1994, when air temperature was relatively warmer. Heat stress reduces the duration of vegetative growth (Saini, 1988) and therefore reduces leaf area (Warrington *et al.*, 1977) and leaf number (Acevedo *et al.*, 1990).

3.3. Crop Growth Rate

Sharma *et al.* (1991) have observed significant positive association of crop growth rate (CGR) with relative growth rate (RGR) and dry weight per ear and negative association of RGR with tillers number per plants. Karimi and Siddique, (1991) considered that higher grain yields of modern cultivars of wheat was achieved with higher RGR during the negative phase and greater CGR from ear emergence to harvest. Delayed sowing decreased crop growth rate and net assimilation rate during the late grain filling period in wheat cultivars and decreased biomass (Takahashi and Nakaseko, 1992). Simane *et al.* (1993) observed that drought resistant wheat cultivars had high RGR in favorable period of the growing season and a low RGR during moisture stress.

Plant growth substances play an important role in regulation of such production and partitioning processes. Furthermore, yield stability could be obtained through incorporation of tolerance to abiotic stresses (Sairam and Srivastava, 2000). Crop plants have also to be tailored for changing climate caused by increase in greenhouse gases which may alter temperature and precipitation (Ghildiyal and Sharma-Natu, 2000; Uprety and Mahalaxmi, 2006).

3.4. Root System

Any effects of thermal stress on the morphology and physiology of the root system may influence water movement through the plant (Mahan, McMichael and Wanjura, 1995). Morover roots are an important sink for assimilates in wheat. Since remobilization of assimilates occurs after anthesis, assimilates from roots may supplement primary sources from the leaf and stem (Cook and Evans, 1978; Hay and Walker, 1989).Wardlay, Solfield and Cartwright (1980) found in pot-grown wheat exposed to higher temperatures from anthesis that a greater movement of ^{14}C-labelled photo-synthate from the flag leaf to the ear was due to a change in the pattern of distribution, with less going to roots.

4. Physiological and Biochemical Parameters

4.1. Chlorophyll Content

Decreases in chlorophyll content due to high temperatures have been observed in wheat leaves (Liv and Su, 1985). Bhullar and Jenner (1983) estimated chlorophyll content in the control and warmed ears and reported that chlorophyll content in the pericarp can be taken as a indicator of developmental stage. They further concluded that warming accelerates chlorophyll degradation by about 4th day. Buttery *et al.* (1981) reported that the variability in photosynthetic rate, nearly 44 per cent was due to variability in chlorophyll content. Xu (1991) reported that photosynthetic rate, thylakoid chlorophyll and protein content of whole, wheat cv. Len, plants were nearly constant at 15/10°C but decreased rapidly at 35/30°C during maturation. He further observed that chlorophyll content of the flag leaf of wheat cv. Len was also reduced under stress environment (high temperature and low moisture), which accelerated the senescence processes and ultimately reduced the grain growth period. Studies of Reynolds *et al.* (1994) showed significant correlation among yield, photosynthetic rate and stomatal conductance and this was probably due to premature loss of chlorophyll associated with heat sensitivity.

4.2. Water Status

Heat stress injury involves water deficit and cell turgor loss (Ahmad *et al.*, 1989). Maintenance of favorable water status is essential for plant tolerance to heat stress (Jiang and Huang, 2000). Drought preconditioning-enhanced heat tolerance may be related to the maintenance of plant water relations by reducing water loss and or increasing water uptake capacity. Osmotic adjustment is well known to be an important physiological mechanism of water retention and cell turgor maintenance (Turner and Jones, 1980 and Morgan, 1984). Relative water content (RWC) has been considered as a better indicator of water status as it reflects the balance between water supplied to leaves and transpiration rate through its relation to cell volume (Sinclair and Ladlow, 1985). Due to high temperature induced higher transpiration, situation similar to water stress is created and RWC becomes important in heat stress also. Spomer (1985), while discussing the different conventional parameters for measuring plant water status, advanced the opinion that water potential is probably the best single measure of the water status in plants. Water potential, ψ, describes water status in terms of its availability for plant processes. NMR relaxation times have been used widely to describe the water status as well as heat the freezing injury in plant tissues (Abass and Rajashekar, 1991). Nuclear magnetic resonance (NMR) offers a non-invasive method for characterization water status in biological tissues (Lewa and Lewa, 1990). Membrane disorganization leads to changes in molecular mobility and other biophysical properties of tissue water (Maheswari *et al.*, 1999). The ^{1}H NMR transverse relaxation time T_2 of the tissue water has been shown to be related to the water content of the plant tissue, properties of water in different parts of the tissue and interactions with macromolecules (Hills and Duce, 1990, Ratcliffe, 1994). It has been shown that proton spin-lattice relaxation time T_1 of leaves of barley (Colire *et al.*, 1988) and wheat (Nagarajan *et al.*, 1993) is a good indicator of plant water status than conventional water relation. Sairam *et al.* (2000) reviewed that extent of oxidative injury and activity of antioxidant enzymes in relation to heat stress induced by manipulation dates of sowing the increase in temperature by late sowing significantly decreased leaf relative water content, ascorbic acid content and increased H_2O_2 content at 8 and 23 days after anthesis. Paulsen and Machado (2001) observed that high temperature interacts with drought to affect water relations and the effect was greater in heat sensitive wheat (*Triticum aestivum*) than in sorghum. Mohamed *et al.* (2002) reported that tapary bean genotypes with less reduction in relative water content could cope-up better under severe stress conditions.

4.3. Membrane Thermostability

The two major sites potentially affected by high temperatures are enzymes and membrane. Membrane disruption may alter water, ion and organic solute movement, photosynthesis and respiration (Christiansen, 1978). Martineau *et al.* (1979) observed that when leaf tissue was injured by high temperature, membrane damage was increased and electrolytes diffused out of cells. Blum and Ebercon (1981) estimated drought and heat tolerance of sorghum by measuring the electrical conductivity of aqueous media containing the leaf disc that were previously treated with polyethylene glycol and concluded that relationship between the degree of leakage and tolerance to drought and heat stress did not exist. Other workers have also used conductometers measurements of solute leakage from cells to estimate heat damage to plasma membranes (Singh *et al.*, 1991). Blum and Ebercon (1981) observed genetic variation in membrane thermostability using conductometric measurements in various field crops including spring wheat. The heat susceptibility of plasma membrane has been shown by ion leakage studies in many crop plants. The increased leakage of solutes is an indication of damage to membrane (Chaisompongopan *et al.*, 1990). Heat injury to the plasmalemma may be measured by ion leakage (Chaisompongpan *et al.*, 1990, Hall, 1993). However, stable cell membrane that remains functional during stress appears to control adaptation to high temperature and found related to heat and drought tolerance (Sullivan and Ross, 1979, Raison *et al.*, 1980.) Blum (1988) reported that membrane is one of the sites of primary physiological injury by heat. A proportional change in ion leakage with increasing temperature in tissues of wheat (Navari- izzo *et al.*, 1993). Deshmukh *et al.* (1985, 1991, 1996) have also suggested the use of ion leakage as a measurements index for screening genotypes against heat and drought stress in wheat. They observed that with increase in temperature there was a proportional increase in ion leakage. Nagarajan and Rane (1997) also emphasized the use of membrane thermostability as one of the simple parameter for screening wheat genotypes against high temperature stress. The ability of plants to acclimate to higher temperatures was conducted on plants adapted to high temperature growth (Berry and Bjorkman, 1980). The plasmalemma and membranes of cell organelles play a vital role in the functioning of cells. Any adverse effect of temperature stress on the membranes leads to disruption of cellular activity or death. Injury to membranes from a sudden heat stress event may result from either denaturation of the membrane proteins or from melting of membrane lipids which leads to membrane rapture and loss of cellular contents (Ahrens and Ingram, 1988). Membrane fatty acids of plants from temperate environments show similar trends in response to temperature, an observation that suggests that alterations in membrane lipids generally contribute to the ability of plants to acclimate to different temperatures (Williams *et al.*, 1988 and Harwood *et al.*, 1994).

The susceptibility of biomembranes to sudden heat stress may be due to:

1. Denaturation of membrane proteins
2. Melting of lipids in membrane which leads to breakage of membrane and loss of cellular contents. Long term acquisition of heat tolerance in several plants is accompanied by a decrease in fatty acid unsaturation resulting in a decrease in the lipid fluidity and an increase in the phase transition temperature. This leads to the stability of membrane during heat stress. Lipid unsaturation is a component of heat tolerance in FAD mutants of Arabidopsis in which a pronounced decrease in trienoic fatty acids of chloroplast membrane results in heat tolerance while large increase in the lipid saturation had only minor effects on the rate of photosynthetic electron transport (Somerville and Browse, 1991).

The chloroplast membrane of higher plants contains an unusually high concentration of trienoic fatty acids. Plants grown in colder temperatures have a higher content of trienoic fatty acids. Transgenic tobacco plants in which the gene encoding chloroplast omega-3 fatty acid desaturase, which synthesizes trienoic fatty acids, was silenced contained a lower level of trienoic fatty acids than wild-type plants and were better able to acclimate to higher temperatures (Murakami *et al.*, 2000). Ibrahim and Quick (2001) suggested that heat tolerance based on membrane thermal stability can be improved using the existing genetic variability available within germplasm. Talwar (2002) and Singh *et al.* (2004) observed that membrane stability is one of the important parameters to evaluate genetic variability for heat strees.The low membrane injury index and high relative water content also essential for tolerant genotypes. The close association (0.76) of membrane injury index with per cent reduction in biomass has clearly indicated that this trait could be used for screening large number of genotypes under late sown conditions. On the basis of results obtained it is concluded that tolerant genotypes should possess a set of characteristics, *i.e.* low membrane injury index and high relative water content which enable them to maintain relatively low leaf temperature under stress condition. It is further suggested that the simple traits like membrane injury index and relative water content which are relatively simple and require less sophisticated instruments could be used easily in a breeding programme for screening tolerant types and for improving productivity of wheat under late planting conditions in any part of world where rice-wheat cropping system is popular (Deshmukh *et al.*, 2006). High-temperature induced changes in the membrane composition may therefore play a role in the stability of such proteins leading to positive effects on whole plant growth at elevated temperature. Almeselmani *et al.* (2006) observed significant increase in membrane injury index in all genotypes under late and very late planting compared to normal planting of wheat.

4.4. Photosynthesis

The important response of photosynthesis, among other physiological processes, to changes in temperature, was clearly shown by Bjorkman *et al.* (1974, 1975) working with plants from different ecology. It is well documented that high temperatures inhibit photosynthetic CO_2 fixation and adjustments in the level of unsaturation of thylakoid membranes may affect the capacity of plants to adapt to elevated temperature conditions in order to avoid a reduction in photosynthetic efficiency (Berry and Bjorkman, 1980). While photosynthetic rates were found to be temperature sensitive in other crops, wheat and rice appear to be different. In wheat, no measurable differences were found in photosynthetic rates per unit flag leaf area or on a whole plant basis in the temperature range from 15 to 35°C (Bagga and Rawson, 1977), but in rice, there is little temperature effect on leaf carbon dioxide assimilation from 20 to 40°C (Egeh *et al.*, 1994). Al-Katib and Paulsen (1990) and Shah (1992) reported that heat stress decreased mean photosynthetic rates by 32 and 11 per cent in seedlings and mature plants respectively. The influence of high temperature even in the course of short time results in structural and functional changes of photosynthetic apparatus of higher plants and algae. As for extreme high temperatures, the thylakoid is a distinct sensitive membrane which is inactivated in its function at a lower temperature compared to other cellular components, like the mitochondria (Thebud and Santarius, 1982), and PSII has been recognized as the most sensitive component of the complete photosynthetic system (Berry and Bjorkman, 1980) and heat stress damage photosynthetic electron transport, particularly at the site of PSII (Havaux and Tardy, 1996). The reaction center of PSII is considered to be the most sensitive component of the thylakoid membrane to thermal breakdown, and the function of the water-splitting D1 protein within PSII has been implicated as the most readily damaged by high temperature as well as being the primary target for photoinhibition (Aro *et al.*, 1993 and 1994). Studies conducted on the ability of plants to acclimate to elevated temperature have mainly focused on components most likely to affect the stability of photosynthetic electron transport, particularly PSII. In this respect, the composition of the chloroplast thylakoids is expected to be important in the thermal tolerance of photosynthetic electron transport (Berry and Bjorkman, 1980).

The heat susceptibility of PSII is studied by measuring the chlorophyll fluorescence in wheat (Mishra and Singal, 1992,) and is related to peroxidation of thylakoid lipids. Chlorophyll (Chl) fluorescence is one of the few physiological parameters that have been shown to correlate with thermo and salinity tolerance. Measurement of PSII activity during leaf heating was used here as a sensitive indicator of the thermostability that might be conferred by the different membrane compositions (Falk *et al.*, 1996). One of the effects of high temperature on thylakoids, specifically PSII, is the observed sharp rise of the basal fluorescence (F_o fluorescence emitted by PSII when all PSII complexes are open). The increase in F_o occurs in plants over a small range of temperatures (40-45°C), and has been used as a signal for the irreversible damage of the photosynthetic apparatus. Such increase in the basal fluorescence has been interpreted to be a result of a functional disconnection of the LHC II from the reaction centre and also from a block in the reducing side of PSII (Bukhov *et al.*, 1990).

In common bean (*Phaseolus vulgaris L.*) the PSII heat susceptibility studies by measuring O_2 evolution and chlorophyll fluorescence (Chaisompongpan *et al.*, 1990), and it was correlated with heat acclimation potential of different bean genotypes. The quantum yield of PSII depends on its heat stability. The ratio of $\hat{o}H/\hat{o}CO_2$ ($\hat{o}H$: electron transported via PSII per quantum absorbed, $\hat{o}CO_2$: true quantum yield of CO_2 assimilation) was not affected by differences in the biochemical mechanism of CO_2 concentration in bundle sheath cells among five C_4 plants *viz.*, *Flaveria trinervia* (NADP-ME), *Zea mays* (NADP-ME), *Amaranthus cruentus* (NAD-ME). *Panicum maximum* (PEP-CK) and *P. milliaceum* (NAD-ME), and the ratio is lower than a C_3 plant *Elaveria pringlei*. The $\hat{o}H/\hat{o}CO_2$ ratio of C_3 plant increases with raising temperature (Oberhuber *et al.*, 1993). Al-Katib and Paulsen, (1999) compared between Wheat (*Triticum aestivum L.*), a C_3 species adapted to cool environments: rice (*Oryza Sativa L*), a C_3 species adapted to warm environments; and millet (*Pennesitum glaucum L.*) a C_4 species adapted to hot environments. Photosynthesis was measured in plants grown at 22, 32, or 42°C, and light reactions were measured in protoplasts, chloroplasts, and thylakoids isolated from seedlings grown at 27°C and treated in vitro at 22, 32, and 42°C. Leaf photosynthesis of millet and rice increased from 22 to 32°C and then decreased as temperature increases to 42°C, whereas in wheat it was highest at 22°C and decreased as temperature increased. Photosynthetic rates of protoplasts and chloroplasts from all species decreased after being treated *in vitro* over the same temperature range.

PSII activity declined steadily in protoplasts, chloroplasts and thylakoids of millet and rice from 22 to 42°C but decreased abruptly in organelles of wheat from 32 to 42°C. The results suggest that differences in photosynthetic responses to high temperature are associated with light reactions and extreme sensitivity of wheat may be attributable to injury to PSII. A higher or lower degree of membrane lipid saturation is beneficial for high temperature tolerance (Klueva *et al.*, 2001). The loss of PSII activity is associated with changes in the distribution of excitation energy between the two photosystems and partly reversible after transfer the leaves and chloroplasts to low temperatures. The rate of oxygen evolution/consumption, *in vivo* electron transport activity, overall photosynthetic capacity were studied after treatment of intact barley seedlings at 40°C for 3 hours either in presence of low white light (100 $\mu mol\ m^{-2}.s^{-1}$) or in the dark. High temperature impact in the dark resulted in lowering of water splitting capacity, photosynthetic electron transport rate, suppression of non-photochemical energy dissipation and membrane energization (qE), caused the photoinhibition of PSII at high PFD and indicate impairment of CO_2 assimilation rate in old (11 days old) leaves. Chen *et al.* (2004) found that high temperature and strong light stress induced a striking inactivation of PSII particle electron transport and degraded OEC 33 KD polypeptide components.

Inhibition of photosynthesis at high leaf temperature could be markedly reduced by artificially increasing C_i. A high C_i also stimulated photosynthetic electron transport and resulted in reduced non photochemical fluorescence quenching (Haldimann and Feller, 2004). The capacity of a plant to acclimate and maintain photosynthesis under high temperatures is a critical factor in heat-tolerance (Jason *et al.*, 2004). Gupta and Gupta (2005) reported that the high temperature adversely affects the photosynthetic functioning and the thylakoid membrane particularly the PSII complexes located on the membrane, which are most heat sensitive part of the PSII mechanism. In addition, Rubisco and other enzymes of carbon metabolism are also adversely influenced by high temperature. Due to above traits in the tolerant genotypes the rate of photosynthesis was relatively high. High temperatures are known to have deleterious effects on photosynthesis, respiration and reproduction (Evans *et al.*, 1975). At molecular level, these effects are brought about by altered gene expression and manifested at the biochemical and metabolic level, membrane stability, and production of heat shock proteins (HSPs)(Abrol *et al.*, 1996). A 5°C increase results in selective expression of HSPs, with continued synthesis of normal cellular proteins.

4.5. Rubisco

Weis (1981b) reported that light-dependent activation of Rubisco in spinach (*Spinacia oleracea* L.) chloroplasts was inhibited by moderately elevated temperatures and that inhibition was closely correlated with reversible inhibition of CO_2 fixation. A similar effect of temperature on Rubisco activation and CO_2 fixation was reported for wheat (*Triticum aestivum* L.) leaves (Kobza and Edwards, 1987). The temperature induced inhibition of carbon assimilation cannot be explained solely by increased photorespiration via the enhanced specificity of Rubisco for oxygen at higher temperatures (Jordan and Ogren, 1984). The synthesis of RuBPcase/oxygenase, the most abundant protein on the earth is susceptible to heat stress. In green suspension cultures of soybean, when the temperature is raises 38-40°C the amount of RuBP decreased 20-30 per cent (Nover *et al.*, 1989). The small subunit synthesis is more heat susceptible than larger subunit in field grown soybean. There exist genetic differences even within a species for heat tolerance in Rubisco SSU synthesis. At 34°C, the heat tolerant wheat variety mustang was able to maintain higher level of SSU synthesis than the heat susceptible variety (Krishnan *et al.*, 1989). The sensitivity of SSU mRNA to heat shock suggests that decreased synthesis of chloroplast proteins produced in the cytoplasm may be an important causal factor of heat damage to plants (Nover *et al.*, 1989). However, after one-hour heat acclimation at 37°C, wheat varieties did not show any difference in SSU synthesis. This may be results of HSP induction at 37°C and protection of transcriptional and translational machinery by HSPs (Krishnan *et al.*, 1989). In soybean leaves the degree of Rubisco activation did not show a clear correlation with either CO_2 concentration or with air temperature, whereas the concentration of RuBP in the upper canopy leaves decreased with increasing air temperature from 26 to 36°C (Campbell *et al.*, 1990). The reduction in the RuBP concentration at high temperature may be due to the increased rate of oxygenation reaction catalyzed by the Rubisco at high temperature (Ogren, 1984). Sharkova (1994) observed that Rubisco activity was not significantly altered even after the heat treatment which induced severe damage to photosynthesis and photosynthetic electron transport. Increase in temperature from 20°C to 25°C increased photosynthesis with a Q_{10} value of 1.90. As the temperature increased from 25°C to 30°C the Q_{10} decreased possibly following a restriction in the regeneration of the acceptor for CO_2 and as temperature increased from 30 to 35°C Q_{10} was further decreased, as the result of the inability of the thylakoid to maintain adequate supply of NADH (Pastenes and Horton, 1995).

Feller *et al.* (1998) showed that light activation of ribulose-1, 5-bisphosphate Rubisco is inhibited by moderately elevated temperature through an effect on Rubisco activase. When cotton (*Gossypium hirsutum* L.) or wheat (*Triticum aestivum* L.) leaf tissue was exposed to increasing temperatures in the light, activation of Rubisco was inhibited above 30 and 35°C, respectively, and the relative inhibition was greater for wheat than for cotton. The temperature induced inhibition of Rubisco activation was fully reversible at temperatures below 40°C. In contrast to activation state, total Rubisco activity was not affected by temperatures as high as 45°C. Non-photochemical fluorescence quenching increased at temperatures that inhibited Rubisco activation, consistent with inhibition of Calvin cycle activity. Initial and maximal chlorophyll fluorescence was not significantly altered until temperatures exceeded 40°C. Thus electron transport, as measured by Chl fluorescence, appeared to be more stable to moderately elevated temperatures than Rubisco activation. Moderately elevated temperatures have a direct effect on the distribution and form of activase, the biochemical component controlling Rubisco activation. A sensitivity of activase to inactivation at moderately elevated temperatures is consistent with the poor thermal stability of the isolated enzyme (Robinson and Portis, 1989, Crafts-Brandner *et al.*, 1997; Eckardt and Portis, 1997).

For both cotton and wheat, a 5 min exposure of leaf tissue to temperatures of 30 to 35°C had a rapid and readily reversible effect on light activation of Rubisco Weis (1981a). Weis (1981a, 1981b) and Kobza and Edwards (1987) observed a close relationship between loss of Rubisco activation and inhibition of CO_2 fixation in response to increasing temperature. Salvucci and Ogren, (1996) suggested that increase in qN that were observed upon exposure of cotton and wheat to elevated temperatures were attributable to the consequent effects of activase inhibition on Rubisco activation and CO_2 fixation. Weis (1981a) reported an increase in thylakoid energization (a major component of qN) in response to loss of Rubisco activation at elevated temperature. Similarly, Salvucci *et al.* (1987) showed than qN and light scattering were both increased under ambient conditions in a mutant of Arabidopsis that is unable to activate Rubisco because of a lack of activase.

Crafts-Brandner *et al.* (1997) reported that high temperature physically perturbs activase, leading to an inhibition of enzyme activity and the consequent effect on light activation of Rubisco. The effect of high temperature stress on the expression of ribulose-1,5-bisphosphate Rubisco activase was examined in wheat (*Triticum aestivum* L.) leaves, which normally possess 46 and 42 kDa activase forms. Heat stress at 38°C significantly reduced total activase mRNA levels compared to controls, and recovery of activase transcription was only marginal 24 h after alleviating heat stress. In contrast to transcript abundance, immunoblot analysis indicated that heat stress increased the accumulation of the 42 kDa activase and induced a putative 41 kDa form. Heat stress did not affect the amounts of the 46 and 42 kDa activase forms (present as 51 and 45 kDa preproteins) recovered after their immunoprecipitation from in *vitro* translation products. *De novo* protein synthesis *in vivo* in the presence of [35 S] Met/Cys showed an increase in the amount of newly synthesized 42 kDa subunit after 4 h of heat stress, and synthesis of the putative 41 kDa activase was apparent. In contrast to activase, heat stress led to a rapid and large reduction in the *de novo* synthesis of the large and small subunits of Rubisco. Long term (48 h) heat stress further increased the amounts of *de novo* synthesized 42 and 41 kDa activase forms. After 24 h of recovery from heat stress, *de novo* synthesis of the 42 kDa activase returned to control levels, while a small amount of 41kDa protein was still expressed. Southern analysis suggested the presence of a single activase gene. This indicated that heat stress alters activase expression, most likely post transcriptionally, and suggest that the heat induced expression of the 42 and 41 kDa subunits of wheat leaf Rubisco activase may be related to the maintenance and acclimation of photosynthetic CO_2 fixation during high temperature stress in wheat (Law and Crafts. 2001).

Rubisco activation in tobacco protoplasts decreased at temperatures higher than 30°C. The same point in the temperature response where the activities of isolated Rubisco and activase began to deviate. The first indications of thermal denaturation of activase both *in vitro* and *in vivo* occurred at temperatures near the optimum for ATP hydrolysis (42°C – 44°C) and denaturation was extensive at temperatures above the optimum. Because activase physically interacts with Rubisco, minor changes in its structural integrity or oligomeric state could affect its ability to interact productively with Rubisco. A disruption of activase-Rubisco interactions would explain the inability of activase to maintain Rubisco in an active state at temperatures between 30°C and 44°C, *i.e.* elevated temperatures that are at or below the temperature optimum for ATP hydrolysis. In an alternate manner, activase activity may simply be inadequate to offset the faster rate of Rubisco deactivation at these temperatures (Crafts-Brandner and Salvucci, 2000).

At temperatures higher than the optimum for ATP hydrolysis, thermal denaturation of activase was extensive. As a consequence, the marked decrease in Rubisco activation at temperatures greater than 44°C is almost certainly caused by loss of activase activity per second and disruption of activase Rubisco interactions. Compared with other chloroplast proteins, activase was extraordinarily sensitive to thermal denaturation. In Arabidopsis plants subjected to abrupt heat stress, both forms of activase exhibited similar patterns of thermal aggregation. Thus, species variability may exist in the relative sensitivities of the two-activase polypeptides to thermal denaturation. When heat stress is imposed rapidly, activase rapidly loses structural integrity, probably overwhelming the constitutive chaperonin system.

However, if heat stress is imposed slowly photosynthesis can acclimate, requiring higher temperatures for inhibition of CO_2 fixation (Berry and Bjorkman, 1980, Weis and Berry, 1988) and for deactivating Rubisco (Law and Crafts-Brandner, 1999).

The mechanistic basis for photosynthetic acclimation is unknown, but could involve stabilization of activase structure by chaperonins to prevent activase from participating in unproductive associations with other activase molecules or with Rubisco or other chloroplast proteins. In an alternate manner, *de novo* synthesis of more thermally stable forms of activase (Sanchez de Jimenez *et al.*, 1995, Law and Crafts-Brandner, 2001), modifications that improve the thermal stability of existing forms of activase or changes in expression that increase the ratio of the longer, more thermally stable form of activase (Crafts-Brandner *et al.*, 1997) could also provide mechanisms for photosynthetic acclimation to high temperature. A more permanent way of increasing the thermal stability of photosynthesis may be to transform plants with additional copies of activase or with an activase that is engineered to be more thermally stable. Modification of activase structure to improve its thermal stability will require changes that do not compromise the ability of activase to interact with Rubisco. Salvucci *et al.* (2004) reported that the formation of high-molecular-weight aggregates of activase after exposure of intact leaf tissue to high temperature provided direct evidence that the physical structure of the enzyme was perturbed by the temperature treatment. Activase from wheat was much more susceptible to structural damage compared with cotton, similar to the results for Rubisco activation, and Chl fluorescence. The temperature required perturbing the distribution and form of activase was much higher than the temperature that caused reversible inhibition of Rubisco activation, occurring at temperatures that caused irreversible damage.

4.6. Respiration

Respiration is a process, which conserves energy for cellular function, growth and differentiation. In wheat, dark respiration rate at 20°C is about 1/3 of net photosynthetic rate at each stage of growth and it increases upon temperature upshift. For most plant species the Q_{10} for respiration between 5-28°C is 2 to 2.5°C. Upto 35°C the rate of respiration increases and Q_{10} begins to decrease. Therefore, heat stress can reduce the net carbon fixed by doubling the dark respiration, even within the physiological range. Dark respiration rate of wheat increases from 0.3 mg CO_2 dm^{-2} h^{-1} to 2.5 mg CO_2 $dm^{-2}h^1$ upon temperature upshift from 15°C to 35°C and thus contributes to considerable yield loss (Evans *et al.*, 1973). The electron transport and oxidative phosphorylation was less affected at 32°C in *Phaseolus acutifolius*. This lower mitochondrial efficiency of *P. vulgaris* resulted in reduced plant growth at 32°C. Therefore, thermostability of mitochondrial functions may determine the high temperature tolerance in plants (Lin and Markhart, 1990).

4.7. Assimilate Supply

Since the number of endosperm cells is decided by the assimilate supply (Brocklehurst *et al.*, 1978), the availability of assimilate is a major factor that determines the final grain weight. Heat stress affects the assimilate availability mainly through its effects on the photosynthesis, rate of respiration and translocation of photosynthates to the grain (Acevedo *et al.*, 1990). There are two sources of assimilate supply for grain growth: photosynthesis of flag leaf and ear after anthesis, and translocation of materials assimilated before anthesis (Gallagher *et al.*, 1976). Due to the high temperature the rates of flag leaf photosynthesis decreased (Al-katib and Paulsen, 1984) and respiration rate increased approximately 7 times upon temperature upshift from 15 to 25°C. In wheat ear photosynthesis may contribute to the final grain dry weight upto 50 per cent (Evans *et al.*, 1973) and presence of awns doubles the rate of net photosynthesis by ear. Presence of awns is an important component of thermotolerance, because of its photosynthetic stability at high temperatures (Blum, 1986).Under irrigated warm condition, the reserves of carbohydrates are not mobilized to grain (Jenner, 1991) as well as harvest index of most wheat cultivars is reduced at high temperature (Ruwali and Prasad, 1991), which may be due to:

1. Inhibitory effect of heat on translocation of assimilates.
2. Inactivation of starch synthase at high temperatures (Rijven, 1986; Mohabir and John, 1988). Hence the sugars accumulate which reduce the phloem unloading and thus reduce the sink demand resulting in decreased partitioning to the grain at high temperature. The sink limitation in turn may also reduce the source strength (Evans *et al.*, 1973, Sinha and Khanna, 1975) and thus contribute to lower yield.

4.8. Starch Synthesis

The rate and duration of starch deposition in wheat grains determines the final grain weight and is determined by the capacity of plant to provide photoassimilates and the capacity of grains to synthesize starch (Jenner, 1991). Among the environmental stresses high temperature is most deleterious, which reduces the rate as well as the duration of starch synthesis (Gallagher *et al.*, 1976). Among the enzymes involved in starch synthesis, following two are most susceptible to high temperature:

1. Starch synthase of wheat (Rijven, 1986, Bhullar and Jenner, 1986), potato (Mohabir and John, 1988).
2.
   Sucrose synthase of wheat (Jenner, 1991) and barley (McLeod and Duffus, 1988). Sucrose synthase converts the sucrose to UDP-G and fructose. UDPG is then converted to ADPG, which is then converted to starch by starch synthase. Both enzymes from wheat and potato (cool season crops) are highly susceptible to high temperatures (35°C) whereas the starch synthesis in rice is stable even at 39/34°C (Tashiro and Wardlaw, 1991). Between wheat cultivars, there is variation in tolerance of starch deposition to elevated temperatures (Wardlaw *et al.*, 1989). Therefore, there is opportunity to manipulate heat tolerance by conventional breeding or by transferring genes coding for more heat tolerance enzymes, *e.g.* rice by genetic engineering (Jenner, 1991).

Wallwork *et al.* (1998) studied the effect of high temperature during grain filling on starch synthesis in the developing barley grain. Plants of malting barley were exposed to 5 days of high temperature (up to 35°C) during mid grain filling under controlled environmental conditions. Grains from heat treated plants accumulated 30 per cent less starch than grains from control plants (21/16°C, 14 hrs day). Their findings suggested 11-75 per cent reduction in activity of the enzymes sunder investigation followed by high temperature exposure. In addition, ADP glucose pyrophosphorylase, branching enzyme and granule bound starch synthase showed increased activity during exposure to moderate temperatures (28-32°C), but reduced activity at high temperatures, while soluble starch synthase showed an immediate loss of activity, even at moderate temperatures. Sucrose synthase and UDP glucose pyrophosphorylase showed the greatest reduction in catalytic activity after plants was returned to cooler conditions.

Wilhelm *et al.* (1999) studied the effect of extended period of high temperature during grain filling on kernel growth, composition, and starch metabolism of seven maize genotypes. Reduction in kernel growth rate resulted in an average mature kernel dry weight loss of 7 per cent. Proportionally similar reductions occurred for starch, protein and oil contents of the kernel. A survey of 11 enzymes of sugar and starch metabolism extracted from developing endosperm revealed that ADP glucose pyrophosphorylase, glucokinase, sucrose synthase and soluble starch synthase were most sensitive to the high temperature treatment. Prakash *et al.* (2003) observed that high temperature (+5-7°C) during grain development phase decreased grain growth in wheat (*Triticum aestivum*) cultivars. Associated with a decrease in grain growth under elevated temperature (ET), there was a decrease in grain starch content and soluble starch synthase (SSS) activity. SSS activity decreased by ET, 20 DAA, a stage beyond which both grain growth and starch accumulation in grains began to decrease at ET compared to control. Relatively tolerant cultivars showed less decrease compared to susceptible types in grain growth, starch accumulation and SSS activity. Out of soluble and granule bound starch synthesis, activity of SSS was found to be more sensitive to high temperature. Prasad *et al.* (2004) also observed that when excised developing grains were pre exposed to gradual rise in temperature from 15 to 45°C, the decrease in SSS activity was less compared to direct exposure to 45°C.

Zehedi *et al.* (2003) reported that in two wheat cultivars differing in tolerance to high temperature, final grain weight was reduced by 33 per cent in the least sensitive (cv. Kavko) and by 40 per cent in the most sensitive (cv. Lyallpur) cultivar as post anthesis temperature was raised from 20/15°C (day/night) to 30/25°C. The difference in the response of the two cultivars was mainly due to changes in the rate of grain filling at high temperature. The response of the rate of grain filling at high temperature and the differential effects on the two cultivars did not seem to be explained by an effect of temperature on the supply of assimilate (sucrose) or on the availability of the substrate (ADPG) for starch synthesis in the grains. *In vitro* but not *in vivo*, the differential responses of the efficiency (V_{max}/K_m) of SSS in the two cultivars to an increase in temperature were associated with differences in the temperature sensitivity of grain filling. *In vivo*, the most remarkable difference between the two varieties was in the absolute values of the efficiency of SSS, with the most tolerant cultivar having the highest efficiency.

4.9. Grain Growth Rate and Duration

To get 30 days of grain filling duration, anthesis must occur by 15 December to early January in central India and in south India. Temperature increase from 15/

10°C to 21/16°C reduced the duration of grain filling from 60 to 36 days. Growth rate per grain depends on floret position within the ear and growth rate varies between cultivars and increases with rise in temperature (Sofield *et al.*, 1977). Temperature has pronounced effect on duration of grain filling. At low temperature, grain development continues for a longer period. This results in higher final grain weight (Warrington *et al.*, 1977). The heat degree days (HDD) required for wheat grain to attain 95 per cent of the maximum harvest weight from date of anthesis is around 400 heat degree days for HD 4152, Kalyansona and Sonalika. Therefore, high temperatures during this phase will provide 400 HDD in less days and thus the duration of grain development in reduces (Saini and Dadhwal, 1986). The response of grain filling to high temperature appears to have a direct effect on the grain itself as the response is largely independent of the source sink balance (Wardlaw *et al.*, 1989). Rate and duration of grain filling determine final grain weight (Duguid and Brule, 1994). Kernel dry weight at maturity is positively correlated with the rate of kernel filling (Wardlaw and Moncur, 1995). Higher temperatures affect all phases of crop growth, accelerate floral initiation, reduce the period of spike development, resulting in shorter spike with lower number of spikelets, and adversely affecting pollen development. The duration of grain growth in the post-anthesis period is considered the most significant determinant of yield in wheat (Evans *et al.*, 1975). Both the day and night temperatures have pronounced effect on the duration of grain-filling. Higher temperatures further associated with limitation of water cause rapid shrinkage of grain volume. Grain growth in wheat takes place under a rapidly ascending temperature when hot winds are frequent. High temperature during grain filling stage is an important yield limiting factor in wheat. The main effect of high temperature after anthesis is the reduction in grain size. Physiologically the rate of grain filling reflects the rate of biochemical reactions involved in the synthesis of reserves. Since, starch constitutes around 70 per cent of dry matter in cereal grains, the synthesis and deposition of starch is therefore, an important determinant of the size of the grain. Attempt was therefore, made to elucidate the high temperature effect on starch synthesizing capacity of wheat grain. Starch in grains is deposited in amyloplasts involving ADP glucose pyrophosphorylase (AGPase), starch synthases and branching enzymes. Prakash *et al.* (2003) showed that decrease in grain growth under high temperature during grain development in wheat was due to a decrease in soluble starch synthase (SSS) activity. A parallelism in the effect of high temperature on grain growth, starch accumulation and SSS activity was observed. Relatively tolerant cultivars showed less decrease compared to susceptible types. Out of soluble starch synthase and granule bound starch synthase, activity of SSS was found to be more sensitive to high temperature (Prakash *et al.*, 2004). Study indicated that SSS could be an important traits for improving terminal high temperature tolerance for grain growth in wheat.

4.10. Source-Sink Relationship

Crop productivity depends primarily on photosynthesis (source) and its efficient partitioning to economically important parts (sink) to obtain high yield Physiological approaches to yield improvement, therefore, attempt to target these key constrains limiting crop productivity particularly under changing environment of rising CO_2

concentration and temperature. An understanding of such key constraints helps in developing plants best suited for such conditions. The projected increase in CO_2 concentration and accruing high temperature as a result of greenhouse effect will have its impact on key enzymes of sources-sink biosynthetic pathways. It is in this context studies on the regulation of photosynthesis and photosynthetic carbon partitioning were attempted to gain insight into the possibilities of improving photosynthetic source-sink efficiency not only under present condition but also in a changed environment of high CO_2 and temperature. Studies conducted on wheat revealed that potential photosynthetic efficiency is not fully utilized because of end product inhibition and yield remains sink limited (Sharma-Natu and Ghildiyal, 2005). There could be two possibilities for enhancing photosynthetic productivity: (1) To overcome end product inhibition of photosynthesis at leaf level itself, (2) To improve the efficiency of sink particularly the storage sink capacity.

Photosynthetic responses of wheat cultivars to long term exposure of elevated temperature evaluated by plots of photosynthesis versus temperature, internal CO_2 concentration (Ci) and photosynthetically active radiation (PAR) revealed Rubisco limitation, which was found to be due a decrease in activation state of Rubisco. This was further substantiated by observed decrease in Rubisco activase activity in elevated temperature grown plants. The decrease in photosynthesis under moderate heat stress thus appeared to be mainly due to a decrease in activation state of Rubisco catalysed by Rubisco activase. Identification and incorporation of a thermostable Rubisco activase in to crop plants would, therefore, improve thermotolerance of photosynthesis (Pushpalatha *et al.*, 2008).

4.11. Improving Sink Efficiency

One of the important components of storage sink is grain number. An increase in grain number per ear in wheat, results in lower individual grain weight. This is because wheat spike structure does not allow sufficient flow of assimilates to distally located grains in spikelets (Kumari and Ghildiyal, 1997). This was also shown by ^{14}C-sucrose translocation studies by Ravi and Ghildiyal (2001). A change in ear structure whereby grain number should be increased by increasing spikelet number per ear instead of increasing grain number per spikelet was therefore, suggested to minimize the inverse relationship between grain number and grain size. A new plant type of wheat having higher grain number with better grain size has been developed by IARI breeders by increasing spikelet number per ear.

4.12. Yield and Yield Attributes

Zhong-Hu, HE and Rajaram, S. (1994) observed that seeds yield per spike, biomass, and plant height are more thermo-sensitive than spike number per square meter, 1000 kernel weight, and test weight. The grain-filling rate was more temperature-sensitive than days to anthesis and duration of grain-filling. Simple phenotypic correlation analysis indicated that yield was highly and positively correlated with seeds per spike, biomass, and harvest index (HI) independent of seasons and genotypes under high temperatures. Effect of earliness on the yield under high temperature was highly dependent on the temperature regime during the heading stage. Grains per spike, biomass, HI, and test weight could be considered potential selection criteria for yield under high temperature.Primitive 'Khapli wheat (*Triticum dicoccoides*) grown in parts of Maharashtra and Karnataka have a large phytomass and very low HI and grain yield compared to the modern wheat cultivars. Wheeler *et al.* (1996 *b*) showed in study with spring wheat, the majority of the reduction in grain yield can be attributed to sink limitation given the similarities of the declines in grain number, grain yield and harvest index with an increase in the maximum temperature Warmer maximum temperatures over 4 consecutive days close to anthesis directly reduced grain number and, as a consequence, grain yield at harvest maturity, while higher mean temperatures reduced root biomass immediately after anthesis. If climates become both warmer and more variable, the occurrence of high temperature during anthesis could reduce wheat grain yield substantially. Deleterious effects of higher temperatures on root growth could pose a further threat to crop production. Ferris *et al.* (1998) observed that the number of grains per ear at maturity declined with increasing maximum temperature recorded over the mid-anthesis period (76-79 DAS) and, more significantly, with maximum temperature 1 degree after 50 per cent anthesis (78 DAS). Grain yield and harvest index also declined sharply with maximum temperature at 78 DAS. Grain yield declined by 350 gm^{-2} at harvest maturity with a 10°C increase in maximum temperature at 78 DAS and was related to a 40 per cent reduction in the number of grains per ear. Grain yield was also negatively related to thermal time accumulated above a base temperature of 31°C (over 8 d of the treatment from 5d before to 2d after 50 per cent anthesis). Thus, grain fertilization and grain set was most sensitive to the maximum temperature at mid-anthesis. These results confirm that wheat yields would be reduced considerably if, as modelers suggest, high temperature extremes become more frequent as a result of increased variability in temperature associated with climate change.

4.12.1. Biomass Weight

Sharma *et al.* (1991) observed significant positive association of aerial biomass with CGR, LAI, RGR, NAR and LAD. Al-Katib and Paulsen (1990) measured various parameters and productivity in ten genotypes from major world wheat productivity regions under moderate (22°C/17°C day/night) and high (32°C/27°C day/night) temperature for two weeks as seedlings or from anthesis to maturity. They found that heat stress decreased mean total biomass by 32 per cent and 15 per cent in seedlings and maturing plants, respectively. Wheat plants grown in silt loam soil in a greenhouse were subjected to day/night temperature of 15/10, 25/20 or 35/30°C and high temperature exposure decreased shoots dry weight (Shah, 1992). For plants experiencing a 3°C increase in day and night temperatures relative to local mean temperatures, dry matter yields were reduced by 18 per cent compared to control in wheat (Moot *et al.*, 1996). There are many other reports which emphasized that unfavorable temperature exposures upon sowing early of late by deviating from recommended time reduced total plant biomass in wheat (Singh and Rajat de 1978, Randhawa *et al.*, 1981, Singh and Singh, 1985, Bhanu-prakash, 1997 and Chaturvedi *et al.*, 1985). Biomass production is a pre-requisite for its desirable partitioning to economically important parts to obtains higher yield. Limitation in yield could therefore, due to limitation in biomass production or its partitioning and in nutrient uptake and utilization (Abrol *et al.*, 1999; Sharma-Natu P and Ghildiyal, 2005).

4.12.2. Tiller Numbers

Jain *et al.* (1974) suggested that number of ear bearing tillers at the time of harvest is an important yield component contributing to higher grain yield. Singh and Srivastava (1988) suggested that in wheat grain yield was dependent on tillers per plant during vegetative phase. Bansal and Sinha (1991) reported that maintenance of tillers was considered in maintaining the number of spikes across environment. Positive correlation of reproductive tillers number and grain yield have been observed (Sharma, 1995). No significant influence on tillering by increasing day temperatures from 21 to 27°C or night temperatures from 13 to 21°C was observed (Campbell and Read, 1968). In the post earing phase temperature had great effect on number of ear/ plant through number of tillers and percent fertile tillers. Owen, (1971) observed that plant given higher day temperatures continued to produce fertile tillers over a longer period. Mean temperature of 25°C and above was always unfavorable for tillering and vegetative growth in October sown wheat (Bhardwaj, 1978). High temperatures during emergence to double ridge decreased number of spike bearing tillers (Shpiler and Blum, 1986 and Acevedo *et al.*, 1990). High mean temperature in early planting compared to normal planting reduced tiller number in various wheat genotypes (Singh and Rajat *et al.*, 1978 and Randhawa *et al.*, 1981). Yield reduction of wheat under stress environment was attributed to reduced productive tillers per plant (Pal, 1992). Sharma, (1995) observed that reproductive tiller number was positively correlated with grain yield. Decrease in tiller number under field condition due to high temperature in wheat was also reported by Waines (1994).

4.12.3. Grain Number

One of the most important yield attributes affected by heat stress is grain number. In wheat the number of grains that develop in an ear is dependent on the number of viable florets that are formed and the effective fertilization of these after anthesis (Evans and Wardlaw, 1976). Grain number in wheat was reduced by warmer weather prevailing between floral initiation and anthesis (Owen, 1971). Good evidence has been found for a reduction in grain number per ear associated with high temperature during the stage of booting *i.e.*, the stage of pollen and embryonic mother cell meiosis (Saini and Aspinall, 1982). High temperature upto 3°C late in the development of the ear of wheat can result in a considerable reduction in the number of grain set at anthesis with the most sensitive stage at the time of pollen meiosis when the ear is still enclosed by the flag leaf sheath (Dawson and Wardlaw, 1989). Chaturvedi *et al.* (1985) suggested that tillers are affected more for number of spikelets and grains per spike and spike weight and grain weight per spike as compared to main shoot in late planting. Wardlaw *et al.* (1989) reported that reduction in grain yield associated with high temperature following anthesis results from variation in kernel size and not due to changes in kernel number. Stone and Nicolas (1994) observed that short period of very high temperature (max > 35°C) in the post anthesis period can significantly reduce grain yield in wheat. Most important yield component affecting yield variation among cultivars under heat stress in kernel number per spike. Under hot conditions heat tolerant cultivars sustain relatively more kernel per spike that heat susceptible cultivars (Shpiler and Blum, 1991). Shah *et al.* (1994) suggested that delayed sowing reduced grain number per ear, while Abdelghani *et al.* (1994) concluded that number of grains per ear were higher due to late sowing. Kelly *et al.* (1994) found that grain number per square meter was most closely correlated with grain yield followed by grain number per spike. Rawson *et al.* (1996) observed that increasing mean temperature from 12°C to 20°C resulted in 16 per cent reduction in grain number per spike. Many researchers have concluded that the increases in yield and harvest index in wheat and barley have been almost entirely due to an increase in grain number per unit area, rather than to an increase in grain weight (Austin *et al.*, 1980, Mc Caig and DePauw, 1995 and Sayre *et al.*, 1996). Fisher, (1996) concluded that yield potential increase was correlated with higher harvest index and higher kernels per square meter and not with change in days to anthesis or biomasss.

4.12.4. 1000-Grain Weight

Chinoy (1947) compared grain weight per plant and 1000-grain weight of varieties grouped into eight classes according to flowering time, which ranged from 90-100 to 160-170, days with delay in flowering, grain developed at increasingly higher temperatures and lower relative humidity, with the consequences that both 1000-grain weight and grain yield per plant diminished progressively. Asana and Williams (1965) concluded that increase in the grain weight was depressed beyond 25°C due to increase in the respiration rate of grain. Kolderop, (1979) and Bhullar and Jenner (1985) reported that high temperature imposed 10 days after anthesis reduced yield of wheat through effects on individual kernel mass. Bhullar and Jenner (1983) also observed reduction in grain weight on brief warming of wheat ears. Grain weight was reduced due to reductions in individual grain weight and in grain number.Wardlaw *et al.* (1989) observed that under controlled temperature conditions kernel weight was less susceptible to heat stress than kernel number. Delayed sowing decreased 1000-grain weight (Shah *et al.* (1994). Stone and Nicolas (1995) concluded that mature individual kernel mass was most sensitive to heat stress applied early in the grain filling and became progressively less sensitive through out grain filling. Hu and Fang (1995) found grain weight per ear was the main factor for improving grain weight per ear. Setter *et al.* (1997) observed that stem carbohydrate which are accumulated before and during the early period after anthesis usually account for 10-30 per cent of the stem dry weight in wheat, and in some cereals exposed to environmental stresses during grain filling this may contribute 70 per cent or more of the grain weight.

4.12.5. Grain Yield

Late planting of wheat result in reduction of grain yield due to reduced crop duration and high temperature prevalent at the time of grain filling. Shah *et al.* (1994) observed that delayed sowing reduced grain yield. Crop sown in November showed 8-24 per cent more grain yield that sown in December (Singh *et al.*, 1995, Sarker and Torofder, 1992 and Singh and Verma, 1990). Asana and William (1965) observed that decrease in grain yield with increasing day temperature was remarkably uniform. The mean percentage reduction in yield for the 3°C from 25°C to 28°C was 8.4 and 16.4 for 6°C rise in temperature. Wardlaw and Wrigley (1994) concluded that in the high temperature range wheat yields decline approximately 3-4 per cent for each 1°C rise in average temperature above 15°C. Anthesis and the end of grain filling, period were advanced and grain and dry matter yield were reduced by 27 and 18 per cent respectively in the plant experiencing a 3°C increase in day and night temperatures (Moot *et al.*, 1996).Guha-Sarkar *et al.* (2001) reported that early maturing genotypes like HD 2285, HD 2307, Sonalika, UP 2338, Lok 1 and C 306 showed better performance under high temperature stress conditions with less reduction in grain yield and have relatively higher grain growth rates with more tolerance to heat stress for most of the yield attributing characters. Guedira and Paulsen (2002) showed that high whole plant temperatures (30/30°C Shoot/root temperatures) accelerated linear grain growth in wheat but diminished the duration. Shoots as well as roots were responsive to thermal signals and elevated temperatures of both organs result in similar effects on the development and metabolism of the grain. Tahir and Nakata (2005) indicated that heat stress negatively affected grain yield, its components and nitrogen remobilization while it increased total non structural carbohydrates remobilization because of the increasing demand for resources. Harvest maturity was defined as that stage when grain moisture content declined naturally to 15-18 per cent.

5. Mineral Nutrition and Quality

Soil fertility, availability of macro- and micronutrient, and organic matter to support microbial population should provide an environment conducive for full expression of the genetic yield potential. With the predicted climate changes and global warming, terminal heat stress of the wheat crop is likely to increase in the near future. External energy inputs in the form of irrigation, nutrients and pest-control measures help to capture a larger harvest of solar energy-a free resource. The various options for increasing grain productivity, energetic cost and N requirements have been examined earlier (Bhatia *et al.*, 1981).

Nitrogen economy and N-use efficiency are altered under heat stress. Increasing soil N during the period of grain-filling at 20/15°C enhanced (Jahedi *et al.*, 2004) grain weight, grain protein and grain protein per cent. However, at higher temperatures (30/25°C), these attributes were not enhanced by increasing N supply. High N at thermal stress reduced the duration of protein deposition in the grains due to inadequate supply of protein deposition in the grains due to inadequate supply from the soil, 80 per cent of grain N comes from remobilization from the leaves. Carbohydrates in wheat grain are derived from post-anthesis CO_2 assimilation, of which the flag-leaf contributes a large share (Evans *et al.*, 1975). Rubisco constitutes 60 per cent of the soluble protein in the flag-leaf. In limiting soil N supply, Rubisco becomes the source of N for the developing grains (Hirel and Gallais 2006). Degradation of Rubisco further enhances flag-leaf senescence, contributing to reduction in the supply of assimilates to developing grains. The present yield gap between the mean productivity and that realized in the AICWIP trials is 50 per cent (5400 − 2700 = 2700 kg/ha). The current yield gap in Punjab is (5400 − 4200 = 1200 kg/ha).Another possibility is by enhancing nitrogen use efficiency. This has been recently reported in rice by transferring alanine aminotrasferase (AlaAT) c-DNA from barley into rice along with a root epidermal-specific promoter (Shrawat *et al.*, 2007). Transgenic plants had significantly higher biomass and seed yield compared to the control. The reduced nitrogen technology seems to work for wheat and other crops.

Cereal grains, although low in protein content and deficient in protein nutritial quality (low levels of the essential amino acid lysine), constitute the major source of energy and nutrients in the world (Roderic and Fox, 1987). Wheat is the principal food item in most developing countries for it provides the population with more energy and nutrients than any other single food source (Pomeranz, 1987). Although genetically controlled, gluten quality can be greatly infuenced by grain productivity factors such as soil fertility and crop management practices (grain yield is negatively correlated to protein contents), and by environmental conditions such as frost heat and rainfall during grain development (Sander _et al._, 1987). Production Value (PV) is defined as the weight of the end-product by the weight of the substrate required for carbon skeletons and energy production. Its unit is g/g. Wheat grain protein is deficient in lysine, methionine, threonine, tryptophane, isoleucine, leucine, phenylalanine and valine, and is predominantly rich in glutamic acid and its amide. Hence PV of wheat grain protein is comparatively higher than HSPs and Rubisco.

6. Molecular Approaches

The general complexity of terminal heat tolerance and the difficulty in phenotypic selection, MAS is considered an effective approach to improve this kind of tolerance. Breeding for heat tolerance is still in its infancy stage and warrants more attention in future (Ortiz _et al._, 2008; Ashraf 2010). Although, significant variation for heat tolerance exists among wheat germplasm (Joshi _et al._, 2007a, b), no direct criteria are available for selection to heat tolerance. Phenotypic selection for heat tolerance has been performed using indirect selection for GFD (Yang _et al._, 2002), thousand grain weight (TGW) and canopy temperature depression (CTD) (Reynolds _et al._, 1994; Ayeneh _et al._, 2002). Crop physiological studies on cereals under drought environments have identified several indirect traits that can be considered for physiological breeding, radiation and water use efficiency, green leaf duration, harvest index, rate of senescence, grain fill duration, leaf area index, deep roots, vigorous crop establishment, stem-reserve utilization, and maintaining cellular hydration (Araus _et al._, 2008; Reynolds and Tuberosa, 2008). The challenge and difficulty that breeders face is in the manner of how these traits should be combined to design cultivars ready for drought stress that can potentially hit it at any development stage. Drought resistance was found to be associated with some molecular markers and the combination of the physiological and molecular research tools have generated promising genetic material for the Mediterranean drylands (Nachit and Elouafi, 2004). However, in last two decades, enormous efforts have been made to develop mapping populations and to identify genetic markers associated with heat tolerance in wheat. The creation of suitable mapping populations and the development of molecular markers have enabled linkage studies in wheat and many QTLs have been identified for yield under drought environments (Varshney _et al._, 2006). QTL for heat tolerance in wheat were reported using different traits like GFD, CTD and yield (Yang _et al._, 2002; Mason _et al._, 2010; Pinto _et al._, 2010) and senescence related traits (Vijayalakshmi _et al._, 2010). Paliwal _et al._ (2012) used the values from controlled and stressed trials for four different traits including heat susceptibility index (HSI) of thousand grain weight (HSITGW);

HSI of grain fill duration (HSIGFD); HSI of grain yield (HSIYLD); and canopy temperature depression (CTD) to determine heat tolerance and identified three major QTLs for heat stress. Despite the available maps, populations, and marker technology, advances in transferring knowledge from QTL studies on yield under drought to breeding remains slow.The advances in technology in genomics, omics, quantitative genetics, and bioinformatics offer us enormous opportunity to study the molecular regulation of quantitative traits in crops. It is only a matter of time for this to be exploited fully in wheat research. Further, studies focused on phenotypic flexibility and assimilate partitioning under heat stress and factors modulating crop heat tolerance are imperative. Such studies combined with genetic approaches to identify QTLs conferring thermotolerance will not only facilitate marker-assisted breeding for heat tolerance but also pave the way for cloning and characterization of underlying genetic factors which could be useful for engineering plants with improved heat tolerance.

7. Climatic Temperature Variations and Adaptation

Climate change could strongly affect the wheat crop the accounts for 21 per cent of food and 200 million hectares of farmland worldwide. Future climate scenarios suggest that global warming may be beneficial for the wheat crop in some regions, but could reduce productivity in zones where optimal temperatures already exist. Wheat geneticists and physiologists are also assessing wild relatives of wheat a potential sources of genes with inhibitory effects on soil nitrification. (Ortiz *et al.*, 2008).The mega-environment (ME) zonation 1 is a favourable, irrigated low rainfall environment with high yield potential whereas mega-environment 5 is a heat-stressed environment (early and late season heat stress) with available irrigation but in its humid and hot areas, the fungus *Bypolaris sorokiniana* is the causal agent of *Helminthosporium* leaf blight. Using site classification data, in combination with long-term normal climate data and irrigated area data, potential mega-environment zones were updated and delineated on an agro-climatic basis. These two major wheat mega-environments in the sub-continent have been differentiated on the basis of coolest quarter minimum temperature ranges (3-11°C for ME-1 and 11-16°C for ME-5). In some of the mega-environment 5 areas poorer infrastructure, socio-economic factors and crop management coupled with the stresses brought by *Helminthosporium* leaf blight and the shortened vegetative phase ensuing from heat stress, particularly at grain filling, lead to low yield in wheat, whose quality may be also affected by grain shriveling (White *et al.*, 2001).

7.1. Canopy Temperature Depression (CTD)

Shimshi and Ephrat (1975) showed positive correlations between grain yield, photosynthetic rate, and stomatal conductance (g_s) in irrigated short spring wheats. One such instrument for rapid indirect determination of g_s is the hand-held infrared thermometer, used for instantaneous measurement of canopy temperature. The results are usually expressed as canopy minus air temperature, but here we will refer to canopy temperature depression (CTD), usually a positive number with well irrigated wheat. Researchers developed the application of infrared thermometry to irrigated wheat and demonstrated that when other variables are constant, CTD increased as g_s

increased (smith *et al.*, 1986; Pinter *et al.*, 1990). While net radiation, air temperature, and wind speed have minor effects (Smith *et al.*, 1986). In the late 1980s, CIMMYT commenced CTD measurements on various irrigated experiments in northwest Mexico. On occasions, but not always, phenotypic correlations of CTD with grain yield were positive. New evidence derived from studies of carbon-13 discrimination was suggestive of yield-g_s relationships across wheat cultivars (Condon *et al.*, 1987; 1990). Nagarajan and Rane (2002) suggested that canopy temperature depression is a good indicator of high temperature stress damage and could be used as criteria for classification of genotypes. The canopy temperature depression measurement is indirectly related with canopy and leaf cooling.

Choose early afternoon cloudless periods for all activity measurements; some times late morning or midday readings were also made. Air temperatures ranged from 22 to 32°C. Radiation use efficiency (g MJ^{-1}) was calculated on the basis of CGR divided by the estimated daily absorbed photosynthetically active radiation, obtained from measured incident solar radiation, and assuming 50 per cent radiation is photosynthetically active and 90 per cent absorbed. The air flow porometer, used in 1995 as an indirect but rapid measure of stomatal opening, electronically timed a given amount of air movement through the leaf under pressure. Each measurement used the portion of the flag leaf exposed to full sunlight, with the cuvette positioned normal to the sun to give a light intensity between 1800 and 2000 μmol m^{-2} s^{-1}. Each year at anthesis, a hand-held meter (SPAD 502, Minolta, Spectrum Technologies Inc., Plainfield, IL) was used to measure flag leaf greenness in arbitrary absorbance or SPAD units. These units have, at a given stage of development, been found to be linearly related to chlorophyll concentration in several situations (Yadav, 1986). Four measurements were made on each of four leaves per plot. Leaf area of 10 representative flag leaves, sampled from each plot around anthesis, was measured using a LI-COR Portable Area Meter (LI3000). Leaf dry weight was then determined and specific leaf weight calculated. Flag leaf N concentration was measured by micro Kjeldahl analysis of the dried leaves. Mean grain yield was not correlated with flag leaf area. In fact, CTD measured in central measured in central Mexico, was generally highly significantly correlated with grain yield of the same cultivars at irrigated sites in Egypt, India, and Sudan (Reynolds *et al.*, 1994). Although CTD can be affected by leaf angle and the presence of spikes, and by canopy height in small plots. With CTD, airborne infrared sensing systems offer exciting possibilities for cost reduction, and such rapid measurement has the advantage of reducing error due to subtle weather fluctuations. Besides it again seems that cultivar differences can be detected in small plots. Balota *et al.* (2008) observed that greater daytime CTD of the drought tolerant line TX86A8072 are consistent with its smaller, thicker leaves. Data showed that the smaller- and thicker-leaved line had higher A, A: [CO$_2$]$_i$, and grain yield than larger-leaved TX86A5606 under similar conditions of plant water availability. Gas exchange and carbon-13 data suggest that smaller leaves also lead to greater CTD during the day.Balota *et al.* (2008) reported that wheat (*Triticum aestivum* L.) cultivars with high canopy temperature depression (CTD) tend to have higher grain yield under dry, hot conditions. Therefore, CTD has been used as a selection criterion to improve adaptation to drought and heat. The CTD is a result of the leaf's energy balance, which includes terms determined by environment and physiological traits.

7.2. Day and Night Temperature

Terminal heat stress is a common abiotic factor for reducing the yield in certain areas of West Asia and North Africa (Ferrera et al., 1993). Heat tolerance thus should be essential characteristic of wheat cultivars to be developed. Stay green is a trait that has been used to indicate heat tolerance in hot environment (Kohli et al., 1991). Photosynthetic rate is maximum at 20-22°C and decreases abruptly at 30-32°C (Al Khatib and Paulsen 1999, Murata and Lyama 1963). Heat stress injuries of the photosynthetic apparatus during reproductive growth of wheat diminish source activity and sink capacity which results in reduced productivity (Harding et al., 1990). Source activity is damaged by heat because both leaf area (Slovacek and Hind 1981 and Spiertz 1974) and photosynthesis is reduced (Al Khatib and Paulsen 1984, Kuroyanagi and Paulsen 1985). Heat injury limits sink growth potential particularly when stress is imposed during early sink developmental stages (Nocolas et al., 1984). Grain yield was negatively related to the thermal time accumulated above the base temperature of 31°C (Ferris et al., 1998, Mian et al., 2007). High temperature above 32°C has been reported reducing grain yield and grain weight (Bluementhal et al., 1995, Gibson and Paulsen, 1999, Wardlaw et al., 2002).

Night-time T is generally not associated with carbon fixation in C_3 plants, it has been linked to improved supply of water-soluble nutrients carried in the transpiration stream (Caird et al., 2005), especially nitrate (McDonald et al., 2002; Conroy and Hocking, 1993). However, in glass-house experiments with Helianthus species, Howard and Donovan (2007) found that nighttime T did not increase with induced nitrate deficiency, but it did decrease with induced water stress. Similarly, Zweifel et al. (2007) found that nighttime T of Scots pine (Pinus silvestris L.) in Switzerland was reduced by drought conditions. Daley and Phillips (2006) concluded that nighttime T of paper birch (Betula papyrifera L.) was responsible for over 10 per cent of the total daily sap flux during the growing season. Caird et al. (2006) pointed out that strong correlations exist for nighttime and daytime gas exchange patterns. Thirty-two accession/cultivars of Triticum and its related species belonging to diploid, tetraploid and hexaploid group were evaluated for heat stress tolerance in the field under full irrigation by late sown technique. Wide variability was observed for grain yield component stability in Triticum species under heat stress. Hexaploidy conferred the productive and adaptive advantages as it combined high yield and stability when compared to the tetraploid and diploid groups. Heat tolerant genotype were indentified. Stability in grian number and biomass conferred yield stability in all ploid levels (Khanna-Chopra and Viswanathan, 1998; Viswanathan and Khanna Chopra, 1996 and 2001).

7.3. Biochemical Changes

7.3.1. Antioxidant Enzymes Activity

Environmental stresses induce production of active oxygen species such as superoxide, hydrogen peroxide (H_2O_2), hydroxyl radical (OH), and singlet oxygen ($^1O_2$). They are highly reactive and can damage many important cellular components such as lipids, proteins, and nucleic acids in living cells (Smirnoff, 1993, Foyer et al.,

1994). Plants have developed enzymatic and nonenzymatic scavenging systems to quench active oxygen species. When plants are subjected to stresses such as high temperatures, the scavenging system may lose its function and the balance between producing and quenching active oxygen species can be disturbed, resulting in oxidative damage (Price *et al.*, 1989, Bowler *et al.*, 1992, Zhang and Kirkham, 1994). Induction of HSPs in flag leaves of a broad range of winter wheat varieties under field condition has already been reported (Nguyen *et al.*, 1994).Tolerance to high temperature stress in crop plants have been associated with an increase in antioxidant enzymes activity (Rui *et al.*, 1990, Gupta *et al.*, 1993, Badiani *et al.*, 1994 and Zhau *et al.*, 1995). Several enzymatic and non enzymatic antioxidant defense systems maintain AOS concentration under tight control to protect cells from damage (Noctor and Foyer, 1998). The activity of enzymes associated with the antioxidant defense system, especially ascorbate peroxidase has been shown to increase rapidly under heat stress in mustard (Dat *et al.*, 1998). High temperature induced oxidative stress in various higher and lower plants (Upadhyaya *et al.*, 1990, Jagtap and Bhargava, 1995, Davidson *et al.*, 1996) and heat injury in cool season grasses has been associated with oxidative damage (Jiang and Huang, 2001 and Liu and Huang, 2000). Plants protect cell and subcellar systems from the cytotoxic effects of the active oxygen radicals using antioxidant enzymes such as superoxide dismutase, ascorbate peroxidase, glutathione reductase, catalase and metabolites like glutathione, ascorbic acid, tocopherol and carotenoids (Sairam *et al.*, 2000). Physiological injury due to heat stress has been associated with increases in oxidative damage in perennial grasses (Liu and Huang, 2000) and other plant species (Larkindale and knight, 2002). Heat stress has been shown to cause oxidative stress due to over production of active oxygen species; such reactive molecules cause cellular damage, particularly to cell membrane, and Oxidative protection considered as an important component in determining the survival of plant during heat stress (Larkindale and Huang, 2004).

Water and salinity stresses induce an increase in reactive oxygen species (ROS), *i.e.* superoxide radical (O_2), hydrogen peroxide (H_2O_2) and hydroxyl radical (OH), which resulted in increased lipid peroxidation and decline in membrane stability and chlorophyll content. The activity of antioxidant enzymes like superoxide dismutase (SOD), ascorbate peroxidase (APX), peroxidase (POX), glutathione reductase (GR) and catalase (CAT) also increased under abiotic stresses, while contents of antioxidant metabolites like, ascorbic acid and carotenoids decreased. Our results suggest that tolerance of wheat genotypes to drought (C 306) and salinity (Kharchia 65) was associated with higher activity of antioxidant enzymes, ascorbic acid and carotenoid contents and lower oxidative stress (H_2O_2, TBARS contents). Moisture stress tolerance was also associated with higher ABA accumulation in tolerant (C 306 and HW2024) than susceptible type (Hira) (Chandrasekhar *et al.*, 2000, Sairam *et al.*, 2001b).

7.3.2. Heat Shock Protein

Under heat stress conditions, the normal cellular protein synthesis will decrease with concomitant increase in HSP synthesis in all organisms. However different organs in an organism as well as different species have different responses to heat stress. Heat shock causes a drastic decrease of normal cellular proteins synthesis in soybean and maize leaves but causes very subtle change in maize root tips. However, in wheat and rice grains, protein synthesis is largely unaffected although an increase in grain protein may result at high temperatures (Bhullar and Jenner, 1985). One of the most prevalent environmental challenges encountered by plants is the exposure to a broad range of temperatures. Plants use a variety of anatomical, metabolic and cellular strategies to deal with changing environmental temperatures. Acclimation to elevated temperatures is a mediated at the cellular level in part by the induction of general stress responses, which include the increased expression and activity of heat shock proteins (Gallie *et al.,* 1998). Heat stress (5-10°C above the normal growth temperature of organisms) induces expression of specific gene families called heat shock genes (HSPs).which lead to synthesis of heat shock proteins (HSPs), in every organism in which it has been sought, from unicellular prokaryote to highly evolved homosapiens. The only exception to this is the germinating pollen of higher plant, which does not synthesize HSPs during heat stress (Farova *et al.,* 1989). All the species examined so far are able to acquire tolerance to lethal temperatures, if the cell previously exposed to non lethal hsp inducing heat shock (Schlesinger, 1990, Vierling, 1991). This acquired thermotolerance occurs very raipidly, with concomitant synthesis of HSPs, which require of few minutes of heat shock induction. Hsps, express from within 1-3 hours of heat shock and decay 12-48 h after (Nover *et al.,* 1989). The HSPs, can be classified based on their molecular weight into following families:

1. HSP 110 (110 to 110 KDa)
2. HSP 90 (85 to 95 KDa)
3. HSP 70 (70 to 80 KDa)
4. HSP 60 (50 to 68 KDa)
5. HSP 20 (15 to 28 KDa)

This argues for their roles in normal functioning of cell during optimal and stressed environment. Constitutive HSPs (HSCs) have weak ATPase activity, which may be essential for its role in protein transport, protein folding and assembly (Pelham, 1986). Some members of HSP 90, HSP 70 and HSP 60 are expressed not only during normal cell growth and development but also induced by other stresses such as toxic metals, ethanol etc. (Howarth, 1990).The high molecular weight (HMW) HSPs ranging from 68-110 KDa are ubiquities among all organisms, where as the low molecular weight (LMW) HSPs (HSP 20 family) predominate in higher plants (Vierling, 1991) ATP dependent "unflodase" activity of HSPs is essential and regulates its function as molecular chaperone (Miernyk *et al.,* 1992). HSP 70s also have calmodulin binding site (Schlesinger, 1990). Therefore, during the growth of an organism in its optimal environment, the HSPs are playing an essential role in DNA replication (Craig and Cross, 1991), transcription, translation, transport, folding, unfolding and assembly of protein and regulation of activities of proteins. Besides this, HSP 70 regulates the HSPs synthesis upon heat stress (Craig and Cross, 1991 and Morimot, 1993). Induction of HSPs in flag leaves of a broad range of winter wheat varieties under field condition has already been reported (Nguyen *et al.,* 1994). Mitra and Bhatia (2008) object that heat shock proteins (HSPs) synthesized to cope with the heat stress in different organisms are known to provide protection and repair the cellular damage caused by heat. Incorporation of terminal heat tolerance into high-yielding cultivars will have an energetic cost, and would require additional carbon assimilates and N inputs. Rehman *et al.* (2009) Recorded that the effects of heat stress were lesser in shorter period exposure and more drastic in prolonged exposure of the genotypes to heat. The ability of lines to stay green for longer period in heat shock had no direct relationship with seed setting. To develop heat tolerant lines, selection should be made on the basis of yield grain weight and grain number per spike. High temperature effects on yield and grain weight had also been reported by Bluementhal *et al.* (1995), Wardlaw *et al.* (2002) and Main *et al.* (2007).

A parallelism between catalytic efficiency of SSS and heat shock protein (HSP 100) level in the grians was observed in wheat cultivars differing in their sensitivity to high temperature (Sumesh *et al.*, 2008). Temperature tolerant cultivars showed high efficiency of SSS and increased HSP 100 levels than sensitive cultivars. HSP100, however, did not increase with increase in temperature and was constitutively present in higher amount in thermotolerant cultivars. Low molecular weight-HSP (HSP18), however, was found to be induced by high temperature. The relatively tolerant cultivars showed a greater expression of HSP 18 compared to susceptible types in response to moderate heat stress. The studies indicated the possibility of improving thermotolerance for grain growth in wheat either through incorporation of thermostable from of SSS or through over expression of HSP 18 in an otherwise high yielding cultivar of wheat.

7.3.3. Role of HSPs in Thermotolerance

The phenotypic characterization of HSP mutants has provided strong evidence that the HSPs are an essential component of thermotolerance. The hin mutant of *E. coli* (Yamamori and Yura, 1982), HSP 104 mutant of yeast (Sanchez and Lindquist, 1990) and SSA1 and SSA2 mutant of yeast were unable to survive in lethal temperatures, even after giving a prior hsp inducing non-lethal heat shock. But the wild types could survive. The cells could not acquire thermotolerance, where HSP synthesis is either blocked at transcriptional (Johnston and Kucey, 1988) and/or translational level (Yamamori and Yua, 1982) or inactivated by antibodies (Riabowel *et al.*, 1998). HSPs induced by other stresses can also provide thermotolerance (Howarth, 1990), which provides further evidence for the involvement of HSPs in thermotolerance.HSPs are essential component of acquired thermotolerance of germinating seeds (Abernethy *et al.*, 1989) and seedling growth of wheat (Blumenthal *et al.*, 1990, Krishnan *et al.*, 1989) soybean (Lin *et al.*, 1984) sorghum and pearlmillet (Ougham and Stoddart, 1986, Howarth, 1990). However, the exact mechanism by which HSPs provide thermoprotection is still obscure. HSPs may give thermotolerance to an organism by protection of nucleus, ribosome synthesis (Schlesinger, 1990) protein synthesis (Krishnan *et al.*, 1989), soluble proteins (Hsich *et al.*, 1992), thylakoid membranes of chloroplast (Schuster *et al.*, 1988) and electron transport chain of mitochondria (Chon *et al.*, 1989) during heat stress. The role of HSPs in thermotolerance is not incontrovertible (Apuya and Zimmerman, 1992, Ramsay, 1988, Susek and

Lindquist, 1989). However, if these studies are closely examined, only a group of HSPs appear to be involved in thermotolerance. Since in plants and other organisms, there are 6 major HSP families, each of which consists of upto 15 hsps, all of them may not be required for thermotolerance at all stages of growth.

7.4. Evaluation and Utilization of PGR for Climate Resilient Agriculture

The Indian Council of Agricultural Research (ICAR) launched a mega network project on 'National Initiative on Climate Resilient Agriculture' (NICRA) in February, 2011, with the objective to enhance resilience of Indian agriculture to climate change and vulnerability, through strategic research and technology demonstration. Under this, NBPGR has been awarded a research project on 'Acquisition, evaluation and identification of climate resilient wheat and rice genetic resources for tolerance to heat, drought, and salt stresses.' Under the project wheat and rice germplasm is screened for abiotic stresses for identifying the stable donor parents for use in future breeding programmes aimed to developing resistant/tolerant varieties. About 22,000 accessions of wheat *(Triticum aestivum, T. dicoccum* and *T. durum)* were sown in augmented block design at CCS Haryana Agriculture University (CCSHAU), Hisar, to characterize for agro-morphological traits and to develop a core set of wheat germplasm conserved in the National Genebank. Another set of the same accessions of wheat was sown at Issapur Farm, NBPGR, New Delhi, for evaluation of the wheat germplasm for heat and drought tolerance (NBPGR Newsletter October-December. 2011). The development of core set will enable breeders to utilized the wheat germplasm in a comprehensive way. In addition, deployment of Genebank accessions for future climate changes is also being envisaged using climatic analogs in collaboration with various international organizations. Utilization of wild species and land races for genetic base broadening through pre- breeding is also being attempted in some selected crops as an insurance against climatic changes.

7.5. Adaptation to Heat Prone Environments

Late planting of wheat in India is common due to the intensive cropping system, often delays the sowing of the crop up to the middle of January, particularly in North West India where it is generally sown after harvest of paddy, sugarcane, pigeonpea and potato. As a result, a portion of maturity period of the crop pushed forward and thus has to face higher temperature of the summer with hot spells often occurring at the time of maturity (Abrol *et al.,* 1991) the late sown wheat is more affected by high temperature stress leading to reduced yield and quality (Wardlaw and Veringley, 1994). Morphologically, similar wheat varieties have showed different degree of tolerance to post anthesis high temperature stress. Hence breeding for heat stress tolerance can open new insights significant in Indian agriculture. There are several physiological traits that are associated with heat tolerance. Photoassimilation, chlorophyll retention, chlorophyll a:b ratio, canopy temperature depression, stomatal conductance, membrane stability are some of the examples (Shukla *et al.,* 1997). Photosynthesis, respiration, conversion of sugar into starch in developing grain has been found to be affected by high temperature (Rawson, 1992). It has been reported that high temperature causes membrane damage resulting in electrolyte leakage and this leakage has been shown to be related to temperature and drought tolerance of the wheat genotypes (Deshmukh *et al.*, 1996).

Clearly, wheat yield in lower latitudes may decrease as per the above global warming forecast, which may be further affected by water scarcity or drought. About 9 million hectares of wheat grow in tropical and subtropical areas of the developing world with temperatures above 17°C in the coolest month of the growing season. In heat stress already affects wheat plant senescence and photosynthesis (Al-Khatib and Paulsen, 1984), thereby influencing grain filling (Wardlaw *et al.*, 1980). Wheat cultivars capable of maintaining high 1000-kernel weight under heat stress appear to possess higher tolerance to hot environments (Reynolds *et al.*, 1994). Physiological traits that are associated with wheat yield in heat-prone environments are canopy temperature depression, membrane thermo-stability, leaf chlorophyll content during grain filling, leaf conductance and photosynthesis (Reynolds *et al.*, 1998). Amani *et al.* (1996) used canopy temperature depression to select for yield under a hot, dry, irrigated wheat environment in Mexico, whereas Hede *et al.* (1999) found that leaf chlorophyll content was correlated with 1000-kernel weight while screening Mexican wheat landraces. Such sources of alleles coupled with some of the above traits can provide means for genetically enhanced wheat by design in heat-prone environments.

Multidisciplinary research involving genetic resources enhancement and crop physiology at CIMMYT have led to a physiological trait-based approach to breeding for abiotic stress which has merit over breeding for yield *per se* by increasing the probability of successful crosses resulting from additive gene action. Advances have already been made in the drought-breeding program (Reynolds and Borlaug, 2006), and this strategy will be used to breed wheat for the high temperature-stressed environments. Since high temperature episodes appear to be more severe around anthesis these may affect the pollination process. Grain set is reduced by temperatures warmer than 30°C during the period from the onset of meiosis in the male generative tissue to the completion of anthesis (Smika and Shawcroft, 1980). The production and transfer of viable pollen grains to the stigma, germination of the pollen grains and growth of the pollen tubes down the style, and fertilization and development of the zygote are necessary for successful seed set. Although all these phases are temperature sensitive, some are more sensitive than others and high temperatures can cause both male and female sterility in wheat (Saini and Aspinall, 1982).

8. Summary, Conclusions and Recommendations

Wheat is the most important staple food crop in the world and second important crop in India after rice and is now grown extensively in countries with tropical and subtropical climates. It is cultvatied in about 28 million ha area with a production of 94 million tonnes and productivity of 3140 kg/ha (2011-12). Global reports indicate a loss of 10-40 per cent in crop production by 21st century because of climate change. The productivity and yield of wheat is significantly influenced by selection of suitable varieties, soil and environmental conditions as well as the management factors. The environmental stresses like waterlogging and drought, heat and chilling and salinity are the major problems of wheat growing areas which substantially reduce the yield and quality. The wheat crop in UK and NW European countries grows at much lower temperatures all through the crop duration with no water stress, and the grain-filling takes place over an extended period of 60 days at temperatures below 20°C. In India, the lower productivity is due to shorter crop duration and period of grain-filling and higher temperatures during crop growth, particularly during grain-filling. Crop duration varies from 150 days in the north and goes down to 100 days in Maharashtra and further south, with corresponding decline in yield.

Drought and heat are major abiotic stresses that decrease quality and productivity of wheat. Continual heat stress is a problem in about 7 m ha area, while terminal heat stress is a problem in about 40 per cent of the irrigated wheat growing areas of the world. High temperature during grain filling reduces the number of grain per ear, grain weight and subsequently the harvest index, resulting in reduced grain yields. Late planting of wheat in India is common due to the intensive cropping system, often delays the sowing of the crop up to the middle of January, particularly in North West India where it is generally sown after harvest of paddy, sugarcane, pigeonpea and potato. As a result, a portion of maturity period of the crop pushed forward and thus has to face higher temperature of the summer with hot spells often occurring at the time of maturity. The late sown wheat is more affected by high temperature stress leading to reduced yield and quality.

High temperature causes membrane damage resulting in electrolyte leakage and related to temperature and drought tolerance of the wheat genotypes. Wheat is a sink limited crop and high temperature during grain filling causes the production of shriveled grains due to forced maturity. Late sown crop gets exposed to mean maximum temperature of about 35°C during grain growth and causes yield reduction of 270 kg ha^{-1} per degree rise in temperature. Temperature adversely alters the growth and development of wheat during the early phase of panicle emergence, grain set and grain development. High temperature reduces the yield drastically due to its deterimental effect on metabolism and duration of phenological phases. Wild relatives of crops which have survived under strong natural selection pressure can be particularly useful as a source of genes for specific adaptive traits. Modern wheat varieties are well adapted to controlled cultural practices, but they are generally not highly tolerant to extreme environmental stresses such as high temperature. The physiological approaches of breeding for heat tolerance can provide a tool by identifying the traits that can be easily measured and used as a selection criterion for identifying heat tolerant genotypes. Therefore there is a wide range in the temperature tolerance both between species and amongst genotypes within a species. This review will be essentially confined to the latest understanding of the physiological, biochemical and molecular responses of organisms to heat stress and the effect of heat stress on molecular function, cellular processes, plant growth and development and crop production. Abiotic stress adversely affect almost all major field-grown plants belonging to varied ecosystems. In marginal or arid lands, environmental factors such as drought, high salinity, high temperature and flooding are serious problems. These stresses cause a great amount of loss that lead to instability in crop production. To address these problems, identification of traits leading to development of stress tolerant plants is of paramount importance.

In India, the lower productivity of wheat is due to shorter crop duration and period of grain-filling and high temperatures during grain-filling. Morphologically, similar wheat varieties have showed different degree of tolerance to post anthesis high temperature stress indicating that breeding for heat stress tolerance can open new insights into Indian agriculture. The physiological traits associated with heat tolerance are: photoassimilation, chlorophyll retention, chlorophyll a:b ratio, canopy temperature depression, stomatal conductance, membrane stability and relative water content and photosynthesis, respiration, conversion of sugar into starch in developing grain has been found to be affected by high temperature.

The environmental factors such as drought, high salinity, high temperature and flooding are serious problems causing a great amount of loss that lead to instability in crop production. To address these problems, identification of traits leading to development of stress tolerant plants is of paramount importance. A strong correlation between water use efficiency (WUE) and carbon isotope discrimination (CID) was observed in wheat genotypes suggesting that CID has the potential as a tool in selection for improved WUE in wheat breeding programme. Physiological traits like membrane injury index and RWC have also been suggested as simple techniques of selection of stress tolerance for use in breeding programme. New Wheat cultivars are needed to adapt the crop to changing environments and meet the nutritional needs of people, particularly those in the developing world, where farmers, increasingly adopt resource-conserving practices. In resource limited areas short duration/early maturing and latest released varieties are useful for minimising the yield losses by heat stress.

9. Future Research Areas

Variability is needed to further increase wheat's yield potential by providing new sources of biotic and abiotic resistance and to maintain the yield levels, develop germplasm adapted to more marginal environments, and to improve quality. The introduction of wheat cropping into marginal areas will present many abiotic stress challenges. Mineral ion deficiencies and toxicities, drought, wind, salinity and temperature extremes are some of the factors that will limit wheat production in these environments. Primitive wheats and wild relatives that originated in such environments can be expected to provide genes for tolerance to these abiotic stresses. The new varieties to be bred for having better physiological traits such as canopy temperature depression, membrane thermo-stability, leaf chlorophyll content during grain filling, leaf conductance and photosynthesis. Molecular biology is also likely to aid conventional breeding in changing the quality of wheat grain by developing it for novel industrial uses and improving its nutritional structure in ways that would clearly benefit consumers. This would be especially important if climate change accelerates.

10. References

Abass, M. and Rajashekhar, C. B. (1991). Characterization of heat injury in graps using ^{1}H Nuclear Magnetic Resonance methods: Changes in transverse relaxation times. *Plant Physiology*, 96: 957-961.

Abdelghani, A.M., Abdelshafi, A.M. and El-Menofi, M.M. (1994). Performance of some wheat germplasm adapted to terminal heat stress in upper Egypt. *Assiut. J. Agric. Sci.*, 25: 59-67.

Abernethy, R.H., D.S. Thiel, N.S. Peterson, and Helm, K. (1989). Thermotolerance is developmentally dependent in germinating wheat seed. *Plant Physiol.*, 89: 569-576.

Abrol, Y.P. and Ingram, K.T. (1996). Effect of higher day and night temperatures on growth and yield of some crop plants. In *Global Climate Change an Agricultural Production* (eds Bazzaz, F. and Sombroek, W.), John Wiley, NY and FAQ, Rome, 1996, pp. 123-140.

Abrol, Y.P., Bagga, A.K., Chakravorty, N.V.K. and Wattal, P.N. (1991). Impact of rise in temperature on productivity of wheat in India. In: Impact of Global Climatic Change on Photosynthesis and Plant Productivity, Y.P. Aborl *et al.* (eds.). Oxford and IBH Publishers, New Delhi, pp. 787-798.

Abrol, Y.P., Chatterjee, S.R., Ananda Kumar, P. and Jain, V. (1999). Improvement in nitrogen use efficiency: Physiological and molecular approaches. *Current Science*, 76: 1357-1364

Acevedo, E.M., Nachit, and Ortiz, G. (1990). Effect of heat stress on wheat and possible selection tools for use in breeding for tolerance. In: Wheat for non traditional warm Areas- A proceedings of the international conference July 29-Aug. 3, 1990 (ed.). saunders. D.A., pp. 401-420. CIMMYT, Mexico, D.F.

Aebi, H. (1984). Catalase *in vitro. Meth. Enzymol.*, 105: 121-126.

Ahmad, S., Ahmad, N., Ahamd, R. and Hamid, M. (1989). Effect of high temperature stress on wheat productive growth. *J. Agric. Re. Lahore*, 27: 307-313.

Ahrens, M.J. and Ingram, D.L. (1988). Heat tolerance of citrus leaves. *Hort. Si.*, 23: 747-748.

Alfonso, M., Yruela, I., Almarcegui, S., Torrado, E., Perez, M.A., Picorel, R. (2001). Unusual tolerance to high temperatures in a new herbicide-resistant D1 mutant from *Glycine max* (L.) Merr. cell cultures deficient in fatty acid desaturation, *planta*, 212: 573-582.

Al-Katib, K. and Paulsen, G.M. (1984). Mode of high temperature injury to wheat grain development. *Physiol. Plant.*, 61: 363-368.

Al-Katib, K. and Paulsen, G.M. (1990). Photosynthesis and productivity during high temperature stress of wheat genotypes from major world regions. *Crop Sci.*, 30: 1127-1132.

Al-Katib, K. and Paulsen, G.M. (1999). High temperature effects on photosynthetic processes in temperate and tropical cereals. *Crop. Sci.*, 39: 119-125.

Al-Khatib, K. and Paulsen, G.M. (1990). Photosynthesis and productivity during high temperature stress of wheat genotypes from major world regions. *Crop Sci.*, 30, 1127-1132.

Almeselmani,M.;Deshmukh, P.S.; Sairam, R.K.; Kushwaha,S.R. and Singh, T.P. (2006). Protective role of antioxidant enzymes under high temperature stress. *Plant science* 171: 382-8.

Amani, I.; Fisher, R.A and Reynolds, M.P. (1996). Canopy temperature depression association with yield of irrigated spring wheat cultivars in hot climate. *J. Agron. Crop Sci.*, 176: 119-229.

Amon, D.I. (1949). Copper enzyme in isolated chloroplasts polyphenol oxidase in Beta vulgaris. *Plant Physiol.*, 24: 1-15.

Anonymous (2005). Economic survey: Government of India, Ministry of Finance and company affairs. Economic Division, New Delhi.

Apuya, N.R and Zimmerman, J. L. (1992). Heat stock gene expression is controlled primarily at the translational level in carrot cells and somatic embryos. *Plant Cell*, 4: 657-665.

Araus, J.L., Slafer, G., Royo, C. and Serret, M.D. (2008). Breeding for yield potential and stress adaptation in cereals. *Critical Reviews in Plant Sciences* 27: 377–412.

Aro, E.M., McCaffery, S. and Anderson, J. M. (1994). Recovery from photoinhibition in peas (*Pisum sativum* L.) acclimated to varying growth irradiances. *Plant Physiol.*, 104: 1033-1041.

Aro, E.M., Virgin, I. and Andersson, B. (1993). Photoinhibition of Photosystem II. Inactivation, protein damage and turnover. *Biochem. Biophys. Acts*, 1143: 113-134.

Asada, K. (1996). Radical production and scavenging in chloroplasts. In: Baker N. (ed.) Photosynthesis and the environment. Kluwer Academy Press. Dordrecht, pp. 123-150.

Asana, R.D. and William, R.F. (1965). The effect of temperature stress on grain development in wheat. *Aust. J. Agri, Res.*, 16: 1-13.

Ashraf, M. (2010). Inducing drought tolerance in plants: recent advances. *Biotechnol. Adv.* 28(1): 169–183.

Austin, R.B., Bingham, I., Blackwell, R.D., Evans, L.T., Ford, M.A., Morgan, C.L. and Taylor, M. (1980). Genetic improvement in winter wheat yield since 1900 and associated physiological changes. *J. Agric. Sci.*, 94: 675-689.

Ayeneh, A., Van, G.M., Reynolds, M.P. and Ammar, K. (2002). Comparison of leaf, spike, peduncle, and canopy temperature depression in wheat under heat stress. *Field Crops Res* 79: 173–184.

Badiani, M., Schenone., G., Paolacci, A.R. and Fumagalli (1994). l: Daily fluctuations of antioxidant in bean (*Phaseolus vulgaris* L.) leaves. The influence of climatic factors. *Agrochimica*, 38: 25-36.

Bagga, A.K. and Rawson, H.M. (1977). Contrasting responses of morphologically similar wheat cultivars to temperatures appropriate to warm temperate climates with hot summers: a study in controlled environment. *Aust. J. Plant Physiol.*, 4: 877-887.

Balota, Maria., Payne, William A., Evett, Steven R., Peters, Troy R. (2008). Morphological and physiological traits associated with canopy temperature depression in three closely related wheat lines. *Crop science*, 48: 1897-1910.

Bansal, K.C. and Sinha, S.K. (1991). Assessment of drought resistance in twenty accessions of *Triticum aestivum* L. and related species. 1l. Stability in yield components. *Euphytica*, 56: 15-26.

Bansal, K.C., Dutta, M., Kumari Jyoti, Pandey, A.C., Singh, T.P., Kumar Sandeep, Trivedi, A.K., Phogat, B.S., Kumar Sandeep, Viswanathan, C., Kumar Neeraj, Sharma Pankaj, Singh, V.K., Aftab Tariq, Sharma, R.K., Tyagi, R.K., Bisht, I.S., Gangopadhyay, K.K., Prasad,T.V., Singh Mohar, Panwar, B.S., Yadav Mamta, Kumari Jyotisna, Jacob Sherry, Srinivasan Kalyani, Karale, M., Katiyar Amit, Smita Shuchi, Archak Sunit, Singh, A.K., Arya Lalit, Verma Manjusha, Bhat, K.V., Bhandari, D.C., Tyagi Vandana and Rathi, Y.S. (2013). Evaluation of 21000 wheat accessions conserved in the national Genebank in India for tolerance to terminal heat stress. Abstract book 12th International wheat genetic symposium, September 8-14, Pacifico Yokahama, Japan, pp. 62.

Barrow Em, Hulum M. (1996). Changing probabilities of daily temperature extremes in the UK related to future global warming and changes in climate variability. *Climate Research* 6: 21-31.

Barrs, H.D. and Weatherley (1962). A re-examination of the relative turgidity technique for estimating water deficit in leaves. *Aust. J. Biol. Sci.*, 15: 413-428.

Bartosz, G. (1997). Oxidative stress in plants. *Acta Physicol.* 19: 47-64.

Bates, L.S., Waldran, R.P. and Teare, I.D. (1973). Rapid determination of free praline for water stress studies. *Plant Soil*, 39: 205-208.

Berry, J. and Bjorkman, O. (1980). Photosynthetic response and adaptations to temperature in higher plants. *Ann. Rev. Plant Physiol.*, 31: 491-532.

Bhanu Parkash (1997). Influence of moisture and temperature stress in tolerant and susceptible wheat genotypes. PhD. Thesis, IARI, New Delhi, India.

Bhardwaj. R.B.L. (1978). New agronomic practices. In: Wheat research in India. 1966-1976, ICAR, New Delhi, pp. 79-98.

Bhatia, C.R. and Mitra, R. (1987). Bioenergetic considerations in genetic improvement of crop plants. In Biotechnology in Tropical Crop Improvement: Proceedigns of the International Biotechnology Workshop, ICRISAT Centre, Patancheru, 12-15 January, pp. 109-118.

Bhatia, C.R. and Rabson, R. (1976). Bioenergetic considerations in cereal breeding for protein improvement. *Science*, 194: 1418-1421.

Bhatia, C.R., R. and Rabson, R. (1981). Bioenergetic and energy constraints in increasing cereal productivity. *Agric. Syst.* 7: 105-111.

Bhullar, S.S. and Jenner, C.F. (1983). Responses to brief periods of elevated temperature in ears and grains of wheat. *Aust. J. Plant Physiol.*, 10: 549-560.

Bhullar, S.S. and Jenner, C.F. (1985). Differential responses to high temperature of starch and nitrogen accumulation in the grain of four cultivars of wheat. *Aust. J. Plant Physiol.*, 12: 313-325.

Bhullar, S.S. and Jenner, C.F. (1986). Effecty of a brief episode of elevated sucrose. *Aust. J. Plant Physiol.*, 13: 617-626.

Bjorkman O., Mahall B., Nobs M., Ward W. and Money, F.N.H. (1974). Growth response of plants from habitats with contrasting thermal environments: An analysis of the temperature dependence under controlled conditions. *Camegie Inst. Washington Yearb.*, 73: 757-767.

Bjorkman O., Mooney H.A. and Ehleringer, J. (1975). Photosynthetic responses of plants from habitats with contrasting thermal environment: comparison of photosynthetic characteristics of intact plants. *Carnegie Inst Washington Yearb.*, 74: 743-748.

Bluementhal C, Bekes F, Gras PW, Barlow EWR and Wrigley CW (1995). Influence of wheat genotypes tolerant to the effects of heat stress on grain quality. *Cereal Chem* 72:539-544.

Blum, A. (1986). Hat tolerance. Plant breeding for stress environment CRC Press, Boca Raton, Florida, P. 223.

Blum, A. (1986). The effect of heat stress on wheat leaf and ear photosynthesis. *J. Exp. Bot.*, 17: 37-47.

Blum, A. (1988). The effect of heat stress on wheat leaf and ear photosynthesis. *J.Exp.Bot.*, 17: 37-47.

Blum, A. and Ebercon, A. (1981). Cell membrane stability as a measure of drought and heat tolerance in wheat. *Crop Sci.*, 21: 43-47.

Blumenthal, C., Akes, F.B., Wrigley, C.W. and Barlow, E.W.R. (1990). The acquisition and maintenance of thermotolerance in Australian wheat. *Aust. J. Plant Physiol.*, 17: 37-47.

Bowler, C., Van Montagu, M. and Inze, D. (1992). Superoxide dismutase and stress tolerance. Ann. Rev. *Plant Physiol. Plant Mol. Biol.*, 43: 83-116.

Bradford, M.M. (1976). A rapid and sensitive method for quantification of proteins utilizing the principle of protein dye binding. *Anal. Biochem.*, 72: 248-254.

Braun, H.J., Rajaram S. and Ginkel, M.V. (1996). CIMMYTs approach to breeding for wide adaptation. *Euphytica.*, 92: 175-183.

Brocklehurst, P.A., Moss, J.P. and Williams, W. (1978). Effects of irradiance and water supply on grain development in wheat. *Ann. Appl. Biol.* 90: 265-276.

Bukhov, N.G., Sabat, S.C. and Mohanty, P. (1990). Analysis of chlorophyll 'a' fluorescence changes in weak light in heat treated Amaranthus chloroplasts. *Photosynth. Res.*, 23: 81-87.

Buttery, B.R., Buzzel, R.J. and Findlay, W.L. (1981). Relationship among photosynthetic rate, bean yield and other characters in field grown cultivars of soybean. *Can. J. Plant Sci.*, 61: 191-198.

Caird, M., J. Richards, J. James, F. Ludwig, L. Donovan, and K. Snyder. (2005). Nighttime transpiration and nutrient acquisition: Is there a benefit of losing water at night> Ecology Society of America Abstracts. 90th Annual Meeting. Ecology Society of America, Washington, DC.

Caird, M., J.H. Richards, and L.A. Donovan. (2006). Overview of nighttime stomatal conductance and transpiration in C_3 and C_4 plants. Abstract Proceedings of "The biology of transpiration: From Guard Cells to Globe." Snowbird Mountain Resort, UT. October 10-14, 2006. American Society of Plant Biologists, Rockville, MD.

Campbell, C.A. and Read, D.W.L. (1968). Influence of air temperature, light intensity and soil moisture on the growth and some growth analysis characteristics of Chinook wheat grown in the growth chamber. *Can. J. Plant Sci.*, 48: 294-315.

Campbell, G.S., and J.M. Norman. (1998). An introduction to environmental biophysics. 2nd ed. Springer-Verlag, Inc., New York.

Campbell, W.J., Allen, L.H. and Bowes, G. (1990). Response of soybean canopy photosynthesis to CO_2 concentration, light and temperature. *J. Exp. Bot.*, 41: 427-433.

Castillo, F.I., Penel, I. and Greppin, H. (1984). Peroxidase release induced by ozone in Sedum album leaves. *Plant physiol.*, 74: 846-851.

Chairompongopan, N.L., Davis, D.W. and Mackhast, A.H. (1990). Photosynthetic response to heat stress in common bean genotypes differing in heat acclimation potential. *Crop Sci.*, 30: 100-104.

Chandrasekar, V., Sairam, R.K. and Srivastava, G.C. (2000). Physiological and biochemical respones of hexaploid and tetraploid wheat to drought stress. *Journal of Agronomy and Crop Science*, 185: 219-227.

Chaturvedi, G.S., Aggarwal, P.K., Singh, A.K. and Sinha, S.K. (1985). Effect of date of sowing on the relative behavior of mother shoot and tillers of wheat. *India J. Agric. Sci.*, 55: 87-98.

Chen, G.X. and Zhang, R.X. (2004). *Scientia Agriculturia Sinica*, 37: 36-42.

Chinoy, J.J. (1947). Correlation between yield of wheat and temperature during ripening of grains. *Nature*, 159: 442-444.

Chon, M., Chen, Y.M. and Lin, C.Y. (1989). Thermotolerance of isolated mitochondria associated with heat shock proteins. *Plant Physiol.*, 89: 617-621.

Chowdhury, S.R. and Choudhuri, M.A. (1985). Hydrogen peroxide metabolism as an index of water stress tolerance in jute. *Physiol. Plant.*, 65: 503-507.

Christiansen, M.N. (1978). The physiology of Plant tolerance to temperature extremes. In: Crop Tolerance to Suboptimal Land Conditions (Ed. G.A. Jung). American Soc. Agron. Wis., pp. 173-191.

Colire, C., Rumeor, E.L., Gallier, J. J, Certaines, J.D. and Larhar, F. (1988). An assessment of proton nuclear magnetic resonance as an alternative method to describe water status of leaf tissues in wilted plants. *Plant Physiol Biochem.*, 26: 764-776.

Condon, A.g., R.A. Richards, and G.D. Farquhar. (1987). Carbon isotope discrimination is positively correlated with grain yield and dry matter production in field-grown wheat. *Crop Sci.* 27: 996-1001.

Condone, A.G., Farquhar, G.D. and Richards, R.A. (1990). Genotype variation in carbon isotope discrimination and transpiration efficiency in wheat. Leaf gas exchange and whole plant studies. *Aust. J. Plant Physiol.*, 17: 9-22.

Conroy, J., and P. Hocking. (1993). Nitrogen nutrition of C_3 plants at elevated atmospheric CO_2 concentrations. *Physiol. Plant.* 89: 570-576.

Conroy, J.P., Seneweera, S., Basra, A.S. Rogers, G. and Nissen-Wooller, B. (1994). Influence of rising atmospheric CO_2 concentrations and temperature on growth, yield and grain quality of cereal crops. *Aust. J. Plant Physiol.*, 21: 741-758.

Cook MG, Evans LT. (1978). Effect of relative size and distance of competing sinks on the distribution of photosynthetic assimilates in wheat. *Australian Journal of Plant Physiology* 5: 495-509.

Crafts-Brandner, S.J. and Law, R.D. (2000). Effect of heat stress on the inhibition and recovery of ribulose-1, 5-bisphosphate carboxylase/oxygenase activation state. *Planta,* 212: 6774-6779.

Crafts-Brandner, S.J. and Salvucci, M.E. (2000). Rubisco activase the photosynthetic potential of leaves at high temperature and CO_2. *Proc. Natl. Aca. Sci. USA,* 97: 13430-13435.

Crafts-Brandner, S.J. and Salvucci, M.E. (2002). Sensitivity of photosynthesis in the C_4 plant maize, to heat stress. *Plant Physiol.*, 129: 1773-1780.

Crafts-Brandner, S.J., Van-de-Loo, F.J. and Salvucci, M.E. (1997). The two forms of ribulose-1, 5-bisphosphate carboxylase/oxygenase activase differ in sensivity to elevated temperature. *Plant Physiol.*, 114: 439-444.

Craig, E.A. and Gross, C.A. (1991). Is HSP 70 the cellular thermotolerance? TIBS, 16: 135-140.

Daley, M.J., and N.G. Phillips, (2006). Interspecific variation in nighttime transpiration and stomatal conductance in a mixed New England deciduous forest. *Tree Physiol.* 26:411-419.

Das, D.K. and Rao, T.V. (1993). Growth and spectral response of wheat as influenced by varying nitrogen levels and plant densities. *Ann. Agric. Res.*, 14: 421-428.

Dat, J.F., Lopez-Delgado, H., Foyer, C.H. and Scott, I.M. (1998). Parallel changes in H_2O_2 and catalase during thermotolerance induced by salicylic acid or heat-acclimation in mustard seedlings. *Plant Physiol.*, 116: 1351-1357.

Davidson, J.E., Whyte, B., Bissinger, P.H. and schiestl, R.H. (1996). Oxidative stress is involved in heat induced cell death in Saccharomyces cerevisae. *Proc. Nat. Acad. Sci. USA,* 93: 5116-5121.

Dawson, I.A. and Wardlaw, I.F. (1989). The tolerance of wheat to high temperatures during reproductive growth lll. Booting to anthesis. *Australian J. Agric. Res.*, 40: 965-980.

Deshmukh, P.S., Hill, R.D., Shukla, D.S. and Wasnik, K.G. (1985). Influence of high temperature stress on ion leakage and nitrate reductase activity in crop plants. In: Proc. National Seminar, Plant Physiology from December 15-17, Imphal, Manipur, India, pp. 1-6.

Deshmukh, P.S., Sairam, R.K. and Shukla, D.S. (1991). Measurement of ion leakage as a screening technique for drought resistance in wheat genotypes. *Indian J. Plant Physiol.*, 34: 89-91.

Deshmukh, P.S., Sairam, R.K., Shukla, D.S. and Chaudhary H.B. (1996). Screening wheat genotypes for high temperature stress tolerance. Processing 2nd International Science Congress, 17-24 November, New Delhi, India, p. 119.

Deshmukh, P.S.; Sairam, R.K.; Kushwaha, S.R.; Singh, Tej Pal; Chaudhary, H.B. and Almeselmani, Moaed (2006). Physio-genetic approaches for increasing wheat (*Triticum aestivum*) productivity under rice (*Oryza sativa*) – wheat cropping system. *Indian Journal of Agricultural Sciences* 76 (11): 667-9.

Dewey, D.R.(1984).The genomic system of classification as a guide to intergeneric hybridization with the perennial Tr4iticeae. pp. 209-279 *in* Gene Manipulation in Plant improvement (J.P. Gustafson, ed.). Plenum Press, New York.

Dhindsa, R.S., Dhindsa, P.P. and Thorpe, T.A. (1981). Leaf senescence: Correlated with increased level of membrane permeability and lipid peroxidation, and decreased levels of superoxide dismutase and catalase. *J. Exp. Bot.*, 32: 93-101.

Donovan, G.R., Lee, J.W., Longhurst, T.J. and Martin, P. (1983). Effect of temperature on grain growth and protein accumulation in cultured wheat ears. *Aust. J. Plant Physiol.*, 10: 445-450.

Dubin, H.J., Fishcer, R.A., Mujeeb-Kazi, A., Pena, R.J., Sayre, K.D., Skovmand, B. and Valkoun, J. (1997). Wheat. Book biodiversity in trust conservation and use of plant Genetic Resources in CGIAR centres, edited by Dominic Fuccillo, Linda sears and Paul Stapleton, pp. 309-320.

Duguid, S.D. and Brule-Babel, A.L. (1994). Rate and duration of grain filling in five spring wheat (Triticum aestivum L.) genotypes. *Can. J. Plant Sci.*, 74: 681-686.

Dutta, M., Phogat, B.S., Kumar Sandeep, Kumar Naresh, Kumari Jyoti, Pandey Avinash, Singh, T.P., Tyagi, R.K., Jacob Sherry, Srinivasan Kalyani, Bisht, I.S., Karale, M., Sharma Pankaj, Aftab Tariq, Rathi, Y.S., Singh Amit, Archak Sunil, Bhat, K.V., Bhandari, D.C., Solanki, Y.P.S., Singh Dheeraj, Katiyar Amit, Smita Shuchi and Bansal, K.C. (2013). Development of core set of wheat germplasm conserved in the National Genebank in India. Abstract book 12th International wheat genetic symposium, September 8-14, Pacifico Yokohama, Japan, p. 52.

Eckardt, N.A. and Portis, A.R. (1997). Heat denaturation profiles of ribulose-1, 5-bisphosphate carboxylase/oxygenase (Rubisco) and Rubisco activase and the inability of Rubisco activase to restore activity of heat-denatured Rubisco. *Plant Physiol.*, 113: 243-248.

Egeh, A.O., Ingram, K.T. and Zamora, O.B. (1994). High temperature effects on leaf exchange. *Philippine J. Crop Sci.*, 17: 21-26.

Evans, L.T. and Wardlaw, I.F. (1976). Aspects of the comparative physiology of grain yield in cereals. *Adv. Agron.* 28: 301-359.

Evans, L.T., Wardlaw, I.F. and Fisher, R.A. (1975). Wheat. In Crop Physiology (ed. Evans, L.T.), University Press, Cambridge, pp. 101-149.

Evans, L.T., Wardlaw, I.F. and Fisther, R.A. (1973). Wheat. In: Crop Physiology. Some case histories (Eds. L.T. Evans). Cambridge University Press, London, pp. 101-149.

Fair, P., Rew, J. and Cresswell, C.F. (1973). Enzyme activity associated with CO_2 exchange in illuminated leaves of *Hordeum vulgare* L.l. Effect of light period, leaf age and position on CO_2 compensation point. *Ann. Bot.*, 37: 831-844.

Falk, S., Maxwell, D.P., Laudenbach, D.E. and Huner, N.P.A (1996). In Advances in Photosynthesis, V.5, Photosynthesis and the Environment (Neil, R. Baker, ed), pp. 367-385. Kluwer Academic Publishers Correct/Boston/London.

Farova, C., Taramino, G. and Binelli, G. (1989). Heat shock proteins during pollen development in maize. *Dev. Genet.*, 10: 324-332.

Farrar, T.C. and Becker, E.D. (1971). Pulse and Fourier Transform NMR-Introduction to theory and Method. Academic Press, New York.

Fedak, G. (1985). Alien species as sources of physiological traits for wheat improvement. *Euphytica.*, 34: 673-680.

Feierabend, J. and Engel, S. (1986). Photoinactivation of catalase in viro and in leaves. *Arch. Biochem Biophys.*, 251: 567-576.

Feirabend, J. (1977). Capacity for chlorophyll synthesis in heat bleached 70s ribosme deficient rye leaves. Planta, 135: 83-88.

Feller U., Crafts-Brandner, S.J. and Salvucci, M.E. (1998). Moderately high temperatures inhibit ribulose-1.5-bisphosphate carboxylase/oxygenase (Rubisco) activase-mediated activation of Rubisco. *Plant Physiol.*, 116: 539-546.

Ferrera, O.G., Rajaram, S. and Mosaad, M. G. (1993). Breeding strategis for improving wheat in heat stressed environments. Proc. Int. Conf. Wad Medani, Sudan and Dinajpur, *Bangladesh.* pp. 24-32.

Ferris, Rachel., Ellis, R.H., Wheeler, T.R. and Hadley, P. (1998). Effect of high temperature stress at anthesis on grain yield and biomass of field-grown crops of wheat. *Annuals of botany* 82: 631-639.

Fisher, R.A. (1985). Number of kernels in wheat crops and the influence of solar radiation and temperature. *J. Agric. Sci.* (Cambridge) 105: 447-461.

Fisher, R.A. (1996). Wheat physiology at CIMMYT and raising the yield potential. In: Increasing yield potential in wheat breaking the barriers. M.P. Reyonlds, S. Rajaram and A. McNab (Eds.) Proceeding of a workshop held in Civclad Obregon, Sonora, Mexico., p. 195.

Fisher, R.A., and Maurer, R. (1978). Drought resistance in spring wheat cultivars. I: Grain yield responses. *Aust. J. Agric Res.*, 29: 897-912.

Foster, J.G. and Hess, J.L. (1980). Responses of superoxide dismutase and glutathione reductase activities in cotton leaf tissue exposed to an atmosphere of enriched oxygen. *Plant Physiol.*, 66: 482-487.

Foyer, C.H., Descourvieres, P. and Kunert, K.J. (1994). Protection against oxygen radicals: an important defense mechanism studied in transgenic plants. *Plant Cell Environ*, 17: 507-523.

Foyer, C.H., Nector, G. (2000). Oxygen processing in photosynthesis regulation and signaling. *New Phytol.*, 146: 359-388.

Frank, A.B. and Bauer, A. (1997). Temperature effects prior to double ridge on apex development and phyllochron in spring barley. *Crop Sci.*, 37: 1527-1531.

Frederick, J.R., Camberato, J.J. (1995). Water and nitrogen effects on winter wheat in the south-eastern coastal plain. II- physiological responses, *Agron. J.* (USA), 87: 527-533.

Gallagher, J.N., Biscoe, P.V. and Hunter, B. (1976). Effect of drought on grain growth. Nature, 264: 541-542.

Gallie D.R., Le, H., Tanguay, R.L. and Browning, K.S. (1998). Translation initiation factors are differently regulated in cereals during development and following heat shock. *Plant J.*, 14: 715-722.

Ghildiyal, M.C. and Sharma-Natu, P. (2000). Photosynthetic acclimation to rising atmospheric carbon dioxide concentration. *Indian Journal of Experimental Biology*, 38: 961-966.

Gibson LR, and Paulsen GM (1999). Yield component of yield grown under high temperature stress during reproductive growth. *Crop Sci* 39: 1841-1846

Graves, W.R., Joy, R.J. and Dana, M.N. (1991). Water use and growth of honey locust and tree-of- heaven at high root-zone temperature. *Hort. Sci.*, 26: 1309-1312.

Gregroy, F.G. (1926). The effect of climate conditions on the growth of barley. *Ann. Bot.*, 40: 1-26.

Groden, D. and Beck, E. (1979). H_2O_2 destruction by ascorbate-dependent system from chloroplast. Bichim. *Biophys. Acta*, 546: 426-435

Guedira, M., Paulsen, G.M. (2002). Accumulation of starch in wheat grain under different shoot/root temperatures during maturation. Fun. *Plant Biol.*, 29: 495-503.

Guha Sarkar, C.K., Srivastava, P.S.L. and Deshmukh, P.S. (2001). Grain growth rate and heat susceptibility index : Traits for breeding genotypes tolerant to terminal high temperature stress in bread wheat (*Triticum aestivum* L.). *Indian J. Genet.*, 61: 209-212.

Gupta, A.S., Webb, R.P., Holaday, A.S. and Allen, R.D. (1993). Over expression of superoxide dismutase protect plants from oxidative stress. Induction of ascorbate peroxidase in superoxide dismutase over expersion plants. *Plant Physiol.*, 103: 1067-1073.

Gupta, S. and Gupta, N.K. (2005). High temperature induced antioxidative defense mechanism in seedlings of contrasting wheat genotypes. *India J. Plant Physiol.*, 10: 73-75.

Gutiman, M.R., Arnozis, P.A. and Barneix, A.J. (1991). Effect of source sink relations and nitrogen remobilization in the flag leaf of wheat. *Physiol. Plant*, 82: 278-284.

Haldimann, P. and Feller, U. (2004). Inhibition of photosynthesis by high temperatures in oak (*Quercus pubescens* L.) leaves grown under natural conditions closely correlates with a reversible heat-dependent reduction of the activation state of ribulose-1,5-bisphosphate carboxylase/oxygenase. *Plant Cell Environ.*, 27: 1169.

Hall, A.E. (1993). Breeding for heat tolerance. *Plant Breed. Res.*, 10: 129-168.

Halliwell, B. (1974). Superoxide dismutase, ctalase and glutathione peroxidase: solutions to the problems or living with oxygen. *New Phytol.*, 73: 1075-1086.

Haolbrook, G.P., Galasinski, S.C. and Salvucci, M.E. (1991). Regulation of 2-carboxyarabinitol 1-phosphatase. *Plant Physiol.*, 97: 894-899.

Harding, S.A., Guikema, G.A. and Paulsen, G.M. (1990). Photosynthetic decline from high temperature stress during maturation of wheat. II Interaction with source and sink processes. *Plant Physiol.* 92: 654-658

Harding, S.A., Gurikema, G.A. and Paulsen, G.M. (1990). Photosynthetic decline from high temperature stress during maturation of wheat. I- Interaction with senescence processes. *Plant Physiol.*, 92: 648-653.

Harwood, J.L., Jones, A.L., Perry, H.J., Rutter, A.J., Smith, K.M. and Williams, M. (1994). Changes in plant lipids during temperature adaptation. In: Temperature adaptation of biological membranes (Ed. Cossins, A.R.). London: Portland Press, pp. 107-117.

Haung, B., Fry, J.D. and Wang, B. (1998). Water relations and canopy characteristic of tall fescue cultivars during and after drought stress. *Hort. Sci.*, 33: 837-840.

Havaux. M. and Tardy, F. (1996). Temperature-dependent adjustment of the thermal stability of photosystem ll in vivo: possible involvement of xanthophyll-cycle pigments. *Planta*, 198: 324-333.

Hawker, J.S., Jenner, C.F. (1993). High temperature affects the activity of enzymes in the committed pathway of starch synthesis in developing wheat endosperm. *Australian Journal of Plant Physiology* 20: 197-209.

Hay, R.K.M, Walker, A.J. (1989). Dry matter partitioning. An introduction to the physiology of crop yield. Harlow: Longman Scientific and Technical, 107-156.

He, Z.H., Rajaram, S. (1994). Differential responses of bread wheat characters to high temperature. *Euphytica.*, 72: 197-203.

Hede, A., Skovmand, B., Reynolds, M.P., Crossa, J., Vilhelmsen, A.L., Stølen, O. (1999). Evaluating genetic diversity for heat tolerance in Mexican wheat landraces. *Genet. Resour. Crop Evol.* 46: 37-45.

Herzog, H. (1985). Relation of source and sink during the grain filling period in wheat and some aspects of its regulation. *Physio Plant* 56:155-166

Hill, B.P. and Duce, S.L. (1990). The influence of chemical and diffusive exchange on water proton transverse relaxation in plant tissues. Magnetic Resonance imaging, 8: 321-331.

Hirel, B. and Gallais, A. (2006). Rubisco synthesis, turnover and degradation: Some new thoughts on an old problem. *New Physiol.*, 169: 445-448.

Hiscox, J.D. and Israelstem, G.F. (1979). A method for extraction of chlorophyll from leaf tissue without maceration using dimethyl sulfoxide. *Can J. Bot.*, 57: 1332-1334.

Hoagland, C.R. and Arnon, D.I. (1950). The solution-culture method for growing plants without soil. Calif. *Agric. Exp. Circ.* p. 347.

Howard, A.R., and L.A. Donovan. (2007). Helianthus nighttime conductance and transpiration respond to soil water but not nutrient availability. *Plant Physiol.* 143:145-155.

Howart, C.J. (1990). Heat shock proteins in sorghum and Pearlmillet: Ethanol, sodium arsenite, sodium malonate and the development of thermotolerance. *J. Exp. Bot.*, 41: 877-883.

Howarth, C.J. and Ougham, H.J. (1993). Gene expression under temperature stress. *New Physiol.*, 125: 1-26.

Hsich, M.H., Chen, J.T., Jinn, T.L., Chen, Y.M. and Lin, C.Y. (1992). A class of soybean low molecular weight HSPs. *Plant Physiol.*, 99: 1279-1284.

Hu, Y.J., and Fang, T. (1995). Studies on the effect of grain weight in breeding high yielding wheat. *Acta Agronomica Sinica*, 21: 671-678.

Ibrahim, A.M.H. and Quick, J.S. (2001). Genetic control of high temperature tolerance in wheat as measured by membrane thermal stability. *Crop sci.* 41: 1405-1407.

Inze, D, and Van Montagu, M. (1995). Oxidative stress in plants. Curr. Opin. *Biotechnol.*, 6: 153-158.

Irina, I.P., Roman, A.V. and Schoffi, F. (2002). Heat stress and heat shock transcription factor dependent expression and activity of ascorbate peroxidase in Arabiodopsis. *Plant Physiol.*, 129: 833-853.

Jagtap, V. and Bhargava, S. (1995). Variation in the antioxidant metabolism of drought tolerant and drought susceptible varieties of Sorghum bicolor (L.) Moench. Exposed to high light, less water and high temperature stress. *J. Plant Physiol.*, 145: 195-197.

Jain, H.K., Sinha, S.K., Kulshresta, V.P. and Mathur, V.S. (1974). Breeding for yield in dwarf wheats. In: Proc. ll. International Wheat Genet. *Symposium Columbia*, USA, p. 527.

Jason J.G. Thomas G.R. and Mason D.P. (2004). Heat and drought influence photosynthesis water relations, and soluble carbohydrates of two ecotypes of red bud. *Journal of American Society of Horicultural Sciences*, 129: 497-502.

Jenner, C.F. (1991). Effects of exposure of wheat ears to high temperature on dry matter accumulation and carbohydrate metabolism in grain of two cultivars. I. Immediate response. *Aust. J. Plant Physiol.*, 18: 165-177.

Jenner, C.F. (1994). Starch synthesis in the kernel of wheat under high temperature conditions. *Aust. J. Plant Physiol.*, 21: 791-806.

Jiang, Y. and Huang, B. (2000). Effects of calcium on antioxidant activities and water relation associated with heat tolerance in two cool-season grasses. *J. Exp. Bot.* (In press).

Jiang, Y. and Huang, B. (2001). Effects of calcium on antioxidant activities and water relations associated with heat tolerance in two cool season grasses. *J. Exp. Bot.*, 52: 341-349.

Johnston, R.N. and Kucey, B.L. (1988). Competitive inhibition of hsp 70 gene expression causes thermosensitivity. *Sci.*, 242: 1551-1553.

Jordan, D.B. and Orgen, W.L. (1984). The CO_2/O_2 specificity of ribulose-1, 5-bisphosphate carboxylase/oxygenase: dependence on ribulose bisphosphate concentration and temperature. *Planta*, 161: 308-313.

Jorgenson, B.B., Isakeen, M.F. and Jannasch, H.W. (1992). Bacterial sulfate reduction above 100°C in deep sea hydrothermal vent sediments. *Sci.*, 258: 1756-1757.

Joshi, A.K., Ferrara, O., Crossa, J., Singh, G., Sharma, R., Chand, R. and Parsad, R. (2007b). Combining superior agronomic performance and terminal heat tolerance with resistance to spot blotch (Bipolaris sorokiniana) in the warm humid Gangetic Plains of South Asia. *Field Crops Res.* 103:53–61.

Joshi, A.K., Mishra, B., Chatrath, R., Ferrara, G.O. and Singh, R.P. (2007).Wheat improvement in India: Present status, emerging challenges and future prospects. *Euphytica* 157: 431-446

Kaku, S. (1993). Monitoring stress sensitivity by water proton NMR relaxation times in leaves of azaleas that originated in different ecological habitats. *Plant Cell Physiol.*, 34: 535-541.

Kalituho, L.N., Pshybytko, L.F., Kabashnikova, and Jahns, P. (2003). Photosynthetic apparatus and high temperature. Role of light. *Bulg. J. Plant Physiol.* (Special Issue): 281-289.

Karimi, M.M. and Siddique, K.H.M. (1991). Crop growth and relative growth rates of old and modern wheat cultivre. *Aust. J. Agric. Res.*, 42: 13-20.

Karus, T.E., McKerise, B.D. and Fletcher, R.A. (1995). Paclobutrazol induced tolerance of wheat leaves to paraquat may involve increased antioxidant enzyme activity. *J. Plant Physiol.*, 145: 570-576.

Kaur, J., Sheogan, I.S. and Nainawatee, H.S. (1989). Effect of heat stress on photosynthesis and respiration in wheat (*Triticum aestivum* L.) mutant. In: G.S. Singhal, J. Barber, R. A. Dilley, R. Govindjee, Haselkorn and P. Mohanty (eds.). Photosytnesis, Molecular Biology and Bio-energetics, Narasa Publishing House, New Delhi, pp. 297-303.

Kelly, J.T., Bacon, R.K. and Gbur, E.E. (1994). Relationship of grain yield and test weight. *Cereal Research Communications*, 23: 53-57.

Khanna-Chopra, R. and Viswanathan, C. (1998). Evaluation of heat stress tolerance in irrigated environment of *T. aestivum* and related species I. Stability in yield and yield components. *Euphytica*, 106, 169-180.

Kirschbaum, M.U.F. and Farquhar, G.D. (1984). Temperature dependence of whole-leaf photosynthesis in *Eucalyptus pauciflora* Sieb. ex Spreng. *Aust. J. Plant Physiol.* 11: 519-538.

Kislyuk, I.M., Bubolo, L.S. and Vaskovskii, M.D. (1997). Heat shock induced increase in the length and number of thylakoids in wheat leaf chloroplasts. *Russion J. Plant Physiol.*, 44: 30-35.

Kler, D.S. and Bains, D.S. (1989). Relations among plant height chlorophyll content, leaf area index and grain yield of durum wheat (*Triticum durum* Desf.) under different sowing patterns. *Haryana J. Agron.*, 5: 87-92.

Klueva, N.Y., Maestri, E., Marmioli, N. and Nguyen, H.Y. (2001). Mechanisms of thermotolerance in corp. In: A.S. Basra (ed.), Crop Response and Adaptations to Temperature Stress, Food Production Press, Binghamton, New York, pp. 177-217.

Kobza, J. and Edwards, G.E. (1987). Influences of leaf temperature on photosynthetic carbon metabolism in wheat. *Plant Physiol.*, 83: 69-74.

Kohli, M.M., Mann,C. and Rajaram, S. (1991).Global status and recent progress in breeding wheat for the warmer areas. Pp 225-241, DA Saunders ed. Wheat for the nontraditional warmer areas. Mexico, D.F: CIMMYT.

Kolderop, F. (1979). Application of different temperatures in three growth phases of wheat. I. effect on grain and straw yields. *Acta Agric. Scandinavia*, 29: 6-10.

Krishnan, M., Nguyen, H.T. and Burke, J.J. (1989). Heat shock protein synthesis and thermal tolerance in wheat. *Plant Physiol.*, 90: 140-145.

Kumari, Sunita and Ghildiyal, M.C. (1997). Availability and utilization of assimilates in relation to grain growth within the ear of wheat. *Journal Agronomy Crop Science*, 178: 245-249.

Kunst, L., Browse, J. and Somerville (1989). Enhanced thermal tolerance in a mutant of Arabidopsis deficient in palmitic acid unsaturation. *Plant Physiol.*, 91: 401-408.

Kuroyanagi, T. and Paulsen, G.M. (1985). Mode of high temperature injury to wheat. II. Comparisons of wheat and rice with and without inflorescence. *Plant Physiol.*, 65: 203-208.

Larkindale, J. and Huang, B. (2004). Thermotolerance and antioxidant systems in *Agrostis stoloifera*: Involvement of salicylic acid, abscisic acid, calcium, hydrogen peroxide, and ethylene. *J. Plant Physiol.*, 161: 405-413.

Larkindale, Jane and Marc R. Knight (2002). Protection against Heat Stress-Induced Oxidative Damage in Arabidopsis Involves Calcium, Abscisic Acid, Ethylene, and Salicylic Acid. *Plant Physiol.*, 128: 682-695.

Law R.D. and Crafts-Brandner S.J. (2001). High temperature stress increases the expression of wheat leaf ribulose-1, 5-bisphosphate carboxylase/oxygenase activase protein. *Arch. Biochem. Biophys.*, 386: 261267.

Law, R.D. and Crafts-Brandner, S.J. (1999). Inhibition and Acclimation of Photosynthesis to Heat Stress Is Closely Correlated with Activation of ribulose-1,5-bisphosphate carboxylase/oxygenase. *Plant Physiol.*, 120: 173-182.

Lee, P.C., Bochner, B.R. and Ames, B.N. (1983). A heat shock stress and cell oxidation. Proc. Natl. Acad. Sci., USA, 80: 7496-7500.

Lehman, V.G. and Engelke, M.C. (1993). Heritability of creeping bent-grass shoot water content under soil dehydration and elevated temperature. *Crop Sci.*, 33: 1061-1066.

Lewa, C.J. and Lewa, M. (1990). Temperature dependence of ^1H NMR relaxation time T_2 for intact and neoplastic plant tissues. *J. Magnetic Resonance.*, 89: 219-226.

Lin, C.Y., Roberts, J.K. and Key, J.L. (1984). Acquisition of thermotolerance in soybean seedlings. *Plant Physiol.*, 74: 152-160.

Lin, T.Y. and Markhart, A.H. (1990). Temperature effect on mitochondrial respiration in *Phaseolus acutifolius*. A Gray and *P. vulgaris* L. *Plant Physiol.*, 94: 54-58.

Liu, X. and Huang, B. (2000). Heat stress injury in relation to membrane lipid peroxidation in creeping bentgrass. *Crop Sci.*, 40: 503-510.

Liv, Z.C. and Su, D.Y. (1985). Effect of high temperature on chloroplast proteins in wheat. *Acta Botanica Sinica*, 27: 63-67.

MacLeod L.C. and Duffus, C.M. (1988). Reduced starch content and sucrose synthase activity in developing endosperm of barley plants grown at elevated temperatures. *Aust. J. Plant Physiol.*, 15: 367-375.

Mahan, J.R., McMichael, B.L.and Wanjura, D.F. (1995). Methods of reducing the adverse effects of temperature stress on plants: a review. *Environmental and Experimental Botany* 35: 251-258.

Maheswari, M., Joshi, D.K., Saha, R., Nagarajan, S. and Gambhir, P.N. (1999). Transverase relaxation time of leaf water protons and membrane injury in wheat (*Triticum aestivum* L.) in response to high temperature. *Ann. Bot.*, 84: 741-745.

Martineau, J.R., Specht, J.E. and Morciere, I.J.S. (1979). The role of abscisic acid and fernesol in the alleviation of water stress. Phil. Trans. Res. Soc. London, Ser., 284: 471-482.

Mason, R.E., Mondal, S., Beecher, F.W., Pacheco, A., Jampala, B., Ibrahim, A.M.H. and Hays, D.B. (2010). QTL associated with heat susceptibility index in wheat (*Triticum aestivum* L.) under short-term reproductive stage heat stress. *Euphytica* 174:423–436.

McCaig, T.N. and De Pauw, R.M. (1995). Breeding hard red spring wheat in western Canada: Historical trends in yield and related variables. *Canadian J. Plant Sci.,* 75: 387-393.

McDonald, E.P., J.e. Erickson, and E.L. Kruger. (2002). Can decreased transpiration limit plant nitrogen acquisition in elevated CO_2? *Funct. Plant Biol.* 29:1115-1120.

Mian,M.A., Mahamood, A., Ihsan, M. and Cheema,N.M. (2007).Response of different wheat genotypes to post anthesis temperature stress. *J Agric Res* 45:269-276

Michael, E.S., Katherine, W.O., Steven, Crafts-Brandner, J. and Elizabeth V. (2001). Exceptional Sensitivity of Rubisco Activase to Thermal Denaturation in Vitro and *in vivo. Plant Physiol.,* 127: 1053-1064.

Miernyk, J.A., Duck, N.B., Shattes, R.G. and Folk, W.R. (1992). The 70 KDa heat shock cognate can act as a molecular chaperone during the membrane translocation of a plant secretary protein precursor. *Plant. Cell,* 4: 821-829.

Mishra, R.K. and Singhal, G.S. (1992). Function of photosynthetic apparatus of intact wheat leaves under high light and heat stress and its relationship with thylakoid lipids. *Plant Physiol.,* 98: 1-6.

Mishra, V., Mishra, R.D., Singh, M. and Verma, R.S. (2003). Dry matter accumulation at pre and post anthesis and yield of wheat (*Triticum aestivum*) as affected by temperature stress and genotypes. *Indian J. Agron.,* 48: 227-281.

Misra, S.M. (1990). Physiological studies on the interaction of moisture stress and nitrogen in wheat. M.Sc. Thesis, IARI, New Delhi, India.

Mitra, R., and Bhatia, C.R. (2008). Bioenergetic cost of heat tolerance in wheat crop. Current Science, Vol. 94, No. 8, 1049-1053.

Mitra, R.K., Bhatia, C.R. and Rabson, R. (1979). Bioenergetic cost of altering amino acid composition of cereal grains. *Cereal Chem.,* 14: 9-25.

Mohabir, G. and John, P. (1988). Effect of temperature on starch synthesis in potato tuber tissue and in amyloplast. *Plant Physiol.* 88: 1222-1228.

Mohamed M.F. Keutgen N. Tawfik A.A and Noga G. (2002). Dehydration avoidance responses of tepary bean lines differing in drought resistance. *Journal of Plant Physiology,* 159: 31-38.

Monk, L.S., Fagerstedt, K.V. and Crawford, R.M.M. (1989). Oxygen toxicity and superoxide dismutase as an antioxidant in physiological stress. *Plant Physiol.,* 76: 456-459.

Moon, B.Y., Higashi, S., Gombos, Z. and Murata, N. (1995). Unsaturation of the membrane lipids of chloroplasts stabilizes and photosynthetic machinery against low-temperature photoinhibition in transgenic tobacco plants. *Proc. Natl. Acad, Sci.,* 92: 6219-6223.

Moot, D.J., Henderson, A.L, Porter, J.R. and Semenov, M.A. (1996). Temperature, CO_2 and the growth and development of wheat: Changes in the mean and Variability of growing conditions. *Climatic Changes,* 33: 351-368.

Morgan, J.M. (1983). Osmoregulation as a selection criterion for drought tolerance in wheat. *Aust. J. Agric. Res.*, 34: 607-614.

Morgan, J.M. (1984). Osmoregulation and water stress in higher Plants. *Ann. Rev. Plant Physiol.*, 35: 299-319.

Morgunov, A. (1994). Bread wheat breeding for heat tolerance. In: Rajaram, S. and G.P. Hettel (Eds), Wheat breeding at CIMMYT: Commemorating 50 years of research in Mexico for global Wheat improvement. Chapter 3: pp 29-35, Ciudad Obregon, Sonora, Mexico, 21-25 March.

Morimoto, R.I. (1993). Cells in stress: Transcriptional activation of heat shock genes. *Sci.*, 259: 1409-1410.

Mujeeb-Kazi, A. and G.P. Hettel, eds. (1995). Utilizing Wild Grass Biodiversity in wheat Improvement. 15 Years of Wide Cross Research At CIMMYT. CIMMYT Research Report No. 2. CIMMYT, Mexico, D.F.

Mukherjee, S.P. and Choudhuri, M.A. (1983). Implications of water stress induced changes in the levels of endogenous ascorbic acid and hydrogen peroxide in Vigna Seedlings. *Plant Physiol,* 58: 166-170.

Murakami, Y., Tsuyama, M., Kobayashi, Y., Kodama, H. and Iba, K. (2000). Trienoic fatty acids and plant tolerance to high temperature. Sci., 287: 476-479.

Murata, Y. and Lyama, J. (1963). Studies on photosynthesis of forage crops II, Influence of air temperature upon the photosynthesis of forage and grain crops. *Proc Crop Sci Society of Japan*, 31:315-321

Nachit, M.M. and Elouafi, I. (2004). Durum adaptation in the Mediterranean dry land: breeding, stress physiology, and molecular markers. In: Rao SC, Ryan J, eds. Challenges and strategies for dry land agriculture. CSSA Special Publication 32. Madison, Wisconsin, USA: Crop Science Society of America Inc., American Society of Agronomy Inc, 203–218.

Nagarajan, S. (2005). Can India produce enough wheat by 2020? *Curr. Sci.*, 89: 1467-1471.

Nagarajan, S. and Rane, J. (1997). Techniques for quantifying heat tolerance in wheat. International group meeting on "Wheat Research Needs Beyond 2000 A.D." August 12-14, 1997. "Abstracts" D.W.R. ICAR: 30-31.

Nagarajan, S. and Rane, J. (2002). Physiological traits associated with yield performance of spring wheat (*Triticum aestivum* L.) under late sown conditions. *Indian J. Agric. Sci.*, 72: 135-140.

Nagarajan, S., Chahal, S.S. Gambhir, P.N. and Tiwari, P.N. (1993). Relationship between leaf water spin-lattice relaxation time and water relation parameters in three wheat cultivars. *Plant Cell and Environment.* 19: 87-92.

Nagarajan, S., Joshi, D.K., Anjali, A., Verma, A.P.S. and Pathak, C.P. (2005). Proton NMR transverse relaxation time and membrane stability in wheat leaves exposed to high temperature shock. *Indian J. Biochem. Biophysics*, 42: 122-126.

Nagarajan, S.and Rane, J. (2002). Physiological traits associated with yield performance of spring wheat (*Tricticum aestivum* L.) under late sown conditions. *Indian Journal of Agricultural Sciences* 72: 135-40.

Nakano Y. and Asada, K. (1981). Hydrogen peroxide is scavenged by ascorbate specific peroxides in spinach chloroplasts. *Plant cell Physiol.*, 22: 867-880.

Navari-Izoo, F., Milone, M.T.A. and Guartacci, M.I. (1993). Metabolic changes in wheat plants subjected to a water deficit stress programme. *Plant Sci.*, 92: 151-157.

NBPGR, Newsletter, October-December (2011).

Nguyen, H.T., Joshi, C., Klueva, N., Weng, J., Hendershot, K.L. and Blum. A. (1994). The heat shock response and expression of heat shock proteins in wheat under diurnal heat stress and field conditions. *Aust. J. Plant Physiol.*, 21: 857-867.

Nocolas, M.E., Gleadon,R.M. and Dalling, M.J. (1984). Effects of drought and high temperature on grain growth in wheat. *Aust J Plant Physiol.*, 11: 553-566

Noctor, G. and Foyer, C.H. (1998). Ascorbate and glutathione: keeping active oxygen under control. Ann. Rev. *Plant Physiol. Mol. Biol.*, 49: 249-279.

Nover, L., Neuman, D. and Scharf, K.D. (1989). Heat shock and other stress response systems of plants. Sringer-verlag. Berlin., pp. 5-86.

Oberhuber, W., Dai, Z.Y. and Edwards, G.E. (1993). Light dependence of quantum yields of photosysem II and CO_2 fixation in C_3 and C_4 plants. *Photsynth. Res.*, 35: 265-274.

Ogren, W.L. (1984). Photorespiration: pathways, Regulation, and modification. *Annu. Rev. Plant Physiol.*, 35: 415-442.

Olmos, E., Harnandez, J.A., Sevilla, F. and Hellin, E. (1994). Induction of several antioxidant enzymes in the selection of salt tolerant cell line of Pisum Sativum. *J. Plant Physiol.*, 144: 594-598.

Ortiz, R., Sayre, K.D., Govaerts, B., Gupta, R., Subbarao, G.V., Ban, T., Hodson, D., Dixon, J.M., Ortiz-Monasterio, J.I.and Reynolds, M. (2008). Climate change: can wheat beat the heat? *Agric Ecosyst Environ.*, 126: 46–58.

Ortiz, R., Trethowan, R., Ortiz Ferrara, G.F., Iwanaga, M., Dodds, J.H., Crouch, J.H., Crossa, J., Braun, H.J.(2007). High yield potential, shuttle breeding, genetic diversity and new international wheat improvement strategy. *Euphytica* 157: 365-384.

Ortiz, Rodomiro, Sayre, Kenneth D., Govaerts, Bram., Gupta, Raj., Subbarao, G.V., Ban, Tomohiro., Hodson, David., Dixon, John M., Ortiz-Monasterio Ivan J. and Reynolds, Mathew. (2008) Climate change: Can wheat beat the heat? *Agriculutre, Ecosystems and Environment*, 126: 46-58.

Ostapenko, N.V. (1993). The effect of the weather and nitrogen on photosynthetic activity of winter wehat. *Agro Khimiya*, 3: 3-7.

Ougham, H.J. and Stoddart, J.L. (1986). Synthesis of heat shock proteins and acquisition of thermotolerance in high temperature tolerant and high temperature susceptible lines of sorghum. *Plant Sci.*, 44: 163-167.

Owen, P.C. (1971). Response of semi dwarf wheat to temperature representing a tropical dry season. I. Non-extreme temperature. *Experimental Agric.*, 7: 33-41.

Pal, S.K. (1992). Effects of high temperature and water stress on grain yield of wheat. Dissertation-Abstracts-International B. Abstract of Thesis. University of Sackatchewan, Canada, 1990, pp. 212

Paliwal, R., Roder, M.S., Kumar, U. Srivastava, J.P. and Joshi, A.K.(2012). QTL mapping of terminal heat tolerance in hexaploid wheat (*T. aestivum* L.). Theor Appl Genet DOI 10.1007/s00122-012-1853-3.

Panse, V.G. and Sukhatme, P.V. (1967). Statistical methods for agricultural workers. ICAR Publication, New Delhi.

Pastenes, C. and Horton, P. (1995). The effect of high temperature on photosynthesis. In photosynthesis from light to biosphere volume IV. Proceeding of the Xth International Photosynthesis Congress, Mont Pellier, France, 20-25 August (1995). (Edited by Mathis, P.) Dordrcht, Netherlands, Kluwer Academic Publishers (1995), 797-800 ISBN 0-7923-3860. (Vol. IV) ISBNO-7923-3962-6.

Paulsen, G.M. and Machado, S. (2001). Combined effect of drought and high temperature on water relation of wheat and sorghum. *Plant and soil*, 233: 179-187.

Pearcy R.W. (1978). Effect of growth temperature on the fatty acid composition of the leaf lipids in Atriplex lentiformis (Torr.) Wats. *Plant Physiol.*, 61: 484-486.

Pearcy, R.W. (1977). Acclimation of photosynthetic and respiratory carbon dioxide exchange to growth temperature in *Atriplex lentiformis* (Torr.) Wats. *Plant Physiol.*, 59: 795-799.

Pelham, H. (1986). Speculations on the major heat shock and glucose regulated proteins. *Cell*, 46: 959-961.

Pinto, R.S., Reynolds, M.P., McIntyre, C.L., Olivares-Villegas, J.J. and Chapman, S.C. (2010). Heat and drought QTL in a wheat population designed to minimize confounding agronomic effects. *Theor Appl Genet* 121:1001–1021.

Polle, A. (1997). Defense against photooxidative damage in plants. p. 785-813. In: J. Scandalios (ed.), Oxidative stress and the molecular biology of antioxidant defense. Cold Spring Harbor Laboratory Press. Cold Spring Harbor, NY.

Pomeranz, Y. (ed.). (1987). Modern cereal Science and Technology. VCH Publishers, Inc., New York, NY.

Pool, R. (1990). Pushing the envelope of life. *Sci.*, 247: 158-160.

Porter, J.R., Kirby,E.J.M., Day,W., Adam,J.S., Appleyard, M., Ayling, S. Baker, C.K., Beale, P., Belford,R.K., Biscoe,P.V., Chapman, A., Fuller, M.P., Hampson, J., Hay, R.K.M., Hough, M., Matthews,S., Thompson, W.J., Weir, A.H., Willington, V.B.A., and Wood. D.W. (1987). An analysis of morphological development stages in

Avalon winter wheat crops with different sowing dates and at ten sites in England and Scotland. *Journal of Agricultural Science,* Cambridge, 109: 107-121.

Prakash, P., Sharma-Natu, P. and Ghildiyal, M.C (2003). High temperature effect on starch synthase activity in relation to grain growth in wheat cultivars. *Indian J. Plant Physiol.* (Special Issue), 8: 390-398.

Prasad, P.V.V., Kenneth, J., Booke, Joseph, C.V.V.U. and Allen, L.H. (2004). The carbohydrate metabolism enzymes sucrose-P synthase and ADP-pyrophosphorylase in Phaseolus bean leaves are upregulated at elevated growth CO_2 and temperature. *Plant Sci.,* 166: 1565-1573.

Price, A.H., Atherton, N.M. and Hendry, G.A.F. (1989). Plants under drought- stress generated activated oxygen. Free Radical Res. *Commun.,* 8: 61-66.

Printer, P.J. Jr., G. Zipoli, R.J. Reginato, R.d. Jackson, S.B. Idso, and J.P. Hohman. (1990). Canopy temperature as a an indicator of differential water use and yield performance among wheat cultivars. *Agric Water Manage.* 18:35-48.

Pushpalatha, P., Sharma-Natu, P. and Ghildiyal, M.C. (2008). Photosynthetic response of wheat cultivar to long term exposure of elevated temperature, *Photosynthetica,* Accepted, (in press).

Quartacci, M.F. and Navari-Izzo, F. (1992). Water stress and free radical mediated changes in sunflower seedlings. *J. Plant Physiol.,* 139: 621-625.

Raison, J.K., Berry, J.A, Armond, R. A. and Pike C.S. (1980). Membrane properties in relation to the adaptation of plants to temperature stress. In: N.C. Turner and P.J. Kramer (eds.). Adaptation of plants to water and high temperature stress, John Wiley and Sons, New York, pp. 261-273.

Rajashekar, C.B. and Burke, M.J. (1986). Methods to study the freezing process in plants. *Methods in Enzymology,* 127: 761-771.

Ramsay, N. (1988). A Mutant in a major heat shock protein of E. coli continues to show inducible thermotolerance. *Mol. Gen. Genet.,* 211: 332-334.

Randall,P.J., Moss, H.J.(1990). Some effects of temperature regime during grain filling on wheat quality. *Australian Journal of Agricultural Research* 41: 603-617.

Randhawa, A.S., Dhillon, S.S. and Dilip, S. (1981). Productivity of wheat varieties as influenced by the time of sowing. Punjab *Agric. Univ. J.,* 18: 227-233.

Rane, J., Jag, S. and Nagarajan, S. (2000). Heat stress environments and impact on wheat productivity in India; Guestimate of losses. *Indian Wheat Newslett.,* 6: 5-6.

Rao, V.S. (2004). Temperature tolerance in wheat- an overview-29th. All Indian Wheat Workers" Meeting at Indian Agricultural Research Institute, New Delhi.

Rao, V.U.M., Singh, D. and Singh, R. (1999). Heat use efficiency of winter crops in Haryana. *J. Agrometerol.,* 1: 143-148.

Ratcliffe, R. G. (1994). *In vivo* NMR studies of higher plants and algae. *Advances in Botanical Research,* 20: 44-123.

Ravi, I. and Ghildiyal, M.C. (2001). Translocation of ^{14}C-sucrose within the ear in durum and aestivum wheat vaieties. *Journal of Agronomy and Crop Science*, 186, 9-13.

Rawson, H.M. and Bagga, A.K. (1979). Influence of temperature between floral initation and flag leaf emergence on grain number in wheat *Aust. J. Plant. Physiol.*, 6: 391-400.

Rawson, H.M. (1992). Plant responses to temperature under conditions of elevated CO_2. *Aust. J. Bot.*, 40: 473-490.

Rawson, H.M., Zajac, M. and Noppakoonwong, R. (1996). Effect of temperature, light and humidity during the phase encompassing pollen meiosis on floret fertility in wheat. In: Sterility in wheat subtropical Asia, extent, causes and solutions: Proceeding of Workshop 18-21 Sep., 1995. Lumle Agriculture Center, Pokhara, Nepal.

Rehman, Aziz, U.R., Habib, Imran, Ahmad, Nadeem, Hussain, Mumtaz, Arif Khan, M., Farooq, Jehanzeb and Amjad Ali, Muammad (2009). Screening wheat germplasm for heat tolerance at terminal growth stage. *Plant Omics Journal* 2(1): 9-19

Reynolds, M.P., Balota, M., Delgado, M.I.B., Amani, J. and Fischer, R.A. (1994). Physiological and morphological traits associated with spring wheat yield under hot, irrigated conditions. *Aust J Plant Physiol.* 21:717–730.

Reynolds, M.P., Borlaug, N.E. (2006). International collaborative wheat improvement: impacts and future prospects. *J. Agric. Sci.* (Cambridge) 144: 3-17.

Reynolds, M.P., Singh, R.P., Ibrahim, a., Ageeb, O.A.A., Larque-Saavedra, A., Quick, J.S. (1998). Evaluating the physiological traits to complement empirical selection for wheat in warm environments. Euphytica 100: 85-94.

Riabowel, K.T., Mizzen, L.A. and Welch, W.J. (1998). Heat shock is lethal to fibroblasts microinjected with antibodies against hsp 70. *Sci.*, 242: 433-436.

Rijven, A.H.G.C. (1986). Heat inactivation of starch synthase in wheat endosperm tissue. *Plant Physiol.*, 81: 448-453.

Ritchi, S.W., Nguyen, T.H. and Holaday, S.A. (1990). Leaf water content and gas-exchange parameters of two wheat genotypes differing in drought resistance. *Crop Sci.*, 30: 105-111.

Robinson, S.P. and Portis A.R. (1989). Adenosine triphosphate hydrolysis by purified Rubisco activase. *Arch. Biochem. Biophys.*, 268: 93-99.

Roderick, C.E. H. and Fox, (1987). Nutritional value of cereal grains. Pp. 1-10 *in* Nutritional Quality of Cereal Grains: Genetic and Agronomic Improvement. Agronomy Monograph No. 20. ASA-CSSA-SSSA, Madison, WI.

Rui, R.L., Nie, Y.Q. and Tong, H.Y. (1990). SOD activity as a parameter for screening stress tolerant germplasm resources in sweet potato (*Lpomoea batatas* L.). *J. Agric. Sci.*, 6: 52-56.

Ruwali, K.N. and Prasad, L. (1991). Effect of temperature stress on yield production of wheat. In: Proc. Intl. Conf. Plant Physiol., BHU, India, pp. 271-277.

Saadala, M.M., Quick, J.S. and Shanahan, J.F. (1990). Heat tolerance in winter wheat. II. Membrane thermostability and field performance. *Crop Sci.*, 30: 1248-1251.

Saini, A.D. (1988). Impact of temperature, photoperiod and light intensity on flowering, grain number and yield in wheat. In: proceedings of the International Congress of Plant Physiology. New Delhi, India. Feb. 15-20, (eds.) Sinha, S.K., Sane, P.V. Bhargava, S.C and P.K. Agrawal. Volume 1, pp. 101-113.

Saini, A.D. and Dadhwal, V.K. (1986). Influence of sowing date on grain growth and kernel size in wheat. *Indian J. Agric. Sci.*, 56: 439-447.

Saini, H.S., Aspinall,D. (1982). Abnormal sporogenesis in wheat (*Triticum aestivium* L.) induced by short period of high temperature. *Animals of Botany*, 49: 835-846.

Sairam, R.K., Rao, K.V. and Srivastava, G.C. (2001b). Differential response of wheat genotypes to long-term salinity stress in relation to oxidative stress, antioxidant activity and osmolyte concentration. *Plant Science*, 163: 1037-1046.

Sairam, R.K., Srivastava, G.C. and Sexena, D.C. (2000). Increased antioxidant activity under elevated temperature: a mechanism of heat stress tolerance in wheat genotypes. *Biologia Plantarum*, 43: 245-251.

Salvucci, M.E. and Crafts-Brandner, S.J. (2004). Inhibition of photosynthesis by heat stress: the activation state of Rubisco as a limiting factor in photosynthesis. *Plant Physiol.*, 120: 179-186.

Salvucci, M.E. and Orgren, W.L. (1996). The mechanism of Rubisco activase: insights from studies of the properties and structure of the enzyme. *Photosynth. Res.*, 47: 1-11.

Salvucci, M.E., Werneke J.M., Ogren, W.L. and Portis, A.R. (1987). Purification and species distribution of Rubisco activase. *Plant Physiol.*, 84: 930-936.

Sanchez,Y. and Lindquist, S.L. (1990). HSP 104 is required for induced thermotolerance. *Sci.*, 248: 1112-1115.

Sanchez-de-Jimenez, E., Medrano, L. and Martinez-Barajas, E. (1995). Rubisco activase, a possible new member of the molecular chaperone family. *Biochemistry*, 34: 2826-2831.

Sander, D.H., W.H. Allaway and R.A. Olson. (1987). Modification of nutritional quality by environment and production practices. pp. 43-82 *in* Nutritional Quality of Cereal Grains: Genetic and Agronomic Improvement. Agronomy Monograph No. 20. ASA-CSSA-SSSA, Madison, WI.

Sarker, S. and Torofder, M.G.S. (1992). Effect of date of sowing and seed rate on wheat (*Triticum aestivum*) under rainfed conditions. *Indian J. Agron.*, 37: 352-354.

Savin, R. and Slafer, G.A. (1991). Shading effects on the yield of an Argentinean wheat cultivars. *J. Agri. Sci.*, 116: 1-7.

Sayed, O.H., Emes, M.J., Earnshaw, M.J. and Butler, R.D. (1989). Photosynthetic responses of different varieties of wheat to high temperature. *J. Exp. Bot.*, 40: 625-631.

Sayre, K.D., Rajaram, S. and Fisher, R.A. (1996). Yield potential progress in short bread wheats in north- west Maxico. *Crop sci.*, 36: 103-105.

Schlesinger, M.J. (1990). Heat shock proteins. Varieties of wheat to high temperature. *J. Biol. Chem.*, 265: 12111-12114.

Scholander, P.L., Hammel, H.T., Bradstreet, E.D. and Hemmingsen, E.Q. (1964). Hydrostatic pressure and osmotic potential in leaves of mangroves and some other plants. *Proceedings of National Academy of Science*, USA. 52: 119-125.

Schuster, G.D., Even, Kloppsteck, K. and Ohard, I. (1988). Evidence for protection by heat shock proteins against photoinhibition during heat shock. *EMBO J.*, 7: 1-6.

Setter, T.L., Anderson, W., Asseng, S. and Barclay (1997). Review of the impact of high stem carbohydrate concentrations on maintenance of high yields in wheat exposed to environmental stress during grain filling, Physiological approaches for increasing wheat productivity under stress environment. International group meeting on "Wheat Research Needs Beyond 2000 AD", August 12-14, 1997 (Abstract) DWR, ICAR, p. 29.

Shah, N.H. (1992). Responses of wheat to combined high temperature and drought or osmotic stresses during maturation. Dissertation,Abstract, International, Sci. Eng., 52: 3984.

Shah, S.A., Harrison, S.A., Boquet, D.J., Colyer, P.D. and Moore, S.H. (1994). Management effects on yield and yield components of late planted wheat. *Crop Sci.*, 34: 1298-1303.

Sharkova, V.E. (1994). Effects of high temperatures on photosynthesis, Hill reaction and the activity of some chloroplast enzymes in wheat plants. *Russion J. Plant Physiol.*, 41: 641-654.

Sharkova, V.E. and Bubolo, L.S. (1996). Effect of heat stress on the arrangement of thylakoid membranes in the chlorplasts of mature wheat leaves. *Russian J. Plant Physiol.*, 43: 358-365.

Sharma, B.D., Sandha, G.S., Gill, K.S. and Dhingra, G.S.C. (1991). Physiological and morphological determinants of grain yield at different stages of plant growth in triticale. Proceedings of the second international triticale symposium. Mexico, D.F. CIMMYT 1991, pp. 98-104. Mexico.

Sharma, K. Dhingra, K.K. and Dhillon, M.S. (1994). Physiological indices for high yield potential indifferent genotypes of wheat. *Environ. Ecology*, 12: 717-719.

Sharma, R.C. (1995). Tiller mortality and its relationship to grain yield in spring wheat. *Field Crop Res.*, 41: 55-60.

Sharma-Natu, P. and Ghildiyal, M.C. (2005). Potential targets for improving photosynthesis and crop yield. *Current Science*, 88: 1918-1928.

Shimshi, D., and J. Ephrat. (1975). Stomatal behaviour of wheat cultivars in relation to their transpiration, photosynthesis and yield. *Agron. J.* 67:326-331.

Shpiler, L. and Bulm, A. (1986). Differential reactions of wheat cultivars to hot environment. *Euphytica*, 35: 483-492.

Shpiler, L. and Bulm, A. (1991). Heat tolerance for yield and its components in different wheat cultivars. *Euphytica*, 51: 257-263.

Shrawat, A., Carroll, R., Taylor, G.J. and Good, A.G. (2007). Genetic engineering of rice to improve nitrogen-use efficiency. Abstr., In Nitrogen 2007- An International Symposium on Nitrogen Nutrition of Plants http:/boil.lancs.ac.uk/nif2007/book of abstracts.pdf

Shreiber, U. and Armond, P.A. (1978). Heat-induced changes of chlorophyll fluorescence in isolated chloroplasts and related heat damage at the pigment level. *Biochem. Biophys. Acta,* 502: 138-151.

Shukla, D.S., Gupta, N.K., Kapashi, S.B., Deshmukh, P.S., Sairam, R.K. and Pande, P.C. (1997). Relationship between stem reserve and grain development in terminal heat stress susceptible and tolerant wheat genotypes. *Indian Journal of Plant Physiology*, 2: 36-40.

Simane, B., Peacock, J.M. and Struik, P.C. (1993). Differences in developmental plasticity and growth rate among drought resistant and susceptible cultivars of durum wheat. *Plant* Soil, 157: 155-166.

Sinclair, T.R. and Ladlow, M.M. (1985). Who taught plants thermodynamics? The unfilled potential of plant water potential. *Australian J. Plant Physiol.* 12: 213-217.

Singh, A.K. and Singhal, G.S. (2001). Effect of irradiance on the thermal stability of thylakoid membrane isolated from acclimated wheat leaves. *Photosynthetica*, 39: 23-27.

Singh, D.P., Rawson, H.M. and Turner, N.C. (1982). Effect of radiation temperature and humidity on photosynthesis, transpiration and water use efficiency of chickpea (*Cicer arietinum* L.). *Indian J. Plant Physiol.* 25: 32-39.

Singh, G.S. and Srivastava, R.D.L. (1988). Morpho-physiological traits associated with drought tolerance in wheat. International Congress of Plant Physiology. Feb. 15-20, 1988. New Delhi

Singh, K. and Purohit, S.S. (1995). Plant productivity under environmental stress. Agro Botanical Publisher, India Bikaner 1st Eds.

Singh, M., Srivastava, J.P. and Kumar, A. (1991). Cell membrane stability in relation to drought tolerance in wheat genotypes. *J. Agron. Crop Sci.,* 168: 186-190.

Singh, R.K. and Rajat D. (1978). Effects of rates and methods of nitrogen application under different soil moisture regimes and the yield and quality of wheat. *Indian J. Agric. Sci.,* 48: 229-233.

Singh, T.P., Deshmukh, P.S. and Kushwaha, S.R. (2004). Physiological studies on temperature tolerance in chickpea (*Cicer arietinum* L.) genotypes. *Indian J. Plant Physiol.,* 9: 294-301.

Singh, V., Singh, R.P. and Panwar, K.S. (1995). Response of wheat (*Triticum aestivum* L.) to seed rate and date of sowing. *Indian J. Agron.*, 40: 697-699.

Singh, V.P. and Singh, M. (1985). Nitrate reductase activity and its relationship with grain protein and yield of wheat. *J. Plant Physiol.*, 28: 235-242.

Sinha, S.K. and Khanna, R. (1975). Physiological, biochemical and genetic basis of heterosis. *Adv. Agron.*, 27: 123-174.

Slovacek, R.E. and Hind, G. (1981). Correlation between the photosynthesis and the transthylakoid proton gradient. *Biochem Acta* 35: 393-404.

Smika, D.E., Shawcroft, R.W. (1980). Preliminary study using a wind tunnel to determine the effect of hot wind on a wheat crop. *Field Crops Research* 3: 129-135.

Smirnoff, N. (1993). The role of active oxygen in the responses of plants to water deficit and desiccation. *New Phytol.*, 125: 27-58.

Smirnoff, N. and Colombe, S.V. (1988). Drought influences the activity of enzymes of the chloroplast hydrogen peroxide scavenging system. *J. Exp. Bot.*, 39: 1097-1108.

Smith, I.K., Vierheller, T.L. and Thorne, C.A. (1988). Assay of glutathione reducatse in crude tissue homogenates using 5,5'-dithiobis (2-nitobenzioic acid). *Anal. Biochem*, 175: 408-413.

Smith, R.C.G., H.D. Barrs, and J.L. Steiner. (1986). Alternative models for predicting the foliage-air temperature difference of well irrigated wheat under variable meteorological conditions. *Irrig. Sci.* 7: 225-236.

Smith, W.K. (1978). Temperatures of desert plants: Another perspective on the adaptability of leaf size. *Science.*, 201: 614-616.

Sofield, I., Evans, L.T., Cook, M.G. and Wadlaw, I.F. (1977). Factors influencing the rate and duration of grain filling in wheat. *Aust. J. Plant Physiol.*, 4: 785-797.

Somerville, C. and Browse, J. (1991). Plant lipids, metabolism and membranes. *Sci.*, 252: 80-87.

Spiertz, J.H.J. (1974). Grain growth and distribution of dry metter in the wheat plant as influence by temperature, light energy and seed size. *Neth. J. Agri Sci.*, 22: 207-220

Spomer, L.A, (1985). Techniques for measuring plant water. *Hort. Sci.*, 20: 1021-1028.

Srivalli, B., Vishanathan, C. and Renu, K.C. (2003). Antioxidant defense in response to abiotic stresses in plants. *J. Plant Biol.*, 30: 121-139.

Stone, P.J. and Nicolas, M.E. (1994). Wheat cultivars vary widely their responses of grain yield and quality to shorts periods of post anthesis heat stress. *Aust. J. Plant Physiol.*, 21: 887-900.

Stone, P.J. and Nicolas, M.E. (1995). Effect of heat stress during grain filling on two wheat varieties differing in heat tolerance grain growth. *Aust. J. Plant Physiol.*, 22: 927-934

Stone,P.J., Nicolas, M.E. (1994). Wheat cultivars vary widely in their responses of grain yield and quality to short periods of post-anthesis heat stress. *Australian Journal of Plant Physiolog,* 21: 887-900.

Sullivan, C.Y. and Ross, W.M. (1979). Selecting for drought and heat resistance in grain sorghum. In: Stress Physiology in Crop Plants (Eds. H. Mussell and R.C. Staples), Wiley Inter. science, New York, pp. 262-281.

Sumesh, K.V., Sharma-Natu, P. and Ghildiyal, M.C. (2008). Starch synthase activity and heat shock protein in relation to thermal tolerance of developing wheat grains. Biologia Plantarum, 52, Accepted, in press.

Susek, L. and Lindquist, S.L. (1989). HSP 26 of Saccharomyces cerevisiae is related to the super family of small heat shock proteins but is without a demonstrable function. *Mol. Cell. Biol.,* 9: 5265-5271.

Tahir, I.S.A. and Nakata, N. (2005). Remobilization of nitrogen and carbohydrate from stems of bread wheat in response to heat stress during grain filling. *J. Agron. Crop Sci.,* 191: 106.

Takahashi, C.Y. and Nakaseko, K. (1992). Varietal differences in yield response to delayed sowing of spring wheat in Hokkaido. *Japanese J. Crop Sci.,* 61: 22-27.

Talwar, H.S., Chandra, S. and Nageshwara, R.C. (2002). Genotypic variability in membrane thermostability in groundnut. *India. J. Plant Physiol.,* 7: 97-102.

Tashiro, T. and Wardlaw I.F. (1991). The effect of high temperature on the accumulation of dry matter, carbon and nitrogen in the kernel of rice. *Australian J. Plant Physiol.,* 18: 259.

Teranishi, Y. Tanaka, A., Osumi, M. and Fukui, S. (1974). Catalase activity of hydrocarbon utilizing candida yeast. *Agr. Biol. Chem.,* 38: 1213-1216.

Thebud, R. and Santarius, K.A. (1982). Effects of high temperature stress on various bio-membranes of leaf cells *in situ* and *in vitro. Plant Physiol.* 70: 200-205.

Turner, M.C. and Jones, M.M. (1980). Turgor maintenance by osmotic adjustment: a review and evaluation. In: Turner NC, Kramer PJ (Eds). Adaptation of Plants to Water and High Temperature Stress. Johan Wiley, New York, pp. 155-175.

Upadhyaya, A., Davis, T.D. and Sankhla, N. (1990). Epibrassinolide does not enhance heat shock tolerance and antioxidant activity in bean. *Hort. Sci.,* 26: 1065-1067.

Uprety, D.C. and Mahalaxmi, V. (2006). Effect of elevated CO_2 and nitrogen on some physiological characters in *Brassica juncea. Journal of Agronomy and Crop Science,* 184, 271-276.

Varshney, R.K., Balyan, H.S. and Langridge, P. (2006). Wheat. In: Kole C, ed. Genome mapping and molecular breeding in plants, Vol. 1. Cereals and millet. Berlin: Springer, 79–134.

Vierling, E. (1991). The roles of heat shock proteins in plants. *Plant Mol. Biol.,* 42: 579-620.

Vijayalakshmi, K., Fritz, A.K., Paulsen, G.M., Bai, G., Pandravada, S.and Gill, B.S. (2010). Modeling and mapping QTL for senescence-related traits in winter wheat under high temperature. *Mol. Breed* 26:163–175.

Vile, D., E. Garnier, B. Shipley, G. Laurent, M. Navas, C. Roumet, S.Lavorel, S. Diaz, J.G. Hodgson (2005). Specific leaf area and dry matter content estimate thickness in laminar leaves *Ann. Bot.* (Lond.) 96:1129-1136.

Vishwanathan, C. and Khanna- Chopra, R. (2001). Effect of heat stress on grain growth, starch and protein synthesis in grains of wheat (*Triticum aestivum* L.) varieties differing in grain weight stability. *Journal of Agronomy and Crop Science*, 186, 1-7.

Vishwanathan, C. and Khanna-Chopra, R. (1996). Heat shock proteins – Role in thermotolerance of crop plant. *Current Science*, 71, 275-283.

Waines, J.G. (1994). High temperature stress in wild and spring wheats. *Aust. J. Plant Physiol.* 21: 705-715.

Wallwork, M.A.B., Logue S.J., Macleod, I.C. and Jenner, C.F. (1998). Effect of high temperature during grain filling on starch synthesis in the developing barley grain. *Aust. J. Plant Physiol.*, 25: 173-181.

Wardlaw, I.F. and Moncur, L. (1995). The response of wheat to high temperature following anthesis. I. The rate and duration of kernel filling *Aust. J. Plant Physiol.*, 22: 391-397.

Wardlaw, I.F. and Wringley, C.W. (1994). Heat tolerance in temperate cereals: An overview. *Aust J. Plant Physiol.*, 21: 695-703.

Wardlaw, I.F., Blumenthal,C., Larroque, O. and Wrigley,C.W. (2002) Contrasting effects of chronic heat stress and heat shock on grain weight and flour quality in wheat. *Functional Plant Biol.* 29: 25-34.

Wardlaw, I.F., Dawson, I.A. and Munibi, P. (1989). The tolerance of wheat to high temperatures during reproductive growth. II grain development. *Aust. J. Agric. Res.*, 40: 15-24.

Wardlaw, I.F., Solfield, I., Cartwright, P.M. (1980). Factors limiting the rate of dry matter accumulation in the grain of wheat grown at high temperature. *Australian Journal of Plant Physiology* 7: 387-400.

Wardlaw, J.F. Sofield, I., Cartwright, P.M. (1980). Factors limiting the rate of dry matter in the grain wheat grown at high temperature. *Aust. J. Platn Physiol.* 7, 121-140.

Warrington, I.F., Dunstone, R.L. and Green, L.M. (1977). Temperature effects at the developmental stages on the yield of wheat ear. *Aust. J. Agric. Res.*, 28: 11-27.

Watson, D.J. (1952). The physiological basis of variation in yield. *Adv. Agron.*, 4: 101-145.

Weis, E. (1981a). Reversible heat-inactivation of the Calvin cycle; a possible mechanism of the temperature regulation of photosynthesis. *Planta*, 151: 33-39.

Weis, E. (1981b). The temperature-sensitivity of dark-inactivation and light-activation of the ribulsoe-1,5-bisphosphate carboxylase in spinach chloroplasts. *FEBS Lett.*, 129: 197-200.

Weis, E. and Berry, J.A. (1988). Plants and high temperature stress. In: S.P. Long, F.I. woodward (eds.), Plants and Temperature. *Society of Experimental Botany,* Cambridge, UK, pp. 327-346.

Wheeler, T.R., Hong,T.D., Ellis, R.H., Batts, G.R., Morison, J.I.L., Hadley, P. (1996b). The duration and rate of grain growth and harvest index of wheat (*Triticum aestivium*) in response to temperature and CO_2. *Journal of Experimental Botany,* 47: 623-630.

White, J.W., Tanner, D.G., Corbett, J.D. (2001). An Agro-Climatological Characterization of Bread Wheat Producton Areas in Ethiopia. NRGGIS series 01-01. CIMMYT, Mexico, D.F.

Wilhelm, E.P., Mullen, R.E., Keeling, P.L. and Singletary, G.W. (1999). Heat stress during grain filling in maize : effects on kernel growth and metabolism. *Crop Sci.,* 39: 1733-1741.

Willekens, H., Inze, D., Van Montagu, M. and Van Camp, W.(1995). Catalase in plants. Mol. Breed, 1: 207-228.Willey RW, Dent JD, 1969. The supply and storage of carbohydrate in wheat and barley. *Agricultural Progress* 44: 43-55.

Williams, J., Khan, M., Mitchell, K. and Johnson, G. (1988). The effect of temperature on the level and biosynthesis of unsaturated fatty acids in daicylglycerols of Brassica napur leaves. *Plant Physiol.,* 87: 904-910.

Wise, R.R. (1995). Chilling enhanced photo-oxidation: the production, action and study of reactive oxygen species produced during chilling in the light. *Photsynth. Res.,* 45: 79-97.

Xu, Q.A. (1991). Biochemical and ultra structural changes in wheat (*Triticum aestivm*) plant exposed to high temperature and phytohormones. Dissertation-Abstracts-International B. Science and Engineering. U.S.A., p. 171.

Yadav, U.L. (1986). A rapid and nondestructive method to determine chlorophyll in intact leaves. *Hort. Science,* 21: 169-183.

Yamamori, J. and Yura, T. (1982). Genetic control of heat shock protein synthesis and its bearing on growth and thermal resistance in E. coli K-12. *Proc. Natl. Acad. Sci. USA,* 79: 860-865.

Yang, J., Sears, R.G., Gill, B.S. and Paulsen, G.M. (2002). Quantitative and molecular characterization of heat tolerance in hexaploid wheat. *Euphytica,*126: 275–282.

Zahedi, M., McDonald, G. and Jenner, C.F. (2004). Nitrogen supply to the grain modifies the effects of temperature on starch and protein accumulation during grain-filling in wheat *Aust. J. Agric. Res.,* 55: 551-564.

Zahedi, M., Sharma, R. and Jenner, C.F. (2003). Effects of high temperature on grain growth on the metabolites and enzymes in the starch synthesis pathway in the grains of two wheat cultivars differing in their responses to temperature. *Funct. Plant Biol.*, 30: 291-300.

Zhang, J.X. and Kirkham, M.B. (1994). Drought-stress-induced changes in activities of superoxide dismutase, catalase, and peroxidase in wheat species. *Plant Cell Physiol.*, 35: 785-791.

Zhau, R.G., Fan, Z.H., Li, X.Z., Wang, Z.W. and Han, W. (1995). The effect of heat acclimation on membrane thermo-stability and reactive enzyme activity. *Act. Agron. Sin.*, 21:568-572.

Zhong-hu, He and Rajaram, S. (1994). Differential responses of bread wheat characters to high temperature. *Euphytica*, 72:197-203.

Zweifel, R., K. Steppe, and F.J. Sterck. (2007). Stomatal regulation by microclimate and tree water relations; Interpreting ecophysiological field data with a hydraulic plant model. *J. Exp. Bot.* 58: 2113-2131.

Chapter 5

Nitric Oxide in Plants: Sources, Methods of Detection and Role as Signaling Agent in Plant Responses to Biotic Stresses

Padmanabh Dwivedi* and Bansh Narayan Singh

Department of Plant Physiology, Institute of Agricultural Sciences, Banaras Hindu University, Varanasi – 221 005
*E-mail: pdwivedi25@rediffmail.com

CONTENTS

1. Introduction

Nitrogen monoxide or nitric oxide (NO) was discovered by Joseph Priestley in 1772. Initially it was considered toxic, however, over the past 30 years, the free radical NO has been of interest to many scientists worldwide. In 1980, Furchgott and Zawadzki reported that endothelial cells released a substance responsible for relaxation of vascular smooth muscle, and they named it as 'endothelium-derived relaxing factor', EDRF. Later on in 1987, Palmer and co-workers suggested EDRF as NO, produced by oxidation of L-arginine. The NO is an important biologically active molecule with widespread use in animal and plant signaling machinery. It is an intracellular and intercellular signaling molecule with various regulatory functions in the central nervous, cardiovascular and immune systems, as well as in platelet inhibition, programmed cell death and host responses to infection, etc. NO mediates various plant physiological and patho-physiological processes. For example, NO has a regulatory role in the inhibition of catalase, ascorbate peroxidase and aconitase activities (Navarre *et al.*, 2000, Clarke *et al.*, 2000), in cell wall lignifications (Ferrer and Ros-Barceló, 1999), the regulation of ion channels of guard cells (Garcia *et al.*, 2003), the mitochondrial and chloroplastic functionality (Yamasaki *et al.*, 2001, Zottinni *et al.*, 2002, Takahashi and Yamasaki, 2002), cell death (Pedroso *et al.*, 2000), senescence (Hung and Kao, 2003), accumulation of ferritin (Murgia *et al.*, 2002), wound signaling (Orozco-Cardenas and Ryan, 2002), cytokinin induced programmed cell death (Neill *et al.*, 2003) and ABA induced stomatal closure (Neill *et al.*, 2002). NO mediates maturation and senescence and works antagonistic to ethylene (Lamattina *et al.*, 2003).

NO has a low reactivity, reacting predominantly with molecules that have orbitals with unpaired electrons. It has a relatively short half-life in biological systems, in the order of 5-15 seconds (Lancaster, 1997), though *in vivo*, half-life may be much longer. The NO molecule gives rise to a wide array of NO derived compounds, collectively referred as reactive nitrogen species (RNS). Because of its free radical nature, it can adopt an energetically more favorable electron structure by gaining or losing an electron, such that the three interchangeable species includes nitric oxide free radical ($NO^\bullet$), nitrosonium cation (NO^+) and nitroxyl anion (NO^-). The reaction of $NO^\bullet$ with

O_2 results in the generation of NOx compounds (including $NO_2^{\bullet}$, N_2O_3, and N_2O_4), which can either react with cellular amines and thiols or simply hydrolyze to form the end metabolites nitrite (NO_2^-) and nitrate (NO_3^-) (Wendehenne et al., 2001). NO⁻ reacts with di-oxygen to form NO_2^- or with reactive oxygen species (ROS) to form a highly oxidizing agent peroxynitrite (ONOO⁻) that mediates cellular injury. NO⁺ mediates electrophilic attack on reactive sulfur, oxygen, nitrogen and aromatic carbon centers, with thiols being the most reactive groups. This chemical process is referred to as nitrosation. Nitrosation of many enzymes or proteins results in chemical modification and can affect the activity of these entities. Such modifications are reversible and protein nitrosation–denitrosation could represent an important mechanism for regulating signal transduction (Hayat et al., 2010).

NO emission from plants was first reported in soybean by Klepper in 1975, much earlier than in animals (Klepper, 1979). The widespread biological significance of nitric oxide in plants was recognized in the nineties (Gouvea et al., 1997, Lesham et al., 1998). Two milestone publications in 1998 demonstrated a regulatory role for NO as a plant defense signal against pathogen infection which induced a burst in NO research (Delledonne et al., 1998, Durner et al., 1998).

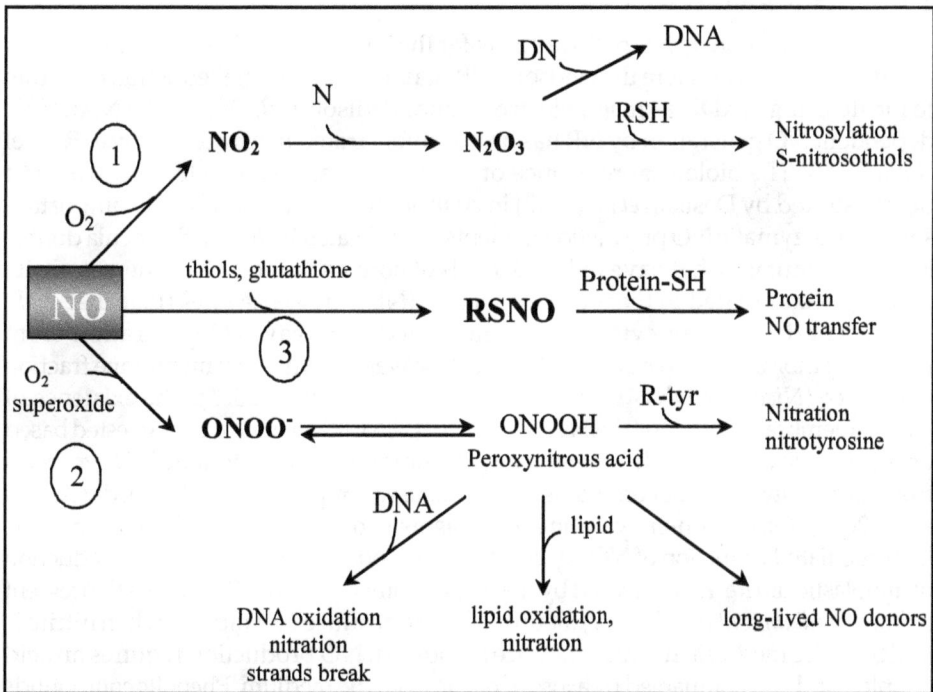

Figure 5.1: Reactions of the Free Radical NO and Reactive Nitrogen Oxide Species.
1: Reactions of NO with O_2 producing N_2O_3 lead indirectly to DNA deamination. N_2O_3 reacts with thiols producing S-nitrosothiols (RSNOs), 2: NO reacts with O_2^- to form ONOO⁻, 3: The bioactive formation of S-nitrosothiols for the storage, transfer and production of NO (*Source*: Durzan, 2002).

This review briefly discusses the role of nitric oxide (NO) in plant hypersensitive response (HR) and various methods available for NO detection. It also emphasizes on involvement of NO in plant-pathogen interaction vis-à-vis its cross talk with reactive oxygen species in triggering HR, leading to various plant defense-related events.

2. NO Sources

The cellular/sub-cellular localization and sources of endogenous nitric oxide synthesis in plant cells are diverse and debated. NO can be produced in plants by non-enzymatic and enzymatic systems depending on plant species, organ or tissue as well as on physiological state of the plant and changing environmental conditions. The best documented NO sources in plant are NO production from nitrite as a substrate by cytosolic (cNR) and membrane bound (PM-NR) nitrate reductases (NR), and NO production by several arginine-dependent nitric oxide synthase-like activities (NOS). The recent studies indicate that mitochondria are an important source of arginine- and nitrite-dependent NO production in plants (Crawford and Guo, 2005, Gupta and Kaiser, 2010). The other potential sources of NO production in plants includes xanthine oxidoreductase, peroxidase, cytochrome P450.

2.1. Nitrate Reductase (NR)

NO production by NR pathway is by far the best characterized pathway for NO production in plants. Here the cytosolic NR mainly catalyzes the reduction of nitrate to nitrite using NADH as major electron donor (Wilson *et al.*, 2008). The NAD(P)H dependent NO generation by NR has been demonstrated *in vitro* and *in vivo* (Rockel *et al.*, 2002). The biological relevance of NR activity as a source for NO was first demonstrated by Desikan *et al.* (2002) in *Arabidopsis* guard cells. NR is an important source of enzymatic NO production in plants, as indicated by NR-deficient nia double knock-out mutants that have reduced levels of both nitrite and NO, whereas nitrite and NO accumulated in the nitrite reductase (NiR) antisense lines (Planchet *et al.*, 2005). Apart from the cytosolic NR mediated pathway NO generation from plasmamembrane bound nitrite:NO reduactase was shown in cell membrane fraction of tobacco (*Nicotiana tabacum*) roots, not leaves (Stohr *et al.*, 2001). The existence of plasmamembrane bound nitrite:NO reduactase (NiNOR) enzyme was suggested based on native size exclusion chromatography technique. Non-enzymatic NO synthesis from nitrites under acidic conditions by ascorbate was reported in barley seeds (Beligni *et al.*, 2002). Other non-enzymatic mechanisms proposed for NO formation is the light-mediated reduction of NO_2 by carotenoids (Cooney *et al.*, 1994) and the reduction of apoplastic nitrite. As measured by mass spectrometry or an NO-reactive fluorescent probe, *Hordeum vulgare* (barley) aleurone layers produce NO rapidly when nitrite is added to the medium in which they are incubated. NO production requires an acid apoplast and is accompanied by a loss of nitrite from the medium. Phenolic compounds in the medium can increase the rate of NO production (Bethke *et al.*, 2004).

2.2. Xanthine Oxidoreductase

Nitrite reduction to NO can also be catalyzed by the peroxisomal enzyme xanthine oxidoreductase (XOR). In pea (*Pisum sativum*) leaves, XOR activity is associated with peroxisomes, and as such, the possibility of an interaction between the production of reactive oxygen and reactive nitrogen species (ROS and RNS, respectively) has been suggested (del Rio *et al.*, 2004).

2.3. Mitochondrial NO

In plants, the first evidence for mitochondrial NO generation was obtained by Tischner *et al.* (2004) who measured NO production under anoxic conditions from the unicellular green alga *Chlorella sorokiniana*. This green alga does not produce NO in the presence of nitrate (NO_3^-), but actively generates NO in the presence of nitrite (NO_2^-). Inhibitors of mitochondrial electron transport also prevented NO production. Shortly thereafter, mitochondrial NO production was also revealed in higher plants. Mitochondrial NO generation was reported in barley plants under anoxic conditions (Gupta and Kaiser, 2010). A tobacco nia 1,2 (nitrate reductase deficient) cell suspension was able to produce NO from exogenous nitrite under anoxic conditions, although nitrate reductase (which can also produce NO from nitrite) was completely absent (Gupta *et al.*, 2011).

2.4. NOS like Activity

NO generation via oxidation of argignine into citrulline catalyzed by the enzyme NO synthase was shown in animals (Palmer *et al.*, 1987). After the discovery of a role for NO in plants in 1998 (Delledonne *et al.*, 1998, Durner *et al.*, 1998), many researchers began to search for NOS activity in plants even though the *Arabidopsis* genome did not reveal any gene with significant homology to that encoding animal NOS (Moreau *et al.*, 2010). Most studies revealing evidence for the existence of arginine-dependent NOS-like biochemical pathway in plants are based on correlation studies between NO production and the supply of L-Arginine or its analogs. L-arginine analogs that are widely used as NOS inhibitors in animals, such as PBITU (S,S0-1,3-Phenylene-bis(1,2-ethanediyl)-bisisothiourea), AET (2-(2 aminoethyl) isothiourea), L-NAME (NG-Nitro-L-arginine methyl ester), L-NMMA (NG-Monomethyl-L-arginine), L-NIL (N6-(1-Iminoethyl)-L-lysine), have also been tested in plants and changes in NO production following their application is interpreted to be indicative of the involvement of NOS-like enzymes in specific physiological or developmental processes (Corpas *et al.*, 2009). However, the concentrations at which these analogs are supplied are sometimes high. For instance, in some studies, up to 300 mM of L-NAME (Lum *et al.*, 2002) was applied. Interestingly such concentrations are several orders of magnitude higher than those regularly used in animal research, which normally uses concentrations in the micromolar range (Legrand *et al.*, 2009). Although these high concentrations might be required because of the low uptake rate of the arginine analogs by plant cells, it also indicates that special care is required when interpreting these studies as evidence for the existence of NOS like activity in plants.

3. NO Detection

Next section of this review briefly describes some of the widely used methods to detect NO and consider their advantages and disadvantages.

3.1. Measurement of NO in Gas Phase

3.1.1. Chemiluminescence

By far the most well-established approach to measure gaseous NO is the chemiluminescent assay which is based on its reaction with O_3 to yield light photons. The concentration of NO can be determined using a simple chemiluminescent reaction involving ozone to produce oxygen and nitrogen dioxide:

$$NO + O_3 \rightarrow NO_2 + O_2; \qquad NO_2 \rightarrow NO_2 + h\nu$$

This is a two stage reaction whereby the reaction of NO with O_3 produces excited-state nitrogen dioxide ($NO_2^{\bullet}$), which emits a photon upon relaxation to the ground state; the emitted light, at >600 nm wavelength, is measured with a photodetector with its intensity proportional to the amount of NO. The results are highly specific for NO. Role of NO during cryptogein induced HR was demonstrated by measuring NO emission from detached tobacco leaves and cell suspension by chemiluminescence methods (Planchet and Kaiser, 2006). NO production was measured with chemiluminescence and DAF-fluorescence. Results from both methods were partially consistent, but kinetics was different (Planchet *et al.*, 2006). The chemiluminescence approach exhibits excellent sensitivity with limits of detection as low as 20–50 pmol and needs only minimal equipment which has contributed to its commercialization as robust platform for NO measurement (Byrnes *et al.*, 1996). However, its major limitation is that it measures only the emitted NO in gas phase (Planchet and Kaiser, 2006). Additionally, it measures only the pure NO emitted from biological samples; while only a small portion of produced NO is emitted from biological samples, the major portion is quenched in the reaction with superoxide (Vanin *et al.*, 2004).

3.1.2. Laser-Based Photo-Acoustic Detection

It uses absorption of rapidly chopped infrared light by NO (Mur *et al.*, 2011). Here, sound is generated during absorption and relaxation which is then detected by the microphone located in photo-acoustic cell. Mur *et al.* (2005) used this method to detect NO from tobacco leaves infected with *Pseudomonas*. Though the method is very precise, it can detect NO only in gas phase and also it does not provide information about NO production in specific cells (Gupta and Igamberdiev, 2013).

3.1.3. Quantum Cascade Laser-Based Spectroscopic Detection

Here the laser is integrated with a thermo-electrically cooled infrared detector (Moeskops *et al.*, 2006, Gupta *et al.*, 2013). The detection sensitivity with this method is 0.03 ppb which is higher by 1-2 folds than in the photo-acoustic detection, as described above.

3.1.4. Membrane Inlet Mass Spectrometry-Based Detection

In this technique the dissolved gases are allowed to diffuse through a capillary and are then identified with a bench top mass spectrometer according to their different mass/charge ratios (m/z) (Conrath *et al.*, 2004). Changes in NO levels are evaluated by changes in the abundance of mass 30, corresponding to NO (m/z = 30). This method, referred to as membrane inlet mass spectrometry (MIMS), was developed in animals using mammalian cell cultures, but has been subsequently adapted to study NO production in plant cell suspensions. This mass spectrometric approach allowed the on-line detection of NO from either tissue cultures or whole plants. In membrane inlet mass spectrometry (MIMS), a membrane separates the mass spectrometer from the tissue culture, but allows the diffusion of small molecular weight gases, such as NO. MIMS was used to detect NO production from tissue cultures of either tobacco or soybean inoculated with HR-eliciting or disease forming strains of *Pseudomonas syringae*. In a restriction capillary inlet MS (RIMS) configuration, NO was sampled in the gaseous phase from cuvettes sprayed with 20 mM $NaNO_3$. This method allows direct, fast and specific NO detection in plant cell suspensions and leaves or entire plant. A particular attractiveness of RIMS/MIMS is that it is able to distinguish between different N isotopes, so that on supplementation of (for example) cultures with likely substrates for NO generating enzymes (for example, N^{15}-labelled nitrate/ nitrites/polyamines/hydroxyl amines), their contribution (if any) to NO production can be estimated.

3.2. Measurement of NO in Liquid Phase

3.2.1. The Oxyhaemoglobin Assay

This is a spectroscopic method where oxyhaemoglobin (HbO_2) reacts with NO to produce methaemoglobin (MetHb) and nitrate (NO_3). The NO production is detected by measuring the shift in absorbance from 415–421 nm (HbO_2) to 401 nm (MetHb). This is a robust and sensitive assay with a predicted detection limit of 1.3–2.8 nM (Murphy and Noack, 1994). However, this method faces a major constraint as it requires production of fresh H_2O_2 which is technically demanding, since it needs hemoglobin oxygenation followed by isolation using chromatography. Another consideration is the false readings that arise from the oxidation of HbO_2 by reactive oxygen species. Also the assay can be affected by plant defense responses manifested as changes in the cytosolic pH and the presence of heme containing proteins.

3.2.2. The Griess Reaction

Given the fact that NO is highly reactive and its small portion is emitted from biological samples while the rest is oxidized, it is not possible to measure oxidized forms of NO like nitrate, nitrite by using gas phase chemiluminescence. However, this problem is negated by using indirect chemiluminescence in which case nitrate and nitrite produced from oxidation of NO are reduced back to NO by injecting sample extracts into boiling acidic vanadium chloride, as suggested by Gupta and Kasier (2010). In case it is not possible to analyze oxidized forms of NO using indirect chemiluminescence, Griess eragent assay can be used instead. The Griess reaction was first reported by Johann Peter Griess in 1879 as a method of analysis of nitrite (NO_2^-). He suggested that nitrite reacts under acidic conditions with sulfanilic acid ($HO_3SC_6H_4NH_2$) to form a diazonium cation ($HO_3SC_6H_4-N\equiv N^+$) which subsequently couples to the aromatic amine 1-naphthylamine ($C_{10}H_7NH_2$) to produce a red–violet coloured ($\lambda max \approx 540$ nm), water-soluble azo dye ($HO_3SC_6H_4-N= N-C_{10}H_6NH_2$) (Ivanov, 2004). This remains the basic reaction except that today sulphanilamide and N-(1-naphthyl) ethylenediamine (NED) are used to react with NO_2^-. The resulting water-soluble azodye is quantified by measuring spectrometric absorption at 540 nm (Ivanov, 2004). Since its discovery the Griess reaction has been used for various biological fluids notably blood and urine to detect derivatives of NO. The Griess assay has been used to measure nitrite content in a number of plants like cucumber, tomato and wine (Tsikas, 2006, Ghafourifar *et al.*, 2008), apart from tobacco and *Arabidopsis*. Importance of Griess assay in plants have been well presented by Vitecek *et al.*, by measuring nitrite content in tobacco cultures inoculated with the cell death elicitor cryptogein (Vitecek *et al.*, 2008).

3.2.3. Diaminofluoresce in (DAF) Fluorescent Dyes

The DAF dyes have been widely used in plants to reveal likely sites of NO generation (Foissner *et al.*, 2000). NO can be visualized via fluorescence microscopy or quantified by spectrofluorometric measurements (Planchet *et al.*, 2006). DAF dyes react with N_2O_3, a by-product of NO oxidation, with a resulting dramatic increase in fluorescence. Diaminofluoresceins (DAF) and their cell permeating diacetate derivatives have been used to measure, visualize and localize NO production inside and outside plant and animal cells and tissues. The fluorescence increase is attributed to the formation of the highly fluorescing DAF-2 triazol (DAF-2T), formed by nitrosation through an NO oxidation product, probably dinitrogentrioxide, N_2O_3 (Jourd'Heuil, 2002). This fluorescent dye was initially commercialized in a diacetate-form, 4, 5-diaminofluorescein diaceate (DAF-2-DA), which allowed ready uptake by living cells. This dye is used for real time bioimaging of NO in plant tissues with fine temporal and spatial resolution (Gould *et al.*, 2003). DAF-2-DA has a detection limit of less than ~5nM (Nakatsubo *et al.*, 1998). Recently, the use of 3-amino-4-(N-methylamino)-2',7'-difluorofluorescein diacetate (DAF-FM-DA) as fluorescent probe has become a very common and sensitive technique in plant systems than DAF-2-DA with detection limit of ~3 nM (Murad, 1999) which fluoresces not only at neutral pH but also at acidic pH (Zhang *et al.*, 2003).

To elucidate the role of NO in plant defense, DAF dyes are being used in combination with NO scavenger, cPTIO and its derivatives to demonstrate the involvement of NO in plant pathogen interaction. The cryptogein induced hypersensitive response was efficiently prevented by the NO scavengers PTIO or cPTIO and it was interpreted as an indication for a requirement of NO. However, cPTI also prevented or weakened cell death and PR-1 expression without scavenging NO (Planchet *et al.*, 2006). Therefore, these results cast serious doubts on specificity of cPTIO as an NO scavenger and its use in implicating NO in plant defense responses. Use of fluorescent dyes has been challenged in various studies. DAF2 reacts with dehydroascorbic acid and ascorbic acid to produce fluorescent products within similar range of fluorescence as DAF-2T (Zhang *et al.*, 2002).

3.2.4. Electron Spin Resonance

Electron spin resonance (ESR), also known as electron paramagnetic resonance (EPR), is based on observing unpaired electrons in magnetic fields which in the microwave region exhibit a 'resonance' between parallel and antiparallel electron spin orientations. ESR instruments will scan the magnetic field strength until resonance between the parallel and antiparallel states is reached at a given microwave frequency (which will be specific to a given radical) until a signal is observed. As EPR only detects free radical species, it is highly selective to NO over all other products of N oxidation. Being a paramagnetic molecule, NO possesses an unpaired electron however, it cannot be studied with simple EPR because of too small relaxation time of excited electron to the ground state (Marples *et al.*, 1999). Therefore, spin straps are used to stabilize the free electron and allow NO measurements. Commonly used spin trapping agents for studying plant-pathogen interactions are diethyldithiocarbamate (DETC) and N-methyl-D-glucamine dithiocarbamate (MGD).

3.2.5. NO Electrodes

The basic structure of NO electrode consists of an Ag/AgCl reference electrode and platinum/teflon coated working electrode. Both these electrodes are encased in a glass micropipette filled with 30 mM NaCl/0.3 nM HCl solution with an open end, covered with an NO-permeable membrane which can be made up of chloroprene rubber, cellulose acetate, collodion/polystyrene, polytetrafluoroethylene (PTFE), and phenylenediamine (Davies and Zhang, 2008). Upon passage of an electric current NO is detected based on its oxidation at +0.8 to +0.9 V compared to the reference electrode (Shibuki, 1990). Reported sensitivities of NO electrodes have been in the order of 10^{-20} mol of NO in single cells (Malinski and Taha, 1992). It was demonstrated that an NO electrode could be used in plants to detect NO simply pushing the electrode into fruits (Lesham, 1996), in tobacco cells (Besson-Bard *et al.*, 2008) and to measure NADH-dependent NO scavenging activity in plant extracts (Igamberdiev *et al.*, 2004). However, several plant organs are not amendable to such intervention, thus electrodes have been most often used in plant tissue culture.

It is recommended that use of two independent methods, particularly one each from gas and liquid phase method, with controls will help in drawing significant information about NO concentration and distribution in plant growth, development and stress response (Gupta and Igamberdiev, 2013).

4. NO in Plant Responses to Biotic Stresses

One of the most powerful weapons in plant's arsenal against pathogen attack is the 'hypersensitive response' (HR), which could be considered as 'programmed cell death' (PCD). The disease resistance to pathogen depends on whether plant is able to recognize the pathogens during infection process. This recognition event leads to a rapid tissue necrosis at the infection site, which is termed as HR. HR restricts the pathogen's access to nutrients thus confining pathogen's growth to a small region of the plant.

4.1. ROS Production during HR

The production of ROS as a by-product of chloroplastic, mitochondrial, peroxisomal and glyoxysomal redox systems is a well known phenomenon. ROS serves as signaling molecules in plant host defense; however, uncontrolled production may also cause damage to the cell. There are anti-oxidant mechanisms in place which scavenges ROS from the plant cell (Noctor and Foyer, 1998), such as SODs, catalase, ascorbate peroxidases. There is plethora of information published on the role of ROS during plant-pathogen interaction (Dat *et al.*, 2000, Neill *et al.*, 2003, Gara *et al.*, 2003). ROS, particularly, H_2O_2, crosses cellular membranes and act as diffusible signaling molecules for inducing defense-related genes (Desikan *et al.*, 1998). ROS production in plants is detected within minutes of elicitor or microbial treatment. (Draper, 1997).There are two phases during ROS production: a weak non-specific phase which follows within minutes of pathogen addition, and a secondary phase which is dependent on recognition of incompatible pathogens by the host begins 1 – 3 h after initial burst (Levine *et al.*, 1994, Baker and Orlandi, 1995). This two phase kinetics of ROS production is characteristic of incompatible plant-pathogen interactions that involve HR. Besides oxidative burst, HR is characterized by other metabolic events such as ion flux across plasma membrane, changes in pH and NO production.

4.2. ROS and NO Cross talk during HR

The involvement of NO in plant disease resistance was first reported during resistance (R) gene-mediated protection in soybean suspension cultures against *Pseudomonas syringae pv. glycinea* expressing the AvrA avirulence gene (Delledonne *et al.*, 1998) and N-mediated recognition of tobacco mosaic virus (TMV) in *Nicotiana tabacum* (Durner *et al.*, 1998). This led to a widespread interest in the role of NO in plant signaling, specially, in plant responses to biotic stress. It has been demonstrated that plant cells accumulate NO in response to infection with bacterial, fungal and viral pathogens. Along with NO, ROS (H_2O_2) is also involved in theprogrammed cell death (PCD) which also limits pathogen spread. HR is associated with oxidative burst in which there is an increased generation of reactive oxygen species (ROS), PCD and activation of defense related genes. During HR, there is induction of defense related genes that restrict pathogen's growth, either indirectly to reinforce cell walls, or directly by providing anti-microbial enzymes and toxic secondary metabolites, such as phytoalexins, killing pathogens. A local HR is often associated with the onset of systemic acquired resistance (SAR) (Ryals *et al.*, 1996), and due to this, plant becomes resistant to secondary infection.

Reactive oxygen species (ROS) interacts synergistically with NO during plant HR (Delledonne *et al.*, 1998, de Pinto *et al.*, 2002, Lum *et al.*, 2002). NO emission was quantified by chemiluminescence in tobacco from roots, leaves, cell suspensions or anti-oxidant enzyme solutions (Planchet *et al.*, 2005, P. Dwivedi *et al.*, unpublished data). The requirement of NO for HR was further proved by Clarke *et al.* (2000) and Foissner *et al.* (2000). This was supported by the inhibition of cryptogein induced HR formation by treating tobacco cell suspensions with NO scavengers PTIO (2-phenyl-4,4,5,5-tetramethylimidazolinone-3-oxide-1-oxyl) or cPTIO (2-(4-carboxy-phenyl)-4,4,5,5-tetramethylimidazoline-1-oxyl-3-oxide) (Planchet *et al.*, 2005). Application of NO donors induced phenylalanine ammonia lyase (PAL) activity and CHS genes. Treatment of soybean cotyledons with NO triggered the biosynthesis of phytoalexins (Modolo *et al.*, 2002). Also, treatment of potato tubers with exogenous NO stimulated rishitin accumulation, and upon application of an NO-scavenger, this effect of inhibitory synthesis of rishitin was observed; this also demonstrates how NO influences phytoalexin accumulation in plant disease resistance (Able, 2003). Using cell suspension cultures of *Arabidopsis*, Clarke *et al.* (2000) proposed that NO acts independently of ROS production and that NO-induced cell death did not result from the production of highly reactive free radical ONOO⁻ formed by the reaction with O_2^- and NO. Nevertheless, NO also appears to act independently from ROS in the induction of various defense genes including PR proteins and enzymes of phenylpropanoid metabolism which are involved in production of lignin, antibiotics and salicyclic acid (Delledonne *et al.*, 1998). A model was proposed by Delledonne *et al.* (2001) wherein it was shown that interaction between NO and ROS determines if HR (PCD) would take place; the ratio of NO and O_2^- determines HR/PCD. Similarly, de Pinto *et al.* (2002) demonstrated a correlation between H_2O_2, NO and antioxidant levels. In animals, NO is synthesized from the enzyme nitric oxide synthase (NOS) along with other alternative pathways. The presence of NOS like activity has been demonstrated in plants by the use of NOS inhibitors that block the emission of NO (Foissner *et al.*, 2000, Pedroso *et al.*, 2000). Commonly used NOS inhibitors are L-NAME (N^w-nitro-L-arginine methyl-ester hydrochloride), L-NIL (L-N6-(1-Imminoethyl)-lysine acetate) and L-NMMA (*NG*.-monomethyl-L-arginine monoacetate) (Planchet *et al.*, 2005). A NOS like activity was detected in plants, although, with limited sequence homology with mammalian NOS (The *Arabidopsis* genome initiative, 2000). Furthermore, NOS inhibitors inhibited elicitor induced phytoalexin formation, implicating NOS as source of NO in plant defense responses. Infection of tobacco plants with HR-inducing varieties of tobacco mosaic virus (TMV) induced NOS activity that was inhibited by NOS inhibitors. A plant gene, AtNOS1, has been identified that exhibited significant sequence similarity to a snail gene which encoded a NOS-like activity (Guo *et al.*, 2003). Although AtNOS1 does not share any features with archetypal mammalian NOS enzymes, this protein was reported to show NOS activity (Guo *et al.*, 2003). More recent data, however, seem to suggest that recombinant AtNOS1 protein does not exhibit NOS activity (Crawford *et al.*, 2006, Zemojtel *et al.*, 2006). Nevertheless, loss of AtNOS1 function reduced *in vivo* NO levels in response to abscisic acid (ABA) (Guo *et al.*, 2003). The absence of AtNOS1 activity also compromised the nitrosative burst induced in response to bacterial lipopolysaccharide (LPS) (Zeidler *et al.*, 2004). While AtNOS1 is unlikely to encode a NOS-like activity this gene product may well be required for NO production. Thus, AtNOS1 may operate directly or indirectly to regulate NO synthesis. AtNOS1 has been renamed *Arabidopsis thaliana* nitric oxide associated 1 (AtNOA1) (Crawford *et al.*, 2006). One significant discovery of nitric oxide synthase (NOS) in plant kingdom was reported recently in two green algae species of the *Ostreococcus* genus, *O. tauri* and *O. lucimarinus* (Foresi *et al.*, 2010). The amino acid sequence of *O. tauri* NOS was found to be 45 per cent similar to that of human NOS. *Escherichia coli* cells expressing recombinant *O. tauri* NOS have increased levels of NO and cell viability.

5. Conclusion

Nitric oxide is a bioactive signaling molecule. Various methods are available for NO detection in plants along with their merits and demerits. It has to be accepted that it is impossible to arrive at a conclusion by using any single method for NO detection. DAF dyes seem to be indispensible fluorescent marker for NO detection until some other alternative dyes are available even with its various limitations. Its availability and ability to detect NO generation in various cell types makes it unparalleled. When using DAF dyes the background fluorescence of the treated materials should be taken into consideration and the increase in fluorescence be reported as a factor of the background. Results obtained by using NO and NOS inhibitors should not be the only basis for arriving at a definite conclusion. Independent NO detection methods should be used to check similar trends. Methods like EPR, chemiluminescence, etc. that are costly and are not available to many groups may be substituted with the use of NO electrodes and the Griess reagent assay. In summary, a combination of two or more methods simultaneously will be a better way for NO measurements with some degree of certainty. The plant HR is characterized by rapid accumulation of ROS and NO, triggering expression of resistance genes and mediates a systemic signal network that is involved in establishment of plant immunity. In the light of these recent discoveries, the simultaneous use of combinatorial knock outs and the next generation sequencing technologies to further elucidate the role of NO in various plant responses could be taken up.

Ever since the role of nitric oxide has been established in plant hypersensitive responses (HR), there has been a growing interest in deciphering the mechanism of nitric oxide action in plant patho-physiological responses. The involvement of NO in HR was demonstrated by using NO scavengers, PTIO and cPTIO, and NOS inhibitors, L-NAME, L-NIL and L- NMMA. In the process, a number of methodologies were tried for NO detection in plant system. Due to its ready availability and ease the DAF dyes have been widely used in plants to detect NO generation. However, the use of these fluorescent dyes has been controversial mainly because of its specificity and contradictions with more sensitive chemiluminescence assays. The plant HR is characterized by rapid accumulation of ROS and NO triggering expression of resistance genes, and mediates a systemic signal network that is involved in establishment of plant immunity. In the light of these recent discoveries, the simultaneous use of combinatorial knock outs and the next generation sequencing technologies to further elucidate the role of NO in various plant responses could be taken up.

6. References

Able, A.J. 2003. Role of reactive oxygen species in the response of barley to necrotrophic pathogens. *Protoplasma*, 221: 137-173.

Baker, C.J. and Orlandi, E.W. 1995. Active oxygen in plant pathogenesis. *Ann. Rev. Phytopath.*, 33: 299-321.

Beligni, M.V., Fath, A., Bethke, P.C., Lamattina, L. and Jones, R.L. 2002. Nitric oxide acts as an antioxidant and delays programmed cell death in barley aleurone layers. *Plant Physiol.*, 129: 1642–1650.

Besron-Bard, A., Griveans, S., Bedioui, F. and Werdehenne, D. 2008. Real time electrochemical detection of extracellular nitric oxide in tobacco cells exposed to cryptogein an elicitor of defense response. *J. Exp. Bot.*, 89: 3407-3414.

Bethke, P.C., Badger, M.R. and Jones, R.L. 2004. Apoplastic synthesis of nitric oxide by plant tissues. *Plant Cell*, 16: 332-341.

Byrnes, C.A. and Bush, A. and Shinebourne, E.A. 1996. Measuring expiratory nitric oxide in humans. *Nitric Oxide Pt B*, 269: 459–473.

Clark, D., Durner, J., Navarre, D.A. and Klessig, D.F. 2000. Nitric oxide inhibition of tobacco catalase and ascorbate peroxidase. *Mol. Plant Microbe Interact.*, 13: 1380–1384.

Conrath, U., Amoroso, G., Kohle, H. and Sultemeyer, D.F. 2004. Non-invasive online detection of nitric oxide from plants and some other organisms by mass spectrometry. *Plant J.*, 38: 1015–1022.

Cooney, R.V., Harwood, P.J., Custer, L.J. and Franke, A.A. 1994. Light mediated conversion of nitrogen dioxide to nitric oxide by carotenoids. *Environ. Health Persp.*, 102: 460–462.

Corpas, F.J., Palma, J.M., delRio, LA and Barroso, J.B. 2009. Oxides supporting the existence of L-arginine dependent nitric oxide synthase activity in plants. *New Phytol.*, 184: 9-14.

Crawford, N.M. and Guo F.Q. 2005. New insights into nitric oxide metabolism and regulatory functions. *Trends Plant Sci.*, 10(4): 195-200.

Crawford, N.M., Galli, M., Tischur, R., Heimer Xon, Okamoto, M. and Mack, A. 2006. Response to Zemojtel *et al*: plant nitric oxide synthase: back to square. *Trends Plant Sci.*, 11: 526-527.

Dat, J., Vandenbeele, S., Vranova, E., Van Montagu, M., Inze, D. and Van Breusegem, F. 2000. Dual activation of the active oxygen species during plant stress responses. *Cell Mol. Life Sci.*, 57: 779-795.

Davies, I.R. and Zhang, X.J. 2008. Nitric oxide selective electrodes. *Globins and other nitric oxide-reactive proteins, Pt A*, 436: 63–95.

de Pinto M.L., Tommasi, F. and De Gara, L. 2002. Changes in the antioxidative system as part of the signaling pathway response for the program cell death activated by nitric oxide and reactive oxygen species in tobacco bright yellow 2 cell. *Plant Physiol.*, 130: 698-708.

Del Rio, L.A., Lorpas, F.J. and Barroso, J.B. 2004. Nitric oxide and nitric oxide synthase activity in plants. *Phytochem.*, 65: 783-792.

Delledonne, M., Xia, Y.J., Dixon, R.A. and Lamb, C. 1998. Nitric oxide functions as a signal in plant disease resistance. *Nature*, 394: 585–588.

Delledonne, M., Zeier, J., Marocco, A. and Lamb, C. 2001. Signal interaction between nitric oxide and vanadium oxide intermediates in the plant hypersensitive disease resistance response. *Proc. Natl. Acad. Sci. USA*, 98: 13454-13459.

Desikan, R., Griffith, R., Hancock, J. and Neill, S. 2002. A new role for an old enzyme: nitrate reductase-mediated nitric oxide generation in required for abscisic acid induced stomatal closure in *Arabidopsis thaliana*. *Proc. Natl. Acad. Sci. USA*, 99: 16314-16318.

Draper, J. 1997. Salicylate, superoxide synthesis and cell suicide in plant defense. *Trends Plant Sci.*, 2: 162-165.

Durner, J., Wendehenne, D., Klessig, D.F. 1998. Defense gene induction in tobacco by nitric oxide, cyclic GMP and cyclic ADP ribose. *Proc. Nat. Acad. Sci. USA*, 95: 10328–10333.

Durzan, D.I. 2002. Stress-induced nitric oxide and adaptive plasticity in conifers. *J. Forest Sci.*, 48: 281-291.

Ferrer, M.A. and Ros-Barcelo´, A. 1999. Differential effects of nitric oxide on peroxidase and H_2O_2 production by the xylem of *Zinnia elegans*. *Plant Cell Environ.*, 22: 891–897.

Foissner, I., Wendehenne, D., Langebartels, C. and Durner, J. 2000. *In vivo* imaging of an elicitor-induced nitric oxide burst in tobacco. *Plant J.*, 23: 817-824.

Foresi, N., Correa- Aragunde, N., Parisi, G., Cato, G., Salerno, G. and Lamattina, L. 2010. Characterization of a nitric oxide synthase fom the plant kingdom: NO generation from the green alga *Ostreococcus tausi* in light, irradiation and growth phase dependent. *Plant Cell*, 22: 3816-3830.

Furchgott, R.F. and Zawdzki I.V. 1980. The obligatory role of endothelial cells in the relaxation of arterial growth muscle by acetylcholine. *Nature*, 288: 373-376.

Gara, L.D., de Pinto, M.C. and Tommasi, F. 2003. The acti-oxidative synthase vis-à-vis reactive oxygen species during plant pathogen interaction. *Plant Physiol. Biochem.*, 41: 863-870.

Garcia, M.C., Gay, R., Sokolovski, S., Hills, A., Lamattina, L. and Blatt, M.R. 2003. Nitric oxide regulates K^+ and Cl^- channels in guard cells through a subset of abscisic acid-evoked signaling pathways. *Proc. Natl. Acad. Sci. U.S.A.*, 100: 11116–111121.

Ghafourifar, P., Parihar, M.S., Nazarewicz, R., Zenebe, W.J. and Parihar, A. 2008. Detection assays for determination of mitochondrial nitric oxide synthase activity: advantages and limitations. *Methods Enzymol.*, 440: 317–334.

Gould, K.S., Lamotte, O., Klinguer, A., Pugin, A. and Wendehenne, D. 2003. Nitric oxide production in tobacco leaf cells: a generalized stress response? *Plant Cell Environ.*, 26: 1851–1862.

Gouvea, C.M.C.P., Souza, J.F. and Magalhaes, M.I.S. 1997. NO-releasing substances that induce growth elongation in maize root segments. *Plant Growth Reg.*, 21: 183–187.

Guo, F.Q. and Crawford, N.M. 2005. *Arabidopsis* nitric oxide synthase is targeted to mitochondria and protects against oxidative damage and death induced senescence. *Plant Cell*, 17: 3436-3450.

Guo, F.Q., Okamoto, M. and Crowford, N.M. 2003. Identification of a plant nitric oxide synthase gene involved in hormonal signaling. *Science*, 302: 100-103.

Gupta, K.J. and Igamberdiev, A.U. 2013. Recommendation of using at least two different methods for measuring nitric oxide. *Frontier Plant Sci.*, 4: 1-4.

Gupta, K.J. and Kaiser, W.M. 2010. Production and scavenging of nitric oxide by barley root mitochondria. *Plant Cell Physiol.*, 51: 576-584.

Gupta, K.J., Brostman, X., Seger, S., Zeier, T., Zeier, J., Persiin, S.T. *et al.*, 2013. The form of nitrogen nutrients affects resistance against *Pseudomonas syringae* pv. *Phasecolicola* in tobacco. *J. Exp. Bot.*, 64: 557-568.

Gupta, K.J., Fernie, A.R., Kaiser, W.M. and Van Dongen, J.T. 2011. On the origin of nitric oxide. *Trends Plant Sci.*, 16: 160-168.

Hayat, S., Hasan, S.A., Mori, M., Fariduddin, Q. and Ahmad, A. 2010. Nitric oxide: chemical biosynthesis and physiological value. In: Nitric oxide in Plant Physiology. S. Hayat, M. Moni, J. Pichtel and A. Ahmad (eds). Wiley Vch Verlag Gmb H and Co, Kga A, Weinheim.

Hung, K.T. and Kao, C.H. 2003. Nitric oxide counteracts the senescence of rice leaves induced by abscisic acid. *J. Plant Physiol.*, 160(8): 871–879.

Igamberdiev, A.U., Seregelyes, C., Mana'h, N. and Hill, R.D. 2002. NADH-dependent metals of nitric oxide in alfalfa root culture expression barley hemoglobin. *Planta*, 219: 95-102.

Ivanov, V.M. 2004. The 125th anniversary of the Griess reagent. *J. Anal. Chem.*, 59: 1002–1005.

Jourd'Heuil, D. 2002. Increased nitric-oxide dependent nitrosylation of 4,5-diaminofluorescein by oxidants. Implication for the measurement of intracellular nitric oxide. *Free Rad. Biol. Med.*, 33: 676-684.

Klepper, L.A. 1979. Nitric oxide (NO) and nitrogen dioxide (NO_2) emissions from herbicide-treated soybean plants. *Atmos. Environ.*, 13: 537–542.

Lamattina, L., Garcý´a-Mata, C., Graziano, M. and Pagnussat, G. 2003. Nitric oxide: the versatility of an extensive signal molecule. *Annu. Rev. Plant Biol.*, 54: 109–136.

Lancarter, J.R. 1997. A tutorial on the diffusibily and reactivity of free nitric oxide. *Nitric Oxide Biol. Chem.*, 1: 18-30.

Legrand, M., Almac, E., Mik, E. G., Johannes, T., Kandil, A., Bezemer, R., Payen, D. and Incel, C. 2009. L-NIL prevents renal microvascular hypoxia and increase of renal oxygen consumption after ischemia-reperfusion in rats. *Am. J. Renal Physiol.*, 296: 1109-1117.

Lesham, X.Y., Wills, R.B.H. and Ku, V.W. 1998. Evidence for the function of the free radical gas nitric oxide as an endogenous regulating factor in higher plants. *Plant Physiol. Biochem.*, 36: 825-833.

Leshem, Y.Y. 1996. Nitric oxide in biological systems. *Plant Growth Regul.*, 18: 155–159.

Leshem, Y.Y., Wills, R.B.H. and Veng-Va Ku, V. 1998. Evidence for the function of the free radical gas-nitric oxide (NO) as an endogenous maturation and senescence regulating factor in higher plants. *Plant Physiol. Biochem.*, 36: 825–833.

Levins, A., Tenhaken, R., Dixon, R.A. and Lamb, C.J. 1994. H_2O_2 from the oxidative burst orchestrates the plant hypersensitive response. *Cell*, 79: 583-593.

Lum, H.K., Butt, Y.K.C. and Lo, S.C.L. 2002. Hydrogen peroxide induced rapid production of nitric oxide in mung bean (*Phaseolus aureus*). *Nitric Oxide Biol. Chem.*, 6: 205-213.

Magalhaes, J.R., Monte, D.C. and Durzan, D. 2000. Nitric oxide and ethylene emission in *Arabidopsis thaliana*. *Physiol. Mol. Biol. Plants*, 6: 117-127.

Malinski, T. and Taha, Z. 1992. Nitric-oxide release from a single cell measured *in situ* by a porphyrinic-based microsensor. *Nature*, 358: 676–678.

Maples, K.R., Sandstrom, T., Su, Y.F. and Henderson, R.F. 1999. The nitric oxide/ heme protein complex as a biologic marker of exposure to nitrogen dioxide in humans, rats and *in vitro* models. *Am. J. Respir. Cell Mol. Biol.*, 4: 538–543.

Modolo, L.V., Cunha, F.Q., Braga, M.R. and Salgado, I. 2002. Nitric Oxide Synthase-mediated phytoalexin accumulation in soybean cotyledons in response to the *Diaporthe phaseolorum* f. sp. *Meridionalis* elicitor. *Plant Physiol.*, 130(3): 1288-1297.

Moeskops, B.W., Cristescu, S.M. and Harren, F.J. 2006. Sub part per billion monitoring of nitric oxide by use of wavelength modulators spectroscopy in combination with a thermoelectrically cooled, continuous wave quantum cascade laser. *Opt. Lett.*, 31: 823-25.

Morean, M., Linder mayr, C., Dwerver, J. and Klessig, D.F. 2010. NO synthesis and signaling in plants- where do we stand. *Physiol Plant.*, 138: 372-383.

Mur, L.A.J., Mandon, J., Cristescu, S.M., Harrens, F.J. and Prats, E. 2011. Methods of nitric oxide detection in plants: a commentary. *Plant Sci.*, 181: 509-519.

Mur, L.A.J., Santosa, I.E., Laarhoven, L.J.J., Holton, N.L., Harren, F.I.M. and Smith, A.R. 2005. Laser phytoacoustic detection allows in planta detection of nitric oxide in tobacco following challenge with avirulent and virulent *Pseudomonas syringae pathovars*. *Plant Physiol.*, 138: 1247-1258.

Murad, F. 1999. Discovery of some of the biological effects of nitric oxide and its role in cell signaling. *Biosci. Rep.*, 19: 133–154.

Murgia, I., Delledonne, M. and Soave, C. 2002. Nitric oxide mediates iron-induced ferritin accumulation in *Arabidopsis*. *Plant J.*, 30: 521–528.

Murphy, M.E. and Noack, E. 1994. Nitric oxide assay using hemoglobin method. *Methods Enzymol.*, 233: 240–250.

Nakatsubo, N., Kojima, H., Kikuchi, K., Nagoshi, H., Hirata, Y., Maeda, D., Imai, Y., Irimura, T. and Nagano, T. 1998. Direct evidence of nitric oxide production from bovine aortic endothelial cells using new fluorescence indicators: diaminofluoresceins. *FEBS Lett.*, 427: 263–266.

Navarre, D.A., Wendehenne, D., Durner, J., Noad, R. and Klessig, D.F. 2000. Nitric oxide modulates the activity of tobacco aconitase. *Plant Physiol.*, 122, 573–582.

Neill, S.J., Desikan, R. and Hancock, J.T. 2003. Nitric oxide signalling in plants. *New Phytol.*, 159: 11–35.

Neill, S.J., Desikan, R., Clarke, A., Hancock, J.T. 2002. Nitric oxide is a novel component of abscisic acid signalling in stomatal guard cells. *Plant Physiol.*, 128: 13–16.

Neill, S.J., Desikan, R., Clarks, A., Hurst, R.D. and Hancock, J.T. 2002. Hydrogen peroxide and nitric oxide as signaling molecule in plants. *J. Exp. Bot.*, 53: 1237-1247.

Noctor, G. and Foyer, C.H. 1998. Ascorbate and glutathione - keeping active oxygen under control. *Annu. Rev. Pl. Physiol. Pl. Mol. Biol.*, 49: 249-279.

Orozco-Cardenas, M.L. and Ryan, C.A. 2002. Nitric oxide negatively modulates wound signaling in tomato plants. *Plant Physiol.*, 130: 487–493.

Palmer, R.M.J., Ferrige, A.G. and Moncada, S. 1987. Nitric oxide accounts for the biological activity of endothelium derived relaxing factor. *Nature*, 307: 524-526.

Pedroso, M.C., Magalhaes, J.R. and Durzan, D. 2000. A nitric oxide burst proceeds apoptosis in angiosperm and gymnosperm callus cells and foliar tissues. *J. Exp. Bot.*, 51: 1027-1036.

Planchet, E. and Kaiser, W.M. 2006. Nitric oxide (NO) detection by DAF fluorescence and chemiluminescence: A comparison using abiotic and biotic NO sources. *J. Exp. Bot.*, 57: 3043–3055.

Planchet, E., Sonoda, M., Zeier, J., Kaiser, W.M. 2006. Nitric oxide (NO) as an intermediate in the cryptogein induced hypersensitive response – a critical re-evaluation. *Plant Cell Environ.*, 29: 59–69.

Planchet, E., Gupta, J.K., Sonoda, M. and Kaiser, W.M. 2005. Nitric oxide (NO) emission from tobacco leaves and cell suspensions: rate limiting factors and evidence for the involvement of mitochondrial electron transport. *Plant J.*, 41: 732–743.

Rockel, P., Strube, F., Rockel, A., Wildt, J. and Kaiser, W.M. 2002. Regulation of nitric oxide (NO) production by plant nitrate reduction *in vivo* and *in vitro*. *J. Exp. Bot.*, 53: 103-110.

Ryals, J.A., Neuensch Wander, U.H., Willits, M.G., Molina, A., Steiner, H.Y. and Hunt, M.D. 1996. Systemic acquired resistance. *Plant Cell*, 8: 1809-1819.

Shibuki, K. 1990. An electrochemical microprobe for detecting nitric-oxide release in brain-tissue. *Neurosci. Res.*, 9: 69–76.

Stohr, C., Strube, F., Marx, G., Ullrich, W.R. and Rockel, P. 2001. A plasma membrane bound enzyme of tobacco roots catalyses formation of nitric oxide from nitrite. *Planta*, 212: 835-841.

Takahashi, S. and Yamasaki, H. 2002. Reversible inhibition of photophosphorylation in chloroplasts by nitric oxide. *FEBS Lett.*, 512 (1-3): 145–148.

The *Arabidopsis* genome initiative. 2000. Analysis of the genome sequence of the flowering plant *Arabidopsis thaliana*. *Nature*, 408: 796-815.

Tischner, R., Planchet, E. and Kaiser, W.M. 2004. Mitochondrial electron transport as a source for nitric oxide in the unicellular green alga *Chlorella sorokiniana*. *FEBS Lett.*, 576: 151–155.

Tsikas, D. 2006. Analysis of nitrite and nitrate in biological fluids by assays based on the Griess reaction: Appraisal of the Griess reaction in the l-arginine/nitric oxide area of research. *J. Chromatography B*, 851: 51–70.

Vanin, A.F., Svistonenko, D.A. Mikoyan, V.D., Serezhenkon, V.A., Fryer, M.J., Baker, N.R. *et al.* (2004). Endogenous superoxide production and the nitrite/nitrate ratio control the concentration of bioavailable free nitric oxide in leaves. *J. Biol. Chem.*, 279: 24100–24107.

Vitecek, J., Reinohl, V. and Jones, R.L. 2008. Measuring NO production by plant tissues and suspension cultured cells. *Mol. Plant*, 1: 270–284.

Wilson, I.D., Neill, S J. and Hancock, J.T. 2008. Nitric oxide synthesis and signaling in plants. *Plant Cell Environ.*, 31: 622-631.

Yamasaki, H., Shimoji, H., Ohshiro, Y. and Sakihama, Y. 2001. Inhibitory effects of nitric oxide on oxidative phosphorylation in plant mitochondria. *Nitric Oxide Biol. Chem.*, 5: 261–270.

Zeidler, D., Zahringer, U., Herber, I., Dubery, I., Hartung, T., Bors, W., Hutjler, P. and Durner, J. 2004. Innate immunity in *Arabidopsis thaliana*: lipopolysaccharides activate nitric oxide synthase (NOS) and induce defense genes. *Proc. Natl. Acad. Sci. USA*, 101: 15811-15816.

Zemojtel, T., Frochlich, A., Palmieri, M.C. *et al.*, 2006. Plant nitric oxide system: a never- ending story. *Trends Plant Sci.*, 11: 524-525.

Zhang, C., Czymmek, K.J. and Shapiro, A. D. 2003. Nitric oxide does not trigger early programmed cell death events but may contribute to cell-to-cell signaling governing progression of the Arabidopsis hypersensitive response. *Mol. Plant Microbe Interact.*, 16: 962–972.

Zottini, M., Formentin, E., Scattolin, M., Carimi, F., LoSchiavo, F. and Terzi, M. 2002. Nitric oxide affects plant mitochondrial functionality *in vivo*. *FEBS Lett.*, 515: 75–78.

Chapter 6

Salicylic Acid: A Key to Regulate Drought Stress in Chickpea

A. Hemantaranjan[1] and Pradeep Kumar Patel[2]

[1]*Department of Plant Physiology, Institute of Agricultural Sciences,*
Banaras Hindu University, Varanasi – 221 005, U.P.
E-mail: hemantaranjan@gmail.com
[2]*Division of Crop Improvement, Indian Institute of Vegetable Research,*
Shahanshahpur, Varanasi – 221305, U.P.

CONTENTS

1. Introduction

Crop yields are more dependent on an adequate supply of water than on any single environmental factor. Even plants with an optimum water supply experience transient period of shortage, where water absorption cannot compensate for water loss by transpiration, a situation that largely depends on environmental factors such as temperature, relative humidity and wind velocity. In addition, many other environmental stresses, such as cold, salt and high temperature, have a water stress component. Passioura (2002) pointed out, drought may have very different cannotations and time scales for different practitioners, the latter ranging from years to decades for meteorologists and farmers, to a few hours for molecular biologist (with agronomist and crop and plant physiologist somewhere in between time frame of days to week, and most rarely months). The real significance of drought can also be quite different depending upon the level of study.

Drought is undoubtedly one of the most important environmental factors regulating plant growth and development (Manivannan *et al.*, 2007). Therefore, it is a major threat affecting the life of plants and is responsible for limiting crop yields globally (Kavar *et al.*, 2008). There are many studies which demonstrate that drought stress induces numerous metabolic, biochemical and physiological changes in plants (Levitt, 1980). These include water status, growth, membrane integrity, pigment content, osmotic adjustment and photosynthetic activity (Zhang *et al.*, 2007; Efeoglu *et al.*, 2009; Praba *et al.*, 2009). The sequence of events that occurs where drought stress develops starts with cellular growth as most sensitive response followed by a wide range of biochemical and physiological events that more negative water potential becomes. This picture is further complicated because of the sensitivity of some response is highly dependent on the plant species. Grain legumes as compared to cereals appear to have more sensitivity towards drought and chickpea is one of them.

Chickpea (*Cicer arietinum* L.) is a leguminous plant known to contain high nutritional properties, due mainly to its seed protein content. In addition to being an important source of food, feed and fodder (Singh, 1997), this annual legume plays a significant role in agricultural sustainability. The legume is utilized both as a rotation crop, allowing for the diversification of agricultural production systems, and for its N_2-fixing properties (Gan *et al.*, 2006). Although it is recognized that vegetative and/ or reproductive growth and productivity of chickpea is adversely affected by global climate change and the increasing shortage of water resources (Leport *et al.*, 2006; Canci and Toker, 2009), there is very limited information in the literature on the tolerance of chickpea cultivars to drought stress at the vegetative and/or reproductive stage (Kalefetoglu Macar and Ekmekci, 2008). Plants tend to cope with water deficit stress via a common response, known as osmotic adjustment. This includes the synthesis and accumulation of compatible (non-toxic) solutes, such as proline (Szira *et al.*, 2008) and total soluble sugars. In addition, up-regulation of antioxidant defense system. Comparatively a little work has been reported to overcome drought stress injuries in this important pulse crop through salicylic acid (SA), because these compounds appear to have innate potentiality for increasing activity of antioxidant enzymes and antioxidant molecules in oxidative stressed plants (Singh and Usha, 2003; Hayat *et al.*, 2008).

Salicylic acid (SA) is a naturally existing phenolic compound and is considered to be a potent plant growth regulator because of its diverse regulatory role in plant metabolism. Phenolic compounds have strong free radicals scavenging capacity (Hall and Cuppett, 1997). Evidences put forward that externally applied SA increased plant's tolerance to several abiotic stresses, including osmotic stress (Wang *et al.*, 2010), heavy metal stress (Moussa and El-Gamel, 2010) and also influence a range of diverse processes in plants, including seed germination, stomatal closure, ion uptake and transport, membrane permeability, photosynthesis, and plant growth rate (Aftab *et al.*, 2010). The review summarizes the recent advances in the understanding of the physiological functions of SA, and relevant insights regarding SA mechanisms that control these events are highlighted.

2. Salicylic Acid (SA) Responses

2.1. SA in Growth Regulation

The first and foremost effect of drought is impaired germination and poor stand establishment (Harris *et al.*, 2002). Cell growth is considered as one of the most drought sensitive physiological processes due to the reduction in turgor pressure. Growth is the result of daughter-cell production by meristematic cell divisions and subsequent massive expansion of the young cells. Under severe water deficiency, cell elongation of higher plants can be inhibited by interruption of water flow from the xylem to the surrounding elongating cells (Nonami, 1998). Drought caused impaired mitosis; cell elongation and expansion resulted in reduced growth and yield traits (Hussain *et al.*, 2008). Plant height and fresh weight may decrease because of suppression of cell expansion and cell growth, due to low turgor pressure (Jaleel *et al.*, 2008). Water deficits reduce the number of leaves per plant and individual leaf size, leaf longevity by decreasing the soil's water potential. Leaf area expansion depends on leaf turgor, temperature, and assimilating supply for growth. Drought-induced reduction in leaf area is ascribed to suppression of leaf expansion through reduction in photosynthesis (Rucker *et al.*, 1995). Drought stress may cause leaf abscission (Rouhi *et al.*, 2007). Osmotic regulation can enable the maintenance of cell turgor for survival or for assisting plant growth under severe drought conditions in pearl millet (Shao *et al.*, 2008). However, Kalefetoglu Macar and Ehmekci (2009) reported that the number of leaves of chickpea cultivars decreased to a larger extent is due to a decline in leaf production rather than abscission under drought. Recently in chickpea genotypes, we noticed highest leaf area in DCP 92-3, while it was lowest in ICC 4958 and Tyson both at 50 per cent flowering and 50 per cent podding phase. Genotype DCP 92-3 attained highest leaf area but not produced highest yield under drought (Patel and Hemantaranjan, 2012). It was found that plasticity in leaf area is an important means by which drought-stressed crop maintains water-use. The results suggest that crops need to adjust their transpiring surface through reducing leaf growth (as a mechanism for reducing water loss) (Patel and Hemantaranjan, 2012).

Salicylic acid and other salicylates are known to affect various physiological and biochemical activities of plants and may play a key role in regulating their growth and productivity (Arberg, 1981). SA and its close analogues enhanced the leaf area and dry mass production in corn and soybean (Khan *et al.*, 2003). Enhanced germination and seedling growth were recorded in wheat, when the grains were subjected to pre-sowing seed-soaking treatment in salicylic acid (Shakirova, 2007). Fariduddin *et al.* (2003) reported that the dry matter accumulation was significantly enhanced in *Brassica juncea*, when lower concentrations of salicylic acid were sprayed. However, higher concentrations of SA had an inhibitory effect. In another study, the leaf number, fresh and dry mass per plant of wheat seedlings increased significantly when raised from the grains soaked in lower concentration (10^{-5} M) of salicylic acid (Hayat *et al.*, 2005). Similar growth promoting response was generated in barley seedlings sprayed with salicylic acid (Pancheva *et al.*, 1996). Growth-stimulating effects of SA have been reported in soybean (Gutie´rrez-Coronado *et al.*, 1998), wheat (Shakirova *et al.*, 2003), maize (Gunes *et al.*, 2007), and chamomile (Kova´cik *et al.*, 2009). In soybean plants treated with 10 mM, 100 mM, and up to 10 µM SA increased shoot and root growth ~ 20 per cent and 45 per cent, respectively, 7 days after application. Wheat seedlings treated with 50 µM SA develop larger ears, and enhanced cell division was observed within the apical meristem of seedling roots (Shakirova *et al.*, 2003). Likewise, 50 µM SA stimulates the growth of leaf rosettes and roots of chamomile plants by 32 per cent and 65 per cent, respectively, but higher concentrations (250 µM) have the opposite effect (Kova´cik *et al.*, 2009). This suggests that the growth-promoting effects of SA could be related to changes in the hormonal status (Abreu and Munne´-Bosch, 2009) or by improvement of photosynthesis, transpiration, and stomatal conductance (Stevens *et al.*, 2006). Interestingly in chickpea, the dry matter partitioning during drought stress was found significantly more to different plant parts for growth and maintenance @ 1.5mM SA, which was relatively more in genotype ICC4958 than DCP 92-3 and JG 315 (Patel and Hemantaranjan, 2012).

2.2. SA-Regulated Plant Water Relation

Relative water content (RWC), leaf water potential, stomatal resistance, rate of transpiration, leaf temperature and canopy temperature are important characteristics that influence plant water relations. Relative water content is considered a measure of plant water status, reflecting the metabolic activity in tissues and used as a most meaningful index for dehydration tolerance. RWC of leaves is higher in the initial stages of leaf development and declines as the dry matter accumulates and leaf matures. RWC related to water uptake by the roots as well as water loss by transpiration. A decrease in the relative water content (RWC) in response to drought stress has been noted in wide variety of plants as reported by Nayyar and Gupta (2006) that when leaves are subjected to drought, leaves exhibit large reductions in RWC and water potential. Exposure of plants to drought stress substantially decreased the leaf water potential, relative water content and transpiration rate, with a concomitant increase in leaf temperature (Siddique *et al.*, 2001). The inhibition of plant growth under dehydrated conditions is associated with altered plant water relations (Dichio *et al.*, 2003). Dehydration was found to decrease the relative water contents of plant leaves (Sanchez-Blanco *et al.*, 2002). Dehydration is also implicated in causing damages to membranes. The critical feature of tolerance to dehydration depends on the abilities to limit membrane damage during dehydration and to regain membrane integrity and membrane bound activities quickly upon rehydration

(Tripathy *et al.*, 2000). The changes in membrane permeability following exposure to drought stress can also be used to estimate drought tolerance in crop plants (Blum and Ebercon, 1981).

The osmotic adjustment in chickpea has been reported to be ranged from 0 to 1.3 MPa (Lecoeur *et al.*, 1992). Osmotic adjustment increases water absorption, maintains cell turgor, photosynthesis and leaf area duration, helps stomatal opening, delays senescence and death (Hsiao *et al.*, 1984); reduce flower abortion (Morgan and King, 1984) and improves root growth with the development of water deficit (Morgan *et al.*, 1991). The greater osmotic adjustment leads to higher growth rate and dry matter production in pigeonpea under drought (Subbarao *et al.*, 2000). Reduction in leaf area is expected to reduce water loss. Saxena (2003) reported two chickpea accessions ICC 5680 and ICC10448, with a smaller leaf area. ICC 5680 has fever leaflet trait in ICC 5680 reduced transpiration loss of water by 30 per cent compared to ICC 4958 at ICRISAT.

It is a well-established fact that salicylic acid potentially generates a wide array of metabolic responses in plants and also affects the photosynthetic parameters and plant water relations. Exogenous application of SA was found to enhance the net photosynthetic rate, internal CO_2 concentration, water use efficiency, stomatal conductance and transpiration rate in *B. juncea* (Fariduddin *et al.*, 2003). Further, Khan *et al.* (2003) reported an increase in transpiration rate and stomatal conductance in response to foliar application of SA and other salicylates in corn and soybean. In another study carried out in soybean, foliar application of salicylic acid enhanced the water use efficiency, transpiration rate and internal CO_2 concentration (Kumar *et al.*, 2000). However, contrary to these results, the transpiration rate decreased significantly in *Phaseolus vulgaris* and *Commelina communis* after the foliar application of SA and this decrease in transpiration rate was attributed to the fact that salicylic acid induced the closure of stomata (Larque-Saavedra and Martin-Mex, 2007). Maintenance of favourable plant water relations is vital for the development of drought resistance in crop plants (Blum, 1996; Lilley and Ludlow, 1996; Passioura, 2002).

Adopting Levitt's (1972) definition, the capability of genotypes to maintain relatively higher water content is an avoidance mechanism. When plant adjusts to lower relative water content, it is considered a tolerance mechanism. RWC was low at 50 per cent flowering compared to 50 per cent podding in four chickpea genotypes at drought stress conditions. Genotypes JG 315 and DCP 92-3 showed lower RWC as compared to Tyson and ICC4958 at 50 per cent flowering stage. Tyson and ICC4958 had maintained higher RWC at both the stages. RWC (per cent) was in the range of 71.6 - 74.4 (unstressed) and 67.9 - 71.6 (stressed), whereas SA @1.5 mM increased leaf RWC values close to the control (unstressed plant). Results showed lowest RWC value (68 per cent) in genotype JG 315 and the maximum RWC (71.17 per cent) was remarkably maintained under stress in Tyson (Patel and Hemantaranjan, 2012). Thus, the results indicate that water availability plays a primary role in leaf production processes. The ability of plants to cope with drought may depend on different mechanisms; including the capacity to maintain high relative water content.

2.3. SA-Regulated Biochemical Responses

2.3.1. Lipid Peroxidation (LPO)

Drought induces oxidative stress in plants by generation of reactive oxygen species (ROS) (Farooq *et al.*, 2009). The ROS such as $O_2^{\cdot}$, H_2O_2 and $\cdot OH$ radicals, can directly attack membrane lipids and increase lipid peroxidation (Mittler, 2002). Drought-induced overproduction of ROS increases the content of malondialdehyde (MDA). Malondialdehyde (MDA) is considered as a suitable marker for membrane lipid peroxidation. The content of MDA has been considered an indicator of oxidative damage (Moller *et al.*, 2007). A decrease in membrane stability reflects the extent of lipid peroxidation caused by ROS. Furthermore, lipid peroxidation is an indicator of the prevalence of free radical reaction in tissues. Moreover, oxygen uptake loading on the tissues as both processes generate reactive oxygen species, particularly H_2O_2 that produced at very high rates by the glycolate oxidase reaction in the peroxisomes in photorespiration.

Salicylic acid (SA) has been proved to induce diverse stress tolerance in plants (Senaratna *et al.*, 2000; Yang *et al.*, 2004). However, certain evidences elucidate that SA could cause heavy lipid peroxidation and oxidative damage in plants (Rao *et al.*, 1997; Ganesan *et al.*, 2001). Effects of exogenous SA on plant stress tolerance depended on the concentrations applied. Moderate concentrations of SA could alleviate the stress injury which at high concentrations cause more damage to plants under stress conditions (Kang *et al.*, 2003). Lipid peroxidation (MDA content) was significantly higher in JG 315 at pre-anthesis drought stage over control among chickpea genotypes. Nevertheless, through SA treatments LPO reduction was highest @1.5 mM SA at both the critical stages (*i.e.* pre- and post- anthesis) (Patel *et al.*, 2011).

Plants accumulate different types of organic and inorganic solutes in the cytosol to lower osmotic potential thereby maintaining cell turgor (Rhodes and Samaras, 1994). Under drought, the maintenance of leaf turgor may also be achieved by the way of osmotic adjustment in response to the accumulation of sucrose, soluble carbohydrates, glycine betaine, proline and other solutes in cytoplasm improving water uptake from drying soil.

2.3.2. Sugar Accumulation

Drought-induced decrease in starch and an increase in soluble sugars have been observed in many crops (Huber *et al.*, 1984; Timpa *et al.*, 1986; Mukane *et al.*, 1996). Increased non structural carbohydrates (starch and sugar) with enhanced remobilization of sucrose from stem to leaves in pigeonpea have been reported to be associated with leaf osmotic adjustment in pigeonpea under drought (Subbarao *et al.*, 2000). Accumulation of pinitol derived from sucrose and proline as osmolytes has been reported in pigeonpea under water deficit (Keller and Ludlow, 1993). Osmotic solute in leaves increase gradient for water fluxes in to the cell and maintains turgor (Serraj and Sinclair, 2002). Further, in soybean plants increased soluble sugar content exhibited a favourable effect along with some ions antagonized or ameliorated the inhibitory effect of drought stress by application of salicylic acid (Hakimi, 2006).

Chickpea accumulated hexose sugars and sucrose in leaves when water stress occurred and starch was degraded (Basu *et al.*, 2007). Sucrose was accumulated during OA of chickpea as a minor solute component and with malic acid and malonic acid contributing to 75 per cent with osmotic potential changes under water limiting conditions (Lecoeur *et al.*, 1992). In addition, leaf starch degrades even under the mild stress at about Ψ -1.6 MPa and leads to an increase in hexose (reducing sugars) in four chickpea lines M51, M93, M129 and Tyson subjected to moderate or high water stress (Basu *et al.*, 2007). Like other C3 species, starch occupies major pre-stored photosynthetic product in chickpea leaves. It has been reported that starch breaks down to release hexose when photosynthesis declines below the threshold level (Foundy *et al.*, 1989) or drought could be the causal factor initiating starch breakdown and accumulation of soluble sugar (Huber *et al.*, 1984). It appears that both reducing sugar and sucrose might contribute solute for osmotic adjustment or lowering OP_{100} in chickpea. Correspondingly, an increase in the sucrose content through transient increase in the activation state of sucrose phosphate synthase (SPS) has been reported in spinach leaves to drought (Quick *et al.*, 1989). Accumulation of sucrose appeared to be associated with increase in RWC but decrease in OP_{100}, *i.e.*, sucrose accumulation promoted osmotic adjustment. Subsequently, sucrose as osmolyte increases the gradient for water flux in to the cell and maintained turgor by adjusting LWP and RWC. Patel and Hemantaranjan (2012) revealed that total soluble sugar level differed significantly at 50 per cent flowering and podding stages among SA treated chickpea genotypes under drought. In our experiment, chickpea genotypes Tyson and ICC 4958 though showed more soluble sugar contents under drought stress that increased further by SA @ 1.5 mM in all four genotypes.

2.3.3. Proline Accumulation

Proline can act as a signaling molecule to modulate mitochondrial functions, influence cell proliferation or cell death and trigger specific gene expression, which can be essential for plant recovery from stress (Szabados and Savouré, 2009). Accumulation of proline under stress in many plant species has been correlated with stress tolerance, and its concentration has been shown to be generally higher in stress-tolerant than in stress-sensitive plants. The accumulation of proline in plant tissues is also a clear marker for environmental stress, particularly in plants under drought stress (Routley, 1966). It influences protein recovery and preserves the quaternary structure of complex proteins, maintains membrane integrity under dehydration stress and reduces oxidation of lipid membranes or photoinhibition (Demiral and Turkan, 2004). Furthermore, it also contributes to stabilizing sub-cellular structures, being a sink of energy or reducing being a source for carbon and nitrogen and scavenging free radicals acting as a hydroxyl radical scavenger (Hare *et al.*, 1998) and buffering cellular redox potential under stress conditions (Ashraf and Fooland, 2007). Proline content increased under drought stress in pea (Sanchez *et al.*, 1998; Alexieva *et al.*, 2001) and can also be observed under different stresses such as high temperature and under starvation (Sairam *et al.*, 2002). Proline might confer drought stress tolerance to wheat plants by increasing the antioxidant system rather than as osmotic adjustment) (Vendruscolo *et al.*, 2007). Osmotic adjustment has been considered as one of the crucial mechanisms in plant adaptation to various stresses.

Soluble sugars and proline are major constituents of osmoregulation in the expanded leaves of many species (Morgan, 1994). The protective action of SA during water deficit was demonstrated by the enhanced proline production in stressed plants. Accumulation of proline has been advocated as a parameter of selection for stress tolerance (Yancy *et al.*, 1982; Jaleel *et al.*, 2007). Salicylic acid during water deficit enhanced proline production in stressed tomato and amaranth plants (Umebese *et al.*, 2009). Patel *et al.* (2011) recently reported more proline accumulation in chickpea genotypes Tyson and ICC 4958 as compared to JG 315 and DCP92-3 in response to SA. Proline content in plants treated with SA @1.5 mM under drought stress was maximum in Tyson and minimum in DCP 92-3.

2.3.4. Nitrate Reductase Activity

Nitrate reductase (NR) activity is vital for the metabolic and physiological status of plants and can be used as a biomarker of plant stress including drought. NR activity decreases in plants exposed to water limiting conditions because of a lower flux of nitrate from the soil to the root (Azco'n and Go'mez, 1996). NR activity mediates the reduction of nitrate to nitrite, which is regarded as a rate limiting step in plant growth and development (Solomonson Barber, 1990) and it provides a good estimate of the nitrogen status of plants and is correlated with growth and plant yield. In particular, activities of nitrate reductase (NR; EC 1.7.1.3) and sucrose-phosphate synthase (SPS; EC 2.4.1.14), two key enzymes of nitrogen and carbon assimilation pathways, respectively, are strongly modulated during drought. Under water stress, leaf NR activity decreases (Garg *et al.*, 2001; Correia *et al.*, 2005). NR activity is a very labile enzyme to water stress; therefore under the condition the activity needs being protected by some means. Sing and Usha (2003) reported that SA treatment, under water stress, protected nitrate reductase (NR) activity and maintained, especially at 3 mM SA concentration, when the protein and nitrogen content of wheat leaves compared to water sufficient seedlings. A concentration (0.01-1.0 mM) of $Ca(NO_3)_2$ in association with SA activated the uptake of nitrogen and the activity of nitrate reductase (NR) both in the leaves and roots of maize plants, although higher concentration (5 mM) proved inhibitory (Jain and Srivastava, 1981). Likewise, SA increased the activity of NR in the presence of NO_3 and also favoured protection of the enzyme against proteinase, trypsin (Rane *et al.*, 1995). The plants resulting from the wheat grains, soaked in aqueous solution (10^{-5} M) of SA, exhibited high NR activity (Hayat *et al.*, 2005) and also in mustard leaves whose foliage was fed with SA (10^{-5} M) (Fariduddin *et al.*, 2003). In the former, 10^{-5} M of SA enhanced the activity of the enzyme by 36 per cent and by 13 per cent in the latter, as compared to their respective controls. Moreover, in both these cases the maximum concentration (10^{-3} M) of SA proves to be inhibitory, that decreased the activity of NR by 14 per cent in wheat seedlings and 10 per cent in mustard plants. Higher NR activity, under the influence of SA, in *Glycine max* was coupled with protein content (Kumar *et al.*, 1999) but SA decreased the level of soluble proteins in barley (Pancheva *et al.*, 1996). Nonetheless the level of sugars, starch and phenols exhibited a decrease in *Vigna mungo* cultivars in response to SA (10-50 µM) treatment (Anandhi and Ranjeva, 1997). Further, in seedlings of *Amaranthus hybridus* cv. NHAC-3 (large green, amaranth) and *Lycopersicum esculentum* cv. Roma (tomato) were subjected to 7 days water stress at Early Vegetative (EV), Late Vegetative (LV), Early Flowering (EF) and Late Flowering (LF) stages of both plants. NRA was significantly (p = 0.05) reduced by stress treatment at the LV stage of amaranth, EF stage of tomato and LF stage of both plants. As a result, the reduction of NRA was more pronounced at the reproductive stage of both plants, as 3 mM SA was effective in maintaining NRA at levels similar to the control in both plants (Umebese *et al.*, 2009).

Nitrate reductase (NR) activity was inhibited at both the critical stages of development (*i.e.* pre- and post- anthesis) in chickpea. Since water content of plants is higher at the pre- anthesis stage than the post anthesis stage, a further decrease in water content as a result of deficit treatment could have reduced NR activity markedly at the post -anthesis stage in control as well as stressed plants in chickpea genotypes. SA @1.5 mM was more effective than @1.0 mM in improving NR activity (Patel *et al.*, 2012). Furthermore, efficient N assimilation might have been favoured by a high rate of CO_2 assimilation but reduced stomatal aperture as a result of water stress that lowers the carbon gain of the plant and reduces CO_2 assimilation (Forbes and Watson, 1992).

2.3.5. SA-Regulated Antioxidants and Antioxidant System

Plants appear to have strong defensive mechanism to overcome harmful effects of environmental stresses *i.e.*, plants have an internal protective enzyme-catalyzed clean up system, which is fine and elaborate enough to avoid injuries of active oxygen, thus assuring normal cellular function (Horváth *et al.*, 2007). The balance between ROS production and activities of antioxidative enzyme determines whether oxidative signaling and/or damage will occur (Moller *et al.*, 2007). To minimize the effects of oxidative stress, plants have evolved a complex enzymatic and non-enzymatic antioxidant system, such as low-molecular mass antioxidants (ascorbic acid, glutathione, carotenoids) and ROS scavenging enzymes (superoxide dismutase (SOD), peroxidase (POX), catalase (CAT) and ascorbate peroxidase (APX) (Apel and Hirt, 2004). Non-enzymatic antioxidants cooperate to maintain the integrity of the photosynthetic membranes under oxidative stress. The enzymatic components may directly scavenge ROS or may act by producing a non-enzymatic antioxidant. Efficient destruction of $O_2^{"}$ and H_2O_2 in plant cells requires the concerted action of antioxidants. $O_2^{"}$ can be dismutated into H_2O_2 by SOD in the chloroplast, mitochondrion, cytoplasm and peroxisomes. POX plays a key role in scavenging H_2O_2 which was produced through dismutation of O_2^- catalyzed by SOD. CAT is a main enzyme to eliminate H_2O_2 in the mitochondrion and microbody (Shigeoka *et al.*, 2002) thus helps in ameliorating the detrimental effects of oxidative stress. It is found in peroxisomes, but considered indispensable for decomposing H_2O_2 during stress. Maintaining a higher level of antioxidative enzyme activities may contribute to drought induction by increasing the capacity against oxidative damage (Sharma and Dubey, 2005).

Besides these, salicylic acid (SA) was found to enhance the activities of antioxidant enzymes, CAT, peroxidase (POX) and superoxide dismutase (SOD), when sprayed exogenously to the drought stressed plants of *L. esculentum* (Hayat *et al.*, 2008) or to the salinity stressed plants of *B. juncea* (Yusuf *et al.*, 2008). Krantev *et al.* (2008) reported the exogenous application of SA enhanced the activities of antioxidant enzymes ascorbate peroxidase (APX) and SOD with a concomitant decline in the activity of CAT in maize plants. The role of SA in plant is not limited to pathogen defence signalling but is also a foremost player in light acclimation and in regulation of the redox homeostasis (Mateo *et al.*, 2006; Muhlenbock *et al.*, 2008). Additionally, plant developmental process such as the induction of flowering or seed germination is also dependent on SA (Martinez *et al.*, 2004; Xie *et al.*, 2007). When applied exogenously at suitable concentrations, SA was found to enhance the efficiency of antioxidant system in plants (Knorzer *et al.*, 1999). The priming of seeds with lower concentrations of SA, before sowing, lowered the elevated levels of ROS due to cadmium exposure and also enhanced the activities of various antioxidant enzymes (CAT, guaiacol peroxidase, glutathione reductase and SOD) in *Oryza sativa*, thereby protecting the plants from oxidative burst (Panda and Patra, 2007). Pre-soaked seeds in SA increased the activities of some antioxidant enzymes, namely ascorbate peroxidase and guaiacol peroxidase in pea (Szalai *et al.*, 2011). SA may regulate the synthesis of ascorbic acid (AsA) under senescence and stress conditions (Huang *et al.*, 2008). A high level of endogenous AsA is essential to maintain the antioxidant capacity that protects plants from oxidative stresses (Zhou *et al.*, 2009). Interestingly, increase in AsA content play an important role in preserving APX activity. Patel *et al.* (2011) reported that SA sustained antioxidant system under drought stress in chickpea in which superoxide dismutase (SOD) showed maximum response to SA and also increased activity of APX but not as marked as for SOD. In plants treated with SA @ 1.5 mM AsA contents, SOD and POX activity increased significantly at post- anthesis stress condition as compared to pre- anthesis drought stress.

3. Conclusions

Salicylic acid dilutes the effects of drought to a great extent. Retention of chlorophyll content, minimizing lipid peroxidation, maintenance of relative water content (RWC), and triggering sugar accumulation, proline content in leaves through SA in drought stressed chickpea plants contributes to increased growth and development.

The plants treated with salicylic acid acquire higher photosynthesis, healthy growth of the plants with increase dry mass production. The salicylic acid prevents the changes at the level of membrane organization of the cells and evidently improves plant metabolism and growth under drought stress. It may also be an expression of the involvement of phenols induced by salicylic acid especially in the efficient mobilization of photosynthate to the sink for increasing the sink strength followed by sink efficiency and eventually contributing towards increase in yield potentiality of chickpea crop under drought stress. As a final point, it may be the enhanced antioxidant activity that helps protecting plants from oxidative damage and, hence, favours apposite growth of chickpea under drought stress.

Salicylic acid (SA) is an endogenous plant growth regulator of phenolic nature exhibits a high potential in controlling losses of agricultural crops to adverse environmental conditions, in general, and drought, in particular. Grain legumes as compared to cereals appear to have more sensitivity towards drought. Chickpea

(*Cicer arietinum* L.) is one of them. Salicylic acid (SA) could alleviate the deleterious effects of drought in chickpea genotypes to a great extent, if not up to100 per cent.

4. Summary and Future Perspectives

Salicylic acid plays diverse physiological roles in plants and potentially alleviates the devastating effects generated by drought stress. Overall literature on SA's relevance in mitigating abiotic stresses are motivating for widespread research especially on how this plant hormone interacts and being regulated by the cross-talk in harmony with other established phytohormones and plant growth regulators. It is clear that SA may directly influence the activity of certain enzymes, as described for superoxide dismutase (SOD), peroxidase (POX), ascorbate peroxidase (APX) and catalase (CAT), or may directly or indirectly induce genes responsible for protective mechanisms. However, several questions remain unanswered at both the theoretical and applied levels. First, it is yet to explore extensively whether the effects of exogenous SA are direct or are connected with that of endogenous SA. In future, the exogenous application of this phytohormone might be the powerful tool in enhancing growth, productivity and also in combating the ill effects generated by drought stresses in plants. The future applications of this plant hormone holds a great promise as a management tool for providing tolerance to our agricultural crops against the aforesaid constrains consequently aiding to accelerate potential crop yield in near future.

☆ In view of the immense potential, enough scope is left to work with salicylic acid with crop plants under abiotic stresses. Extensive research is needed on how this plant hormone interacts and being regulated with other established phytohormones and plant growth regulators working at long range (auxins, cytokinins, gibberellins, ethylene etc.), short range (NO, jasmonates, brassinosteriods etc.) and very short range (ROS, H_2O_2).

☆ One could also argue how the regulated doses of these short range phytohormones mostly produced in-vicinity to biotic infestation and then transported systemically to play their role during broad range abiotic stresses.

☆ It is also worthwhile to elucidate the role of aforesaid phytohormone in tissue-specific differentiation and growth of plant parts during growth and development.

☆ Biochemical inhibitors of key enzymes of pathways and mutant study might incident some light on such aspects.

5. References

Aberg B (1981). Plant growth regulators XLI. Monosubstitue benzoic acid. Swed. J. Agric. Res. 11: 93-105.

Abreu ME, Munne´-Bosch S (2009). Salicylic acid deficiency in NahG transgenic lines and *sid2* mutants increases seed yield in the annual plant *Arabidopsis thaliana*. J. Exp. Bot. 60: 1261–1271.

Aftab T, Khan MMA, Idrees M, Naeem MM (2010). Salicylic acid acts as potent enhancer of growth, photosynthesis and artemisinin production in *Artemisia annua* L. J. Crop Sci. Biotech. 13:183–188.

Alexieva V, Sergiev I, Mapelli S, Karanov E (2001). The effect of drought and ultraviolet radiation on growth and stress markers in pea and wheat. Plant Cell Environ. 24: 1337–1344.

Anandhi S, Ramanujam MP (1997). Effect of salicylic acid on black gram (*Vigna mungo*). cultivars. Indian J. Plant Physiol. 2: 138–141.

Apel K, Hirt H (2004). Reactive oxygen species: metabolism, oxidative stress, and signal transduction. Annu. Rev. Plant Biol. 55: 373-399.

Ashraf M, Foolad MR (2007). Roles of glycine betaine and proline in improving plant abiotic stress resistance. Environ Exp. Bot., 59: 206-216.

Azco´n R, Go´mez M, Tobar RM (1996). Physiological and nutritional responses by *Lactuca sativa* L. to nitrogen sources and mycorrhizal fungi under drought conditions. Biol. Fertil. Soils, 22: 156–161.

Basu PS, Masood A, Chaturvedi SK (2007). Osmotic adjustment increases water uptake, remobilization of assimilates and maintains photosynthesis in chickpea under drought. Indian J. Exp. Biol. 45: 261-267.

Blum A, Ebercon A (1981). Cell membrane stability as a measure of drought and heat tolerance in wheat. Crop Sci. 21:43–47.

Blum A (1996). Crop responses to drought and the interpretation of adaptation. Plant Growth Regul. 20: 135–148.

Canci H, Toker C (2009). Evaluation of yield criteria for drought and heat resistance in chickpea (*Cicer arietinum* L.). J. Agron Crop Sci. 195: 47–54

Correia MJ, Fonseca F, Azedo-Silva J, Dias C, David MM, Barrote I, Oso´rio ML, Oso´rio J (2005). Effects of water deficit on the activity of nitrate reductase and content of sugars, nitrate and free amino acids in the leaves and roots of sunflower and white lupin plants growing under two nutrient supply regimes. Physiologia Plantarum 124: 61-70.

Demiral T, Turkan I (2004). Does exogenous glycine betaine affect antioxidative system of rice seedlings under NaCl treatment? J Plant Physiol. 161: 1089-1110.

Dichio B, Xiloyannis C, Angelopoulos KV, Nuzzo SA, Bufo Celano G (2003). Drought-induced variations of water relations parameters in *Olea europaea*. Plant Soil 257: 381-903.

Efeoglu B, Ekmekci Y, Cicek N (2009). Physiological responses of three maize cultivars to drought stress and recovery. S. Afr. J. Bot. 75:34-42.

Fariduddin Q, Hayat S, Ahmad A (2003). Salicylic acid influences net photosynthetic rate, carboxylation efficiency, nitrate reductase activity and seed yield in *Brassica juncea*. Photosynthetica 41: 281–284.

Farooq M, Wahid A, Kobayashi N, Fujita D, Basra SMA (2009). Plant drought stress: effects, mechanisms and management. Agron. Sustain. Dev. 29: 185-212.

Forbes JC, Watson RD (1992). Plants in Agriculture. University Press, Cambridge, ISBN: 0521427916: 355.

Foundy BR, Geiger DR, Servaites JC (1989). photosynthesis, carbohydrate metabolism and export in *Beta vulgaries* L. and *Phaseolus vulgaris* L. during square and sinusoidal light regimes. Plant Physiol. 89: 396.

Gan YT, Siddique KHM, MacLeod WJ, Jayakumar P (2006). Management options for minimizing the damage by ascochyta blight (*Ascochyta rabiei*). in chickpea (*Cicer arietinum* L.). Field Crops Res. 97:121-134.

Ganesan V, Thomas G (2001). Salicylic Acid Response in Rice: Influence of Salicylic Acid on H_2O_2 Accumulation and Oxidative Stress. Plant Sci. 160: 1095-1106.

Garg BK, Kathju S, Burman U (2001). Influence of water stress on water relation, photosynthetic parameters and nitrogen metabolism of moth bean genotype. Biologia Plantarum 44:289-292.

Gunes A, Inal A, Alpaslan M, Eraslan F, Bagci EG, Cicek N (2007). Salicylic acid induced changes on some physiological parameters symptomatic for oxidative stress and mineral nutrition in maize (*Zea mays* L.). grown under salinity. J. Plant Physiol. 164:728–736.

Gutie´rrez-Coronado MA, Trejo-Ló´pez C, Larque´-Saavedra A (1998). Effects of salicylic acid on the growth of roots and shoots in soybean. Plant Physiol. Biochem. 36: 563-565.

Hakimi-Al, AMA (2006). Counteraction of drought stress on soybean plants by seed soaking in salicylic acid. J. Bot. 2:421–426.

Hall CA, Cuppett SL (1997). Structure-activities of natural antioxidants.141–170 In: Auroma O.I. and Cuppett S.L. (eds), Antioxidant methodology in vivo and in vitro concepts. AOCS Press, Champaign, IL.

Hare PD, Cress WA, Van SJ (1998). Dissecting the roles of osmolytes accumulation during stress. Plant Cell Environ. 21: 535-553.

Harris D, Tripathi RS, Joshi A (2002). On-farm seed priming to improve crop establishment and yield in dry direct-seeded rice, in: Pandey S., Mortimer M., Wade L., Tuong T.P., Lopes K., Hardy B. (Eds.), Direct seeding: Research Strategies and Opportunities, International Research Institute, Manila, Philippines, pp. 231-240.

Hayat S, Fariduddin Q, Ali B, Ahmad A (2005). Effect of salicylic acid on growth and enzyme activities of wheat seedlings. Acta Agron. Hung. 53: 433–437.

Hayat S, Hasan SA, Farriduddun Q, Ahmad A (2008). Growth of tomato (*Lycopersicon esculentum*). in response to salicylic acid under water stress. J. Plant Int. 3 (4): 297-304.

Horváth E, Pál M, Szalai G, Páldi E, Janda T (2007). Exogenous 4- hydroxybenzoic acid and salicylic acid modulate the effect of short term drought and freezing stress on wheat plants. Biol. Plant, 51: 480-487.

Hsiao TC, O'Toole JC, Yambao EB, Turner NC (1984). Influence of osmotic adjustment on leaf rolling and tissue death in rice (*Oryza sativa* L.). Plant Physiol. 95: 338–341.

Huang R, Xia R, Lu Y, Xu Y (2008). Effect of Pre- harvest salicylic acid spray treatment on post-harvest antioxidant in the pulp and peel of 'Cara cara' navel orange (*Citrus sinensis* L. Osbeck). J. Sci. Food Agr., 88: 229-236.

Huber SC, Rogers HH, Mowry FL (1984). Effect of water stress on photosynthesis and carbon partitioning in soybean (*Glycine max* (L.). Merr.). plants grown in the field at different CO_2 levels. Plant Physiol. 76: 244–249.

Hussain M, Malik MA, Farooq M, Ashraf MY, Cheema MA (2008). Improving drought tolerance by exogenous application of glycine betaine and salicylic acid in sunflower. J. Agron. Crop Sci. 194: 193-199.

Jain A, Srivastava HS (1981). Effect of salicylic acid on nitrate reductase activity in maize seedlings. Physiol. Plant 51: 339-342.

Jaleel CA, Manivannan P, Lakshmanan GMA, Gomathinayagam M, Panneerselvam R (2008). Alterations in morphological parameters and photosynthetic pigment responses of *Catharanthus roseus* under soil water deficits. Colloids Surf. B. Biointerfaces 61:298–303.

Kalefetetoglu Macar T, Ekmekci Y (2009). Alternations in photochemical and physiological activities of chickpea (*Cicer arietinum* L.). cultivars under drought stress. J. Agron. Crop Sci. 195: 335-345.

Kalefetoglu Macar T, Y Ekmekci (2008). PSII photochemistry and antioxidant responses of a chickpea variety exposed to drought. Z. Naturforsch 63c: 583–594.

Kang GZ, Wang ZX, Sun GC (2003). Participation of H_2O_2 in Enhancement of Cold Chilling by Salicylic Acid in Banana Seedlings. Acta Bot. Sinica. 45: 567–573.

Kavar T, Maras M, Kidric M, Sustar-Vozlic J, Meglic V (2008). Identification of genes involved in the response of leaves of Phaseolus vulgaris to drought stress. Mol. Breeding 21: 159–172.

Keller F, Ludlow MM (1993). Carbohydrate metabolism in drought- stress leaves of pigeonpea (*Cajanus cajan*). J. Exp. Bot. 44: 1351.

Khan W, Prithiviraj B, Smith D (2003). Photosynthetic response of corn and soybean to foliar application of salicylates. J. Plant Physiol. 160:485–492.

Knorzer OC, Lederer B, Durner J, Boger P (1999). Antioxidative defense activation in soybean cells. Physiol. Plant 107: 294–302.

Kova´ cik J, Gru´ z J, Backor M, Strnad M, Repca´k M (2009). Salicylic acid-induced changes to growth and phenolic metabolism in *Matricaria chamomilla* plants. Plant Cell Rep. 28: 135–143.

Krantev A, Yordanova R, Janda T, Szalai G, Popova L (2008). Treatment with salicylic acid decreases the effect of cadmium on photosynthesis in maize plants. J. Plant Physiol. 165: 920–931.

Kumar P, Dube SD, Chauhan VS (1999). Effect of salicylic acid on growth, development and some biochemical aspects of soybean (*Glycine max* L. Merrill). Indian J. Plant Physiol. 4: 327-330.

Kumar P, Lakshmi NJ, Mani VP (2000). Interactive effects of salicylic acid and phytohormones on photosynthesis and grain yield of soybean (*Glycine max* L. Merrill). Physiol. Mol. Biol. Plants 6: 179-186.

Larque-Saavedra A, Martin-Mex F (2007). Effects of salicylic acid on the bio-productivity of the plants. In: Hayat S, Ahmad A (Eds.), Salicylic Acid, A Plant Hormone. Springer Publishers, Dordrecht, The Netherlands.

Lecoeur J, Wery J, Turc O (1992). Osmotic adjustment as a mechanism of dehydration postponement in chickpea (*Cicer arietinum* L.). leaves. Plant Soil 144: 177-189.

Leport L, Turner NC, Davies SL, Siddique KHM (2006). Variation in pod production and abortion among chickpea cultivars under terminal drought. Eur. J. Agron. 24: 236-246.

Levitt J (1972). Responses of Plants to Environmental Stresses. Academic Press, McGraw Hill, New York.

Levitt J (1980). Responses of Plants to Environmental Stresses, 2nd edn. Vol. II. Water, Radiation, Salt and other Stresses, pp. 25–280. Academic Press, New York.

Lilley JM, Ludlow MM (1996). Expression of osmotic adjustment and dehydration tolerance in diverse rice lines. Field Crops Res. 48:185-197.

Manivannan P, Abdul Jaleel C, Sankar B, Kishorekumar A, Somasundaram R, Lakshmanan GMA, Panneerselvam R (2007). Growth, biochemical modifications and proline metabolism in *Helianthus annuus* L. as induced by drought stress. Colloids Surfaces B: Biointerfaces 59: 141–149.

Martinez C, Pons E, Parts G, Leon J (2004). Salicylic acid regulates flowering time and links defense responses and reproductive development. Plant J. 37: 209- 217.

Mateo A, Funck D, Muhlenbock P, Kular B, Mullineaux PM, Karpinski S (2006). Controled levels of salicylic acid are required for optimal photosynthesis and redox homeostasis. J. Exp. Bot. 57: 1795-1807.

Mittler R (2002). Oxidative stress, antioxidants and stress tolerance. Trends Plant Sci. 7: 405-410.

Moller IM, Jensen PE, Hansson A (2007). Oxidative modifications to cellular components in plants. Annu. Rev. Plant Biol. 58: 459-481.

Morgan JM, Rodriguez B, Knights EJ (1991). Adaptation to water deficits in chickpea breeding lines by osmoregulation: relationship to grain-yield in the field. Field Crop Res. 27: 61.

Morgan JM, King RW (1984). Association between loss of leaf turgor, abscisic acid levels and seed set in two wheat cultivars. Aust. J. Plant Physiol. 11: 143-150.

Morgan JM (1994). Osmoregulation and water-stress in higher-plants. Annu Rev Plant Physiol. Plant Mol.Biol. 35:299–319.

Moussa HR, El-Gamel SM (2010). Effect of salicylic acid pre-treatment on cadmium toxicity in wheat. Biol. Plant 54: 315–320.

Muhlenbock P, Szechynska- Hebda M, Plaszczyca M, Baudo M, Mullineaux PM, Parker J E, Karpinska B, Karpinski S (2008). Chloroplast signaling and LESION SIMULATING DISEASE 1. Regulate crosstalk between light acclimation and immunity in Arabidopsis. Plant Cell 20: 2339- 2356.

Mukane MA, Desai BB, Naik RM, Chavan U (1996). Biochemical markers for water stress in pigeonpea genotypes. J Maharashtra Agri Univ 21: 140–141.

Nayyar H, Gupta D (2006). Differential sensitivity of C_3 and C_4 plants to water deficit stress: association with oxidative stress and antioxidants. Environ. Exp. Bot. 58: 106-113.

Nonami H (1998). Plant water relations and control of cell elongation at low water potentials. J. Plant Res. 111: 373-382.

Pancheva TV, Popova LP, Uzunova AM (1996). Effect of salicylic acid on growth and photosynthesis in barley plants. J. Plant Physiol. 149: 57-63.

Panda SK, Patra HK (2007). Effect of salicylic acid potentiates cadmium-induced oxidative damage in *Oryza sativa* L. leaves. Acta Physiol. Plant. 29: 567–575.

Passioura JB (2002). Environmental plant biology and crop improvement. Funct. Plant Biol. 29, 537–546.

Patel PK, Hemantaranjan A (2012). Salicylic Acid Induced Alteration in Dry Matter Partitioning, Antioxidant Defence System and Yield in Chickpea (*Cicer arietinum* L.). under Drought Stress. Asian J. Crop Sci. 4: 86-102.

Patel PK, Hemantaranjan A, Sarma BK, Singh Radha (2011). Growth and antioxidant system under drought stress in Chickpea (*Cicer arietinum* L.). as sustained by salicylic acid. J. Stress Physiol. Biochem. 7: 130-144.

Patel PK, Hemantaranjan A. Sarma, BK (2012). Effect of salicylic acid on growth and metabolism of chickpea (*Cicer arietinum* L.). under early and late drought stress. Indian J. Plant Physiol. 17: 170-176.

Praba ML, Cairns JE, Babu RC, Lafitte HR (2009). Identification of physiological traits underlying cultivar differences in drought tolerance in rice and wheat. J. Agron. Crop Sci. 195: 30–46.

Quick P, Siegel G, Neuhaus E, Fiel R, Stitt M (1989). Short-term water stress leads to a stimulation of sucrose synthesis by activating sucrose-phosphate synthase. Planta 177: 535–546.

Rane J, Lakkineni KC, Kumar PA, Abrol YP (1995). Salicylic acid protects nitrate reductase activity of wheat leaves. Plant Physiol. Biochem. 22: 119-121.

Rao MV, Paliyath G, Ormrod P, Murr DP, Watkins CB (1997). Influence of salicylic acid on H_2O_2 production, oxidative stress, and H_2O_2 - metabolizing enzymes. Plant Physiol. 115:137-149.

Rhodes D, Samaras Y (1994). Genetic control of osmoregulation in plants. In cellular and molecular physiology of cell volume regulation. Strange K Boca Raton: CRC Press, pp. 347-361.

Rouhi V, Samson R, Lemeur R, Van Damme P (2007). Photosynthetic gas exchange characteristics in three different almond species during drought stress and subsequent recovery. Environ. Exp. Bot. 59: 117–129.

Routley DG (1966). Proline accumulation in wilted ladino clover leaves. Crop Sci 6: 358-361

Rucker KS, Kvien CK, Holbrook CC, Hook JE (1995). Identification of peanut genotypes with improved drought avoidance traits. Peanut Sci. 24: 14-18.

Sairam RK, Veerabhadra Rao K, Srivastava GC (2002). Differential response of wheat genotypes to long term salinity stress in relation to oxidative stress, antioxidant activity and osmolyte concentration, Plant Sci. 163: 1037–1046.

Sanchez FJ, Manzanares M, de Andres EF, Tenorio JL, Ayerbe L (1998). Turgor maintenance, osmotic adjustment and soluble sugar and praline accumulation in 49 pea cultivars in response to water stress. Field Crops Res. 59: 225–235.

Sanchez-Blanco MJ, Rodriguez P, Morales MA, Ortuo MF, Torrecillas A (2002). Comparative growth and water relations of *Citrus albidus* and *Citrus monspeliensis* plants during water deficit conditions and recovery. Plant Sci. 162: 107-113.

Saxena NP (2003). Management of drought in chickpea—a holistic approach. In: Saxena NP, ed. Management of agricultural drought—agronomic and genetic options. New Delhi: IBH Publishing, 103–122.

Senaratna T, Touchell D, Bunn E, Dixon K (2000). Acetyl salicylic acid (aspirin). and salicylic acid induce multiple stress tolerance in bean and tomato plants. Plant Growth Regul. 30: 157-161.

Serraj R, Sinclair TR (2002). Osmolyte accumulation: can it really help increase crop yield under drought conditions? Plant Cell Environ. 25: 333–341.

Shakirova FM (2007). Role of hormonal system in the manisfestation of growth promoting and anti-stress action of salicylic acid. In: Hayat S, Ahmad A (Eds.), Salicylic Acid, A Plant Hormone. Springer, Dordrecht, Netherlands.

Shakirova FM, Sakhabutdinova AR, Bezrukova MV, Fatkhutdinova RA, Fatkhutdinova DR (2003). Changes in the hormonal status of wheat seedlings induced by salicylic acid and salinity. Plant Sci. 164: 317-322.

Shao HB, Chu LY, Shao MA, Abdul Jaleel C, Hong-Mei M (2008). Higher plant antioxidant and redox signaling under environmental stresses. Comp. Rend. Biol. 331: 433-441.

Sharma P, Dubey RS (2005). Lead toxicity in plants. Braz. J. Plant Physiol. 17: 35–52.

Shigeoka S, Ishikawa T, Tamoi M, Miyagawa Y, Takeda T, Yabuta Y, Yoshimura K (2002). Regulation and function of ascorbate peroxidase isoenzymes. J. Exp. Bot. 53: 1305-1319.

Siddique MRB, Hamid A, Islam MS (2001). Drought stress effects on water relations of wheat. Bot. Bull. Acad. Sin. 41: 35-39.

Singh KB, Malhotra RS, Saxena MC, Bejiga G (1997). Superiority of winter sowing over traditional spring sowing of chickpea in the Mediterranean region. Agron. J. 89: 112-118.

Singh B, Usha K (2003). Salicylic Acid Induced Physiological and Biochemical Changes in Wheat Seedlings under Water Stress. Plant Growth Regul. 39: 137–141.

Solomonson LP, Barber MJ (1990). Assimilatory nitrate reductase functional properties and regulation. Annu. Rev. Plant Physiol. Plant Mol. Biol. 41: 225-253.

Stevens J, Senaratna T, Sivasithamparam K (2006). Salicylic acid induces salinity tolerance in tomato (*Lycopersicon esculentum* cv. Roma): associated changes in gas exchange, water relations and membrane stabilisation. Plant Growth Regul. 49: 77–83.

Subbarao GV, Nam NH, Chauhan YS, Johansen C (2000). Osmotic adjustment, water relations and carbohydrate remobilization in pigeonpea under water deficits. J. Plant Physiol. 157: 651–659.

Szabados L, Savoure´ A (2009). Proline: a multifunctional amino acid. Trends Plant Sci. 15: 89-97.

Szalai G, Horgosi S, Soós V, Majláth I, Balázs E, Janda T (2011). Salicylic acid treatment of pea seeds induces its de novo synthesis. J. Plant Physiol. 168:213–219.

Szira F, Ba´lint AF, Bo¨rner A, Galiba G (2008). Evaluation of drought-related traits and screening methods at different developmental stages in spring barley. J. Agron. Crop Sci. 194: 334–342.

Timpa JD, Burke JJ, Quinsenberry JE, Wendt CW (1986). Effect of water stress on the organic acid and carbohydrate composition of cotton plants. Plant Physiol. 82: 724–728.

Tripathy JN, Zhang J, Robin S, Nguyen Th T, Nguyen HT (2000). QTLs for cell-membrane stability mapped in rice (*Oryza sativa* L.). under drought stress. Theo. App. Gen. 100: 1197-1202.

Umebese CE, Olatimilehin TO, Ogunsusi TA (2009). Salicyclic acid protects nitrate reductase activity, growth and proline in amaranth and tomato plants during water deficit. Am. J. Agri. Biol. Sci. 4: 224-229.

Vendruscolo ECG, Schuster I, Pileggi M, Scapim CA, Molinari HBC, Marur CJ, Vieira LGE (2007). Stress-induced synthesis of proline confers tolerance to water deficit in transgenic wheat. J. Plant Physiol. 164(10): 1367-1376.

Wang LJ, Fan L, Loescher W, Duan W, Liu GJ, Cheng JS (2010). Salicylic acid alleviates decreases in photosynthesis under heat stress and accelerates recovery in grapevine leaves. BMC Plant Biol. 10: 34.

Xie Z, Zhang ZL, Hanzlik S, Cook E, Shen QJ (2007). Salicylic acid inhibits gibberellin-induced alpha-amylase expression and seed germination via a pathway involving an abscisic-acid inducible WRKY gene. Plant Mol. Biol. 64: 293–303.

Yancy PH, Clark ME, Hand SC, Bowlus RD, Somero GN (1982). Living with water stress: evolution of osmolytes systems. Science 217: 1214–1223.

Yang YN, Qi M, Mei CS (2004). Endogenous salicylic acid protects rice plants from oxidative damage caused by aging as well as biotic and abiotic stress. Plant J. 40: 909-919.

Yusuf M, Hasan SA, Ali B, Hayat S, Fariduddin Q, Ahmad A (2008). Effect of salicylic acid on salinity induced changes in *Brassica juncea*. J. Integrative Plant Biol. 50 (8): 1–4.

Zhang LX, Zhang SX, Li H, Liang ZS (2007). Nitrogen rates and water stress affects on production, lipid peroxidation and antioxidative enzyme activities in two maize (*Zea mays* L.). genotypes. J. Agron. Crop Sci. 193: 387–397.

Zhou SZ, Guo K, Elbaz AA, Yang ZM (2009). Salicylic acid alleviates mercury toxicity by preventing oxidative stress in roots of *Medicago sativa*. Environ. Exp. Bot. 65: 27.

Chapter 7
Production of Secondary Metabolites from Medicinal Plants

Manish Das

Principal Scientist (Plant Physiology), Krishi Anushandhan Bhawan-II
Indian Council of Agricultural Research, New Delhi – 110 012
E-mail: manishdas50@gmail.com

CONTENTS

Plants have the ability to synthesize a wide variety of chemical compounds that are used to perform important biological functions, and to defend against attack from predators such as insects, fungi and herbivorous mammals. Many of these phytochemicals have beneficial effects on long-term health when consumed by humans, and can be used to effectively treat human diseases. At least 12,000 such compounds have been isolated so far; a number estimated to be less than 10 per cent of the total

(Tapsell *et al.*, 2006; Lai and Roy, 2004). These phyto-chemicals are divided into (1) primary metabolites such as sugars and fats, which are found in all plants; and (2) secondary metabolites – compounds which are found in a smaller range of plants, serving a more specific function (Meskin, 2002). For example, some secondary metabolites are toxins used to deter predation and others are pheromones used to attract insects for pollination. It is these secondary metabolites and pigments that can have therapeutic actions in humans and which can be refined to produce drugs— examples are inulin from the roots of dahlias, quinine from the cinchona, morphine and codeine from the poppy, and digoxin from the foxglove (Meskin, 2002). Chemical compounds in plants mediate their effects on the human body through processes identical to those already well understood for the chemical compounds in conventional drugs; thus herbal medicines do not differ greatly from conventional drugs in terms of how they work. This enables herbal medicines to be as effective as conventional medicines, but also gives them the same potential to cause harmful side effects (Tapsell *et al.*, 2006; Lai and Roy, 2004).

Most cultures have a tradition of using plants medicinally. In Europe, apothecaries stocked herbal ingredients for their medicines. In the Latin names for plants created by Linnaeus, the word *officinalis* indicates that a plant was used in this way. For example, the marsh mallow has the classification *Althaea officinalis*, as it was traditionally used as an emollient to soothe ulcers (William, 2004). Ayurvedic medicine, herbal medicine and traditional Chinese medicine are other examples of medical practices that incorporate medical uses of plants. Pharmacognosy is the branch of modern medicine about medicines from plant sources.

Modern medicine now tends to use the active ingredients of plants rather than the whole plants. The phytochemicals may be synthesized, compounded or otherwise transformed to make pharmaceuticals. Examples of such derivatives include Digoxin, from digitalis; capsaicine, from chili; and aspirin, which is chemically related to the salicylic acid found in white willow (William, 2004). The opium poppy continues to be a major industrial source of opiates, including morphine. Few traditional remedies, however, have translated into modern drugs, although there is continuing research into the efficacy and possible adaptation of traditional herbal treatments.

1. Secondary Metabolites

Secondary metabolites are organic compounds that are not directly involved in the normal growth, development, or reproduction of an organism (Fraenkel, 1959). Unlike primary metabolites, absence of secondary metabolites does not result in immediate death, but rather in long-term impairment of the organism's survivability, fecundity, or aesthetics, or perhaps in no significant change at all. Secondary metabolites are often restricted to a narrow set of species within a phylogenetic group (Stamp, 2003). Secondary metabolites often play an important role in plant defense against herbivory (Samuni-Blank *et al.*, 2012) and other interspecies defenses (Chizzali *et al.*, 2012). Humans use secondary metabolites as medicines, flavorings, and recreational drugs.

Studies on plant secondary metabolites have been increasing over the last 50 years. These molecules are known to play a major role in the adaptation of plants to their environment, but also represent an important source of active pharmaceuticals. Plant cell culture technologies were introduced at the end of the 1960s as a possible tool for both studying and producing plant secondary metabolites (Bourgaud *et al.*, 2001). Different strategies, using *in vitro* systems, have been extensively studied with the objective of improving the production of secondary plant compounds. Undifferentiated cell cultures have been mainly studied, but a large interest has also been shown in hairy roots and other organ cultures. Specific processes have been designed to meet the requirements of plant cell and organ cultures in bioreactors. Despite all of these efforts of the last 30 years, plant biotechnologies have led to very few commercial successes for the production of valuable secondary compounds. Compared to other biotechnological fields such as microorganisms or mammalian cell cultures, this can be explained by a lack of basic knowledge about biosynthetic pathways, or insufficiently adapted reactor facilities (Harborne, 1999). More recently, the emergence of recombinant DNA technology has opened a new field with the possibility of directly modifying the expression of genes related to biosyntheses. It is now possible to manipulate the pathways that lead to secondary plant compounds. Many research projects are now currently being carried out and should give a promising future for plant metabolic engineering.

2. Plant Secondary Compounds

Plant secondary compounds are usually classified according to their biosynthetic pathways (Harborne, 1999). Three large molecule families are generally considered: phenolics, terpenes and steroids, and alkaloids. A good example of a widespread metabolite family is given by phenolics: because these molecules are involved in lignin synthesis, they are common to all higher plants. However, other compounds such as alkaloids are sparsely distributed in the plant kingdom and are much more specific to defined plant genus and species. This narrower distribution of secondary compounds constitutes the basis for chemotaxonomy and chemical ecology (Bourgaud *et al.*, 2001).

Due to their large biological activities, plant secondary metabolites have been used for centuries in traditional medicine. Nowadays, they correspond to valuable compounds such as pharmaceutics, cosmetics, fine chemicals, or more recently nutraceutics. Recent surveys have established that in western countries, where chemistry is the backbone of the pharmaceutical industry, 25 per cent of the molecules used are of natural plant origin (Payne *et al.*, 1991). A good example could be aspirin (acetylsalicylate) which derives from salicylate. The genuine molecule can be isolated in large quantities from many plants (*Spiraea ulmaria, Betula lenta*..), but the chemical is synthesized as an acetyl-derivative in order to lower secondary effects (stomachaches). Chemical synthesis apart, the production of plant secondary metabolites has, for a long time, been achieved through the field cultivation of medicinal plants (Bourgaud *et al.*, 2001). However, plants originating from particular biotopes can be hard to grow outside their local ecosystems. It also happens that common plants do not withstand large field cultures due to pathogen sensitiveness (anthracnose on *Hypericum perforatum* or *Arnica montana*). This has led scientists and biotechnologists to consider plant cell, tissue and organ cultures as an alternative way to produce the corresponding secondary metabolites (Payne *et al.*, 1991).

3. Need for Breakthroughs for the Production of Secondary Compounds

Industrial interest emerged quite early in the story of plant *in vitro* cultures and secondary metabolites. Indeed, at least six large-scale industrial productions were engaged between 1976 and 1986 elsewhere in the world (Zryd, 1988). However, since that time, this technology has led to only a few applications for the production of commercial compounds. This lack of success can be attributed to several bottlenecks (Steward *et al.*, 1999a).

First of all, it is obvious that industrial processes were started at a time when basic knowledge was severely lacking on both plant vitro cultures and secondary compounds. It may be surprising to realize that a key-problem like cell viability, and its assessment, has been studied only recently with plant cells (Steward *et al.*, 1999a; Steward *et al.*, 1999b), whereas, this topic has been considered for a long time with animal cell cultures (Cook and Mitchell, 1989). Also, secondary metabolites are produced following long biosynthetic pathways that can involve dozens of enzymes. This synthesis is much more complex than for recombinant proteins produced with mammalian or prokaryotic biotechnologies, which usually involve one or two genes. This can partially explain the lack of success of plant cell and tissue cultures compared to other expression systems (Cook and Mitchell, 1989).

A second bottleneck has to do with the economic feasibility of plant cell and organ cultures. Indeed, this technology requires high-cost bioreactors associated with aseptic conditions that are expensive to maintain (Curtis, 1999). Besides, unlike mammalian cell cultures which produce high-value molecules, plant vitro cultures have also been considered for the production of low or moderate-value molecules such as food ingredients. As a consequence, economic feasibility is even more difficult to reach (Cook and Mitchell, 1989).

4. Strategies to Use Plant *in vitro* Culture Technologies

There have been many attempts in the last decade to address the problem of cost-effectiveness of plant vitro culture technologies. Several new routes have been investigated. The first one is the design of low-cost bioreactors associated with a reduction of online controls and probes, and a simplified sterilization process (Curtis, 1999). Another possible strategy could be the use of plastic bags as already in use for the cultivation of microalgae (Borowitzka, 1999). These plastic bags are obviously much cheaper to use than culture reactors. However, they do not address all the problems encountered with cell cultures such as cell autotrophy or sedimentation. Others have tried to change more radically the concept of plant cell tissue and organ cultures in bioreactors without the use of a traditional culture chamber. A new system has been proposed by Borisjuk and co-workers (Borisjuk *et al.*, 1999) although it was designed for the production of recombinant proteins from genetically modified plants. The basic concept is to grow entire plants in a system where roots are maintained in sterile conditions and separated from the aerial parts. The recombinant proteins are recovered in the nutrient solution that surrounds the roots with the help of the rhizospheric excretion from the plant. This concept is interesting because it uses whole plants. It addresses the problem of biomass production encountered with cell suspension cultures that are hard to develop.

Another system has been recently proposed by Gontier and co-workers (Gontier *et al.*, 2000). This system has been designed specifically for the production of secondary compounds. The basic principle is to get rid of sterility but to keep a culture medium that allows the manipulation of secondary compound production (elicitation, addition of precursors, etc.). Sterility is essential because a traditional *in vitro* culture medium is constituted of organic compounds, especially sugars, that would be altered by microorganisms if maintained in non-sterile conditions. If we use whole plants that are photosynthetically active, it is possible to avoid the use of organic compounds in a mineral-based medium. In this case, the medium is less amenable to contamination in open conditions. The other basic concept of this device is to yield secondary compounds in a non-destructive manner for the plants, therefore allowing repeated and regular harvests from the same biomass. In practice, this is achieved by growing the plants in hydroponic or aeroponic non-sterile conditions and by treating the medium in order to elicit the synthesis and to allow the excretion of the desired compounds, out of the roots. The last step is the recovery of the compounds from the medium with the help of purification processes. As a conclusion, this system uses the plants for what they can do best in natural conditions: they grow quite quickly when they are photosynthetically active. It also uses the concept of culture medium for what it is most useful in reactor culture technology: it allows a close control of secondary compound production (Gontier *et al.*, 2000). This technology has been successfully used at pilot scale on three different model plants: *Datura innoxia* for tropane alkaloids, *Ruta graveolens* for furocoumarins, and *Taxus baccata* for paclitaxel (Borisjuk *et al.*, 1999).

5. Plant Metabolic Engineering and Pharmaceutical Compounds

Plant metabolic engineering has been successfully applied to the production of pharmaceutically useful secondary metabolites. Many pathways are currently being investigated but alkaloids have probably received more attention because of their pharmaceutical relevance (Yun *et al.*, 1992). Pioneer investigations on medicinal plants and metabolic engineering were conducted by Yun *et al.* (1992) on tropane alkaloids. These authors used a hyoscyamine 6β-hydroxylase (*H6H*) gene from *Hyoscyamus niger*, controlled by a 35s promoter, and succeeded in overexpressing h6h activity in *Atropa belladonna*. Transgenic *Atropa* plants displayed an enhanced conversion of hyoscyamine into scopolamine, which is a more pharmaceutically useful compound. These results clearly demonstrate that it is possible to considerably modify secondary metabolite patterns, playing on enzymes located downstream to the synthesis (Bourgaud *et al.*, 2001). Other experiments based on single gene transformations have been carried out by the same group on various alkaloid producing plants (Yun *et al.*, 1992). Overexpression of tobacco putrescine-*N*-methyl-transferase (PMT) in *Atropa belladonna* led to unmodified alkaloid profiles in the transformants whereas the same transformation carried out on *Nicotiana sylvestris* gave a 40 per cent increase in leaf nicotine. Unlike H6H, PMT is placed upstream in alkaloid synthesis. PMT experiments demonstrate that it is possible to increase the metabolite flux so that it can benefit a downstream compound (nicotine). However, this strategy is not always successful, as in the case of *Atropa belladonna*, as other enzymes possibly limit the synthesis. Another strategy has been put forward by Sato and co-workers (Sato *et al.*, 2001) with experiments on scoulerine 9-*O*-methyltransferase (SMT). It consists in playing with pathways that derive from the same branching point. SMT converts scoulerine in successive compounds that lead to berberine, but scoulerine can also be transformed into sanguinarine by a parallel pathway (Bourgaud *et al.*, 2001). SMT overexpression allowed an increase in the berberine content of *Coptis* transgenic cells (20 per cent) to the detriment of the competitive pathway.

Many attempts have been made to invent new devices that could lead to breakthroughs for the production of secondary compounds. However, the major changes in the area of plant secondary metabolites have probably been achieved thanks to the rise of molecular genetics in recent years, through the so-called metabolic engineering approach.

6. Bioactive Natural Products

The discovery of new bioactive natural products is still a fascinating field in organic chemistry as demonstrated by the recent paradigms of the anticancer drug epothilon, the immunosuppressant rapamycin, or the proteasome inhibitor salinosporamide, to name but a few of hundreds of possible examples (Dickschat, 2011). Finding new secondary metabolites is a prerequisite for the development of novel pharmaceuticals, and this is an especially urgent task in the case of antibiotics due to the rapid spreading of bacterial resistances and the emergence of multiresistant pathogenic strains, which poses severe clinical problems in the treatment of infectious diseases (Dickschat, 2011). Biosynthetic aspects are closely related to functional investigations, because a deep understanding of metabolic pathways to natural products, not only on a chemical, but also on a genetic and enzymatic level, allows for the expression of whole biosynthetic gene clusters in heterologous hosts (Ravishankar and Rao, 2000). This technique can make interesting, new secondary metabolites available from unculturable microorganisms, or may be used to optimise their availability by fermentation, for further research and also for production in the pharmaceutical industry.

7. Secondary Metabolites from Medicinal Plants by Plant Tissue Cultures

Plants are a tremendous source for the discovery of new products of medicinal value for drug development. Today several distinct chemicals derived from plants are important drugs currently used in one or more countries in the world. Many of the drugs sold today are simple synthetic modifications or copies of the naturally obtained substances. The evolving commercial importance of secondary metabolites has in recent years resulted in a great interest in secondary metabolism, particularly in the possibility of altering the production of bioactive plant metabolites by means of tissue culture technology (Vanisree *et al.*, 2004). Plant cell culture technologies were introduced at the end of the 1960's as a possible tool for both studying and producing plant secondary metabolites. Different strategies, using an *in vitro* system, have been extensively studied to improve the production of plant chemicals. The focus is now on the application of tissue culture technology for the production of some important plant pharmaceuticals.

Many higher plants are major sources of natural products used as pharmaceuticals, agrochemicals, flavor and fragrance ingredients, food additives, and pesticides (Balandrin and Klocke, 1988). The search for new plant derived chemicals should thus be a priority in current and future efforts toward sustainable conservation and rational utilization of biodiversity (Phillipson, 1990). In the search for alternatives to production of desirable medicinal compounds from plants, biotechnological approaches, specifically, plant tissue cultures, are found to have potential as a supplement to traditional agriculture in the industrial production of bioactive plant metabolites (Rao and Ravishankar, 2002). Cell suspension culture systems could be used for large scale culturing of plant cells from which secondary metabolites could be extracted. The advantage of this method is that it can ultimately provide a continuous, reliable source of natural products (Vanisree *et al.*, 2004).

Discoveries of cell cultures capable of producing specific medicinal compounds at a rate similar or superior to that of intact plants have accelerated in the last few years. New physiologically active substances of medicinal interest have been found by bioassay. It has been demonstrated that the biosynthetic activity of cultured cells can be enhanced by regulating environmental factors, as well as by artificial selection or the induction of variant clones (Vanisree *et al.*, 2004). Some of the medicinal compounds localized in morphologically specialized tissues or organs of native plants have been produced in culture systems not only by inducing specific organized cultures, but also by undifferentiated cell cultures. The possible use of plant cell cultures for the specific biotransformations of natural compounds has been demonstrated (Cheetham, 1995; Scragg, 1997; Krings and Berger, 1998; Ravishankar and Rao, 2000). Due to these advances, research in the area of tissue culture technology for production of plant chemicals has bloomed beyond expectations.

8. Advantages of a Cell Culture System

The major advantages of a cell culture system over the conventional cultivation of whole plants are: (1) Useful compounds can be produced under controlled conditions independent of climatic changes or soil conditions; (2) Cultured cells would be free of microbes and insects; (3) The cells of any plants, tropical or alpine, could easily be multiplied to yield their specific metabolites; (4) Automated control of cell growth and rational regulation of metabolite processes would reduce of labour costs and improve productivity; (5) Organic substances are extractable from callus cultures.

In order to obtain high yields suitable for commercial exploitation, efforts have focused on isolating the biosynthetic activities of cultured cells, achieved by optimizing the cultural conditions, selecting high-producing strains, and employing precursor feeding, transformation methods, and immobilization techniques (Dicosmo and Misawa, 1995). Transgenic hairy root cultures have revolutionized the role of plant tissue culture in secondary metabolite production. They are unique in their genetic and biosynthetic stability, faster in growth, and more easily maintained. Using this methodology a wide range of chemical compounds have been synthesized (Shanks and Morgan, 1999; Giri and Narasu, 2000). Advances in tissue culture, combined with improvement in genetic engineering, specifically transformation technology, has opened new avenues for high volume production of pharmaceuticals, nutraceuticals, and other beneficial substances (Hansen and Wright, 1999). Recent advances in the molecular biology, enzymology, and fermentation technology of plant cell cultures suggest that these systems will become a viable source of important secondary metabolites. Genome manipulation is resulting in relatively large amounts of desired compounds produced by plants infected with an engineered virus, whereas transgenic plants can maintain constant levels of production of proteins without additional intervention (Sajc *et al.*, 2000). Large-scale plant tissue culture is found to be an attractive alternative approach to traditional methods of plantation as it offers a controlled supply of biochemicals independent of plant availability (Sajc *et al.*, 2000). Kieran *et al.* (1997) detailed the impact of specific engineering-related factors on cell suspension cultures. Current developments in tissue culture technology indicate that transcription factors are efficient new molecular tools for plant metabolic engineering to increase the production of valuable compounds (Gantet and Memelink, 2002). *In vitro* cell culture offers an intrinsic advantage for foreign protein synthesis in certain situations since they can be designed to produce therapeutic proteins, including monoclonal antibodies, antigenic proteins that act as immunogenes, human serum albumin, interferon, immuno-contraceptive protein, ribosome unactivator trichosantin, antihypersensitive drug angiotensin, leu-enkephalin neuropeptide, and human hemoglobin (Hiatt *et al.*, 1989; Manson and Arntzen, 1995; Wahl *et al.*, 1995; Arntzen, 1997; Hahn *et al.*, 1997; La Count *et al.*, 1997; Marden *et al.*, 1997; Wongsamuth and Doran, 1997; Doran, 2000). The appeal of using natural products for medicinal purposes is increasing, and metabolic engineering can alter the production of pharmaceuticals and help to design new therapies (Manson and Arntzen, 1995). At present, researchers aim to produce substances with antitumor, antiviral, hypoglycaemic, anti-inflammatory, antiparasite, antimicrobial, tranquilizer and immunomodulating activities through tissue culture technology. Exploration of the biosynthetic capabilities of various cell cultures has been carried out by a group of plant scientists and microbiologists in several countries during the last decade. In the last few years promising findings have been reported for a variety of medicinally valuable substances, some of which may be produced on an industrial scale in the near future.

9. Conclusions and Future Perspectives

In vitro propagation of medicinal plants with enriched bioactive principles and cell culture methodologies for selective metabolite production is found to be highly useful for commercial production of medicinally important compounds. The increased use of plant cell culture systems in recent years is perhaps due to an improved understanding of the secondary metabolite pathway in economically important plants. Advances in plant cell cultures could provide new means for the cost-effective, commercial production of even rare or exotic plants, their cells, and the chemicals that they will produce. Knowledge of the biosynthetic pathways of desired compounds in plants as well as of cultures is often still rudimentary, and strategies are consequently needed to develop information based on a cellular and molecular level. Because of the complex and incompletely understood nature of plant cells in *in vitro* cultures, case-by-case studies have used to explain the problems occurring in the production of secondary metabolites from cultured plant cells. A key to the evaluation of strategies to improve productivity is the realization that all the problems must be seen in a holistic context. At any rate, substantial progress in improving secondary metabolite production from plant cell cultures has been made within last few years. These new technologies will serve to extend and enhance the continued usefulness of higher plants as renewable sources of chemicals, especially medicinal compounds. It is hoped that a continuation and intensification efforts in this field will lead to controllable and successful biotechnological production of specific, valuable, and as yet unknown plant chemicals.

10. References

Arntzen C J. 1997. High tech herbal medicine: plant based vaccines. *Nature Biotechnol.* 15(3): 221-222.

Balandrin M J and J A Klocke. 1988. Medicinal, aromatic and industrial materials from plants. *In* Y.P.S. Bajaj (ed.), Biotechnology in Agriculture and Forestry. *Medicinal and Aromatic Plant*, vol. 4. Springer-Verlag, Berlin, Heidelberg, pp. 1-36.

Borisjuk N V, Borisjuk L G, Logendra S, Petersen F, Gleba Y and Raskin I. 1999. Production of recombinant proteins in plant root exudates, *Nature Biotechnology*, 17: 466–469.

Borowitzka M A. 1999. Commercial production of microalgae: ponds, tanks, tubes and fermenters, *Journal of Biotechnology*, 70: 313–321.

Bourgaud F, Gravot A, Milesi S and Gontier E. 2001. Production of plant secondary metabolites: a historical perspective, *Plant Science*, 161: 839-851.

Cheetham P S J. 1995. Biotransformations: new routes to food ingredients. *Chem Ind.*, pp. 265- 268.

Chizzali L, Cornelia M and Beerhues L. 2012. Phytoalexins of the Pyrinae: Biphenyls and dibenzofurans. *Beilstein J. Org. Chem.* 8: 613–620.

Cook J A and Mitchell J B. 1989. Viability measurements in mammalian cell systems, *Analytical Biochemistry*, 179: 1–7.

Curtis W R. 1999. Achieving economic feasibility for moderate-value food and flavor additives: a perspective on productivity and proposal for production technology cost reduction, *In*: Plant Cell and Tissue Culture for the Production of Food Ingredients, (Eds.), T.J. Fun, G. Singh, W.R. Curtis, *Kluwer Academic, Plenum Publisher*, 1999, pp. 225–236.

Dicosmo F and Misawa M. 1995. Plant cell and tissue culture: Alternatives for metabolite production. *Biotechnol. Adv.* 13(3): 425-453.

Dickschat J S. 2011. Biosynthesis and function of secondary metabolites, *Beilstein J. Org. Chem*, 7: 1620–1621.

Doran P M. 2000. Foreign protein production in plant tissue cultures. *Curr. Opin. Biotechnol.* 11: 199-204.

Fraenkel, G S. 1959. The raison d'Etre of secondary plant substances. *Science,* 129 (3361): 1466–1470.

Gantet P and Memelink J. 2002. Transcription factors: tools to engineer the production of pharmacologically active plant metabolites. *Trends Pharmacol. Sci.* 23: 563-569.

Giri A and Narasu M L. 2000. Transgenic hairy roots: recent trends and applications. *Biotechnol. Adv.* 18: 1-22.

Gontier E, Clement A, Bourgaud F and Guckert A. 2000. Plant metabolite production by hydroponic or aeroponic cultures. *Plant Cell Report,* 20: 177–190.

Hahn J J, Eschenlauer C A, Narrol H M, Somers A D and Srienc F. 1997. Growth kinetics, nutrient uptake, and expression of the *Alcaligenes eutrophus* poly(-hydroxybutarate) synthesis pathway in transgenic maize cell suspension cultures, *Biotechnol. Prog.* 13(4): 347-354.

Hansen G and Wright M S. 1999. Recent advances in the transformation of plants, *Trends Plant Sci.* 4: 226-231.

Harborne J B. 1999. Classes and functions of secondary products, *In*: Chemicals from Plants, Perspectives on Secondary Plant Products (Eds.), N.J. Walton, D.E. Brown, *Imperial College Press*, pp. 1–25.

Hiatt A, Cafferkey R and Boedish K. 1989. Production of antibodies in transgenic plants. *Nature,* 342: 76-78.

Kieran P M, MacLoughlin P F and Malone D M. 1997. Plant cell suspension cultures: some engineering considerations, *J. Biotechnol.* 59: 39-52.

Krings U and Berger R G. 1998. Biotechnological production of flavours and fragrances. *Appl. Microb. Biotechnol,* 49: 1-8.

La Count W, An G and Lee J M. 1997. The effect of PVP on the heavy chain monoclonal antibody production from plant suspension cultures. *Biotechnol. Let.* 19(1): 93-96.

Lai PK and Roy J. 2004. Antimicrobial and chemopreventive properties of herbs and spices. *Curr. Med. Chem.* 11(11): 1451–60.

Manson H S and Arntzen C J. 1995. Transgenic plants as vaccine production system, *Trends Biotechnol,* 3: 388-392.

Marden M C, Dieryck W, Pagnier J, Poyart C, Gruber V, Bournat P, Baudino S and Merot B. 1997. Human hemoglobin from transgenic tobacco, *Nature,* 342: 29-30.

Meskin Mark S. 2002. Phytochemicals in Nutrition and Health. *CRC Press.* p. 123.

Payne G F, Bringi V, Prince C and Shuler M L. 1991. The quest for commercial production of chemicals from plant cell culture, *In: Plant Cell and Tissue Culture in Liquid Systems*, Hanser (Eds.), G.F. Payne, V. Bringi, C. Prince, M.L. Shuler, pp. 1–10.

Phillipson J D. 1990. Plants as source of valuable products. *In:* B.V. Charlwood, and M.J.C. Rhodes (eds.), *Secondary Products from Plant Tissue Culture*. Oxford: Clarendon Press, pp. 1-21.

Rao R S and Ravishankar G A. 2000. Biotransformation of protocatechuic aldehyde and caffeic acid to vanillin and capsaicin in freely suspended and immobilized cell cultures of *Capsicum frutescens. J. Biotechnol*, 76: 137-146.

Ravishankar G A and Rao R S. 2000. Biotechnological production of phyto-pharmaceuticals. *J. Biochem. Mol. Biol. Biophys*, 4: 73-102.

Sajc L, Grubisic D and Vunjak-Novakovic G. 2000. Bioreactors for plant engineering: an outlook for further research, *Biochem. Eng, J.* 4: 89-99.

Samuni-Blank M, Izhaki I, Dearing MD, Gerchman Y, Trabelcy B, Lotan A, Karasov WH, Arad, Z 2012. Intraspecific directed deterrence by the mustard oil bomb in a desert plant. *Current Biology.* 22: 1-3.

Sato F, Hashimoto T, Hachiya A, Tamura K I, Choi K B and Morishige T. 2001. Metabolic engineering of plant alkaloid biosynthesis, *In: Proceedings of the National Academy Science*, USA, 98: 367–372.

Scragg A H. 1997. The production of aromas by plant cell cultures. *In:* T. Schepier (ed.), *Adv Biochem. Eng. Biotechnol.* Vol. 55. Berlin: Springer-Verlag, pp. 239-263.

Shanks J V and Morgan J. 1999. Plant hairy root culture. *Curr. Opin. Biotechnol.* 10: 151-155.

Tapsell LC, Hemphill I and Cobiac L. 2006. Health benefits of herbs and spices: the past, the present, the future. *Med. J. Aust.* 185: 4–24.

Stamp N. 2003. Out of the quagmire of plant defense hypotheses. *The Quarterly Review of Biology*, 78 (1): 23–55.

Steward N, Martin R, Engasser J M and Goergen J L. 1999a. A new methodology for plant cell viability assessment using intracellular esterase activity, *Plant Cell Report*, 19: 171–176.

Steward N, Martin R, Engasser J M and Goergen J L. 1999b. Determination of growth and lysis kinetics in plant cell suspension cultures from the measurement of esterase release *Biotechnology and Bioengineering*, 66: 114–121.

Vanisree M, Lee Chen-Yue, Lo Shu-Fung, Nalawade S M, Lin C Y, and Tsay Hsin-Sheng. 2004. Studies on the production of some important secondary metabolites from medicinal plants by plant tissue cultures. *Bot. Bull. Acad. Sin.*, 45: 1-22.

Wahl M F, An G and Lee J M. 1995. Effects of DMSO on heavy chain monoclonal antibody production from plant cell culture. *Biotechnol. Lett.* 17(5): 463-468.

William S H. 2003. Officina. *Medical meanings: a glossary of word origins*, p. 162.

Wongsamuth R and Doran M P. 1997. Production of monoclonal antibodies by tobacco hairy roots. *Biotechnol. Bioeng.* 54(5): 401-415.

Yun D J, Hashimoto T and Yamada Y. 1992. Metabolic engineering of medicinal plants: transgenic *Atropa belladonna* with an improved alkaloid composition, *In: Proceedings of the National Academy Science*, USA, 89: 11799–11803.

Zrÿd J P. 1988. Cultures in vitro et production de metabolites secondaires, *In: Cultures de Cellules, Tissus, et Organes Végétaux, Fondements Théoriques et Utilisations Pratiques*, J.P. Zrÿd (Ed.), Polytechniques Romandes Presses, pp. 228–234.

Chapter 8
Phytoremediation of Cadmium through Sorghum

*Prasann Kumar and Padmanabh Dwivedi**

*Department of Plant Physiology, Institute of Agricultural Sciences,
Banaras Hindu University, Varanasi – 221 005
E-mail: pdwivedi25@rediffmail.com

CONTENTS

1. Introduction

Sorghum bicolor L. is an important crop due to its wide use as food, feed and energy crop. In addition, Sorghum appears promising as a cereal crop, which has some non-food uses, particularly for bioethanol production (Epelde *et al.*, 2009). It comes originally from North Africa, and therefore, is resistant to heat and drought. It is unique due to specific compounds (phytates, tannins), which affect its nutritional value. Increased concentration of essential elements in agricultural soils means higher plant uptake. However, soil also contains toxic elements such as heavy metals; the higher concentrations, the higher are the uptake of non-essential elements. What is more, the concentrations of these metals are so toxic they can cause death of plants or reduced their production.

Sorghum is able to accumulate large quantities of metals in shoots grown in hydroponic conditions but in field trails these amounts are lower (Epelde *et al.*, 2009). Nevertheless, one advantage is the possibility of some economic returns during the process. Sorghum has high production of biomass in comparison to other crops such as sunflowers or corn (Zhuang *et al.*, 2009). Nowadays, there is a huge energy demand, therefore, the growth of energy crops on polluted sites can be feasible.

Human evolution has led to immense scientific and technological progress. Global development, however, raises new challenges, especially in the field of environmental protection and conservation (Bennett *et al.*, 2003). Nearly every government around the world advocates for an environment free from harmful contamination for their citizens. However, the demand for a country's economic, agricultural and industrial development outweighs the demand for a safe, pure and natural environment. Ironically, it is the economic, agricultural and industrial developments that are often linked to polluting the environment (Ikhuoria and Okieimen, 2000). Since the beginning of the industrial revolution, soil pollution by toxic metals has accelerated dramatically. About 90 per cent of the anthropogenic emissions of heavy metals have occurred since 1900 AD; it is now well recognized that human activities lead to a substantial accumulation of heavy metals in soils on global scale (*e.g.*, 5.6-38 x 10^6 kg Cd yr^{-1}) (Nriagu, 1996). Man's exposure to heavy metals comes from industrial activities like mining, smelting, refining and manufacturing processes (Nariagu, 1996). A number of chemicals, heavy metals and other industries in the coastal areas have resulted in significant discharge of industrial effluents into the coastal water environment and contribute to a variety of toxic effects on living organisms in food chain (Dembitsky, 2003) by bioaccumulation and bio-magnification (Manohar *et al.*, 2006). Heavy metals such as cadmium, copper, lead, chromium, zinc and nickel are important environmental pollutants particularly in areas with high anthropogenic pressure (United State Environmental Protection Agency, 1997).The soil has been traditionally the site for disposal for most of the heavy metals wastes which needs to be treated. Currently, conventional remediation methods of heavy metal contaminated soils are expensive and environmentally destructive (Bio-Wise, 2003; Aboulroos *et al.*, 2006).

2. Sources of Metal Pollution

Geological and anthropogenic activities are sources of heavy metal contamination (Dembitsky, 2003). Source of anthropogenic metal contamination includes industrial effluents, fuel production, mining, smelting processes, military operation, utilization of agricultural chemicals, small-scale industries (including battery production, metal products, metal smelting and cable coating industries), brick kilns and coal combustion (Zhen-Guo *et al.*, 2002). One of the prominent sources contributing to increased load of soil contamination is disposal of municipal wastage. This wastage is either dumped on roadsides or used as landfills, while sewage is used for irrigation. These wastages, although useful as sources of nutrients, are also source of carcinogens and toxic metals. Other source can include unsafe or excess application of pesticides, fungicides and fertilisers (Zhen-Guo *et al.*, 2002). Additional potential sources of heavy metals include irrigation water contaminated by sewage and industrial effluent leading to contaminated soils and vegetables (Bridge, 2004).

3. Metal Toxicity

All plants have the ability to accumulate "essential" metals (Ca, Co, Cu, Fe, K, Mn, Mo, Na, Ni, Se, V and Zn) from the soil solution. Plants need different concentrations for growth and development. This ability also allows plants to accumulate other "non-essential" metals (Al, As, Au, Cd, Cr, Hg, Pb, Pd, Pt, Sb, Te, Tl and U) which have no known biological function (Djingova and Kuleff, 2000). Moreover, metals cannot be broken down and when concentrations inside the plant cells accumulate above threshold or optimal levels, it can cause direct toxicity by damaging cell structure (due to oxidative stress caused by reactive oxygen species) and inhibit a number of cytoplasmic enzymes (Assche and Clijsters, 1990). In addition, it can cause indirect toxic effects by replacing essential nutrients at cation exchange sites in plants (Taiz and Zeiger, 2002). Baker (1981) proposed, however, that some plants have evolved to tolerate the presence of large amounts of metals in their environment by the following three ways:

[A] Exclusion, whereby transport of metals is restricted and constant metal concentrations are maintained in the shoot over a wide range of soil levels.

[B] Inclusion, whereby shoot metal concentrations reflect those in the soil solution in a linear relationship.

[C] Bioaccumulation, whereby metals are accumulated in the roots and upper plant parts at both high and low soil concentrations

Schmidit (2003) reported that elevated heavy metal concentrations in the soil can lead to enhanced crop uptake and negative effect on plant growth. At higher concentrations, they interfere with metabolic processes and inhibit growth, sometimes leading to plant death (Schaller and Diez, 1991). Excessive metals in human nutrition can be toxic and cause acute and chronic disease (Schmidt, 2003).

4. Plant-Metal Uptake

Plant extracts accumulate metals from soil solution. Before the metal can move from the soil solution into the plant, it must pass the surface of the root. This can either be a passive process, with metal ions moving through the porous cell wall of the root cells, or an active process by which metal ions move through the cells of the root. This latter process requires that the metal ions traverse the plasmalemma, a selectively permeable barrier that surrounds cells (Pilon-Smits, 2005). Special plant membrane proteins recognize the chemical structure of essential metals; these proteins bind the metals and are then ready for uptake and transport. Numerous protein transporters exist in plants. For example, the model plant *Arabidopsis thaliana* contains 150 different cation transports (Axelsen and Palmgren, 2001) and even more than one transporter for some metals (Hawkesford, 2003). Some of the essential, non-essential and toxic metals, however, are analogous in chemical structure so that these proteins regard them as the same. For example, arsenate is taken up by P transporter. Abedin *et al.* (2002) studied the uptake kinetics of arsenic species, arsenite and arsenate, in rice plants and found that arsenate uptake was strongly suppressed in the presence of arsenite. Clarkson and Luttge (1989) reported that Cu and Zn, Ni and Cd compete for the same membrane carriers. For root to shoot transport these elements are transported via the vascular system to the above-soil biomass (Shoot). The shoots are harvested, incinerated to reduce volume, disposed of as hazardous waste, or precious metals can be recycled (phytomining). Different chelators may be involved in the translocation of metal cations through the xylem, such as organic acid chelators [malate, citrate, histidine (Salt *et al.*, 1995; von Wiren *et al.*, 1999), or nicotianamine (Stephen *et al.*, 1996; von Wiren *et al.*, 1999)]. Since the metal is complexed within a chelate it can be translocated upwards in the xylem without being adsorbed by the high cation exchange capacity of the xylem (von Wiren *et al.*, 1999).

The pollution of soil and water with heavy metals is an environmental concern today. Metals and other inorganic contaminants are among the most prevalent forms of contamination found at waste sites, and their remediation in soils and sediments are the most techanically difficult. The high cost of existing cleanup technologies led to the search for new clean up strategies that have the potential to be low-cost, low-impact, visually benign and environmentally sound. Phytoremediation is a new clean up concept that involves the use of plants to clean or stabilize contaminated environments.

Phytoremediation is a potential remediation strategy that can be used to decontaminate soils contaminated with inorganic pollutants. Research related to this technology needs to be promoted and emphasized and expanded in developing countries since it is low cost. In situ, solar driven technology makes use of vascular plants to accumulate and translocate metals from roots to shoots. Harvesting the plant shoots can permanently remove these contaminants from the soil. Phytoremediation does not have the destructive impact on soil fertility and structure that some more vigorous conventional technologies such as acid extraction and soil washing. This technology can be applied 'in situ' to remediate shallow soil, ground water and surface water bodies (Dwivedi and Kumar, 2012). Also, phytoremediation has been perceived to be a more environmentally-friendly 'green' and low-tech alternative to more active and intrusive remedial methods.

5. Types of Phytoremediation

5.1. Phytoextraction

This technology involves the extraction of metals by plant roots and the translocation thereof to shoots. The roots and shoots are subsequently harvested to remove the contaminants from the soil. Salt *et al.* (1995) reported that the costs involved in phytoextraction would be more than ten times less per hectare compared to conventional soil remediation techniques. Phytoextraction also has environmental benefits because it is considered a low impact technology. Furthermore, during the phytoextraction procedure, plants cover the soil and erosion and leaching will thus be reduced. With successive cropping and harvesting, the levels of contaminants in the soil can be reduced (Vandenhove *et al.*, 2001). Researchers at the University of Florida have discovered the ability of the Chinese brake fern, *P. vittata* to hyperaccumulate arsenic. In a field test, the fern was planted at a wood-preserving site containing soil contaminated with from 18.8 to 1, 603 parts per million arsenic, and they accumulated from 3,280 to 4,980 parts per million arsenic in their tissue (Ma *et al.*, 2001). Sunflowers have proven effective in the remediation of radionuclides and certain other heavy meals. The flowers were planted as a demonstration of phytoremediation in a pond contaminated with radioactive cesium-137 and strontium-90 as a result of the Chernobyl nuclear disaster in the Ukraine. The concentration of radionuclides in the water decreased by 90 per cent in a two-week period. According to the demonstration, the radionuclide concentration in the roots was 8000 times than that in the water. In a demonstration study performed by Phytotech for Department of Energy, *H. annus* reduced the uranium concentration at the site from 350 parts per billion to 5 parts per billion, achieving a 95 per cent reduction in 24 h (Schnoor, 1997).

5.2. Phytostabilisation

Phytostabilisation, also referred to as in-place inactivation, is primarily used for the remediation of soil, sediment and sludge (United States Protection Agency, 2000). It is the use of plant roots to limit contaminant mobility and bioavailability in the soil. The plants primarily are used to [1] decrease the amount of water percolating through the soil matrix, which may results in the formation of a hazardous leachate, [2] act as a barrier to prevent direct contact with the contaminated soil and, [3] prevent soil erosion and the distribution of the toxic metal to other areas (Raskin and Ensley, 2000). Phytostabilisation can occur through the sorption, precipitation, complexation, or metal valence reduction. It is useful for the treatment of lead (Pb) as well as arsenic

(As), cadmium (Cd), chromium (Cr), copper (Cu) and zinc (Zn). Some of the advantages associated with this technology are that disposal of hazardous material/biomass is not required (United States Protection Agency, 2000) and it is very effective when rapid immobilization is needed to plants also reduced soil erosion and decrease the amount of water available in the system (United States Protection Agency, 2000). Phytoremediation has been used to treat contaminated land areas affected by mining activities and superfund sites. The experiment on phytostabilisation by Jadia and Fulekar (2008) was conducted in a greenhouse, using sorghum (fibrous root grass) to remediate soil contaminated by heavy metals and the developed vermicompost was amended in contaminated soil as a natural fertilizer. They reported that growth was adversely affected by heavy metals at the higher concentration of 40 and 50 ppm, while lower concentrations (5 to 20 ppm) stimulated shoot growth and increased plant biomass. Further, heavy metals were efficiently taken up mainly by roots of sorghum plant at all the evaluated concentration of 5, 10, 20, 40 and 50 ppm. The order of uptake of heavy metals was: Zn>Cu> Cd>Ni>Pb. The large surface area of fibrous roots of sorghum facilitate intensive penetration of roots into soil and capable of immobilizing and concentrating heavy metals in the roots.

5.3. Rhizofiltration

Rhizofiltration is primarily used to remediate extracted groundwater, surface water and wastewater with low contaminant concentration (Ensley, 2000). It is defined as the use of plants, both terrestrial and aquatic, to absorb, concentrate and precipitate contaminants from polluted aqueous source in their roots. Rhizofiltration can be used for Pb, Cd, Cu, Ni, Zn and Cr, which are primarily retained within the roots (United States Protection agency, 2000). Sunflower, Indian mustard, tobacco, rye, spinach, and corn have been studied for their ability to remove lead from water, with sunflower having the greatest ability. Indian mustard has a bioaccumulation coefficient of 563 for lead and has also proven to be effective in removing a wide concentration range of lead (4 mg/L-500 mg/L) (Raskin and Ensley, 2000; United States Protection Agency, 2000). The advantages associated with rhizofiltration are the ability to use both terrestrial and aquatic plants for either in situ or ex situ applications. Another advantage is that contaminants do not have to be translocated to the shoots. Thus, species other than hyperaccumulator may be used. Terrestrial plants are preferred because they have a fibrous and much longer root system, increasing the amount of root area (Raskin and Ensley, 2000). Sunflowers have successfully been implemented for rhizofiltration at Chernobyl to remediate uranium contamination.

An experiment on rhizofiltration by Karkhanis *et al.* (2005) was conducted in a greenhouse, using Pistia, duckweed and water hyacinth to remediate aquatic environment contaminated by coal ash starting from 0, 5, 10, 20, 30 and 40 per cent. Simultaneously the physicochemical parameters of leachate have been analyzed and studied to understand the leachability. The results showed good potential for uptake of these metals next to Pistia. Rhizofiltration of Zn and Cu in case of water hyacinth was lower as compared to Pistia and duckweed. This research showed that Pistia/duckweed/water hyacinth can be good accumulators of heavy metals in aquatic environment.

5.4. Phytovolatilization

Phytovolatilization involves the use of plants to take up contaminants from the soil, transforming them into volatile forms and transpiring them into volatile form and transpiring them into the atmosphere (United States Protection Agency, 2000). Mercuric mercury is the primary metal contaminant that this process has been used for. The advantage of this method is that the contaminant, mercuric ion, may be transformed into a less toxic substance (that is, elemental Hg). The disadvantage is that the mercury released into the atmosphere is likely to be recycled by precipitation and then re-deposited back into lakes and oceans, repeating the production of methyl-mercury by anaerobic bacteria.

In laboratory experiments, tobacco and a small model plant that had been genetically modified to include a gene for mercuric reductase converted ionic mercury (Hg (II)) to the less toxic metallic mercury (Hg (0)) and volatilized it (Meaagher *et al.*, 2000). Similarly transformed yellow poplar plantlets were resistance to, and grew well in, normally toxic concentrations of ionic mercury. The transformed plantlets volatilized about ten times more elemental mercury than did untransformed plantlets (Rugh *et al.*, 1998). Indian mustard and canola may be effective for phytovolatilization of selenium, and, in addition, accumulate the selenium (Banuelos *et al.*, 1997).

6. Cadmium Absorption and Accumulation in different Parts of *Sorghum*

Increasing growth of world population along with the expansion of agricultural and industrial activities for improving supply of nutritional materials and the occurrence of several consecutive years of drought, are the reason why available water resources in most countries located in the dry-region zone have been used with the highest degree, such a condition may naturally exert a great deal of pressure on water resources.

Water supply is considered as one of the most fundamental concern of many governments at present and in the future, and large investments have been done in this area. It should be accepted that one of the main options is using of water with different quality for different consumptions (Carr, 2005). Undoubtedly, wastewater reuse in agriculture sector is one of the most significant parts of the usage-chain. Moreover, having access to wastewater and industrial sewage as an assuring and permanent source of water and nutritional materials can provide the necessary water and fertilizers for agricultural products.

Since, it is possible that an amount of Sorghum may be directly consumed by human beings; therefore, it is necessary to investigate the issue cadmium accumulation in nutritional parts of Sorghum and compare it with the existing standards. From one of the important studies made by Izadiyar and Yargholi, 2010, it is clear that the maximum cadmium accumulation in sorghum was less than the UN standard set for agricultural and nutritional materials which permits at most a daily use of 35 micro-grams of cadmium by human beings, the standard set by the American Food Industries which permits a daily use of at most 92 micro-grams of cadmium by human and the standard set by Miller in his book which has specified a maximum amount of 0.5 ppm to be used by human.

Table 8.1: Comparison of Average Cadmium Concentration in different Parts of Sorghum Species.

Plant Species	Cadmium (mg kg⁻¹)		
	Root	Stem	Leaf
Sorghum	0.507	0.287	0.356

Source: Izadiyar and Yargholi, 2010.

Table 8.2: Comparison of Mean Cadmium Concentration in different Parts of Plant Species in different Treatments.

Treatments	Cadmium (mg kg⁻¹)		
	Root	Stem	Leaf
Control	0.055	0.037	0.043
50 ppm	1.518	0.920	1.108
100 ppm	2.722	1.735	1.983

Source: Izadiyar and Yargholi, 2010.

From the Table 8.2, it is clear that the amount of cadmium accumulation in plant organs (root, shoot and leaf) significantly increased at the level of 1 per cent in third treatment (100ppm), compared with the other two treatments (one with a concentration 50 ppm and the other with no cadmium). Similar results are comparable with the results carried out by Gardiner *et al.* (1995), Ramos *et al.* (2002) and Molla-hoseini *et al.* (2005). Accumulation of cadmium in different plant parts suggests that with a high degree of movement which this metal has in soil, it can be easily absorbed by the root and then transferred to different organs of the plant. It means accumulation of this element in soil and consequently the undesired increase in the amount of this metal in plants in the long term; endanger the health of affected plants and consequently, the health of farm animals and humans.

By considering the concentrations of accumulated cadmium in plant organs corresponding to treatments 50 ppm and 100 ppm, it is found that an increase in cadmium concentration in soil can result in an increase in cadmium accumulation in plants cultivated in such a polluted soil. But, this is not the only necessary condition. In other words, the overall concentration of this metal in soil can not by itself be considered as appropriate criterion for determining the amount of cadmium absorbed by the plant (Sims and Kline, 1991). Moreover, the potential for absorbing heavy metals ions by plants, apart from plant type and ion concentration in root environment, depends on other factors such as its degree of dissolvability in root environment. These factors are dependent on some physical characteristic of soil, like particle size and water-keeping capacity as well as on some chemical characteristics such as acidity, existence of nutritional materials and existence of organic chemicals excreted from plant roots (Alloway, 1990; Robinson, 1997). For this reason, it is important that cadmium accumulation in each of sorghum organs does not yield a significant difference with increasing concentration of cadmium in the soil corresponding to root region.

7. Toxic Metal Cadmium Removal from Soil by AM Fungi Inoculated *Sorghum*

Cadmium is one of the components of the earth's crust and found everywhere in the environment. The natural occurrence of cadmium in the environment results mainly from gradual phenomenon such as rock erosion and abrasion that estimate for 15,000 mt per annum (WHO, 1992). Naturally existing concentration of Cd in atmosphere is 0.1-0.5 ng/m^3, in earth crust 100-500 mg/gm but much higher level may be accumulated in sedimentary rocks and marine phosphates. The wide spade use of Cd is based on its unique physical and chemical properties. It is highly resistant to chemicals, high temperature and ultraviolet light (Morrow and Keatings, 1997). Cd is widely used in special alloys, pigments coatings stabilizers above all (almost 70 per cent of its use) in Ni-Cd batteries (Morrow, 1996). It can enter air from the burning of coal, household waste, and metal mining as well as refining process which may increase the level of Cd in the soil varying from 100-600 mg/kg dry weight (Ernest and Neilson, 2000; Lombi *et al.*, 2000) or more (Meaghler, 2000). The general trends of metal enhancement appears to urban > rural > remote locations.

Some countries have set tolerance limits on heavy metal addition to soil because their long-term effects are unknown. These limits are usually set for plough layer of soil where most of the root activity occurs. The value of potentially toxic elements (PTE) proposed by council of European Economic Committee (Smith, 1996) for Cd concentration in soil is 1.0-3.0 mg/kg of dry soil and maximum annual addition of total cadmium to soils is 150 gm per hectare (Palanaippan, 2002). Phytoremediation of metal contaminated sites offers a low cost method for soil remediation and some extracted metals may be recycled for value. Because cost of growing a crop is minimal compared to those of soil removal and replacement, so the use of plants to remediate the hazardous soils seems to have great promise. Other recent reviews on many aspects of soil phytoremediation are available. Phytoremediation is the use of plants to make soil contaminants non-toxic and is also often referred to as bioremediation, botanical-bioremediation, or green remediation. The idea of using rare plants which hyper-accumulate metals to selectively remove and recycled excessive soil metals was introduced in 1983 (Cheny, 1997), gained public exposure in 1990 (Anonymous 1990), and has increasingly been examined as a potential practice and more cost effective technology than soil replacement, solidifications, or washing strategies recently used (Salt *et al.*, 1995; Cunningham *et al.*, 1996).

In the experiment conducted by Arora and Sharma, 2009, it was found that health hazard posed by the accumulation of toxic metals in the environment accompanied by high cost of removal and replacement of metal polluted soil have prompted effort to develop bioremediation strategies. In their experiment, plants of *Sorghum* were grown in AM and non-AM inoculated substrate and subjected to five soil-[Cd] concentrations (0.1 per cent, 1.0 per cent, 2.0 per cent and 5.0 per cent). The inoculation of AM fungi resulted in significantly better absorption, and accumulation of Cd accumulation was 47.1 per cent, 45.2 per cent, 35.7 per cent, 33.9 per cent and

23.5 per cent for 0.1 per cent, 0.2 per cent, 1.0 per cent, 2.0 per cent and 5.0 per cent, respectively after 80 days of treatments. Table 8.3 deals with the value of soil pH at different concentrations of Cd. The pH of soil having 0.1 per cent concentration of Cd was more or less near to the pH of control for both AMF and non-AMF treatments. For non-AMF treatments, there was a decrease in pH as the concentration of Cd was increased. At 0.2 per cent and 1 per cent of Cd, pH was slightly acidic but at 5 per cent concentration soil was found to be acidic. The AMF treatments didn't show much variation in pH *i.e.* no significant changes in pH was observed. For AMF treatments, pH seemed to be independent of the concentrations of the metal Cd (Pawloska *et al.*, 1996).

Table 8.3: Effects of Cd at different Concentrations on Soil pH.

Days	Control	Concentration of Cd (per cent)									
		0.1		0.2		1		2		5	
		AMF	NAMF	AMF	NAMF	AMF	NAMF	AMF	NAMF	AMF	NAMF
0	7.3	7.08	6.8	7.02	6.36	6.97	6.16	6.95	6.69	6.94	5.2
20	7.36	7.1	6.75	7.06	6.21	7.01	6.11	7	5.79	6.98	5.16
40	7.26	7.09	6.65	7.04	6.18	6.99	6.07	6.96	5.75	6.95	5.15
60	7.2	7.07	6.6	7.03	6.08	6.94	6.02	6.91	5.69	6.93	5.1
80	7.18	7.05	6.45	7	6	6.9	5.98	6.9	5.6	6.89	5.06

Source: Arora and Sharma, 2009.

From the experiment of Arora and Sharma (2009), it is clear that the accumulation of cadmium in plants was maximum at low concentration because as the concentration of the metal increases it becomes toxic to the plant and thus retard its growth. With the inoculation of the AM fungi, the potential of *Sorghum* to accumulate Cd from the soil increases significantly. The efficiency of phytoremediation of heavy metal contaminated site increases with the presence of higher proportion of metal resistant microbial population in the soil, which may likely, confer a better nutritional assimilation and protective effect on plant (Doelman, 1985). There have been reports of significant inhibition of mycorrhizal colonization by the heavy metals like Cd (Griffioen *et al.*, 1994; Leyval *et al.*, 1995). The increase in AM root colonization is likely the results of numerous factors including increase in soil-metal (Cd) concentration and the subsequent decrease in soil pH (Rufyikiri *et al.*, 2003), who assessed root- and hypha-induced substrate. pH groups of soluble proteins and non-proteins thiol operating as tolerating mechanism in root cells (Chaui *et al.*, 1997).

So, finally Arora and Sharma (2009) suggested that, in addition to the metal immobilization in the Mycorrhizosphere, mycorrhizal fungi may also act as effective barrier controlling excessive metal uptake into the root cells. It is also suggested that inoculation with AM fungi can facilitate plant growth and thus increase phytoremediation sufficiency. Plant – mycorrhizal relationship in relation to toxic metal removal from soil can be an effective tool to enhance plant efficiency.

8. Effects of Cadmium on the Growth and Physiology of Sorghum

With the rapid industrial development, soil environmental pollution in India has become increasingly serious. Cadmium (Cd) is a highly toxic heavy metal in the environment (Davis, 1984; Guo, 1994). Cd is a non-essential nutrient for plants, and excessive Cd has not only significant adverse effects (Shamsi *et al.*, 2008), but also endangers human health via food chain (Naidu and Harter, 1998). The alleviation or inhibition of Cd damage in plants has, therefore, caused extensive attention of the whole society (Uraguchi *et al.*, 2009; Wang *et al.*, 2008). Heavy metals have an adverse impact on growth and development of the plants, showing some physiological and biochemical characteristic of damages. To a certain extent, plant growth and physiological characteristic can reflect the adverse impact of heavy metal externally or internally (Zhang and Shu, 2006). The research of the poisoning effect of the heavy metal Cd on plant mainly focuses on food crops such as rice, wheat and maize, but less on *Sorghum* plants, which is known often as animal feed sources.

Sorghum scientists reported some important changes in growth and physiological characteristic of sorghum plant under cadmium stress. In their experiment, the major physiological parameters for observation were like this, [A] effects of cadmium on height of different species of sorghum, [B]effects of cadmium on chlorophyll contents of different species of sorghum plants, [C] Effects of cadmium on root activities in different species of sorghum plants, [D] Effects of cadmium on MDA contents in leaves of different species of sorghum plants. They found several interesting observations and these can be summarized such as, a kind of oxidative stress, heavy metal stress affects the growth of plants. Jalil *et al.* (1994) reported that low concentration of cadmium can promote the growth of hard wheat, while under a relatively high concentration, the growth of wheat and tillering were both inhibited and the degree varied among different varieties. Liu's (2004) research also showed that corn seedling's height under cadmium treatment reduced significantly as the concentration of cadmium increased with prolonged growth period. These studies showed that lower concentration of cadmium stimulates the increase of sorghum height, which may be related to the certain resistance of sorghum plants to Cd, while higher levels of cadmium inhibited height growth of sorghum genus plant. Thus, lower concentration of cadmium stress stimulates the growth of sorghum plants to a certain extent, and higher concentrations inhibited their growth.

Reason for the inhibitory effect of heavy metals to plant growth was probably due to: [A] A series of physical and chemical reactions between excess heavy metal and soil components changes soil properties, thus affecting soil fertility levels (Cieslinski *et al.*, 1996; Chang and Wu, 2005). For example, heavy metal pollution can enhance the fixation of soil phosphorous, which affected the plants absorbing phosphorous, thus influenced the growth of plants (Li *et al.*, 2004; Zhang *et al.*, 2004); [B] Heavy metal poisonous effects caused a reduction in plant photosynthesis, thereby reducing the plant water and nutrient adsorption, which affected the normal growth and development of plants (Qin *et al.*, 2000).

In plant body, photosynthesis is the most fundamental and important physiological and biochemical process, and its initial link is chlorophyll synthesis and function realization. The results from experiments conducted by Liu *et al.* (2004) showed that chlorophyll synthesis was affected by cadmium stress and with the increase of cadmium stress levels, the inhibitory effect was increasingly severe. The reason for change in chlorophyll contents in the leaves of sorghum plants might be that chlorophyll synthesis was formed under the action of a series of enzymes in proplastid and chloroplast (Stobart and Griffith, 1985; Wang, 2000). Cd stress inhibited relevant enzymes activities in the leaves of sorghum plants in the process of chlorophyll synthesis, affecting chlorophyll synthesis process and leaf chlorosis, thus leading to the change in chlorophyll contents.

There have been many reports about heavy metal pollution linked to root activities of *Graminae*. For example, through the hydroponic way, Yang *et al.* (2005) researched the effect of sewage irrigation accelerated the decline of wheat seedlings and root, reducing the root number and the root activities significantly. Jiang *et al.* (2004) research also showed that infected soil made the roots of rice seedling yellow and red, enlarged the rhizome and the root color was brown and yellow.

Under Cd stress, root activities of three kinds of sorghum plants decreased significantly in different growth stages. Underground part and aerial part of plants existed with interdependence and mutual restriction relevance. Roots and leaves of plants not only existed in sink-source relationships in assimilation products, but also in supply-demand relationships between water and inorganic nutrition. Cd stress could directly reduce root activities, impending water and mineral nutrient absorption and influencing of the aerial part growth by showing a drop in height, leaf area and tillering number. It might also influence the root growth by reducing the allocation of photosynthesis products to root, thus influencing the photosynthesis products to root, thus influencing the photosynthetic capacity in leaf (Foy *et al.*, 1978; Kastori *et al.*, 1992).

Under senescence and stress, plant organs undergo lipid membrane peroxidation because of free radical toxicity, and the product, malondialdehyde, damage cell membrane system severely. In normal circumstances, because of the active oxygen scavenging system in plant body, active oxygen in cells exists at very low levels, so it cannot cause damage. When adversity exceeds a certain degree, the active oxygen scavenging system in plant is destroyed, and active oxygen becomes accumulated (Richter and Schweizer, 1997; Shah *et al.*, 2001), deflating or reducing the structure, activities and contents of active oxygen scavengers such as SOD, POD, CAT etc, which lead to further accumulation of active oxygen, thus destroying the oxygen balance. Meanwhile, the increase of reactive oxygen not only causes or aggravates membrane lipid peroxidation (Filek *et al.*, 2009; Tames *et al.*, 2009), but also dehydrogenated protein and produces proteins free radicals, causing damage to chain polymerization and membrane system, with the accumulation of MDA acting as an indicator of the degree of damage of membrane system cells (Phindsa *et al.*, 1981).

9. Changes Induced by Copper and Cadmium Stress in the Anatomy and Grain Yield of *Sorghum*

Despite their potential physiological and economical significance, anatomical alteration induced by heavy metals in plants has so far been grossly overlooked. Furthermore, few studies have been carried out to investigate the impact of heavy metals on one or a few anatomical parameters of the plant and, therefore, coordinated investigations on a broad scale to assess the impact of heavy metal stress on the structure of the plant vegetative body seems crucial (Khudsar *et al.*, 2001; Panou-Filotheou and Bosabalidis, 2004). Similarly, the impact of heavy metals on the yield of economically important crop plants was unduly neglected (Khan *et al.*, 2003; Wu *et al.*, 2003). Plants are seldom exposed in nature to the effect of a single heavy metal since excess metal ions exist mostly in mixtures in variously polluted soils and irrigation water worldwide (Souza and Rauser, 2003). However, only a few studies dealing with the interactive effects of a combination of heavy metals on plants are so far on record (Zeid, 2001; Yang *et al.*, 2004).

In the experiment conducted by Kasim (2006), sorghum plants were grown in sandy soil for two months and irrigated with half-strength Hoagland solution supplemented with 10^{-5} M CuSO$_4$, 10^{-6} M CdSO$_4$ or a mixture of both (1:1, v/v). Kasim found that copper or cadmium applied alone or in combination caused significant reduction in root diameter, width and thickness of leaf midrib, diameter of xylem vessels of all seedlings organs, parenchyma cell area in the stem, leaf midrib and pith and cortex of root, dimensions of stem vascular bundles, number of xylem arms in root and frequency of stomata on abaxial leaf surface.

Table 8.4: The Percentage Reduction Caused by 10^{-5} M Cu, 10^{-6} M Cd or Cu + Cd in Parenchyma Cell Area (PCA) of Stem Ground Tissue, Mesophyll of Leaf Midrib, Root Cortex and Root Pith, Diameter of Metaxylem Vessels (DMX) in Stem, Leaf Midrib and Large Xylem Arms of Root, and in Grain Yield of *Sorghum*.

Criteria	Tissue/Organ	Per cent Reduction		
		Cu	Cd	Cu + Cd
PCA	Stem ground tissue	21.13	23.17	29.67
	Mesophyll of leaf midrib	23.87	31.12	37.68
	Root cortex	35.59	39.81	41.63
	Root pith	33.25	38.17	42.50
DMX	Vascular bundles of midrib	12.80	11.54	10.15
	Vascular bundles of stem	18.25	14.76	15.95
Number	Larger xylem arms in root	20.31	17.95	20.94
Number/plant	All grains	55.31	24.78	23.45
	Filled grains	62.94	40.00	53.48
Weight	1000 intact grains	2.21	5.22	5.99
	1000 chaffed grains	4.40	11.12	15.81

The applied heavy metal concentrations induced a number of measurable alterations in the anatomy of the vegetative organs as well as in the grain yield of sorghum plant. Grain yield was also drastically reduced. Similar alterations in anatomical parameters and grain yield of different crop plants have been earlier reported by several scientists (Setia and Bala, 1994; Kovaceviv *et al.*, 1999; Shalini *et al.*, 1999; Khudsar *et al.*, 2001; Papadakis *et al.*, 2004).

The uptake of heavy metals by root seems to trigger a series of structural alterations with potential functional consequences in the plant. Many scientists demonstrated that a marked decrease in the cell size might be the result of a decrease in the elasticity of cell walls of the roots.

10. Effects of Sewage Sludge and Nitrogen Applications on Grain Sorghum

Sorghum and millets are valuable in the development of year-forage systems, particularly where quality is important. Grain sorghum is grown all around the world under a wide range of climatic conditions. Sorghum is considered very efficient in utilizing nutrients from the soil because of a large fibrous root system.

Nitrogen (N) has traditionally been considered one of the most important nutrients for plants. It usually increases plant growth and crop yield. Because most soils are often deficient in the type of N that plants can readily use, therefore, chemical fertilizers or organic "residuals" such as manure or biosolids are added annually to agricultural soils. The amount of N in biosolids is relatively high, making it an attractive fertilizer. In fact, biosolids application rates must be carefully calculated to avoid adding too much N, which leaches out of the soil in the form of nitrate and degrades the environment.

Sludge land application has proven to be an excellent substitute for inorganic fertilizers and is cost effective for both the municipality applying the sludge and the farmer who accept it. The benefits of using the sewage sludge as fertilizers have been proven by numerous researchers. In most cases the application of sewage sludge improves the physical and chemical properties of infertility soils and increases their fertility. Scientist reported that sludge-treated plots produced corn yields equivalent to those to which commercial fertilizers was applied. In addition to nutrients, however, sludge contains a number of potentially harmful constituents such as heavy metals. The heavy metals in applied sludge create the potential for soil and water contamination, crop damage and accumulation of heavy metals in the food supply. The magnitude of the problem depends on the composition of the sludge, rate of sludge application, soil properties and crop species and cultivation. In general, biosolids increase plant production and improve forage quality when applications are not excessive relative to N requirement of treated plants and metal concentrations are not toxic (Smith, 1996).

Akdeniz *et al.* (2006), in their experiment described that sewage sludge application positively affected grain yield, leaf nitrogen, harvest index, and total N uptake more than chemical fertilizer, except for dry matter yield. Moreover, biosolids applications did not cause the negative effect on heavy metal content of plant, seed as

Table 8.5: Effects of Fertilizer Treatment on Dry Matter, Grain Yield, Plant Length, Number of Panicles, Unit Grain Weight and Whole Plant and Grain, Harvest Index, Total N Uptake, N Harvest Index, Nitrogen Use Efficiency.

Fertilizer Treatment	Dry Matter (t/ha)	Grain Yield (cm)	Plant Length (cm)	Number of Panicle (m²)	Unit Grain Weight	Leaf-N (Per cent)	Whole Plant N (Per cent)	Grain N (Per cent)	Harvest Index	Total N Uptake (kg/ha)	N Harvest Index	N Use Efficiency
Control	5.75	7.25	110.8	52.0	13.9	1.60	0.52	1.13	57.5	113.8	76.0	66.1
N1	5.63	8.72	109.2	50.3	16.6	1.63	0.58	1.16	59.8	129.3	74.7	64.7
N2	6.10	10.03	111.6	53.2	18.9	1.72	0.66	1.21	60.5	166.3	72.9	60.7
S1	5.16	7.84	107.6	54.2	14.5	1.62	0.64	1.15	61.4	124.3	73.4	63.8
S2	5.19	8.30	107.1	51.7	16.0	1.68	0.60	1.08	66.0	121.0	74.3	68.6
S3	6.58	12.82	116.8	57.8	22.2	1.93	0.67	1.20		198.8	77.7	64.8

Source: Akdeniz *et al.*, 2006.

well as soil. The concentration of copper, cadmium, lead and zinc in the soil and Ni and Zn in the leaf increased with increasing sludge application, however, the increase was low. As a result, sewage sludge may be used as nitrogen source for grain sorghum production.

The short-term advantage of applying sewage sludge as a fertilizer is relatively determined based on a level economic return. Although a short-term economic advantage may be realized, the individual farmer may not be aware of the long-term disadvantage of phytotoxicity. Nevertheless, a continuous monitoring of heavy metal accumulation in the soil and plants should be established in area of application of biosolids.

In the experiment of Akdeniz *et al.* (2006), the concentrations of the studied macronutrients and heavy metals in both leaf and grain of sorghum showed no difference with sewage biosolids and nitrogen applications (Tables 8.6 and 8.7). Compared to control, sewage sludge slightly increased plant P, Mg, Cu and Cr concentrations, but this increase was not significant. The Ca, Mg, Cr and Pb concentrations in leaves were relatively higher than the seeds, while the P, K, Fe and Mn concentration in seeds were relatively higher than the leaves. However, there was a significant and positive correlation among P, K, Mg and Ca concentrations in seeds.

In the findings of Akdeniz *et al.* (2006), there was a relationship among Cr, Zn and P concentrations. In the leaf, there was a significant relationship between Ni and Zn, Pb and Cd, but the relation between Mg and Cu concentration was negative. Similarly, there was a significant and positive correlation for P, K, Mg and Ca concentrations in seed. So, on the basis of findings of Akdeniz *et al.* (2006) it can be suggested that sewage biosolids application did not cause any significant increase in heavy metal levels in leaf and seed of grain sorghum. Metals have toxic effects on living organisms when they exceed a certain concentration. However, the heavy metal levels in plants by far did not reach the limits of tolerance for plants (Schmidt, 1997).

11. Municipal Wastewater, Soil Chemical Properties and Heavy Metal Uptake by *Sorghum*

In most arid and semiarid region of the world, including the lands of many countries water crisis is considered as one of the main problems on the path of sustainable development of agriculture. Due to water restrictions and increased water consumption, use of low quality water resources (wastewater) is considered as solution to resolve agricultural water requirements which are pointed out as the largest consumption of water recently. Wastewater can have a positive effects on soil and eventually plant growth, due to richness of organic matter and nutrients such as nitrogen, potassium and phosphorous (Mohammad and Ayadi, 2004; Ghanbari *et al.*, 2007). Recently, one of the issues that attracted the attention of researchers and environmentalists is wastewater chemicals and heavy metals especially those which can penetrate into soil, plant and finally food chain (Ashworth and Alloway, 2003). Heavy metals represented a portion of important environmental pollutants which

Table 8.6: Macronutrient Concentrations (Per cent), Micronutrient and Heavy Metal Concentrations (mg/kg) Dry Weight in Sorghum Leaf.

Fertilizer	P	K	Ca	Mg	Fe	Mn	Cu	Cd	Cr	Ni	Pb	Zn
Control	0.15	1.31	0.50	0.41	13.53	3.40	4.15	0.13	0.93	1.05	0.49	11.63
N1	0.17	1.41	0.47	0.42	13.65	3.35	4.45	0.12	0.80	1.10	0.46	12.15
N2	0.15	1.03	0.51	0.47	13.63	3.15	4.40	0.13	0.82	1.18	0.56	11.80
S1	0.17	1.05	0.71	0.46	13.95	3.08	4.45	0.12	0.86	1.11	0.56	13.03
S2	0.18	1.15	0.64	0.48	13.23	3.20	4.55	0.11	1.06	1.34	0.49	15.80
S3	0.19	1.14	0.57	0.48	14.48	3.60	4.60	0.11	1.01	1.15	0.49	15.18

Source: Akdeniz *et al.,* 2006.

Table 8.7: Macronutrient Concentration (Per cent) Micronutrient and Heavy Metal Concentration (mg/kg) Dry Weight in Sorghum Seed.

Fertilizer	P	K	Ca	Mg	Fe	Mn	Cu	Cd	Cr	Ni	Pb	Zn
Control	0.25	0.41	0.11	0.29	28.40	11.75	4.52	0.13	0.45	1.14	0.36	19.20
N1	0.26	0.43	0.11	0.30	23.95	12.56	4.46	0.12	0.40	1.22	0.41	20.73
N2	0.24	0.44	0.12	0.29	27.40	12.38	4.94	0.12	0.34	1.10	0.36	18.15
S1	0.25	0.39	0.11	0.29	29.28	12.44	4.09	0.13	0.39	1.13	0.44	20.05
S2	0.25	0.40	0.10	0.25	28.33	11.88	4.89	0.13	0.45	1.19	0.46	20.35
S3	0.26	0.38	0.11	0.29	27.15	12.25	4.03	0.11	0.41	1.01	0.45	19.78

Source: Akdeniz *et al.,* 2006.

cause pollution problems by increasing their use in products in recent decades. In spite of gradual accumulation of heavy metals in the soil, the stability of heavy metals in the environment will lead to pollution since they could not be decomposed like organic pollutants by biological or chemical processes (McBridee, 1995). Propagation of heavy metals in biological food chain is one of the important issues of this behaviour, as increasing the amount of several heavy metals in higher stages of food chain is many times more than initial levels (Al-Enezi *et al.*, 2005). The storage of heavy metals severely threatens human health, but due to their long half-life (*e.g.*, 1460 days for lead and 200 days for cadmium), tendency for storing such elements is dramatic (Pescord, 1992). Sorghum crop is important to provide livestock forage, and forage health has a direct effect on human health (Al-Jaloud *et al.*, 1995).

Galavi *et al.* (2010) conducted an experiment in order to investigate the effect of treated municipal wastewater on soil chemical properties and heavy metal uptake by sorghum. The treatments were managed for irrigation; with well water during entire period of growing season as control (T1); wastewater during first half of growing season (T2); wastewater during the second half of growing period (T3); wastewater and well water alternately (T4) and wastewater during entire period of growing season (T5). On the basis of their findings it is clear that wastewater irrigation increased the percentage of organic matter and total nitrogen content of soil. In the irrigation with wastewater in the first half of growing season, nutrients could be used in the beginning of growth. Wastewater irrigation had no significant influence on heavy metal concentration in soil. For total amount of heavy metals present in soil and plant, several factors should be taken into account such as soil, metals, plant species and type of vegetation. Transfer coefficients of molybdenum and lead were high in sorghum (Table 8.9). The absorption of these elements by plants must be reduced with proper management to prevent forage toxicity and consequently poisoning (Table 8.8).

Table 8.8: Comparing of Effects of Irrigation Treatments on Heavy Metal Concentration in Soil and Sorghum Crop.

Parameters		Wastewater				
		T1	T2	T3	T4	T5
Soil	Zn	2.47	2.643	2.60	2.855	2.81
	Cu	0.622	0.633	0.630	0.651	0.677
	Fe	0.80	1.11	1.11	1.1	1.13
	Pb	0.94	0.972	0.967	0.967	0.977
	Mo	0.164	0.200	0.196	0.201	0.208
Sorghum	Zn	0.422	0.431	0.430	0.433	0.432
	Cu	0.040	0.054	0.053	0.052	0.051
	Fe	0.096	0.11	0.123	0.111	0.120
	Pb	0.96	1.00	0.99	0.99	1.00
	Mo	0.242	0.251	0.246	0.253	0.257

Source: Galavi *et al.*, 2010.

Table 8.9: Transfer Coefficient (TC) from Soil to Sorghum.

	Elements				
TC	Zn	Cu	Fe	Pb	Mo
	0.164	0.077	0.107	1.02	1.28

Source: Galavi *et al.*, 2010.

Therefore, on the basis of above findings it is clear that the absorption of these elements by plants must be reduced with proper management to prevent forage toxicity and consequently poisoning.

12. Soil Cadmium Regulates Antioxidants

Cadmium is one of the most highly toxic heavy metals in the atmosphere, soil and water, causing a serious environmental problem and threatening to all organisms (Benavides *et al.*, 2005). Toxic levels of Cd in agricultural soils may be caused by contaminated soil characteristics such as abundant Cd, agricultural manufacturing, mining, waste disposal practices, or use of metal-containing pesticides and fertilizers (Radotic *et al.*, 2000). Although Cd is not an essential element for plants, it is readily taken up and affects directly or indirectly several metabolic activities in different cell compartments (Stobart *et al.*, 1985). Cd can induce the production of reactive oxygen species (ROS) by autoxidation and fenton reaction, block essential functional biomolecules and displace essential metal ions from biomolecules (Schutzendubel and Polle, 2002). It causes oxidative damage to plants through ROS formation (Schutzendubel *et al.*, 2001). ROS are effectively eliminated by non-enzymatic (glutathione, ascorbate, α-tocopherol and carotenoids) and enzymatic defence system such as superoxide dismutase (SOD), catalase (CAT), peroxidase (POD) and glutathione (GSH), which protect plants against oxidative damage (Foyer *et al.*, 1994). The response of antioxidant to Cd varies in different species and tissues (Hassan *et al.*, 2005; Tiryakioglu *et al.*, 2006).

The physiological studies about resistance have indicated that cell membrane damages in the crops is affected by low temperature, drought, salinity and other stresses like heavy metal stress (Wang 1987). Free radical damage theory is widely applied for the mechanism of membrane damage. Active oxygen is mainly removed by some of the enzymes and antioxidants. The main protective enzymes like SOD, CAT and POD could remove excessive ROS and maintain the dynamic balance of ROS generation and quenching, thus inhibit lipid peroxidation process. Therefore, the protective enzymes play an important role in maintaining cell membrane stability (Wang, 2000). The GSH may also be main factor that reduces the stress induced by cadmium and play an important role in the tolerance to heavy metal as non-hyperaccumulator (Rausser 1999).

According to researchers, under the low concentrations of cadmium treatment, the protective enzyme activities of SOD, POD and CAT were induced and showed an upward trend. This indicted that sorghum had generated response to oxidative stress and started removing the peroxide generated by the active oxygen metabolism and reducing damage induced by active oxygen, so as to maintain the normal function of such cells. Such response, however, in sorghum generated protective response to resist heavy metal stress. But these kinds of protective response are limited. When the cadmium stress exceeds the defensive capacity of protective enzymes, the activities of protective enzymes declined rapidly, showing an inhibitory effect, this would result in excessive accumulation of AOS in sorghum plant leading to oxidative effects, damaging the membrane system, and causing cell death when the damage was severe. Further it is suggested that the protective enzyme activities among the sorghum plants were different. The same species also showed a difference at different growth stages. Overall, the protective enzyme activities at the elongation stage were higher than that of seedling and heading stage. According to previous studies the SOD composition in different crops is different. Wu and Zhang (2003) showed that the SOD in leaves and roots of Weiyou 48, 49 and 26 cultivars of rice mainly consisted of Cu/Zn-SOD and Mn-SOD, but Fe-SOD was not found. The SOD in pea seed consists of Cu-Zn SOD, and the former could be inhibited by chloroform-ethanol. These differences may result in different levels of resistance to adversity; perhaps this may be reason that different species show different tolerance to heavy metals. It is necessary to study whether a similar mechanism existed in other enzymes.

13. Influence of Organic Matter on the Uptake of Cadmium by Sorghum Plants

Sorghum is the fifth most important cereal in the world. In South Asia, Central America and greater part of Africa, it is a main staple food, while in the United States, Australia and South America, it is used primarily as animal feed. It is mainly through the consumption of food, in particular, cereals that Cd contamination occurs in animals. It is, therefore, very important to evaluate cadmium accumulation in cereals aimed for food production. The release of trace metal in biologically available forms, as a result of human activity, may impair or alter both natural and man-made ecosystems. Cadmium is one of the most important elements of concern for food-chain contamination. It is a non-essential trace metal pollutant of the environment that results from various agricultural, mining and industrial activities and is also released in the exhaust gases of automobiles. Cadmium is recognized as an extremely significant pollutant due to its high toxicity and large solubility in water.

Knowledge of metal-plant interactions is important for the safety of the environment and for reducing the risks associated with the introduction of trace metals into food chain. Cadmium is of particular interest because it can be accumulated by plants to levels that are toxic to human and animals, but which are not phytotoxic (Prince *et al.*, 2002). Plant species that tend to accumulate large amounts of cadmium include lettuce (Crew and Davies, 1985), celery (Ni *et al.*, 2002), wheat, poppy (Chizzola, 1997) and linseed (Marquard *et al.*, 1990). Cadmium shows a particular affinity for sulfhydryl groups and also reacts with hydroxyl groups and nitrogen-containing ligands. Consequently the metal can inactivate many important enzymes resulting in inhibition of photosynthesis, respiratory rate and other metabolic processes in plants (Torres *et al.*, 2000). In spite of the different mobility of metal ions in plants, the metal content is generally greater in roots than in the above-ground tissue (Ramos *et al.*, 2002). The root acts as a barrier for the uptake and translocation of heavy metals (Vassilev *et al.*, 1998). In general, the content of Cd in plants decreases in the order: roots>stems>leaves>fruits>seeds (Blum, 1997). Some organic molecules with low molecular weight may act as carriers of metal ions including Cd, increase cell membrane permeability or exhibit hormone-like activity (Chen *et al.*, 1998; Clapp *et al.*, 2000; Pizzeghello *et al.*, 2000). Sessi *et al.* (2000) observed increase in the biomass production of corn, soybean, rice and barley developed in nutrient solutions with levels of Cd that varied from 0.25 to 100 mg Cd dm^{-3}.

Typical response curves of biomass production vs. organic matter concentration for plants grown in nutrient solution, have shown an enhanced growth with increasing organic matter levels, followed by a decrease at high concentration. The organic matter concentration responsible for those effects depends on the plant species and experimental parameters (Rauthan and Schnitzer, 1981; Tyler and Mcbride, 1982; Chen *et al.*, 1998). To reduce the risks associated with the introduction of toxic metals in the food-chain, it is very important to know the interaction between trace metal ions, soil organic matter, and plants. An experiment was carried out by Pinto *et al.* (2004) in which they tried to find out the influence of organic matter on the uptake of cadmium, zinc, copper and iron by sorghum plants. This experiment was carried out under controlled environmental conditions, to investigate the effects of a fulvic acid fraction of soil organic matter on growth, cadmium uptake and redistribution by sorghum. Sorghum was grown in nutrient solutions with 0, 0.1, 1 and 10 mg Cd dm^{-3}, in the absence and presence of organic matter, for various periods up to 20 days. Pinto *et al.* (2004) made several observations on the basis of their experiment and these are as under:

13.1. Effect of Organic Matter on Biomass Production and Cd Uptake

Adding organic matter to the nutrient solution promoted an increase in the biomass of roots and shoots, both in the presence or absence of cadmium. Humic substances in nutrient solution have been shown to increase root length and density of secondary roots (Chen *et al.*, 1998; Clapp *et al.*, 2000; Sessi *et al.*, 2000), as well as plant growth. However, high level of organic matter may inhibit plant growth.

Table 8.10: Cd Partioning [Ratio of Cd concentration or (Cd content) in shoot to root] of Sorghum Plants grown in Nutrient Solutions with different Levels of Cd, for 10 days without or with OM Addition (32 mg C dm^{-1}). Content= [Cd] (mg Cd kg^{-1} d. m.) x Biomass (kg)

	Level of Cd (mg Cd dm^{-3})		
	0.1 (Mean ±S.E.)	1 (Mean ±S.E.)	10 (Mean ±S.E.)
Without OM	0.40±0.02 (0.70±0.02)	0.11±0.01 (0.18±0.01)	0.07±0.01 (0.13±0.01)
With OM	0.56±0.02 (1.10 ±0.04)	0.17±0.01 (0.27±0.01)	0.12±0.01 (0.26±0.01)

Source: Pinto *et al.*, 2004.

13.2. Cadmium and Organic Matter Effect on Micronutrient Uptake

The presence of organic matter leads to a decrease in the concentrations of Cu, Zn and Fe in roots suggesting that these ions were also retained by organic matter. Furthermore, in the absence or presence of OM, the general pattern observation for Cu and Zn showed a decrease in their accumulation by plants, with increasing exposure to Cd. This could be consequences of the direct competition between Cu or Zn and Cd. In fact, kinetic data showed that cadmium compete with essential metal ions like Cu^{+2} and Zn^{+2} for the same membrane transporter (Raskin *et al.*, 1994). Zinc and Cd have many physical and chemical similarities as they both belong to group II of the periodic table. They are usually found together in the ores and compete with each other for various ligands. The other plants such as clover, corn, cabbage and ryegrass show decreasing Cu and Zn concentration in the roots and shoots with increasing Cd concentration in nutrient solution (Yang *et al.*, 1996). Reducing Zn uptake in the presence of 1 mg Cd dm^{-3} has also been observed in bean (Wallace *et al.*, 1980), and wheat (Abo-Kassem *et al.*, 1997).

Table 8.11: Cd Partioning [Ratio of Cd concentration in shoot to root] of Sorghum Plants in Nutrient Solutions with different Levels of Cd, for 5 or 20 days. Content = [Cd] (mg Cd kg^{-1} d. m) x Biomass (kg)

Level of Cd (mg Cd dm^{-3})	5 days (Mean ± S.E.)	20 Days (Mean ± S.E.)
0.1	0.32±0.01 (0.58±0.02)	0.41±0.02 (0.69±0.03)
1	0.14±0.0 (0.24±0.01)	0.23±0.01 (0.43± 0.02)
10	0.08±0.01 (0.13±0.01)	0.11±0.01 (0.19±0.01)

Source: Pinto *et al.*, 2004.

14. Conclusion

Sorghum bicolor L. is an important crop with its wide use as food, feed and energy crop and also has some non-food uses, particularly for bioethanol production. It was originated from North Africa, and hence is resistant to heat and drought. It is unique due to specific compounds (phytates, tannins), which affect its nutritional value. Increased concentration of essential elements in soils results in higher plant uptake. However, soil also contains toxic elements such as heavy metals; the higher the concentrations, the higher are the uptake of non essential elements. The concentrations of these metals are so toxic that they can cause death of plants or reduced their production. Sorghum is able to accumulate large quantities of metals in shoots grown in hydroponic conditions but in field trials these amounts are lower. Nevertheless, one advantage is the possibility of some economic returns during the process. Sorghum has high production of biomass in comparison to other crops such as sunflowers or corn. Nowadays, there is a huge energy demand, therefore, the growth of energy crops on polluted sites can be feasible.

Human evolution has led to immense scientific and technological progress. Global development, however, raises new challenges, especially in the field of environmental protection and conservation. Nearly every government around the world advocates for an environment free from harmful contamination for their citizens. However, the demand for a country's economic, agricultural and industrial development outweighs the demand for a safe, pure and natural environmental. Ironically, it is the economic, agricultural and industrial developments that are often linked to polluting the environment. Since the beginning of the industrial revolution, soil pollution by toxic metals has accelerated dramatically. About 90 per cent of the anthropogenic emissions of heavy metals have occurred since 1900 AD; it is now well recognized that human activities lead to a substantial accumulation of heavy metals in soils on global scale (*e.g.* 5.6-38 x 10^6 kg Cd yr^{-1}). Man's exposure to heavy metals comes from industrial activities like mining, smelting, refining and manufacturing processes. A number of chemicals, heavy metals and other industries in the coastal areas have resulted in significant discharge of industrial effluents into the coastal water environment and contribute to a variety of toxic effects on living organisms in food chain by bioaccumulation and bio-magnification. Heavy metals such as cadmium, copper, lead, chromium, zinc and nickel are important environmental pollutants particularly in areas with high anthropogenic pressure. The soil has been traditionally the site for disposal for most of the heavy metals wastes which needs to be treated. Currently, conventional remediation methods of heavy metal contaminated soils are expensive and environmentally destructive. Plants have evolved to tolerate the presence of large amounts of metals in their environment by using ways:

1. Exclusion, whereby transport of metals is restricted and constant metal concentrations are maintained in the shoot over a wide range of soil levels.
2. Inclusion, whereby shoot metal concentrations reflect those in the soil solution in a linear relationship.
3. Bioaccumulation, whereby metals are accumulated in the roots and upper plant parts at both high and low soil concentrations

Since, an amount of sorghum may be directly consumed by human beings, it is necessary to investigate the Cd accumulation in nutritional parts of sorghum and compare it with the existing standards. The maximum Cd accumulation in sorghum was less than the UN standard set for agricultural and nutritional materials which permits at most a daily use of 35 micro-grams of Cd by human beings; the standard set by the American Food Industries permits a daily use of at most 92 micro-grams of cadmium by human while that set by Miller in his book has specified a maximum amount of 0.5 ppm to be used by human.

The changes in growth and physiological characteristic of sorghum plant under cadmium stress were on height of different species of sorghum; chlorophyll contents of, root activities in leaves of different species of sorghum plants; and MDA contents in leaves of different species of sorghum plants. They found several interesting observations including heavy metal stress affects the growth of plants through oxidative stress.

Sorghum phytoremediate Cd from sewage sludge and municipal waste water. Influence of Cd uptake from these polluted sites and its effect on plant metabolism while making such sites cadmium free has been reviewed in this chapter. The basic processes of phytoremediation are still largely not clear and hence require further fundamental or basic and applied research to optimize its field performance. Information gathered from fundamental research at anatomical, physiological, biochemical and genetic level in plants will be helpful in understanding the mechanisms of passive adsorption, ionic balance within the cell, ionic homeostasis, stress tolerance, active uptake, translocation, accumulation and chelation mechanisms. Better knowledge of these biochemical mechanisms may lead to:

1. Identification of novel genes and the subsequent development of transgenic plants with superior remediation capacities
2. Better understanding of the ecological interactions involved (*e.g.* plant-microbe interactions)
3. Appreciation of the effect of the remediation process on ecological interactions
4. Knowledge of the entry and movement of the pollutants in the ecosystem.

15. References

Abedin, M.J., Feldmann, J. and Meharg, A.A. 2000. Uptake kinetics of arsenic species in rice plants. *Plant Physiol.*, 128: 1120-28.

Abo-Kassem, E.M., Sharaf, A. and Mohamed, Y.A.H. 1997. Effect of different cadmium concentration on growth, photosynthesis and ion elation of wheat. *Egypt J. Physiol. Sci.*, 21: 41-51.

Aboulroos, S.A., Helal, M.I.D. and Kamel, M.M. 2006. Remediation of Pb and Cd polluted soils using in situ immobilization and phytoextraction techniques. *Soil Sedi. Contamination*, 15: 199-215.

Akdeniz, H., Yilmaz, I., Bozkurt, M.A. and Keskin, B. 2006. The effects of sewage sludge and nitrogen applications on grain sorghum grown (*Sorghum vulgare* L.) in Turkey. *Polish J. Environ. Stud.*, 15(1): 19-26.

Al-Enezi, G., Hamodam, M. F. and Fawzi, N. 2005. Heavy metals content of municipal wastewater and sludges in Kuwait. *J. Environ. Sci. Health*, 39: 397–407.

Al-Jaloud, A.A., Hussain, G., Al-Saati, A.J. and Karimulla, S. 1995. Effect of wastewater irrigation on mineral composition of corn and sorghum plants in a pot experiment. *J. Plant Nutr.*, 18: 1677-1692.

Alloway, B.J. 1990. *Heavy Metal in Soils*. John Wiley and Sons Inc. New York, pp. 20-27.

Arora, K. and Sharma, S. 2009. Toxic metal (Cd) removal from soil by AM fungi inoculated sorghum. *Asian J. Exp. Sci.*, 23(1): 341-438.

Ashworth, D.J. and Alloway, B.J. 2003. Soil mobility of sewage sludge–derived dissolved organic matter, copper, nickel and zinc. *Environ. Pollut.*, 127: 137-144.

Assche, F. and Clijsters, H. 1990. Effects of metals on enzyme activity in plants. *Plant Cell Environ.*, 24: 1-15.

Axelsen, K.B. and Palmgren, M.G. 2001. Inventory of the superfamily of P-type ion pumps in *Arabidopsis*. *Plant Physiol.*, 126: 696-706.

Baker, A.J.M. 1981. Accumulators and excluders – strategies in the response of plants to heavy metals. *J. Plant Nutr.*, 3: 643-654.

Banuelos, G.S., Ajwa, H.A., Mackey, B., Wu, L.L., Cook, C., Akohoue, S. and Zambrzuski, S. 1997. Evaluation of different plant species used for phytoremediation of high soil selenium. *J. Environ. Qual.*, 26(3): 639- 646.

Benavides, M.P., Gallego, S.M. and Tomaro, M.L. 2005. Cadmium toxicity in plants. *Braz. J. Plant Physiol.*, 17: 21-34.

Bennett, L.E., Burkhead, J.L., Hale, K.L., Terry, N., Pilon, M. and Pilon-smits E.A.H. 2003. Analysis of transgenic Indian mustard plants for phytoremediation of metals-contaminated mine tailings. *J. Environ. Qual.*, 32: 432-440.

Bio-wise 2003. Contaminated Land Remediation: A Review of Biological Technology, London. DTI.

Blum, W.H. 1997. Cadmium uptake by higher plants. Proceeding of extended abstracts from the fourth international conference on the biogeochemistry of trace elements. University of California, Berkely, USA. p. 109-110.

Bridge, G. 2004. Contested terrain: mining and the environment. *Annu. Rev. Environ. Resour.*, 29: 205-259.

Carr, R. 2005. WHO guidelines for safe waste water use: More than just numbers. *J. Iris. Drain*, 54: 103-111.

Chaney, R.L., Malik M., Li, Y.M., Brown, S.L., Brewer, E.P., Angle, J.S. and Baker, A.J.M. 1997. Phytoremediation of soil metals. *Curr. opin. Biotech.*, 8: 279-284.

Chang, Z.M. and Wu, X.H. 2005. Difference comparison of three alfalfa varities resistant to cadmium pollution. *Pratacult. Sci.*, 22(12): 20-23.

Chen, Y., Clapp, C.E., Magen, H. and Cline, V.W. 1998. Stimulation of plant growth by humic substances: effects on iron availability. In: Understanding humic substance: advanced methods, properties and applications. E.A. Ghabbour and G. Davis (eds). Cambridge: The Royal Soc. of Chemistry, pp. 255-263.

Chizzola, R. 1997. Comapative cadmium uptake and mineral composition of cadmium treated *Papaver somniferum*, *Triticum durum* and *Phaseolus vulgaris*. *J. Appl. Bot.*, 71: 147-153.

Cieslinski, G., Neilser, G.H. and Hogue E.J. 1996. Effect of soil cadmium applicat ion and pH on growth and cadmium accumulation in roots, leaves and fruit of strawberry plants. *Plant Soil*, 180: 267-271.

Clapp, C.E., Chen, Y., Cline, V.W., Palazzo, A.J. and Dowdy, R.H. 2000. Plant Growth Stimulation by Humic Substances. Proceedings of 10th International Meeting of the International Humic Substances Society, Toulouse, pp. 895-896.

Clarkson, D.T., and Luttge, U. 1989. Mineral nutrition: divalent cations, transport and compartmentation. *Propag. Bot.*, 51: 93-112.

Crews, H.M. and Davies, B.E. 1985. Heavy metal uptake from contaminated soils by six varieties of lettuce (*Lactuca sativa* L.). *J. Agric Sci Camb*, 105: 591-595.

Cunningham, S.D. and Ow, D.W. 1996. Promises and prospects of phytoremediation. *Plant Physiol.*, 110(3): 715-719.

Davis, R.D. 1984. Cadmium- a complex environmental problem part II. Cadmium in sludge used as fertilizer. *Experientia*, 40(2): 117-126.

Dembitsky, V. 2003. Natural occurrence of arseno compounds in plants, lichens, fungi, algal species and microorganisms. *Plant Sci.*, 165: 1177-1192.

Djingova, R. and Kuleff, I. 2000. Instrumental techniques for trace analysis. In: *Trace elements: their distribution and effects in the environment*. J.P. Vernet (ed) Elsevier Science Ltd., United Kingdom, pp. 146.

Doelman, P. 1985. Resistance of soil microbial communities to heavy metals In: *Microbial communities in soil*. V. Jensen, A. Kjoelles and L.H. Soerensen (eds). Elsevier, London, pp. 369-384.

Dwivedi, P. and Kumar, P. 2012. Phytoremediation: Tool for *In situ* Risk Reduction. In: *Advances in Plant Physiology* 13th Vol. A. Hementaranjan ed.). Scientific Publishers (India), Jodhpur, pp. 527-549.

Ensley, B.D. 2000. *"Rationale for the Use of Phytoremediation." Phytoremediation of toxic metals: Using plants to clean-up the environment.* John Wiley Publishers: New York.

Epelde, L., Mijangos, I., Becerril, J. M., Garbisu, C., Jones, D., Killham, K. and van Hees, P. 2009. Soil microbial community as bio-indicator of the recovery of soil functioning derived from metal phytoextraction with sorghum. *Soil Biol. Biochem.*, 41(9): 1788-1794.

Ernest, W.H.O. and Nielson, H.J.M. 2000. Life cycle phases of Zn and Cd resistant ecotypes of *in* risk assessment of polymetallic metallic mine soils. *Environ. Pollut.*, 107: 329-338.

Filek, M., Zenbala, M., Hartikainen, H., Miszalski, Z., Komas, A., Wietecka-Posluszny, R. and Walas, P. 2009. Changes in wheat plastid membrane properties induced by cadmium and selenium in presence/absence of 2, 4-dichlorophenoxyacetic acid. *Plant Cell Tissue Organ Cult.*, 96: 19-28.

Foy, C.D., Chaney, R.L. and White, M.C 1978. The physiology of metal toxicity in plants. *Annu. Rev. Plant Physiol.*, 29: 511-566.

Foyer, C.H., Lelandais, M. and Kunert, K.J. 1994. Photooxidative Stress in Plants. *Physiol. Plant.*, 92(4): 696-717.

Galavi, M., Jalali, A., Ramroodi, M., Mousavi, S.R. and Galavi, H. 2010. Effects of treated Municipal wastewater on soil chemical properties and heavy metal uptake by Sorghum (*Sorghum bicolor* L.). *J. Agric. Sci.*, 2: 235-241.

Gardiner, D.T., Miller, R.W., Badamchian, B., Azari, A.S. and Sisson, D.R. 1995. Effects of repeated sewage sludge application on plants accumulations of heavy metals. *J. Agric. Ecosys. Environ.*, 55: 1-6.

Ghanbari, A., Abedikoupai, J. and TaieSemiromi, J. 2007. Effect of municipal wastewater irrigation on yield and quality of wheat and some soil properties in sistan zone. *J. Sci. Tech. Agri. Nat. Res.*, 10: 59-74.

Griffioen, W.A.J., Ietswaart, J.H. and WHO, E. 1994. Mycorrhizal infection of an *Agrostis capullaris* population on a copper contaminated soil. *Plant Soil*, 158: 83-89.

Guo, D.F. 1994. Lead and cadmium source in environment and the harms on human and animal. *Environ. Sci. Progress*, 12(3): 71-76.

Hassan, M.J., Shao, G. and Zhang, G. 2005. Inuence of cadmium toxicity on antioxidant enzymes activity in rice cultivars with dierent grain Cd accumulation. *J. Plant Nutr.*, 28: 1259–1270.

Hawkesford, M.J. 2003. Transporter gene families in plants: the sulfate transporter gene family – redundancy or specialization? *Physiol. Plant.*, 117: 155-63.

Ikhuoria, E.U. and Okieimen, F.E. 2000. Scavenging Cadmium, copper, lead, nickel and zinc ions from aqueous solution by modified cellulosic sorbent. *Int. J. Environ Stud.*, 57(4): 401.

Izadiyar, M.H. and Yargholi, B. 2010. Study of Cadmium Absorption and Accumulation in Different Parts of Four Forages. *Amer.-Eur. J. Agric. and Environ. Sci.*, 9(3): 231-238.

Jadia, C.D. and Fulekar, M.H. 2008. Phytotoxicity and remediation of heavy metals by fibrous root grass (sorghum). *J. Appl. Biosci.*, 10: 491-499.

Jalil, A., Selles, F. and Clark, J.M. 1994. Effects of cadmium on growth and the uptake of cadmium and other elements by durum wheat. *Plant Nutr.*, 17: 1839-1895.

Jiang, Y., Liang, W.J., Zhang, Y.G. and Xu, Y.F. 2004. Research on effects of sewage irrigation on soil heavy metal environmental capacity and rice growth. *China Ecol. Agric. J.*, 12(3): 124-127.

Karkhanis, M., Jadia, C.D. and Fulekar, M.H. 2005. Rhizofilteration of metals from coal ash leachate. *Asian J. Water Environ. Pollut.*, 3(1): 91-94.

Kasim, W.A. 2006. Changes induced by copper and cadmium stress in the anatomy and grain yield of *Sorghum bicolour* (L.) Moench. *Intl. J. Agri. Biol.*, 8(1): 123-128.

Kastori, R., Petrovic, M. and Petrovic, N. 1992. Effects of excess lead, cadmium, copper and zinc on water relations in sunflower. *Plant Nutr.*, 15: 2427-2439.

Khan, H.R., McDonald, G.K. and Rengel, Z. 2003. Zn fertilization improves water use efficiency, grain yield and seed Zn content I chickpea. *Plant Soil*, 249: 389-400.

Khudsar, T., Mahmooduzzafar and Iqbal, M. 2001. Cadmium- induced changes in leaf epidermis, photosynthetic rate and pigment concentration in *Cajanus cajan*. *Biol. Plant.*, 44: 59-64.

Khudsar, T., Mahmooduzzafar, M. Iqbal and Sairam, R.K. 2004. Zinc-induced changes in morpho-physiological and biochemical parameters in *Artemisia annua*. *Biol. Plant.* 48(2): 255-260.

Kovacevic, G., Kastori, R and Merkulov, I.J. 1999. Dry matter and leaf structure in young wheat plants as affected by cadmium, lead and nickel. *Biol. Plant.*, 42: 119-23.

Levyal, C., Singh, B.R and Joner, E.J. 1995. Occurance and infectivity of arbuscular mycorrhizal fungi in some Norwegian soils influenced by heavy metals and soil properties. *Water Air Soil Pollut.*, 84: 203-276.

Li, F., Li, M.Y., Pan, X.H. and Xu, Y.F. 2004. Biochemical and physiological characteristic in seedlings roots of different rice cultivars under low phosphorous stress. *Chinese J. Rice Sci.*, 18(1): 48-52.

Ma, L.Q., Komar, K.M., Tu, C., Zhang, W., Cai, Y. and Kennelley, E.D. 2001. A fern that hyperaccumulates arsenic. *Nature*, 409: 579-579.

Manohar, S., Jadia, C.D. and Fulekar, M.H. 2006. Impact of ganesh idol immersion on water quality. *Indian J. Environ. Prot.*, 27(3): 216-220.

McBride, M.B. 1995. Toxic metal accumulation from agricultural use of sludge: Are the USEPA regulations protective? *J. Environ. Qual.*, 24: 5–18

Meagher, R.B., Rugh, C.L., Kandasamy, M.K., Gragson, G. and Wang, N.J. 2000. Engineered Phytoremediation of Mercury Pollution in Soil and Water Using Bacterial Genes. In: *Phytoremediation of Contaminated Soil and Water*. N. Terry and G. Banuelos (eds). Lewis Publishers, Boca Raton, FL, pp. 201-219.

Mohammad, M. J. and Ayadi, M. 2004. Forage yield and nutrient uptake as influenced by secondary treated wastewater. *J. Plant Nutr.*, 27: 351-364.

Molla-hoseini, H., Harati, M., Akbari, A. and Hariri, M. 2005. Accumulation of heavy metals in Sorghum irrigated with sewage. Proceeding of 9th Iranian Soil Sciences Conference.

Morrow, H. 1996. "Questioning the need to develop alternatives for Cd coatings". Proceedings of second annual Cd alternatives for Cd conference, National Defence Centre or Environment Excellence, Johnstown, Pennsylvania, USA, May 13-15.

Morrow, H. and Keating, J. 1997. "Overview Paper for OECD Workshop on the Effective Collection and Recycling of Nickel-Cadmium Batteries" OECD Workshop on the Effective Collection and Recycling of Nickel-Cadmium Batteries, Lyon, France.

Naidu, R. and Harter, R.D. 1998. Effects of different ligands on cadmium sorption by and extract ability from soils. *Soil Sci. Am. J.*, 62: 644-650.

Ni, W.Z., Yang, X.E. and Long, X.X. 2002. Differences of cadmium absorption and accumulation in selected vegetable crops. *J. Environ. Sci.*, 14: 399-405.

Nriagu, J.O. 1996. Toxic metal Pollution in Africa. *Science*, 223: 272.

Palanaippan, M., Shannumugam, K. and Ponnusamy, S. 2002. Soil degradation due to heavy metal accumulation under long term fertilization. The 17th world congress of soil Science (WCSS), Bangkok. (Thailand).

Panou-Filotheou, H. and Bosabalidis, A.M. 2004. Root structural aspects associated with copper toxicity in oregano (*Origanum vulgare* subsp. Hirum). *Plant Sci.*, 166: 1497-504.

Papadakis, I.E., Dimassi, K.N., Bosabalidis, A.M., Therios, I.N., Patakas, A. and Giannakoula, A. 2004. Effects of B excess on some physiological and anatomical parameters of "Navalina" orange plants grafted on two rootstocks. *Environ. Exp. Bot.*, 51: 247-257.

Pawlowska, T.E., Blaszkowski, J. and Rushling, A. 1996. The mycorrhizal status of plants colonizing a calamine spoil mound in Southern Poland. *Mycorrhiza*, 6: 499-505.

Pescod, M.B. 1992. Wastewater treatment and use in agriculture–FAO irrigation and drainage. Food and agriculture organization of the United Nations, Rome.

Phindsa, R.S., Dhindsaa, P.P. and Thorpa, T.A. 1981. Leaf senescence: correlated with increased levels of membrane permeability and lipid peroxidation and decreased levels of superoxide dismutase and catalase. *J. Exp. Bot.*, 32: 93.

Pilon-Smits, E. 2005. Phytoremediation. *Annu. Rev. Plant. Biol.*, 56: 15- 39.

Pinto, A.P. 2003. Sorghum behaviour in hydroponic solution and soils contaminated by cadmium. Organic matter effects. Ph.D thesis, Instituto Superior Tecnico, Lisboa.

Pizzeghello, D., Sessi, E., Muscolo, A., Albuzio, A. and Nardi, S. 2000. High and low apparent molecular size humic substances affecting plant metabolism. 10th International Meeting of the International Humic Substances Society, Toulouse, pp. 919-922.

Prince, W.S.P.M., Senthil Kumar, P., Doberschutz, K.D. and Subburam, V. 2002. Cadmium toxicity in mulberry plants with special reference to the nutritional quality of leaves. *J. Plant Nutr.*, 25: 689-700.

Qin, T.C., Ruan, J. and Wang, L.J. 2000. Effects of cadmium on plant photosynthesis. *Environ. Sci. Technol.*, 13: 33-35.

Radotic, K., Ducic, T. and Mutavdzic, D. 2000. Changes in peroxidase activity and isozymes in spruce needles after exposure to different concentrations of cadmium. *Environ. Exp. Bot.*, 44: 105-113.

Ramos, I., Esteban, E., Lucena, J.J. and Garate, A. 2002. Cadmium uptake and subcellular distribution in plants of *Lactuca* sp. Cd-Mn interaction. *Plant Sci.*, 162: 761-767.

Raskin, I. and Ensley, B.D. 2000. Phytoremediation of Toxic Metals: Using Plants to Clean Up the Environment. John Wiley and Sons, Inc., New York. Rugh CL.

Raskin, I., Kumar, N., Dushenkov, S. and Salt, D. 1994. Bioconcentration of heavy metals by plants. *Curr. Opin. Biotech.*, 5: 285-290.

Rauser, W.E. 1995. Phytochelatins and related peptides. Structure, biosynthesis and function. *Plant Physiol.*, 109: 1141–1149.

Rauser, W.E. 1999. The structure and function of metal chelators produced by plants. *Cell Biol. Biophy.*, 31: 19-48.

Rauthan, B.S. and Schnitzer, M. 1981. Effects of a soil fulvic acid on the growth and nutrient content of cucumber (*Cucumis sativus*) plants. *Plant Soil*, 63: 491-495.

Richter, C. and Schweizer, M. 1997. Oxidative stress in mitochondria. Cold Spring Harbor Laboratory Press, pp. 169-200.

Robinson, B.H. 1997. The phytoextraction of metals from metalliferous soils, PhD Thesis, Massey University, NZ, pp. 19-30.

Rufyikiri, G., Thirty, Y., Devaux, B. and Declerck, S. 2003. Contribution of hyphae and roots to Uranium uptake and translocation by arbuscular mycorrhizal carrot roots under root-organ culture conditions. *New Phytol.*, 158: 391-399.

Rugh, C.L., Senecoff, J.F., Meagher, R.B. and Merkle, S.A. 1998. Development of transgenic yellow poplar for mercury phytoremediation. *Nat. Biotechnol.*, 16(10): 925-928.

Salt, D.E., Blaylock, M., Kumar, N.P.B.A., Dushenkov, V., Ensley, B.D., Chet, I. and Raskin, I. 1995. Phytoremediation: a novel strategy for the removal of toxic metals from the environment using plants. *Biotech.*, 13: 468-475.

Salt, D.E., Prince, R.C., Pickering, I.J. and Raskin, I. 1995. Mechanisms of cadmium mobility and accumulation in Indian mustard. *Plant Physiol.*, 109: 1427-1433.

Schaller, A. and Diez, T. 1991. Plant specific aspects of heavy metal uptake and comparison with quality standards for food and forage crops (In German.) In: Sauerbeck D, Lu¨ bben S (eds) Der Einfluß von festen Abfa¨ llen auf Bo¨ den, Pflanzen. KFA, Ju¨ lich, Germany, pp. 92-125.

Schmidt, U. 2003. Enhancing Phytoextraction: The effects of chemical soil manipulation on mobility, plant accumulation, and leaching of heavy metals. *J. Environ. Qual.*, 32: 1939-1954.

Schnoor, J.L. 1997. Phytoremediation. University of Lowa, Department of Civil and Engineering, 1: 62.

Schutzendubel, A. and Polle, A. 2002. *Plant responses to abiotic stresses: heavy metal-induced oxidative stress and protection by mycorrhization. J. Exp. Bot.*, 53: 1351–1365.

Schutzendübel, A., Schwanz, P., Teichmann, T., Gross, K., Langenfeld Heyser, R., Godbold, D.L. and Polle, A. 2001. Cadmium induced changes in antioxidative systems, H_2O_2 content and differentiation in pine (Pinus sylvestris) roots. *Plant Physiol.*, 127: 887–892.

Sessi, E., Pizzeghello, D., Gessa, C. and Nardi, S. 2000. Effects of low molecular weight humic fraction on growth and nitrogen metabolism of different plant species. 10th International Meeting of the International Humic Substances Society, Toulouse, pp. 915-918.

Setia, R.C. and Bala, R. 1994. Anatomical changes in root and stem of wheat (*Triticum aestivum* L.) in response to different heavy metals. *Phytomorphol.*, 44: 95-104.

Shah, K., Kumar, R.G., Verma, S. and Dubey, R.S. (2001). Effect of cadmium on lipid peroxidation, superoxide anion generation and activities of antioxidant enzymes in growing rice seedlings. *Plant Sci.*, 161(6): 1135-1144.

Shalini, M., Ali, S.T., Siddiqui, M.T.O and Iqbal, M. 1999. Cadmium-induced changes in foliar response of *Solanum melongena* L. *Phytomorphol.*, 49: 295-302.

Shamsi, I.H., Wei, K., Zhang, G.P., Jilani, G.H. and Hassan, M.J. 2008. Interactive effects of cadmium and aluminium on growth and antioxidative enzymes in soyabean. *Biol. Plant.*, 52: 165-169.

Sims, J.J. and Kline, J.S. 1991. Chemical fractionation and plant uptake of heavy metal in soils amended with co-composted sewage sludge. *J. Environ. Sci. Technol.*, 13: 1255.

Smith, S.R. 1996. Agricultural recycling of sewage sludge and environment. CAB. International, Willingford, UK, p. 382.

Souza, J.F. and Rauser, W.E. 2003. Maize and radish sequester excess cadmium and zinc in different ways. *Plant Sci.*, 165: 1009-1022.

Stephen, U.W., Schmidke, I., Stephan, V.W. and Scholtz, G. 1996. The nicotianamine molecule is made-to-measure for complexation of metal micronutrients in plants. *Biometals.* 9: 84-90.

Stobart, A.K. and Griffiths, W.T. 1985. Effects of Cd^{2+} on the biosynthesis of chlorophyll in leaves of barley. *Physiol. Plant*, 63: 293-298.

Taiz, L. and Zeiger, E. 2002. *Plant Physiology*. Sinauer Associates (eds). Sunderland, U.S.A., p. 690.

Tiryakioglu, M., Eker, S., Ozkutl, F., Husted, S. and Cakmak, I. 2006. Antioxidant defense system and cadmium uptake in barley genotypes differing in cadmium tolerance. *J. Trace Elem. Med. Biol.*, 20: 181-189.

Torres, E., Cid, A., Herrero, C. and Abalde, J. 2000. Effect of cadmium on growth, ATP content, carbon fixation and ultra structure in the marine diatom Phaeodactylum tricornutum Bohlin. *Water Air Soil*, 117: 1-14.

Tyler, L.D. and Mcbride, M.B. 1982. Influence of Ca, pH and humic acid on cadmium uptake. *Plant Soil*, 64: 259-262.

United States Environmental Protection Agency (USEPA) 1997. Cleaning Up the Nation's Waste Sites: Markets and Technology Trends. EPA/542/R-96/005. Office of Solid Waste and Emergency Response, Washington, DC.

Uraguchi, S., Mori, S., Kuramata, M., Kawasaki, A., Arao, T. and Ishikawa, S. 2009. Root-to shoot Cd translocation via the xylem is the major process determining shoot and grain cadmium accumulation in rice. *J. Exp. Bot.*, 60: 2677–2688.

Vassilev, A, Berova, M. and Zlatev, Z. 1998. Influence of Cadmium on growth, chlorophyll content and water relations in young barley plants. *Biol. Plant.*, 41: 601-606.

Von Wiren, N., Klair, S., Bansal, S., Briat, J.F., Khodr, H., Shiori, T., Leigh, R.A. and Hider, R.C. 1999. Nicotianamine chelates both Fe(III) and Fe(II). Implications for metal transport in plants. *Plant Physiol.*, 119: 1107-1114.

Wallace, A., Romney, E.M. and Alexander, G. 1980. Zinc-cadmium interaction on the availability of each to bush bean grown in the solution culture. *J. Plant Nutr.*, 2: 51-54.

Wang, L., Zhou, Q.X., Ding, L.L. and Sun, Y. 2008. Effect of cadmium toxicity on nitrogen metabolism in leaves of *Solanum nigrum* L. as a newly found cadmium hyperaccumulator. *Hazard. Mat.*, 154(1-3): 818-825.

Wang, Z. 2000. *Plant Physiology*. China Agriculture Press. Beijing.

WHO, 1992. Cadmium - environmental aspects. World Health Organization, Geneva, p. 156.

Wu, F. and Zhang, G. 2003. Phytochelatin and its function in heavy metal tolerance of higher plants. *Ying Yong Sheng Tai Xue Bao.*, 14(4): 632-636.

Wu, F., H, Wu., Zhang, G. and Bachir, D.M.L. 2004. Difference in growth and yield in response to cadmium toxicity in cotton genotype. *J. Pl. Nutr. Soil. Sci.*, 167: 85-90.

Yang, J.F., Bu, Y.S. and Guo, X.Y. 2005. Research on effects of soil exogenous cadmium and lead pollution on rape growth. *Shanxi Agric. Sci.*, 3: 26-28.

Yang, X., Baligar, V.C., Martens, D.C. and Clark, R.B. 1996. Cadmium effects on influx and transport of mineral nutrients in plant species. *J. Plant Nutr.*, 19: 643-656.

Yang, X.E., Long, X.X., Ye, H.B., He, Z.L., Calvert, D.V. and Stoffella, P.J. 2004. Cadmium tolerance and hyperaccumulation in new Zn-hyperaccumulating plant species (*Sedum alfredii* Hance). *Plant Soil*, 259: 181-189.

Zeid, I.M. 2001. Response of Phaseolus vulgaris to chromium and cobalt treatments. *Biol. Plant.*, 44: 111-115.

Zhang, E.H., Zhang, X.H. and Wang, H.Z. 2004. Adaptable effects of phosphorus stress on different genotypes of faba-bean. *Acta Ecol. Sinica.*, 24(8): 1589-1593.

Zhang, J. and Shu, W.S. 2006. Mechanisms of heavy metal cadmium tolerance in plants. *J. Plant Physiol. Mol. Biol.*, 32(1): 1-8.

Zhen-Guo, S., Xian-Dong, L., Chun-Chun, W., Huai-Man, Ch. and Hong, Ch. 2002. Lead Phytoextraction from contaminated soil with high biomass plant species. *J. Environ. Qual.*, 31: 1893-1900.

Zhuang, P., Zou, H. and Shu, W. 2009. Biotransfer of heavy metals along a soil-plant insect-chicken food chain: Field study. *J. Environ. Sci.*, 21(6): 849–853.

Chapter 9
Physiology of Floral Malformation in Mango

V.K. Singh

Principal Scientist,
Department Plant Physiology, Central Institute for Subtropical Horticulture,
Lucknow – 226 101, U.P.
E-mail: singhvk_cish@rediffmail.com

CONTENTS

The malformation is a very serious physiological disorder of mango (*Mangifera indica* L.) in subtropics causing up to some time 90 per cent crop loss varying from place to place, cultivar to cultivar (Singh, 2006). It was first recorded at a small harvest of the Indian state Bihar from Dharbhanga district in 1891. Since then its occurrence

Figure 9.1: Type of Malformation.

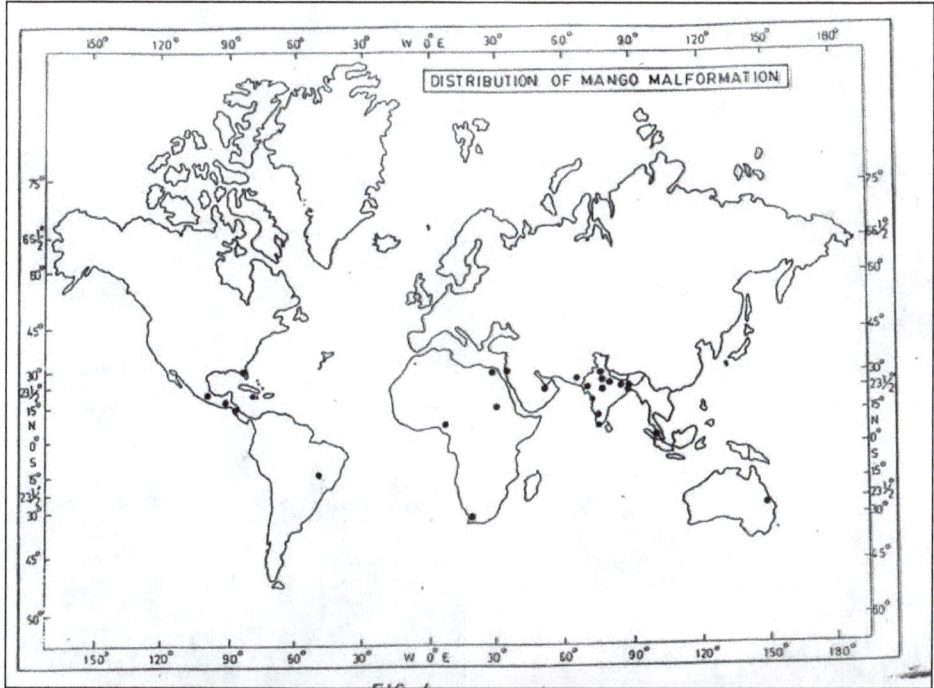

DISTRIBUTION OF MANGO MALFORMATION

World Distribution of Mango Malformation

Vegetative

Floral

Contd...

Figure 9.1–*Contd...*

Loose

Dense

Mixed

is being reported consistently from different parts of the world and hardly any mango growing countries has been left which is not affected by the malformation. Thus, it has been recognized as the plant disorder of International importance.

In recent years no other plant disorder has drawn so much attention from scientists of various disciplines and generated such high-pitched animated debate as mango malformation. The sequence of events that unraveled the confusion in understanding its cause and thereafter stepwise revelation of different aspects of the disorder leading to a common agreement about nature of its causal organism and developing integrated management practices makes a fascinating story. Here attempts have been made to focus the current status of researches on mango malformation (vegetative and floral) and to trace the course of research on this aspect since its first report.

1. Vegetative and Floral Malformation Symptoms

The disease/disorder appears both in young seedling and flowering trees. In the seedlings, the apical and axillary buds get swollen in the initial stages and produce thick short shootlets bearing scaly leaves. Similar shoots later arise from the axils of scaly leaves showing bunch like appearance due to the loss of apical dominance resulting in symptoms called 'bunchy top' disorder the seedlings affected at an early stage remain stunted and may dry up, however, those affected at later stages continue to produce healthy as well as malformed growth.

Floral malformation is characterized by deformation of panicles, loss of apical dominance shortened primary and secondary axes and thickened rachises of panicles giving characteristic clustered appearance of the flowers. The affected panicles have staminate and large flowers which do not set fruits. Several type of malformed panicles were noticed on the trees but loose type are more common.

2. Internal Symptoms

Internal symptoms include the development of hyperplastic and a hypertrophied cells is malformed vegetative and floral parts. The pith cells of malformed panicles are deformed and packed with starch marked anatomical differences have been reported in the rachis of healthy and malformed panicles. The cortex, xylem vessels and pith of the affected panicles contain 50 per cent less number of cells per unit area than the healthy ones (UPCAR final project report, 2003) and resin canal was found increased in affected tissue. Swollen mitochondria in the epithelial cells of malformed part were indicative of increased senescence.

3. Varietal Susceptibility

The incidence of malformation is reported to be variable. Its percentage of incidence varies from time to time, cultivar to cultivar, age of the tree and from season to season the incidence of malformation was started. The great diversity recorded in the cultivars and the variety Elaichi was found field resistance of this malady (Misra *et al.*, 2000), Bhadawaran less susceptible, Beauty Mc-lin most susceptible and Amrapali was found susceptible (Singh, 2006).

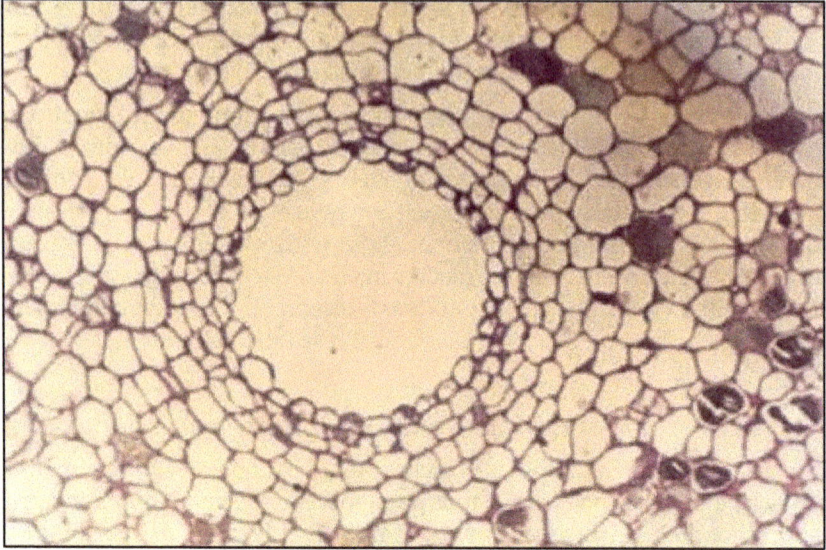

Figure 9.2: Cells of Normal Panicle.

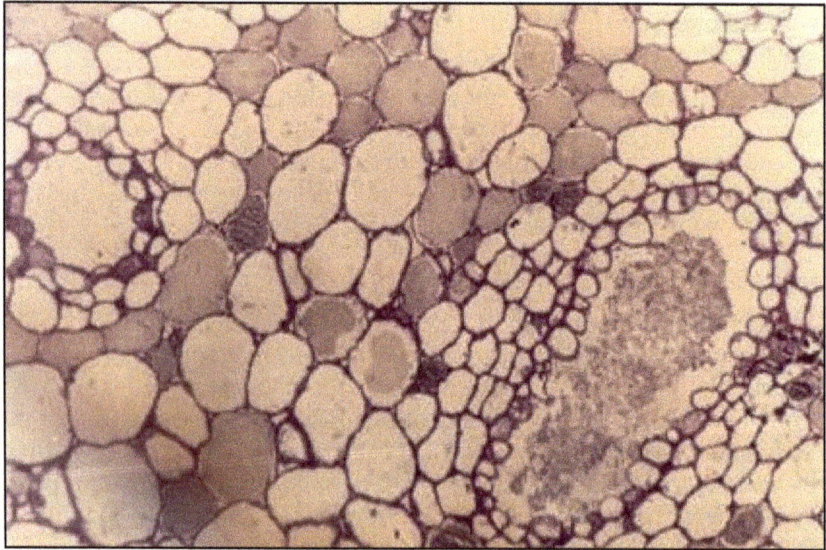

Figure 9.3: Enlarge Cells in Affected Panicle.

Malformation since very beginning has fascinated researchers of various disciplines and every body has attempted to interpret the problem from his particular professional perspective and solve it accordingly. But failure of such type of quick-fix solution suggested that it was not a straight forward problem rather a complex one. Thus, a phase of research with multidisciplinary approach was initiated. The information generated by scientists belonging to various discipline were put together and concerted effort was made.

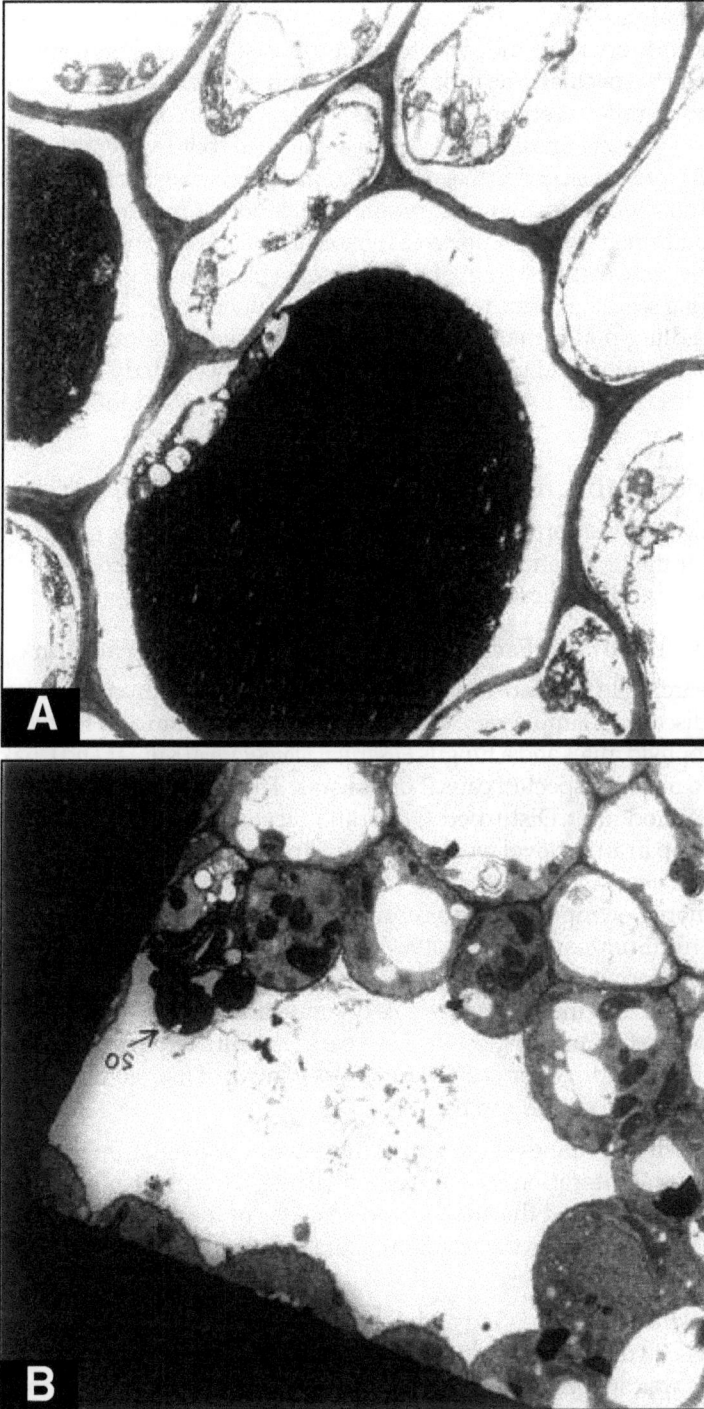

**Figure 9.4: Electron Microscopy of Normal (A) and Malformed Tissue (B)
of Floral Panicle.**

In the initial years *i.e.* upto the fifties, the research was limited to the visual observations and reporting the symptoms of this disorder and severity; attempts were also made to speculate the probable cause and possible remedies. During this period erophyid mites were the prime suspects as causal organism. Besides the hypothesis of virus origin of the disease was mooted, both vegetative and floral malformation were envisaged as the manifestations of the same disorder and attempts were made to reduce it incidence through eradication of malformed plant parts. Initially the malformation was reported as abnormal manifestations on inflorescences. But during this time manifestation of the disease on branches of grown up trees and top of the young seedlings were recorded. The term "Bunchy top" was first time used to denote seedling malformation and is called as vegetative malformation. The association of the eriophyid mite *Aceria mangifera* with this malady was a significant report. The idea that the disease is caused by a virus was also mooted which took long years to clear off.

4. Etiology of Malformation

The etiology of malformation has remained controversial. Diverse claims have been made for the cause and control of malformation. This has been ascribed to a number of biotic and abiotic factors.

4.1. Abiotic and Biotic Factors Associated with Malformation

The researches in the sixties were marked by systemic approach to identify the cause of the disease. Attempts were made to prove Koch's postulates for the first time both with eriophyid mites and a fungus *Fusarium moniliforme* (Kumar and Chakrabarti, 2001) the two major suspected causal organisms. The important research findings during this period are : Disturbed C/N ratio, accumulation of gibberellins and cytokinins, drop in auxin level were implicated to contribute in the development of this disorder. *F. moniliforme* was found to be associated with the malformed tissues and malformation symptoms were reproduced by artificial inoculation with the isolated fungus (Summanwar *et al.*, 1967). On the other hand, a group of scientists reproduced malformed shoots and panicles by inoculating just sprouting buds with mites taken from malformed twigs. The role of temperature in the disease manifestation that has become a favourite aspect in later years was initiated this time (Singh *et al.*, 1998). A remarkable discovery of this time was control of mango malformation by removing malformed shoots and panicles.

The research in seventies witnessed an intensive investigation on the biochemical and physiological alteration in malformed plants. The plant physiologists and horticulturists interpreted the changes as the cause of the malady while the plant pathologists viewed them as the resultant of the pathogenic invasion (*F. moniliforme*) or physiology of pathogenesis.

A control measure consisting of deblossoming and naphthyl acetic acid (NAA) spray which is till date the favourite recommendation for horticulturists and plant physiologists was suggested (Chadha *et al.*, 1979). Differences in growth hormone gibberellins, and protein and DNA contents between healthy and malformed tissues were determined and speculated that these qualitative and quantitative changes of the biochemical components induced the disease symptoms (Chakrabarti *et al.*, 1990).

In seventies a substantial changes in perspectives were recorded. To confirm or disprove the contention of horticulturists and plant physiologists that the aberrant biochemical constituents of the host cause the malady, plant pathologists first inoculated the healthy plants with the pathogen and subsequently reproduced the similar biochemical changes. Thus convinced that the abnormal biochemical constituents were not the cause of the disease; on the contrary, these were the results

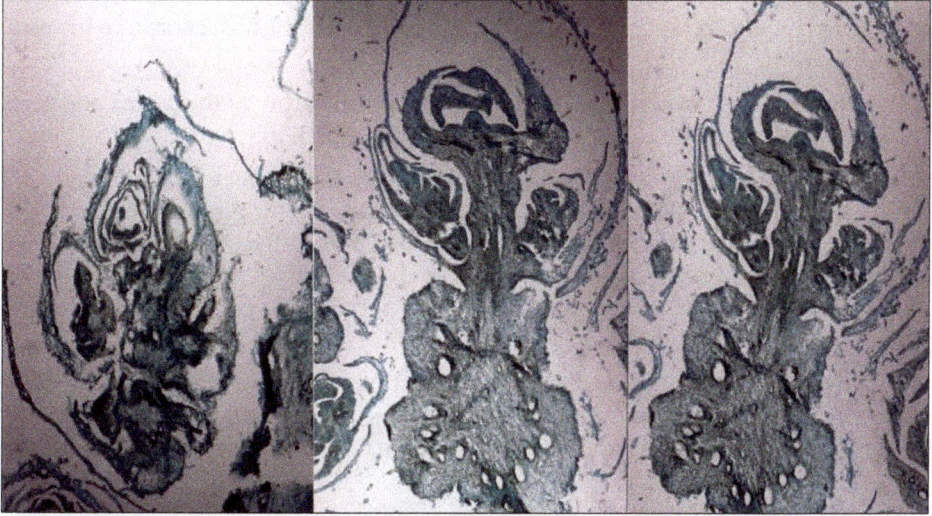

Figure 9.5: Normal Flower Bud.

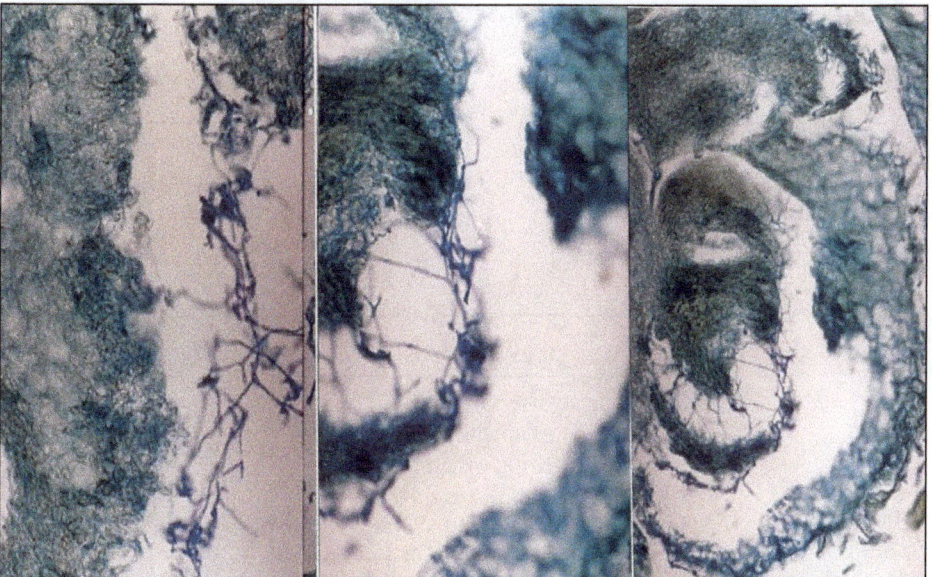

Figure 9.6: Affected Flower Bud.

of the pathogenic invasion. Artificially inoculating the healthy host tissues, biochemical changes in protein and nitrogen content, cell wall components and oxidative enzymes activity similar to that of naturally malformed plants were reproduced (Chakrabarti *et al.*, 1990).

F. moniliforme was renamed as *F. moniliforme* var. *subglutinans* as number of *Fusarium* spp. has been isolated from malformed mango tissue, however, *F. subglutinans* (*F. moniliforme* var. *subglutinans*) was reported to be most commonly associated species with both floral and vegetative malformation (UPCAR final project report, 2003).

Primary indications towards host-specificity of the pathogen and its dependence on external wounding agencies, biotic or abiotic, to enter the host came into the fore.

The presence of toxins of *F. moniliforme* var. *subglutinans* and aberrant host metabolite detected in malformed cells; the chemical nature of the toxins and their role in producing the disease symptoms was established.

Macro and micro nutrients studies in disorder tissue have not yielded valuable information on the possible cause of malformation, but Ca^{++} is generally found lower in malformed tissue which may be responsible for the twisted and deformed tissue, and the death of meristematic zones. Moreover, the effect of selenium on malformation needs investigation as it was known to be more effective substrate for ethylene production than methionine (Liebermann, 1979).

In the eighties, most of the research publications directly or indirectly related to the estimation of growth hormones in malformed tissues and hormonal imbalance in malformed tissues was considered to be one of the important reasons for creating the

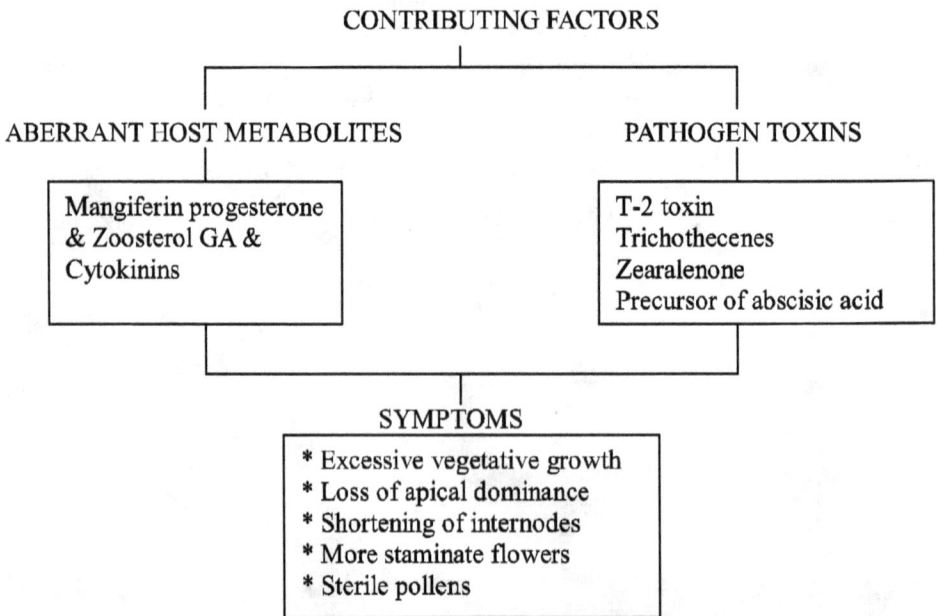

CONTRIBUTING FACTORS

ABERRANT HOST METABOLITES

| Mangiferin progesterone & Zoosterol GA & Cytokinins |

PATHOGEN TOXINS

| T-2 toxin Trichothecenes Zearalenone Precursor of abscisic acid |

SYMPTOMS

* Excessive vegetative growth
* Loss of apical dominance
* Shortening of internodes
* More staminate flowers
* Sterile pollens

Figure 9.7: Development of Malformation in Mango.

disorder of malformation (Majumdar *et al.*, 1970 and Pandey *et al.*, 1974) and treated as physiological disorder. Elaborate authentic experimentation conclusively disproved the virus origin of the disease during this period was concluded.

Information in support of the involvement of the fusarial toxins in the disease manifestation was made available. In this context, a toxic compound Malformin, usually secreted by *Aspergillus niger* is to be mentioned especially. Contrary to this Ram and Bist (1984) reported malformin spp. substances in malformed panicle but absent in healthy ones.

Chakrabarti and Kumar (2002) reported that the physiology of pathogenesis strongly support the contention of the disease manifestation which is the combined effects of the aberrant host metabolites (toxic principles, TP) or (mangiferin, zoosterols, abnormal GA or cytokinins) and the fusarial toxins (malformation inducing principles, MIP) or (trichothecenes, T-2 toxins, zearalenone, abscisic acid). But horticulturist and plant physiologist have different views in this context. Despite the strong evidence supporting *F. moniliforme* var. *subglutinans* as the causal organism of the disease, some publications cast doubt on the fact that the fungus is responsible for mango malformation. Hence, to provide unequivocal evidence, a series of studies using molecular diagnostic tools were undertaken in nineties. This period also witnessed large number of publications on the epidemiology that was so far poorly understood. Besides, mango malformation forecasting models were proposed (Chakrabarti and Chakraborti, 2006).

Improved technologies such as vegetative compatibility (VCG) and use of GUS reporter gene were also used to confirm *F. moniliforme* var. *subglutinans* as the causal organism of the disease. Biochemical evidence was put forward to establish it as a physiological race of *F. moniliforme* var. *subglutinas* and was proposed to name it as *F. moniliforme* var. *subglutinans* sp. *mangiferae*. The mechanism of transformation of the fungus into a physiological race was also reported. Various epidemiological aspects *viz.* structure of the epidemic, seasonal variation of the pathogen vis-à-vis the disease incidence in relation to environmental parameters, mangiferin content and flushing of the host, synergistic role of a new group of mycophagus mite with the pathogen, disease dissemination and variation in symptom expression and virulence of the pathogen under adverse agro-climatic conditions were reported. Several factors like level of genetic resistance/susceptibility of the variety, age of the host plants, bearing nature of cultivars, environment factors including effect of cultural practices that affect the malformation development was reported.

5. Expression Mango Malformation: Association of Mangiferin and other Related Attributes

Role of ethylene in malformation was intensively investigated (Pant, 2000). Significant higher level of ethylene was detected in malformed panicle and could suppress the apical dominance of the panicle, increasing isodiametric growth of the rachides, and thickening the secondary branches of malformed panicle producing overcrowding of flowers. But the nature of stress ethylene either it is due to abiotic or biotic (*Fusarium* spp.) needs further investigation.

Chromosomal anomalies during micro-sporogenesis in malformed flowers resulting into abortive pollens were reported. The PR protein that imparts resistance in the host plant against the development of malformation was identified.

The additional protein band in resistant cultivar Elaichi during flowering was also reported to have association with the expression of mango malformation (Singh *et al.*, 2007). But this addition protein is fungal protein or imparted to the resistance process of this variety needs to be further studied. Study on this resistant source of malformation, (Elaichi variety) has also provided the information on involvement of metabolic processes in resistance to this disorder (Singh, 2006). Mangiferin ((1,3,6,7-terahydroxy xanthone-C_2-β-D-glucoside) was reported to be natural source of resistance to malformation (Singh and Prasad, 2003). Research work on resistant variety and very susceptible variety Beauty Mc-lin and susceptible variety Amrapali and Dashehari at various flower developmental stages suggests that the mangiferin is a potent stimulator of defense related enzymes for the induction of natural defense

Figure 9.8: Necrosis of Affected Plants Due to Excessive Malformation.

Figure 9.9: Induction of Mango Malformation.

system of the plant and could be considered as a good biochemical indicator for screening mango genotypes for resistance to malformation. Along with mangiferin, polyphenol oxidase (PPO), catalase peroxide peroxidase were also found to be imparted in the resistance process of malformation(Singh,2006). However, further studies are needed in to the inheritance of mangiferin content to facilitate its use as a trait that may be used in selecting parental stock of mango. Mangiferin mediated mass production of malformation was developed in first time (Singh and Misra, 2006). The tree which had developed maximum number of malformed buds suffered from die-back later. The necrosis and die-back of malformed tree occurred, may be due to unmetabolized cyanide (Pant, 2003) or oxidation of mangiferin in to polymeric quinine (Chakrabarti and Ghosal, 1989).

6. Molecular Physiology of Mango Malformation

Most of the molecular works are associated *Fusarium* spp to understand the molecular physiology of the malady/disease. Studies using different molecular tools such as nuclear and mitochondrial DNA sequences and isozymes and tests for mating types and compatibility concluded that *Fusarium* isolate from malformed mango tissues represented a new species of *G. fujikuroi* complex and a discrete taxon. It was described as *F. mangiferae*.

The studies on population genetics conducted with the *F. mangiferae* showed very little variations among isolates from different region of the world. *F. mangifeae* in different geographical areas was most probably introduced from India. It was also assumed that the pathogen have originated in India. *Fusarium* spp. associated with the economically diverting mango malformation disease (MMD) was thoroughly studied in mango. Mexico 142 *Fusarium* strains were isolated. MMD associated *Fusarium* spp. were investigated using multilocus DNA sequence data and phylogenetic species recognition. DNA sequence data indicated that at last nine phylogenetically distinct *Fusarium* spp. with in *Gibberella fungikuroi* species complex are associated with MMD. Genetic analysis of *F. mexicanum* and other *Fusarium* spp. associated with MMD, phylogenetic diversity of mango associated *Fusarium* spp. were recently reported in detail (Colina *et al.*, 2010).

7. Management Practices

A computerized expert system has been developed to predict the incidence of this malady in any state of India and to suggest appropriate an IDM strategy. Very viable and effective control measure of mango malformation was developed through pruning of old and malformed shoot and spraying of Chelated Zn^{++} (100 ppm) and Cu^{++} (40 ppm) during flower bud differentiation and flowering stage combined with the spray of Carbendazim (0.1 per cent) during November–December.

Where do we move from here now?

- ☆ Invent ways and means to protect our prized mango varieties from the bane of mango malformation.
- ☆ Exact contribution and role of stress ethylene towards the cause of differential susceptibility of malformation should be worked out.

☆ Contribution of environmental pollution to the disease and its spread, and invasion of various micro-organism may be attended for this cause.

☆ Identification of gene(s) responsible for resistance to the malformation.

☆ Transfer of resistance gene through breeding and genetic engineering for evolving resistance varieties should be taken on priority.

8. Conclusion

We have "miles to go" to completely unwind the mystery of malformation and find out an easy and a single stroke solution. In view of the current enthusiasm among the new generation of scientists and their improved techniques, it may be hoped that it is not far away when the stride to tame the devastating malady will trump the success of century old efforts.

However, once upon a time India was the epicentre of mango malformation research, now-a-days hardly any scientist venture to work on this problem. On contrary, a number of scientists in USA, Israel, South Africa, Australia are vigorously and constantly pursuing the research on malformation. Perhaps we are looking towards these countries to come with technologies to save our mango orchards from the tentacles of malformation.

9. References

Chadha, K.L., Pal, R.N., Prakash, O., Tandon, P.L. and Singh, H. (1979). Studies on mango malformation, its causes and control. *Indian J. Hort.*, 36: 359-368.

Chakrabarti, D. K. and R. Kumar. 2002. Mango malformation : present status and future strategy. In *IPM System in Agriculture*, Vol. 8, (eds. R.K.Upadhyay, D.K.Arora and O.P.Dubey), Aditya Books Pvt. Ltd., N. Delhi, pp. 237-255.

Chakrabarti, D.K. and Ghosal, S. (1989). Disease cycle of mango malformation induced by *Fusarium moniliforme* var. *subglutinans* and the curative effects of mangiferin metal chelates. *J. Phytopathol.*, 125: 238-246.

Chakrabarti, D.K. and Pinaki Chakraborty. 2006. Expert system for management of malformation disease of mango, *ICAR News*, 12(1) : 18.

Chakrabarti, D.K., A. Singh and K. Singh. 1990. Physiological and biochemical changes induced by accumulated mangiferin in *Mangifera indica*, *Journal of Horticultural Sciences*, 65(6) : 731-737.

Colina, G.O., Alvarado, G.R., Pavia, S.F., Maymon, M., Ploetz, R.C., Aoki, T., Donnell, K. and Freeman, S. (2010). Identification and characterization of a novel etiological agent of mango malformation disease in Mexico, *Fusarium mexicanum* sp. nov. *Phytopathology*, 100(11): 1176-1184.

Kumar, J. and Beniwal, S.P.S. (1992). Role of *Fusarium* species in the etiology of mango malformation. In proceedings of the 4th International Mango Symposium, Miami, Fla, USA, p. 133.

Kumar, R. and D.K. Chakrabarti. 1997. Assessment of loss in yield of mango (*Mangifera indica*) caused by mango malformation, *Indian Journal of Agricultural Sciences*, 67(3) : 130-131.

Kumar, R. and D.K.Chakrabarti. 2001.Techniques to reproduce floral malformation of mango. *Proceedings International Conference on Integrated Plant Disease Management for Sustainable Agriculture*, Vol. II, Indian Phytopathological Society, IARI, New Delhi, pp. 1121-1122.

Liebermann, M. (1979). Biosynthesis and action of ethylene. *Ann. Rev. Plant Physiol.*, 30: 533-591.

Majumdar, P.K., Sinha, G.C. and Singh, R.N. (1970). The effect of exogenous application of X-napthyl-acetic acid on mango, *Mangifera indica*. *Indian J. Hort.*, 20: 130-131.

Misra, A.K., Negi, S.S., Rajan, S. and Kumar R. (2000). Cultivar Elaichi – A new source of resistance to mango malformation. Indian Phytopathology Golden Jubilee – Proceedings Vol. II, 751-752.

Pandey, R.M., Rathore, D.S. and Singh, R.N. (1974). Hormonal regulation of mango malformation. *Curr. Sci.*, 43: 694-695.

Pant, R.C. (2000). Is 'Stress-Ethylene' the cause of mango (Mangifera indica L.) malformation? *Physiol. Mol. Biol. Plants*, 6: 7-14.

Pant, R.C. (2003). Final report on Networking Project of Mango Malformation submitted to UPCAR, Lucknow.

Ram, S. and Bist, L.D. (1984). Occurrence of malformin like substances in malformed panicles and control of floral malformation in mango. *Scient. Hort.*, 23: 333-336.

Singh, V. K., Saini, J. P. and Misra, A. K. (1998). Mango malformation in relation to physiological parameters under elevated temperature. *Indian J. Plant. Physiol.*, 3 : 231 – 233.

Singh, V.K. (2006). Physiological and biochemical changes with special reference to mangiferin and oxidative enzymes level in malformation resistant and susceptible cultivars of mango (*Mangifera indica* L.). *Scientia Horticulturae*, Elsevier, 108:43-48.

Singh, V.K. and Misra, A. K. (2006). Reproduction of vegetative malformation in mango. *Indian J. Horticulture*, 63(1) : 36-38.

Singh, V.K. and Prasad, S. (2003). Mangiferin – a natural source of resistance to mango malformation. *Indian J. Plant Physiol.* (Special Issue) : 508-511.

Singh, V.K., Misra, A.K., Rajan, S. and Ansari, M.W. (2007). Protein profile, mangiferin and polyphenol oxidase levels in susceptible and tolerant genotypes of mango against malformation. *Indian J. Hort.*, 64(1): 23-25.

Summanwar, A.S. (1967). Mango malformation, serious economic consequences. *Indian Hort.*, 11: 12-16.

Chapter 10

Recent Advances in Alteration of Fatty Acid Composition and Protein Quality of Major Edible Oil Seed Crops

M.K. Mahatma, A.L. Singh, Lokesh Kumar and J.B. Misra

Directorate of Groundnut Research, Junagadh – 362 001, Gujarat
E-mail: maheshmahatama@nrcg.res.in

CONTENTS

Abbreviations

ALA:	α-linolenic acid
ARA:	Arachidonic acid
DHA:	Docosahexaenoic acid
EPA:	Eicosapentaenoic acid
FA:	Fatty acid
FAD:	Fatty acid desaturase
FFAs:	Free fatty acids
GLA:	γ-Linolenic acid
LEA:	Low erucic acid
MUFAs:	Mono unsaturated fatty acids
O/L:	Oleic/linoleic
PUFAs:	Ployunsaturated fatty acids
SDA:	Stearidonic acid
SFAs:	Saturated fatty acids
TAG:	Triacylglycerol.

1. Introduction

Oils and fat, a major source of calories for human diet, provide fat soluble vitamins and essential fatty acids and contribute significantly to the sensory characteristics of food. Many oils are also used for non-food applications, although current industrial use mostly for biodiesel utilize only a small proportion of the world vegetable oil production. About 80 per cent of edible oils are derived from plant sources. Soybean, rapeseed mustard, cotton seed, groundnut, sunflower, safflower, sesame, linseed and castor are major oil seed crops; coconut and palm are plantation crops. Among the edible oil seed crops, soybean, rapeseed mustard, cottonseed and groundnut contribute about 87 per cent of the world's oil supplies (FAO, 2012) and soybean oil is by far the dominant oil in this category, accounting for over half of the world vegetable oil production. The major fatty acids present in plants oils have chain lengths of 16 or 18 carbons and contain 1–3 cis double bonds (Ohlrogge and Jaworski, 1997).

The fatty acid profile plays a significant role in both the nutritional properties and end-use functionality of edible plant oils. Majority of vegetable oils are oxidatively unstable due to high percentage of polyunsaturated fatty acids, limiting their utility in food applications. Partial hydrogenation can improve the oxidative stability of these plant oils by saturation of fatty acids, and also solid fat functionality, which is especially useful for baking and similar applications. But this improved functionality affects nutritional quality due to the presence of high concentrations of trans fatty acids found in partially hydrogenated oils. Trans fatty acids have been correlated with cardiovascular disease (Korver and Katan, 2006). Intake of other dietary fatty acids have a strong influence on overall health, *i.e.* saturated fatty acids such as palmitic acid, raise harmful low-density lipoprotein (LDL) cholesterol and lower the benecial high density lipoprotein (HDL) cholesterol whereas stearic acid has neutral effect on blood, total and LDL cholesterol levels, unlike other predominant long-chain saturated fatty acids (SFAs) of vegetable oils which are responsible for increasing the cholesterol levels in blood (Minihane and Harland, 2007).

Generally, poly unsaturated fatty acids (PUFAs) lower serum cholesterol and mono unsaturated fatty acids (MUFAs) were neutral, but from 1985 onwards many scientists demonstrated that vegetable oils rich in MUFAs (*i.e.* oleic acid, the most abundant fatty acid in nature) were as effective as PUFA in reducing not only the total blood cholesterol level, but also LDL cholesterol. The oleic acid has additional advantage of neither causing a decrease in benecial HDL cholesterol levels nor an increase in blood triglycerides (Mensink and Katan, 1990).

The linoleic (precursor of ω-6 family PUFAs) and α-linolenic (precursor of ω-3 family PUFAs) are essential fatty acids (EFAs) for human and are required for the production of components of cell membrane, eicosanoid metabolism (ω-6 metabolized to arachidonic acid and ω-3 into eicosapentanoic acid), gene expression and intercellular cell to cell communication (Simopoulos 2000). A high ω-6 content and a high ω-6:ω-3 ratio in the dietary fat intake are atherogenic and diabetogenic. The newly introduced so-called "heart friendly" oils like sunflower, safflower, rice-bran, corn and cottonseed oils have undesirable ratio of PUFAs (very rich in ω-6 with no or negligible amount of ω-3), and the use of single or excess intake of these newer oils are actually detrimental to health. Thus, it is necessary to balance the dietary consumption of ω-6/ω-3 fatty acids for proper health and nutrition. By decreasing the risk of cardiovascular disease (CVD) and other chronic diseases and improve mental health.

The balanced proportion of fatty acids has been a subject of intense research especially for the amount of saturated fats to be consumed. Although world health organization (WHO) recommens less than 10 per cent consumption of saturated fats, several reports have recommend a ratio of 1:1:1 between SFA, MUFA and PUFAs the most suitable for human consumption. Thus, a typical balanced ideal fatty acid intake for human consumption comprises of approximately 30 per cent SFAs, 33 per cent PUFAs (containing essential fatty acids) and 37 per cent MUFAs, but most of the commercial oilseed crops do not have this ideal proportion (Table 10.1). This need attention of researchers (particularly, oil seed biochemists and biotechnologists) to develop designer vegetable oil that can improve functionality and maintain nutritional quality which will have market potential for both the frying and baking industries (Cahoon _et al._, 2009).

The establishment of biotechnological tools for metabolically engineering of plants expressing single genes or multistep biosynthetic pathways and with the understanding of metabolic pathways, the increasing availability of genes that encode enzymes related to fatty acids biosynthesis, the manipulation of fatty acids in plants was stimulated during the last decades (Wu _et al._, 2005; Damude and Kinney, 2008; Lu _et al._, 2010). The manipulation of fatty acids in transgenic plants open up the possibility of plant fatty acid engineering via the transformation of another genome, the plastid genome (Craig _et al._, 2008). The gene has been identified for the synthesis and metabolism of novel fatty acid structures, fatty acids with hydroxy and epoxy residues that are well suited for industrial applications such as lubricants, plasticizers and nylon precursors (Burgal _et al._, 2008; Li _et al._, 2010). Also the incorporation of additional fatty acids of nutritional importance _i.e._ ARA (ω-6 arachidonic acid; 20:4 Δ 5,8,11,14), EPA (ω-3 eicosapentaenoic acid; 20:5 Δ 5,8,11,14,17) and DHA (ω-3 docosahexaenoic acid; 22:6 Δ 4,7,10,13,16,19) in domesticated plant crops are underway (Cahoon _et al._, 2007; Napier, 2007; Walsh and Metz, 2013). In this review, an effort has been made to highlight the advancement made in the fatty acid manipulation and protein quality improvements in soybean, rapeseed/mustard, cotton seed and peanuts crops for better human health (nutritional).

2. Fatty Acid and Triacylglycerol Biosynthesis

The fatty acid (FA) synthesis takes place in plastids (Figure 10.1A), but the triacylglycerol (TAG) molecule synthesize outside the plastid and may be associated with both the endoplasmic reticulum (ER) and the oil body (Figures 10.1B, C) (Ohlrogge _et al._, 1979). In most oil seeds, carbon is delivered to FA synthesis via glycolysis with hexose and/or triose as the predominant carbohydrate entering the plastid. However, green seeds can also use light to supply NADPH and ATP, which allows a 'bypass' of glycolysis via ribulose-1,5-bisphosphate carboxylase activity and pentose phosphate enzymes. Priming and elongation of nascent acyl chains requires acetyl and malonyl-CoA respectively, as direct precursors.

The plastid FA synthesis pathway determines the chain length (up to 18 carbons) and the level of saturated FAs in seed oils. Assembly of FAs occurs on acyl carrier protein (ACP) via a cycle of 4 reactions that elongate the acyl chain by 2 carbons each cycle. After 7 cycles, the saturated 16 carbon acyl-ACP can either be hydrolyzed by

Table 10.1: The Fatty Acid Composition (Per cent) and their Range in Major Oilseed Crops.

Fatty Acids*	$C_{12:0}$	$C_{14:0}$	$C_{16:0}$	$C_{18:0}$	$C_{20:0}$	$C_{18:1}$	$C_{18:2}$	$C_{18:3}$	$C_{20:1}$	$C_{22:1}$
Oilseeds crops										
Groundnut			9-25	1-5	2-10	38-56	16-38		1-3	
Mustard			2-4			7-22	12-24	10-15	6-14	18-49
India-rape			2-3			9-34	11-18	6-12	8-12	24-61
Coconut	44-52	13-20	8-11	1-3		4-8	1-3			
Sunflower			5-6	4-6		22-50	40-67			
Safflower			5-8	1-2	1-4	6-8	75-79			
Sesame			8-12	3-8		35-47	28-47			
Niger			6-10	5-9		13-39	45-66			
Soybean			7-12	2-5		22-34	50-60	2-10		
Palm			32-47	1-6		40-52	2-11			
Cotton seed			22-26	3-4		18-21	50-52			
Corn (Maize)			10-14	3-4		24-30	45-53			
Rice bran			14-18	1-2		42-48	35-39			

Bhattacharya et al., 2012

*Trivial names: $C_{12:0}$ Lauric acid $C_{14:0}$ Myristic acid; $C_{16:0}$ Palmitic acid; $C_{18:0}$ Stearic acid; $C_{18:1}$ Oleic acid; $C_{18:2}$ Linoleic acid; $C_{18:3}$ Linolenic acid; $C_{20:0}$ Arachidic acid; $C_{20:1}$ Ecosenoic acid and $C_{22:1}$ Erucic acid.

the FATB acyl-ACP thioesterase or further elongated by KASII to 18:0-ACP, which is then desaturated to 18:1-ACP and hydrolyzed by the FATA thioesterase (Li-Beisson *et al.*, 2013). The resulting 16:0 and 18:1 free acids are the main products of plastid FA synthesis, and their relative proportions are determined by the activities of FATA, FATB, 18:0-ACP desaturase (SAD) and KASII (Figure 10.1A).

2.1. Export of Acyl Chains from Plastid to Endoplasmic Reticulum (ER)

After FA synthesis, the free FA products of the FATA/B thioesterases are exported from the plastid. After export, it is presumed that long-chain acyl-CoA synthetase (LACS) on the outer plastid envelope forms the acyl-CoA that is the substrate for glycerolipid assembly (Figures 10.1A, B). The esterification of newly synthesized FAs to PC via the acyl-editing cycle can occur at the plastid envelope via acyl-CoA: lysophosphatidylcholine acyltransferase (LPCAT). Acyl editing is a phosphatidylcholine (PC)-deacylation and lysophosphatidyl choline (LPC)-reacylation cycle which exchanges the FA on PC with the acyl-CoA pool without net PC synthesis or degradation (Figure 10.1B, orange arrows). The acyl editing cycle allows first, nascent FA from the plastid to be incorporated into PC while second, FAs that have been desaturated or otherwise modied on PC are released into the acyl-CoA pool where they can participate in the Kennedy pathway or other reactions (Bates *et al.*, 2013).

2.2. Simple and Complex Pathways of TAG Assembly

The *de novo* assembly of TAG from glycerol-3-phosphate and acyl-CoAs (also known as the Kennedy pathway) involves only four enzymatic steps: first, two acylations of G3P by sn -1 glycerol-3-phosphate acyltransferase (GPAT) and lysophosphatidic acid acyltransferase (LPAAT), followed by phosphatidic acid phosphatase (PAP), and a third acylation by diacylglycerol acyltransferase (DGAT) (Figure 10.1C, green arrows). The FA esterified to the sn-2 position of PC is the major site of ER localized FA modification (*e.g.*, desaturation, hydroxylation, etc. (Bates *et al.*, 2012; Vandeloo *et al.*, 1995). In developing seeds, the flux of acyl chains in the ER eventually leads to esterification on all three positions of glycerol to form triacylglycerol (TAG). The low polarity of TAG is believed to result in the accumulation of this lipid between bilayer leaflets leading to the budding of storage organelles termed oil bodies.

3. Approaches of Fatty Acid Alteration

The development of a nutritionally optimized or superior quality edible oil seed crop is a challenging research area and several affords have been made in this regard through conventional breeding and mutation approaches and by biotechnological (Over expression of gene, Gene silencing and insertion of novel gene) approaches. The suitable seed oils to be used for human consumption must satisfy a relatively high level of MUFAs (oleic acid), and an intermediated level of PUFAs with a good balanced ratio (6:1) of ω-6/ω-3. In the food products requiring a solid or semi-solid consistency, it is better to prefer for stearic acid (neutral, saturated fatty acid) instead of conventionally used SFAs (lauric, myristic and palmitic acids) and hydrogenated (trans) fatty acids. Unfortunately, none of the naturally occurring oilseed crops have these balanced fatty acids compositions, which are nutritionally beneficial and also provide the required functional property of the food products.

3.1 Breeding Approaches

3.1.1. Low Erucic Acid Content Rapeseed Varieties

Oil rich in erucic acid (22:1) is considered undesirable for human consumption. Low erucic acid rapeseed (LEAR) cultivar was developed by the introduction of recessive alleles (e^A and e^c) at both loci involved in 22:1 content. A spontaneous mutant of the German "Liho" spring rapeseed is the only source of LEA in rapeseed which was used to breed the first LEAR variety in Canada (Stefansson *et al.*, 1961). Oil from modern LEAR (so called single zero "0" seed quality varieties) lacks nutritionally undesirable long chain FA (less than 2 per cent of 20:1 and 22:1) and is highly appreciated due to its FA profile (5–7 per cent of saturated FA, 58–60 per cent of monounsaturated FA and 30–35 per cent of polyunsaturated FA) that almost its diet recommendations. There is success in selection of seed mutants expressing higher oil yields and a modified fatty acid composition. An example is the virtual elimination of erucic acid, a fatty acid with negative health implications, from rapeseed cultivars destined for edible oil production. These low erucic cultivars are used to produce canola oil, which has become widely consumed cooking oil (Chopra and Vageeshbabu, 1996).

3.1.2. High Oleic and Low Linolenic (HOLL) Rapeseed Varieties

Reduced level of polyunsaturated fatty acids (especially linolenic acid, 18:3) and increased content of monounsaturated FA (oleic acid, 18:1) provide higher oil stability and the resultant product can be used for salad dressings and for food applications requiring high cooking and frying temperatures and hence, breeding rapeseed with high oleic acid and low linolenic acid content is a major goal.

Canadian and European researchers used the combination of x-rays and chemicals mutagens to develop mutants of *B. napus* with <3 per cent linolenic acid and elevated levels of linoleic acid (Robbelen and Thies 1980). The reduced level of linolenic acid observed in these mutants was the result of deactivating the enzymes responsible for desaturation of linoleic acid. In 1988, the cultivar Stellar, with low levels of linolenic acid, was released for commercial production in Canada (Scarth *et al.*, 1988). Mutants with reduced levels of both linoleic and linolenic were also developed (Pleins and Friedt 1989). Auld *et al.* (1992) developed mutants using chemical mutagene in both *B. napus* and *B. rapa* with reduced levels of PUFA and increased levels of oleic acid by blocking oleic acid desaturation, and the most promising mutant identified from M2 seeds of *B. rapa*, was M-30, which had 2.1 per cent linoleic acid and 3.0 per cent linolenic acid, vs. 11.9 per cent linoleic and 8.6 per cent linolenic acid in the original cultivar. This mutant was further crossed with 'Tobin' to derive F4 lines having <6 per cent total PUFA and oleic acid concentrations >87 per cent. The most promising mutant identified from M2 seeds of *B. napus* was X-82, which had 6.6 per cent PUFA, vs. 27.4 per cent PUFA. Several M~ and M4 lines derived from X82 had < 6 per cent total PUFA and > 88 per cent oleic acid.

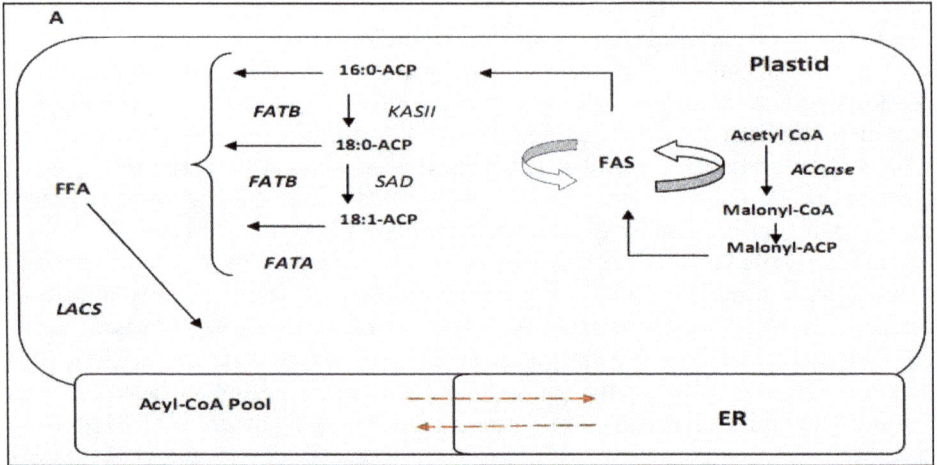

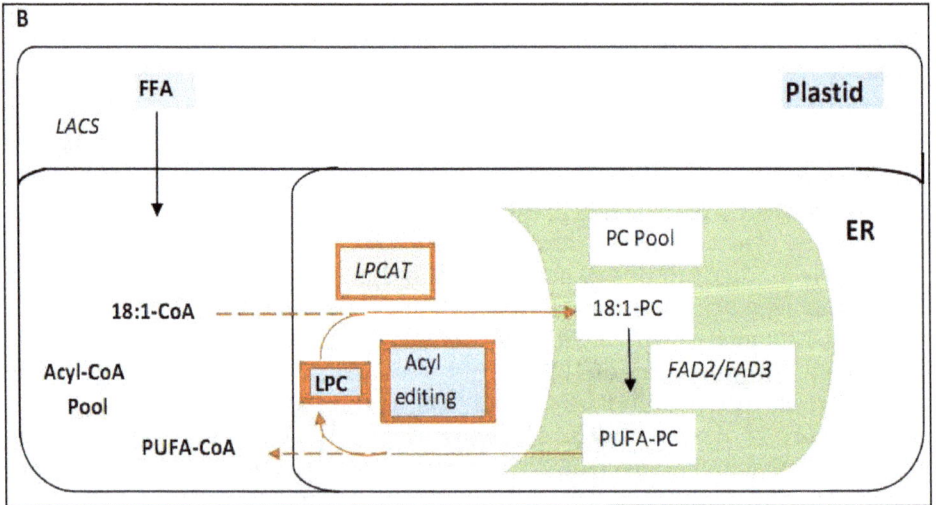

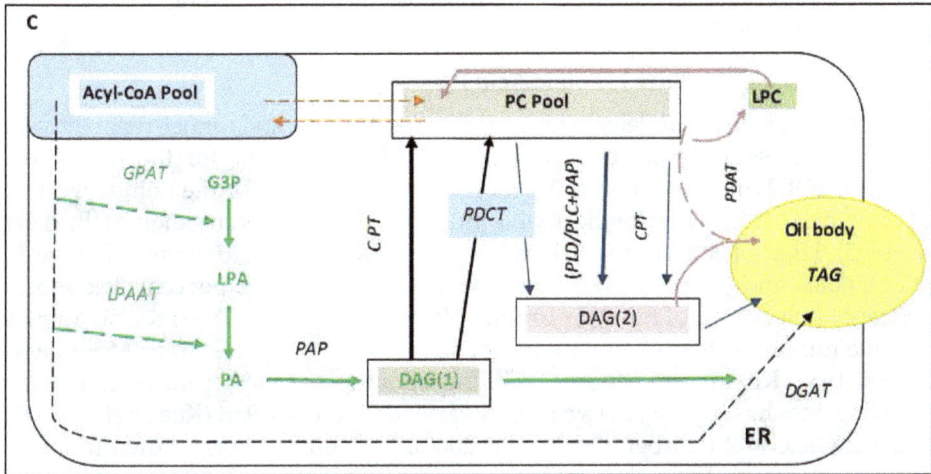

Figure 10.1: Overview of Major Reactions of Fatty Acid and Triacylglycerol (TAG) Synthesis in Plants (Bates *et al.*, 2013).

(A) Fatty acid synthesis in plastid (B) Acyl editing and (C) TAG synthesis. Acyl transfer reactions are dashed lines. Green lines are *de novo* TAG synthesis, blue lines are PC-derived DAG synthesis, orange lines are acyl editing, and pink lines denotes phospholopid: diacylglycerol acyltransferase (*PDAT*) TAG synthesis. DAG (1) is *de novo* synthesized DAG and DAG (2) is PC-derived DAG. Abbreviations: substrate are bold: ACP- Acyl carrier protein; DAG- Diacylglycerol; FFA- Free fatty acid; G3P- Glycerol-3-phosphate; LPA - Lysophosphatidic acid; LPC- Lysophosphatidic acid; MAL- Malonate; PA- Phospahatidic acid; PC- Phosphatidyl choline; PUFA- Polyunsaturated fatty acids; TAG- Triacylglycerol.

Enzymatic reactions are in italics: *ACCase*-Acetyl-CoA carboxylase; *FAS*-fatty acid synthase; *FATA/B*-fatty acyl-ACP thioesterases A/B ; *KASII*-Ketoacyl-ACP synthase; *LACS*-long-chain acyl-CoA synthetase; *SAD*-Stearoyl-ACP desaturase; FAD-fatty acid desaturase; *LPCAT*-Acyl-CoA: lyso phosphatidyl choline acyltransferase; *GPAT*- sn-1 glycerol-3-phosphate acyltransferase; *LPAAT*-lysophosphatidic acid acyltransferase; *PLC/D*- phospholipase C,or phospholipase D; *PDCT*-phosphatidylcholine:diacylglycerol cholinephosphotransferase; *PAP*-Phospahatidic acid phosphatase; *PDAT*-phospholipid:diacylglycerol acyltransferase; *CPT*-CDP-choline:diacylglycerol choline phosphotransferase; *DGAT*-diacylglycerol acyltransferase.

Genetic analysis of high oleic and low linolenic acid containing mutants revealed that one (Schierholt *et al.*, 2007) or two (Falentin *et al.*, 2007) major loci, depending on the mutants and corresponding to the FAD2 (fatty acid desaturase) genes, controlled oleic acid (18:1) content, and two loci corresponding to the FAD3 genes controlled linolenic acid (18:3) content (Barret *et al.*, 1999). Several studies identfiied molecular markers tightly associated with both traits (Jourdren *et al.*, 1996). Recently, the FAD2 and FAD3 genes were mapped to the QTL controlling 18:1 and 18:3 contents respectively. In addition, mutations corresponding to single nucleotide polymorphism (SNP) were identified in the sequences of both FAD2 and FAD3 genes and used to develop markers for direct selection of desirable alleles for breeding high oleic low linolenic varieties (Hu *et al.*, 2006).

3.1.3. High-Oleic and Low-Linoleic Acid Peanuts

In terms of the O/L (oleic/linoleic acid) ratios, Virginia and runner-type cultivars have been classified into four genotypes: OL1OL1OL2OL2 for the wild type, ol1ol1OL2OL2 for homozygous A genome, OL1OL1ol2ol2 for the homozygous B genome, and ol1ol1ol2ol2 for the double mutation in both A and B genomes (Burks *et al.*, 1995). High-oleic peanuts such as SunOleic 97R was developed which have 80 per cent oleic and 2 per cent linoleic fatty acid as compared to 50 per cent oleic and 25 per cent linoleic in oil of normal peanuts (Norden *et al.*, 1987). Two recessive genes (double mutant, ol1ol1ol2ol2) are responsible for the high-oleic trait (Moore and Knauft 1987; Knauft and Moore 1993). Incorporation of these genes into Florida breeding lines has resulted in two peanut varieties, SunOleic 95R (Knauft *et al.*, 1995) and SunOleic 97R (Gorbet and Knauft 2000). This oil has been shown to lower cholesterol levels in hypercholesterolemic women (O'Byrne *et al.*, 1993), decrease the level of monounsaturated fatty acids in swine (Myer *et al.*, 1992), enhance peanut shelf life, and reduce rancidity (O'Keefe *et al.*, 1993; Braddock *et al.*, 1995; Mugendi *et al.*, 1998).

Mutants of FAD2 (Both single and double) contained a higher amount of oleic acid and a lower amount of linoleic acid (Wang *et al.*, 2011). The results obtained by Wang *et al.* (2011) from genotyping the FAD2 functional alleles and the fatty acid profiles of the *FAD2* genotypes indicated that both genes *FAD2A* and *FAD2B* are involved in the conversion of oleic acid to linoleic acid in cultivated peanut. Each gene mutation can lead to a limited amount of oleic acid increase because of the compensation effect on the conversion of oleic acid to linoleic acid from the other homologous gene. Once homologous genes are mutated, the amount of oleic acid is greatly increased, and the amount of linoleic acid is dramatically decreased (Table 10.2). Therefore, the double mutant can significantly impact the O/L ratio (Wang *et al.*, 2011; Isleib *et al.*, 2006).

3.1.4. Fatty Acid Alteration in Soybean

Soybean (*Glycine max*) oil contains five fatty acids: palmitic acid (16:0), stearic acid (18:0), oleic acid (18:1), linoleic acid (18:2), and linolenic acid (18:3), in average of 10, 4, 18, 55, and 13 percent respectively. This fatty acid profile has low oxidative stability that limits the uses of soybean oil in food products and industrial applications.

Soybean breeders have made great strides by exploiting natural variation in oleic acid levels among soybean germplasm and to move an elevated oleic acid phenotype (30-70 per cent) into elite genotypes (Takagi and Rahman 1996; Rahman *et al.*, 2001; Alt *et al.*, 2005). Alteration of fatty acid composition of soybean (*Glycine max* cv. Williams-82) by chemical mutagenesis [N-nitroso-N-methylurea (NMU)] was reported by Hudson (2012) who developed 4566 M3 generation having populations with a wide range of variation in the major fatty acid contents. In this study he identified three low-palmitic acid mutants (two of these had normal levels of oleic acid), six high-palmitic acid mutants, four low-stearic acid mutants and four high stearic acid containing mutants. Six low-oleate mutants with high linoleate, and five of these had high linolenate and marginally low oleate. All 16 high-oleate mutants had less than average levels of linoleate. Two low-linolenate mutants were identified along with high linoleate or oleate. The mutants with high levels of linolenate showed reduced levels of oleate.

Conventional approaches to raise the oleic acid content in soybean has led to the development of "mid-oleic" phenotype, with 30 to 70 per cent oleic acid. Oleic acid is metabolized to linoleic acid by a single desaturation step carried out by a Δ^{12}-desaturase encoded by the FAD2 gene (Heppard *et al.*, 1996). There are at least six FAD2 genes in soybean falling in two classes, *FAD2-1* and *FAD2-2*. The *FAD2-1* class is primarily embryo-specific, while the *FAD2-2* class is generally constitutive and expressed during both vegetative and seed developmental stages (Tang *et al.*, 2005).

In soybean seed the content of palmitate is regulated by a *palmitoyl thioesterase* encoded by *FatB* genes. Low palmitic acid soybean genotypes have been reported (Bubeck *et al.*, 1989; Primomo *et al.*, 2002) which has been associated with allelic variation in *FatB* (Cardinal *et al.*, 2007), but seem to suffer from yield penalty (Cardinal and Burton, 2007).

The concentration of stearate the cardiovascular "neutral" saturated fatty acid, in soybean is controlled by both desaturase (Cheesbrough, 1990) and thioesterase activities (Pantalone *et al.*, 2002). The genetic locus controlling elevated stearic acid levels in soybean is designated *Fas*, and allelic variation at this locus manifests stearic acid levels ranging from 3 per cent in wild type to high stearate germplasm with 35 per cent (Bubeck *et al.*, 1989; Pantalone *et al.*, 2002).

3.1.5. Conventional Approach of Fatty Acid Modification and its Limitations

The conventional approach to develop the mid-oleic phenotype has a few drawbacks. Firstly, the genetics of the phenotypes require the stacking of multiple loci (Alt *et al.*, 2005), which may complicate the breeding process. Secondly, the "mid-oleic" phenotype are affected by environment, requiring warmer climates for stability of the elevated oleic acid trait to be maintained. This is due to the temperature effects the desaturase activity and expression (Heppard *et al.*, 1996; Tang *et al.*, 2005). The third drawback associated with the conventional breeding approach for a "mid-oleic" soybean is the germplasm that expresses this phenotype is associated with yield drag (Primomo *et al.*, 2002).

Significant amount of progress has been made in developing novel soybean germplasm, with altered fatty acid profiles, using conventional breeding strategies.

Table 10.2: Effect of *FAD2* Mutations on Fatty Acids in Peanuts.

Fatty Acid	Genotype	Mean	Tukey Grouping	Fatty Acid	Genotype	Mean	Tukey Grouping
C16:0	Ol1Ol1Ol2Ol2	10.738	A	C20:0	Ol1Ol1Ol2Ol2	1.618	A
	Ol1Ol1ol2ol2	10.279	A		ol1ol1Ol2Ol2	1.529	A
	ol1ol1Ol2Ol2	8.853	B		Ol1Ol1ol2ol2	1.357	A
	ol1ol1ol2ol2	6.103	C		ol1ol1ol2ol2	1.335	A
LSD		0.943		LSD		0.285	
C18:0	Ol1Ol1Ol2Ol2	3.428	A	C20:1	ol1ol1ol2ol2	1.828	A
	ol1ol1Ol2Ol2	3.146	A		Ol1Ol1ol2ol2	1.269	B
	Ol1Ol1ol2ol2	2.76	A		Ol1Ol1Ol2Ol2	1.194	B
	ol1ol1ol2ol2	2.8	A		Ol1Ol1Ol2Ol2	1.015	B
LSD		0.973		LSD		0.363	
C18:1	ol1ol1ol2ol2	80.139	A	C22:0	Ol1Ol1ol2ol2	3.461	A
	ol1ol1Ol2Ol2	57.111	B		Ol1Ol1Ol2Ol2	3.145	A
	Ol1Ol1ol2ol2	52.985	C		ol1ol1ol2ol2	2.973	A
	Ol1Ol1Ol2Ol2	46.075	D		ol1ol1Ol2Ol2	2.886	A
LSD		3.713		LSD		0.636	
C18:2	Ol1Ol1Ol2Ol2	32.054	A	C24:0	ol1ol1ol2ol2	1.613	A
	Ol1Ol1ol2ol2	26.713	B		Ol1Ol1Ol2Ol2	1.606	A
	ol1ol1Ol2Ol2	23.688	B		ol1ol1Ol2Ol2	1.593	A
	ol1ol1ol2ol2	3.211	C		Ol1Ol1ol2ol2	1.491	A
LSD		3.03		LSD		0.284	

Wang et al., 2011, LSD (Least significant difference).

Bates et al., 2013.

However, there tends to be a yield drag associated with these improved oil traits which may be due to the allelic variants selected for altered fatty acid profile during both vegetative and embryogenesis development, thereby increasing the probability of a negative agronomic effect associated with the mutant allele governing the novel oil phenotype. High oleic acid mutants, with 60 to 90 per cent oleic content, have also been developed in corn, peanut, canola and sunflower. But all these mutants have defective FAD2 genes (Perez-Vich *et al.*, 2002; Patel *et al.*, 2004; Hu *et al.*, 2006; Belo *et al.*, 2008).

3.2 Biotechnological Approaches

3.2.1. Low Erucic, High Oleic and Low Linolenic (HOLL) Acid Rapeseed-Mustard

Efforts to develop low erucic acid Indian mustard (*B. juncea*) by back cross breeding, based on the "00" rapeseed (*B. napus*) was not successful, as a result mustard with an acceptably low erucic acid is still not available. However, to produce low erucic acid Indian mustard, a *B. juncea* homologue of the *Arabidopsis thaliana* Fatty Acid Elongation1 (*FAE1*) gene was amplified by polymerase chain reaction (PCR), cloned and characterized. It was found to share 93.6 per cent DNA sequence homology with its *A. thaliana* counterpart (Venkateswari *et al.*, 1999).

Karnar *et al.* (2006) developed a transgenic Indian mustard by cloning of FAE1 homologue from *Brassica juncea* cv. Pusa Bold in a binary vector both in sense and antisense orientations under the control of the CaMV35S promoter. The recombinant binary vectors were used to transform *B. juncea* cv. RLM 198 via *Agrobacterium tumefaciens. The* seed oil from homozygous T4 generation seeds revealed that over-expression of the *FAE1* gene caused a 36 per cent increase while, down-regulation caused an 86 per cent decrease in erucic acid to as low as 5 per cent in the seed oil of transgenic plants.

Increases in oleic acid content can be achieved by reducing the activity of oleate desaturase (oleoyl- phosphatidylcholine Δ^{12}-desaturase), enzyme which converts oleate into linoleate during development of seed. Molecular mechanisms of inactivating antisense and co-suppression genes can be implemented in a tissue-specific manner enabling the inactivation of all copies of Δ^{12}-desaturase in the developing seed without affecting gene expression in other tissues. High-oleic acid containing Australian *B. napus* and *B. juncea* varieties were developed by transformation of plants carrying the Δ^{12}-desaturase co-suppression constructs. Silencing of the endogenous oleate desaturase genes has resulted in substantial increases in oleic acid levels, up to 89 per cent in *B. napus* and 73 per cent in *B. juncea* (Stoutjesdijk *et al.*, 2000).

The transgenic of *B. juncea*, developed by a heterologous gene transfer strategy with the ACP from *Azospirillum brasilense* (AB) (a plant associative aerobic/micro aerophilic bacterium) has an oleic acid (C18:1) rich lipid membrane. The transformed plants showed noticeable changes in seed fatty acid profile, and compared with the untransformed control plants, the transgenic *B. juncea* showed appreciable increase of the oleic acid (C18:1) and linoleic (C18:2) acid by about 23–42 and 18–42 per cent, respectively which was compensated by striking decrease of stearic acid (C18:0) and erucic acid (C22:1) by about 4–28 per cent and 17–37 per cent respectively (Jha *et al.*, 2007).

The reduction of erucic acid content was achieved in Indian mustard using a double stranded or hair-pin (hp) RNA mediated silencing of the endogenous fatty acid elongase (FAE) where diversion of the metabolic flux from the nutritionally undesirable erucic acid towards the production of more desirable fatty acids was obtained (Sinha *et al.*, 2007). The accumulation of erucic acid was reduced by 64–82 per cent in the seed oil and along with it, the loss of erucic acid was compensated by simultaneous increment of health benecial fatty acids stearic, oleic and linoleic acid. However, the double stranded RNA-mediated FAE silencing failed to produce any "zero erucic acid" containing lines suggesting the presence of complex pathway governing the VLCUFA synthesis in the ER of oilseed crops.

Intron containing hpRNA (ihpRNA) silencing construct has been shown to result in much higher degree of silencing compared to co-suppressing, antisense or hpRNA-mediated gene silencing. In a recent study Tian *et al.* (2011) used BnFAE1.1 (*B. napus* FAE1.1) gene as the target for ihpRNA-mediated post-transcriptional silencing under the control of a seed-specic napin promoter. In transgenic *B. napus* lines, expression of endogenous BnFAE1.1 gene was successfully down-regulated in immature seeds, and it resulted in significantly decreased level of both erucic acid and eicosenoic acid as well, in addition to a highly increased per cent of oleic acid in transgenic seed oil.

3.2.2. Changes in Cotton Seed Oil Composition

Cotton (*Gossypium hirsutum*), primarily grown for fibre production, is the world's third largest source of vegetable oil containing about 26 per cent palmitic acid (C16:0), 15 per cent oleic acid (C18:1), and 58 per cent linoleic acid (C18:2). Relatively high level of palmitic acid provides a degree of stability to the oil, that makes it suitable for high-temperature frying, but is nutritionally undesirable because of the low-density lipoprotein cholesterol-raising properties of this saturated fatty acid (Cox *et al.*, 1995). Also, there are evidences that cottonseed oil has been shown to lower total serum cholesterol compared with corn (*Zea mays*) oil (Radcliffe *et al.*, 2001). The cottonseed oil hydrogenated to achieve very high stability for deep frying food service applications or to provide the solidity required for margarine hard stock, but the hydrogenation process results in the production of trans-fatty acids, which have cholesterol raising properties equivalent to those of saturated fatty acids (Ascherio and Willett, 1997).

Thus, there is a growing trend away from the use of palmitic acid and hydrogenated oils in favor of those that are both nutritionally beneficial and can provide the required functionality without hydrogenation. Selective breeding utilizing natural variants or induced mutations has been used to develop a range of improved oils in the major temperate oilseed crops, including high-stearic (HS) soybean (Graef *et al.*, 1985), high-oleic (HO) rapeseed, HO peanut (Norden *et al.*, 1987), and HS (Osorio *et al.*, 1995) and HO (Soldatov, 1976) sunflower (*Helianthus annuus*). However, due to a lack of any genetic variation for fatty acid composition in cottonseed oil and the allotetraploid nature of cultivated cotton, classical breeding techniques and induced mutagenesis have so far been unsuccessful in developing improved cotton seed oil.

To overcome the limitations, genetic engineering techniques have been successfully employed to modify the fatty acid composition in cotton seed oil. The posttranslational gene silencing (PTGS) has been used to down-regulate the activity of the desaturase enzymes that control the synthesis of the major seed oil fatty acids, principally stearoyl-acyl-carrier protein (ACP) Δ^9-desaturase, which converts stearic acid into oleic acid, and oleoyl-phosphatidylcholine (PC) ω 6-desaturase, which converts oleic acid into linoleic acid. However, the antisense and co-suppression strategies used in these cases have variable and unpredictable effectiveness and generally require the production of large populations of transgenic plants to obtain an acceptable number of lines exhibiting sufficient degrees of target gene suppression (Kinney, 1996).

Fatty acid composition, using hairpin RNA-mediated gene silencing to down-regulate the seed expression of two key fatty acid desaturase genes, ghSAD-1encoding stearoyl-acyl-carrier protein 9-desaturase and ghFAD2-1-encoding oleoyl-phosphatidylcholine ω6-desaturase were studied where hairpin RNA-encoding gene constructs targeted against either ghSAD-1 or ghFAD2-1 into cotton (cv Coker 315). The resulting down-regulation of the ghSAD-1 gene, substantially increased stearic acid from the normal levels of 2-3 per cent to as high as 40 per cent and silencing of the ghFAD2-1 gene resulted in elevated oleic acid content, up to 77 per cent as against 15 per cent in seeds of untransformed plants. In addition, palmitic acid was significantly lowered in both high-stearic and high-oleic lines. Similar fatty acid composition phenotypes were also achieved by transformation with conventional antisense constructs targeted against the same genes, but at much lower frequencies than were achieved with the hairpin constructs. By inter crossing the high-stearic and high-oleic genotypes; it was possible to simultaneously down-regulate both ghSAD-1 and ghFAD2-1 to the same degree as observed in the individually silenced parental lines. Silencing of ghSAD-1 and/or ghFAD2-1 at various degrees one can develop cottonseed oils having novel combinations of palmitic, stearic, oleic, and linoleic contents that can be used in margarines and deep frying without hydrogenation and also potentially in high-value confectionery applications (Liu *et al.*, 2002)

3.2.3. Changes in Soybean Seed Oil Composition

Directed modification of fatty acid biosynthesis to alter relative amounts of fatty acids in soybean or to produce novel fatty acids was achieved using biotechnological tools (Damude and Kinney, 2008). Modulation of *FAD2* expression using transgenic approach was carried out by introducing transgenic elements designed to induce post-transcriptional gene silencing (Cerutti, 2003). High oleic acid soybean derived from down-regulating FAD2, concomitantly reduces polyunsaturated fatty acids to below 6 per cent (Kinney and Knowlton 1997), in addition palmitic acid was reduced to approximately 7–8 per cent (a 20 per cent reduction in palmitate). In many uses, such as salad oils, it is essential to reduce saturated fatty acids, especially palmitic acid as much as possible. The FATB class of acyl:*ACP thioesterases* control the release of some saturated fatty acids, including palmitic, into the cytoplasm, making them available for oil biosynthesis (Dormann *et al.*, 2000). Seed-specific silencing of FATB genes in soybean leads to a major reduction in total saturated fatty acids, from about 15 per cent to less than 6 per cent (Kinney, 1996).

An ultra-low linolenic acid in soybean oil was achieved by Flores *et al.* (2008); they utilized a targeted gene silencing approach to down regulate the GmFAD3 family in a seed-specific manner. To elevates oleic acid content of seed storage lipids in soybean the embryo-specific Δ^{-12} fatty acid desaturase FAD2-1 gene was down-regulated by ribozyme-termination of RNA transcripts both ribozyme-terminated antisense and standard antisense constructs were capable of embryo specific gene down- regulation, producing over 57 per cent oleic acid compared with less than 18 per cent in wild-type seed. Eight independent soybean transformants, that harboured standard plus sense or ribozyme terminated FAD2-1 cassette were screened. Two displayed oleic acids levels in the seed storage lipids of over 75 per cent, while none of the lines showed elevated oleic acid phenotypes. The dual constructs targeted FAD2-1 and the Fat B gene encoding a palmitoyl-thioesterase and five transgenic soybean lines harbouring the dual constructs had oleic acid levels, greater than 85 per cent, and saturated fatty acids levels, less than 6 per cent (Buhr *et al.*, 2002). Graef *et al.* (2009) developed transgenic soyabean by down regulation of FAD2-1A and 1B via seed-specific expression of posttranscriptional gene-silencing, which possesses oil with an increased oleic acid (> 85 per cent) and reduced palmitic acid (< 5 per cent) contents.

3.2.4. Production of Novel Fatty Acids and its Importance in Human Health

A vast diversity of fatty acids is found within the Plantae, although the predominant fatty acids present in most plant species are palmitic, stearic, oleic, linoleic, and linolenic acids, at various ratios depending on species and genotypic variation. Many of the relatively rare fatty acids found in plants have commercial applications but due to their low content in native sources they are costly. The ability to combine or "stack" multiple genetic cassettes into a plant genome allows for the assembly of biochemical pathways. Considerable efforts have been made to improve the composition of vegetable oils, by improving our understanding of the seed oil biosynthetic pathway, however terrestrial plants do not have the ability to synthesize ω-3 VLC-PUFA such as EPA and DHA; and hence any approach must involve the introduction of new desaturase and elongase activities. The introduction of the VLC-PUFA biosynthetic pathways into oilseed crops has been successfully demonstrated, but reaching economically viable levels of EPA and DHA is still challenging. In some plant species, a $\Delta 6$ desaturase has been shown to have activity on both linoleic and linolenic acids, leading to the production of γ-linolenic acid (GLA; 18:3Δ6,9,12) and Stearidonic acid (SDA 18:4Δ6,9,12,15), respectively (Sayanova *et al.*, 2003). The GLA, an ω-6 fatty acid, have both pharmacological and nutraceutical properties (Horrobin 1992). However; SDA, an ω-3 fatty acid, is an intermediate in the biosynthesis of the very-long-chain polyunsaturated ω-3 fatty acids EPA and DHA and therefore has nutritional value in food and feed applications (Clemente and Cahoon 2009).

The ALA (α-linolenic acid) from plant sources is converted by the body to EPA (Eicosapentaenoic acid), consuming a high proportion of ω-3 oils, such as flax oil, may be a useful dietary strategy. The first step in the human EPA pathway, the conversion of ALA to stearidonic acid (SDA; a reaction catalysed by a Δ6 desaturase), is rate-limiting in humans (Burdge *et al.*, 2002; James *et al.*, 2003). The ALA in the diet is converted to EPA, with only a fraction of the efficiency of SDA in healthy subjects. Further, Δ6 activity has been shown to decline with age (Bourre *et al.*, 1990) and with a significant number of clinical conditions, such as cancer, making utilization of dietary ALA even less efficient (Lane *et al.*, 2003).

Fortunately, mammals have the necessary enzymes to make VLC-PUFAs from the parent essential Fatty acids, but *in vivo* studies show that only 5 per cent of ALA is converted to EPA and less than 0.5 per cent of ALA is converted to DHA in humans (Williams and Burdge 2006). These low conversion efficiencies and lack of sufficient dietary ω-3 VLC-PUFAs could compromise health (Abeywardena and Patten, 2011), and hence; EPA and DHA considered as 'conditionally essential' Fatty acids. Only the dietary consumption of ω-3 VLC-PUFAs, for example via oil fish, is known to mitigate conditions such as cardiovascular disease, obesity, and metabolic syndrome (Williams and Burdge 2006; Poudyal *et al.*, 2011). However, majority of (vegetarian population) have intakes of EPA and DHA far below the recommended intake amounts (Meyer 2011). Responding to the limited supply and increasing demand for VLC-PUFAs, in particular ω-3 FAs, much research effort has focused on engineering of oilseed plants to synthesize EPA and DHA.

The SDA (Stearidonic acid) as a nutritional fatty acid is largely dependent on the ability of humans and animals to convert SDA to EPA (Eicosapentaenoic acid) and DHA (Docosahexaenoic acid) and diets rich in these very-long-chain ω-3 fatty acids have been widely linked to cardiovascular fitness (Marik and Varon, 2009). The DHA is also important for neuron synthesis and infant brain development (Uauy *et al.*, 2003). The conversion of SDA to EPA requires elongation of the fatty acid chain to C20 and Δ desaturation, while DHA synthesis further elongated to C22 and the introduction of an additional Δ4 double bond.

To increase metabolic flux toward SDA synthesis, Ursin (2003) described the stacking of three transgenic cassettes, harboring the *Mortierella alpina* Δ6 and Δ12 and canola Δ15 desaturase genes, in canola. This transgenic stack resulted in accumulation of 18 per cent and 23 per cent, GLA and SDA respectively. Partial conversion of dietary SDA to EPA, has been shown in human and Atlantic salmon (James *et al.*, 2003; Miller *et al.*, 2007), but, higher levels of EPA accumulation were observed with diets containing oils enriched in EPA rather than SDA (James *et al.*, 2003). Similarly, DHA levels in Atlantic salmon were enhanced with a fish oil-based diet that is rich in EPA and DHA (Miller *et al.*, 2007). These studies clearly indicate that the inclusion of oils containing EPA and DHA in the diet yields greater nutritional efficacy than use of SDA-rich oils. Such findings encourage for the development of soybean seeds that accumulate oils with EPA and DHA, and also can be produced at a lower cost in soybean seeds compared with fish and algae, the current commercial sources of EPA- and DHA-containing oils. In addition, soybean offers a more sustainable production platform than fish. The future use of soybean oil enriched in

EPA and DHA is the preparation of food and feed products, including salad oils, infant formulas, and feed rations for farm fish (Clemente and Cahoon, 2009).

A number of strategies can be used to produce EPA and DHA in soybean seeds. A biosynthetic route described for SDA involves the introduction of a "condensing" enzyme or fatty acid "elongase" to initiate the elongation of SDA to eicosatetraenoic acid (20:4Δ8,11,14,17) and a Δ5 desaturase to generate EPA. The production of DHA through this pathway requires another condensing enzyme to form the C22 docosapentaenoic acid (22:5Δ7,10,13,16,19) and a Δ4 desaturase. Vascular plants do not produce EPA and DHA and therefore lack genes for the synthesis of these fatty acids from SDA. Instead, potential sources of EPA and DHA biosynthetic genes include fungi, marine microalgae, and protists that synthesize very-long-chain polyunsaturated fatty acids (Napier 2007; Damude and Kinney, 2008).

A transgenic approach to make higher levels of ALA in soybean by expressing a bi-functional Δ12/Δ15 desaturase from *Fusarium moniliforme* resulted in ALA concentrations upto 72 per cent in seed (Damude *et al.*, 2006). Transgenic soybean and *B. juncea* have been developed with ~15–20 per cent EPA in seed oils by co-expression of desaturases and elongase. These transgenic plants with relatively high levels of EPA can be further engineered for the synthesis of ω-3 docosahexaenoic acid (DHA; 22:6Δ4,7,10,13,16,19) by adding another elongase and desaturase. However, transgenic DHA levels have been relatively low ranging from 0.5 per cent (*Arabidopsis*) to 1.5 per cent (*B. juncea*) to 3.3 per cent (Kinney 2004; Wu *et al.*, 2005).

The effects of different host species, genes and promoters on EPA biosynthesis was investigated by Cheng *et al.* (2010), where zero-erucic acid *Brassica carinata* was an outstanding host for EPA production, with EPA levels in transgenic seed of this line reaching up to 25 per cent. Two novel genes, an 18-carbon ω3 desaturase (CpDesX) from *Claviceps purpurea* and a 20-carbon ω3 desaturase (Pir-ω3) from *Pythium irregulare*, proved to be very effective in increasing EPA levels in high-erucic acid *B. carinata*. The conlinin1 promoter from flax functioned reasonably well in *B. carinata*, and can serve as an alternative to the napin promoter from *B. napus*.

Liu *et al.* (2001) expressed a fungal Δ6-desaturase from *Mortierella alpina* along with a Δ12-desaturase from the same fungus was expressed in *B. napus* seeds, producing up to 43 per cent ω-6-γ-linolenic acid (GLA; 18:3Δ6,9,12). Similar levels of GLA were produced in *B. juncea* seeds expressing a Δ6-desaturase from the fungus *Pythium irregulare* (Hong *et al.*, 2002) and of *Borago ofcinalis* (Qiu *et al.*, 2003). The Monsanto and Solae LLC companies have developed transgenic soybean producing 15–30 per cent ω-3 stearidonic acid (SDA; 18:4 Δ6,9,12,15) and 5– 8 per cent GLA which can be used in foods and beverages (Hammond *et al.*, 2008).

3.2.5. Modification of Oil for Food Industries

The oil and fats are also used for various food and non-food related industrial applications. The saturated fats are preferred over the unsaturated fats and are widely used as emulsifying agents in cosmetics as well as confectionary industries (Perez-Vich *et al.*, 2004; Zarhloul *et al.*, 2006). The trans-fatty acids are the isomers of unsaturated fatty acids, produced during industrial hydrogenation of vegetable liquid oils, (mostly oleic and linoleic acids) to increase the oxidative stability during storage or frying and to make it solid at room temperature for uses in confectionary industry. However, these *trans*-isomers of unsaturated fatty acids are associated with coronary heart disease (Broun *et al.*1999). For many food applications, vegetable oils with a reduced amount of *trans*-unsaturated fatty acids are desirable to improve human health. This has been achieved using strategies such as cosuppression, antisense, and RNA interference to down-regulate endogenous stearoyl-ACP desaturase genes in soybean, cotton and Brassica oil seeds. In these plants, levels of stearic were increased upto 40 per cent to provide a semi solid margarine feed stock without the need for hydrogenation. An oxidatively stable liquid oil low in saturated fatty acids was also produced in soybean by suppression of the oleoyl desaturase (Knutzon *et al.*, 1992; Kinney 1996; Thelen and Ohlrogge 2002). The oil with oleic acid content increased upto 86 per cent, linoleic acid content reduced from 55 per cent to less than 1 per cent, and saturated fatty acids reduced to 10 per cent has been produced commercially and is extremely stable for high-temperature frying applications (Minihane and Harland, 2007).

Vegetable fats rich in saturates, particularly of stearate a neutral fatty acid towards human health, decreases the need of industrial hydrogenation (hardening, aiming to increase the saturated fatty acids), thereby reducing the cost of hydrogenation as well as lessen the production of trans isomers of fatty acids (Perez-Vich *et al.*, 2004) during the saturate production through hydrogenation process. Soybeans have several Δ9 desaturase (SAD) genes expressed in their seeds and silencing genes (SAD3) in a seed-specific manner resulted in oils with 20–30 per cent stearic acid (Booth *et al.*, 2006). The seeds of mangosteen (*Garcinia mangostana*) contain 45–50 per cent stearate as the predominant fatty acid, a result of a stearate- specific thioesterase and when the mangosteen gene was expressed in transgenic canola seeds, stearate contents of 20–30 per cent were reported (Hawkins and Kridl 1998). Similar increases in stearate have been observed when this gene was expressed in soybean seeds (Kridl, 2002).

By combining the silencing of SAD3 and FAD2 in soybean it has been possible to make oil containing 50– 60 per cent oleic acid and 15–20 per cent stearic acid (Booth *et al.*, 2002). This combination of stearic and oleic acid confers the desired solid fat functionality. Improved functionality has also been obtained by combining increased stearic acid with low linolenic acid oils (Dirienzo *et al.*, 2008). Healthy oils from these types of seeds have the potential to replace solid fats in a wide range of baking and heavy-duty frying applications.

3.2.6. Metabolic Engineering for Increased Total Oil Content

The high demand for soybean oil has sparked an interest in increasing the relative content of oil per seed which accounts for 18-20 per cent of the seed weight and is low relative to most other oilseed crops (*i.e.*, peanut ~45 per cent). The prospects of retooling the metabolism of soybean seeds to yield high content of oil are daunting. Instead, the current focus of both breeding and transgenic efforts has been on increasing the oil content of soybean seeds incrementally without affecting protein content. This strategy addresses the need for more oil yield from soybean without compromising the dual-use nature of this crop.

A number of transgenic approaches have been explored for increasing seed oil content. Though most of these efforts have used *Arabidopsis* as a model for oilseeds, few attempts have been made with oilseed crops. Seed oils are composed almost exclusively of triacylglycerols, and also contain small amounts of other lipidic compounds such as phospholipids and tocopherols. Triacylglycerols consist of three fatty acid chains bound to a glycerol backbone. The metabolic engineering efforts for seed oil enhancement have focused on either increasing the synthesis of fatty acids or increasing their incorporation onto the glycerol backbone. In the former case, this has involved attempts to enhance the partitioning of carbon toward fatty acid synthesis and to increase the pool sizes of substrates for fatty acid synthesis. A variety of approaches have been used, including the overexpression of transcription factors, such as WRI1 (Cernac and Benning 2004; Baud *et al.*, 2007), and key metabolic enzymes, such as acetyl-CoA carboxylase (Roesler *et al.*, 1997). Experiments to increase oil content by enhancing fatty acid sequestration onto the glycerol backbone have centered on acyltransferases, including diacylglycerol acyltransferase (DGAT). This enzyme catalyzes the last step of triacylglycerol biosynthesis, the incorporation of a fatty acyl-CoA onto diacylglycerol. Two structurally divergent classes of membrane associated DGATs (DGAT1 and DGAT2) occur in plants and other organisms (Cases *et al.*, 1998, 2001).

The only transgenic success reported to date for enhancement of soybean oil content was achieved by the introduction of a seed-specific transgene for a DGAT2-type enzyme from the oil-accumulating fungus *Umbelopsis ramanniana* (Lardizabal *et al.*, 2008). In these studies, the oil content was increased from approximately 20 per cent of the seed weight to approximately 21.5 per cent. This increase was stable over three growing seasons in two different geographical locations. Importantly, the increased oil content had little or no impact on the protein levels of seeds and the seed yield per acre.

4. Protein Quality Improvement in Oilseed Crops

Seed proteins in general are deficient in some essential amino acids and hence are of poor nutritional quality. Intensive research is going on to isolate and characterize these proteins and their genes, and to produce transgenic crop plants with modified seed protein genes, with a view to improving their nutritive value as human food and animal feed. Many seed storage protein genes from cereals, legumes and oil seeds have been isolated, sequenced and their regulation has been studied by promoter deletion assay in transgenic plants. The amino acid composition of seed proteins of some transgenic crops has been marginally improved by modifying these genes for more methionine and lysine codons by site-directed mutagenesis or by introducing heterologous genes. Apart from such modifications of proteins and amino acids for quality improvement; enzyme alteration, thermal chemical and pressure treatment and genetic engineering is being extensively in use.

4.1. Rape Seed Protein

The cruciferins (12S globulins), napins (2S albumins) and oleosins (oil body proteins) are the major proteins in rape seeds (Purkrtova *et al.*, 2008). Both cruciferins and napins accounted for up to 70 per cent of seed proteins and display specific features. Napins contain higher levels of sulfur and aromatic amino acids than cruciferins, which make them the most important targets for the improvement of seed protein composition. Several promising attempts for the genetic engineering of 2S albumins have been conducted in rapeseed through introduction of a Brazil nut (*Bertholletia excelsa*) 2S gene (Altenbach *et al.*, 1992) or expression of a cruciferin antisense gene. In all cases, transgenic plants accumulate more napins in their seeds, which led to an increase in cysteine, methionine and lysine contents, the two latter amino acids being essential. In addition, the increase in napins was counter-balanced by a decrease in cruciferin content, suggesting that the 12S/2S balance was tightly controlled (Nesi *et al.*, 2008).

4.2. Soybean Protein

Soybean is an important protein source in human diets and additionally, human consumption of soybean protein is reported to provide specific health benefits. Since low methionine and cysteine contents limit the nutritional value of soybean, efforts involving both traditional breeding and genetic engineering have been employed in attempts to increase the essential amino acids. Traditional breeding has been primarily utilized to increase the total protein content but not to enhance the sulphur amino acid content of soybean. Mutagenesis in conjunction with traditional breeding is a viable approach for enhancing the sulphur amino acid content of soybean. Introduction of methionine-rich heterologous proteins has resulted in a modest increase of this amino acid in soybean. Either elevating the expression of endogenous methionine-rich proteins or introducing synthetic proteins containing a high percentage of essential amino acids are other possible approaches that may increase the nutritional quality of the seed (Krishnan 2005).

Substantial increases in protein levels have been achieved by traditional breeding in soybean (Wilcox and Shibles, 2001) however, lack of variability in methionine content among soybean cultivars (Krober, 1956) has limited the use of conventional methods to increase the sulfur amino acid content. In general, the amount of sulfur amino acids has remained constant regardless of the seed protein (Wilcox and Shibles 2001). Madison and Thompson (1988) identified several methionine-over-producing soybean cell culture lines following treatment with ethionine, a chemical analog of methionine. These lines accumulate >8-fold higher methionine than the parental line, but mature soybean plants could not be generated from these cultures. Mutagenesis of soybean seeds with ethyl methanesulfonate (EMS) and subsequent screening for ethionine resistant plants exhibiting dark-green leaves resulted in the identification of several methionine-over-producing lines and concentration of methionine and cysteine in one mutant was approximately 20 per cent higher than that of the parent line (Imsande 2001). These studies demonstrated the feasibility of increasing the sulfur amino acid content of soybean either by chemical mutagenesis or by ethionine treatment. However, it need to be looked in whether the reported 20 per cent increase in the sulfur amino acid content is adequate to meet the nutritional requirement of livestock and poultry (Krishnan, 2005).

Over the years, several approaches have been employed with varying degrees of success to enhance the level of certain essential amino acids in soy protein to meet the needs of advanced feed formulations. These efforts range from increasing the sulfur content of fertilizers (Holowach *et al.*, 1986) to biotechnological technique transforming soybeans with genes that encode high methionine proteins from the Brazil nut (Saalbach *et al.*, 1994). Another interesting concept is genetic manipulation of the levels of glycinin (11S) and ß-conglycinin (7S), which account for nearly 70 per cent of soybean protein. This idea is based partially on observations that 11S proteins contain 3–4 times more cysteine and methionine than 7S proteins. Therefore, soybean meal with a higher 11S:7S (w/w) ratio should exhibit enhanced nutritional quality. Although information on genetic variation in the 11S:7S ratio among cultivated soybean varieties is limited, the range appears to be narrow between 1.1-2.2 (Ogawa *et al.*, 1989).

4.3. Cotton Seed Protein

In addition to 21 per cent oil, cottonseed is a source of relatively high-quality protein (23 per cent). However, the ability to use this nutrient-rich resource for food is hampered by the presence of toxic gossypol that is unique to the tribe Gossypieae. This cardio and hepatotoxic terpenoid, present in the glands, renders cottonseed unsafe for human and monogastric animal consumption (Risco, 1997). Unfortunately, this toxicity subjugates this abundant agricultural resource to the ranks of a feed for ruminant animals either as whole seeds or as meal after oil extraction. Gossypol and related terpenoids are present throughout the cotton plant in the glands of foliage, floral organs and bolls as well as in the roots, which is induced in response to microbial infections. In addition, these terpenoids are protect the plant from both insects and pathogens (Stipanovic, 1999).

Gossypol and other sesquiterpenoids are derived from (+)-δ-cadinene. The enzyme δ-cadinene synthase catalyzes the first committed step involving the cyclization of farnesyl diphosphate to (+)-δ-cadinene. Thus, tissue-specific RNAi of δ-cadinene synthase expression to disrupt terpenoid biosynthesis offers a possible mechanism to eliminate gossypol from the seed while retaining a full complement of this and related terpenoids in the rest of the plant for maintaining its defensive capabilities against insects and diseases. The RNAi technology coupled with tissue specific promoter was used to disrupt gossypol biosynthesis in cottonseed tissue by interfering with the expression of the δ-cadinene synthase gene during seed development. Cottonseed-gossypol level was significantly reduce in a stable and heritable manner.Furthermore, the levels of gossypol and related terpenoids in the foliage and oral parts were not diminished, and thus their potential function in plant defense against insects and diseases remained untouched. Thus, gossypol-free cottonseed would significantly contribute to human nutrition and health, particularly in developing countries and would help meet the requirements of the predicted 50 per cent increase in the world population in the next 50 years (Sunilkumar *et al.*, 2006).

4.4. Peanut Protein

Peanut proteins are consumed in various form of foods (*i.e.* raw dry nuts, fresh boiled and salted, fried and mixed with sugar syrup, fried and coated with chickpea flour, nuts fermented and fried, roasted and salted, peanut butter, candies and confectionaries). Although peanuts are more popular for their oil content, processing of peanuts for oil extraction yields protein rich co-product which is also used for human consumption. Several peanut proteins have been implicated as allergens (Kang *et al.*, 2007) and of these three proteins considered important peanut allergens belong to legume seed-storage protein families. Ara h 1, a 63.5-kDa glycoprotein, shows homology to vicilin proteins in other legume seeds, Ara h2 belongs to a conglutin storage protein family, and Ara h 3 is a legumin-type storage protein that has high sequence similarity to glycinin, a storage protein of soybean seeds. In peanut plants, these allergenic proteins are detectable only in the embryonic axes and cotyledons of seeds. Of these three allergenic proteins, the Ara h2 (17.5-kDa) glycoprotein is the most potent allergen, with nearly 50-fold greater potency than Ara h1. Overall, allergen proteins comprise 5 per cent of the total cellular protein of peanut seeds (Singh and Bhalla, 2008).

The RNA interference (RNAi) technology for silencing of Ara h 2 in peanut was utilized by Dodo *et al.* (2007). The gene-silenced transgenic peanut plants were produced by *Agrobacterium* -mediated transformation with an intron-spliced Ara h 2 RNAi construct under the control of the CaMV 35S promoter. Crude peanut extract from the transgenic plants showed up to 25 per cent reduction in Ara h 2 content. Immunoassay using individual seed extracts confirmed, however, that about one-quarter of the seeds produced by the transgenic plants were either free of Ara h2 or contained a signicantly reduced amount of the allergen. However, T_0 plants, produce only half of their seeds with reduced allergen content, are not suitable as a source of hypoallergenic pea nuts. Homozygosity of the transgene occurring in T_1 and subsequent generations will ensure that almost all of the seeds produced will inherit the RNAi construct and thus will be allergen free.

Osuji *et al.* (2012) developed allergen free and lower linoleic acid peanut using an approach based on molecular permutation modeling of metabolism at the mRNA level. In this approach it was predicted that induction of GDH isomerization and RNA synthesis by mineral ion treatments of peanut will promote arachin and fatty acid desaturase loss-of-functions leading to production of several metabolic variants that are substantially free of arachins and fatty acid desaturases, because GDH is a target site of mineral ion action in plants. They observed that treatment of peanut with phosphate plus potassium ions induced GDH to synthesize isomeric RNAs that coordinately silenced the mRNAs encoding arachin h 1, and fatty acid desaturase thereby producing arachin-free low linoleic acid (~18 per cent) peanut. This indicated that through ionic treatment (P, K) arachin-free low linoleic acid peanut can be produced.

5. Conclusions

Edible oil seed crops are not only a major source of fatty acids for human and animal nutrition but also considered as a source of amino acids and one of industrial feedstocks. Recent advances in understanding of the basic biochemistry of seed oil and protein biosynthesis, coupled with cloning of the genes encoding the enzymes involved in fatty acid and protein modification, have set the stage for the metabolic engineering of oil seed crops that produce "designer" oil seeds with the improved nutritional values for human being.

Most oilseeds contain only common fatty acids, of 16 or 18 carbons in length with zero to three double bonds at specific positions with varying concentrations. Thus none of the oilseeds is ideal for nutritional and industrial point of view. Modification in various oil seeds crops, achieved through various breeding and biotechnological approaches as per requirement such as low erucic acid containing rapeseed-mustard, high oleic acid containing rapeseed- mustard, soybean, cotton seed and peanut have been developed for table purpose. For food industries stearic acid content was increased in soybean, cottonseed and Brassica oil seeds. Although developed for food uses, the availability of a relatively pure source of oleic acid at commodity prices has stimulated interest in the industrial uses of these oils. Linolenic acid in flax (*Linum usitatissimum* L.) and soybean and erucic acid in rapeseed was increased to making these oils more suitable for industrial feedstocks. Attempts were also made to increase essential amino acids in rapeseed and soybean.

Although genes encoding suitable fatty acid-modifying enzymes are available from many wild species and other sources, still we are far away from alteration of fatty acid in predictable manner. A number of attempts 'failed' to achieve higher accumulation of non-native fatty acids (hydroxy, EPA and DHA); such efforts have equally indicated that plant lipid metabolism is more complicated than previously imagined. However some success stories of fatty acid alteration in the market which encourages utilising genetic engineering techniques with suitable promoters for developing designer oil seed crops with a wide range of applications.

Reduction of gossypol content in cotton, ricin in castor bean and allergen protein from peanut clearly demonstrated that targeted gene silencing can be used to modulate biosynthetic pathways in a specific tissue. Thus, an approach based on the removal or reduction of naturally occurring toxic compounds from the edible portion of the plant not only improves food safety but also provides an additional and potential means to meet the nutritional requirements of the growing world population.

To develop economically viable oilseed crops with modified fatty acid profiles and quality protein there is a requirement to manipulate the gene expression of relevant key constituent steps in the synthetic or modification pathways, *i.e.* to carry out genetic metabolic engineering. This can be achieved either through up or down-regulation of an introduced recombinant gene (transgenic), deletion of endogenous genes (mutagenesis), or by selection of appropriate combinations of the relevant naturally occurring alleles present in the gene pool.

6. Future Prospects and Strategies

There are no fundamental barriers toward making substantial modifications in edible oilseed biosynthesis. In many cases, modification of one or two enzyme and transfer of more than one gene may be essential for efficient production of the desired fatty acids. But for higher production of unusual/non-native fatty acids in plants, multiple factors and genes are responsible and there is no single gene that will allow this in a transgenic oilseed. The prospects for future gene-stacking experiments are strengthened by enhancement of EPA in *B. Juncea* and recinoleic acid content in *Arabidopsis*, which demonstrated clearly that the expression of just one additional enzyme can successfully boost the accumulation of these fatty acids in transgenic plants.

The introduction of novel genes into genetically engineered crops raises issues that require additional safety and regulatory scrutiny. Therefore, some alternative strategies to the transfer of fatty acid biosynthetic genes can be used to improve the characteristics of existing oil crops that produce the target fatty acids. New molecular biology techniques such as TILLING (Targeting Induced Local Lesions in Genomes) or RNA interference can be applied to silence specific genes. More efforts should be concentrated on development of trait specific molecular markers can greatly speed up breeding programs.

The recent techniques of genomics, proteomics and metabolomics will allow researchers to understand carbon flow during embryogenesis which in turn, will allow for the designing of experiments that target manipulation of the metabolic control points. These studies will unravel complications of the relative carbon partitioning between seed storage components and the intricacies of fatty acid metabolism in oil seeds. These approaches may lead to crops with designer fatty acid profile coupled with better quality protein for nutrition.

The emerging science of synthetic biology will provides a new paradigm for plant lipid engineering, allowing re-designing of metabolic pathway or create entirely new biosynthetic pathways *de novo* within cells, thereby enabling production of unusual fatty acids, have the potential for use in industrial applications and nutritional requirements of human and livestock's.

7. Summary

Most of the major oilseed crops have been modified through genetic engineering and majority of the genetically modified (GM) oilseed crops produced to date contain herbicide tolerance or insect resistance input traits. However, genetic manipulation in fatty acids of oilseed crops has the greatest potential. The nutritional and physical qualities of oils are determined by the fatty acid composition of the oil, and in recent years there has been substantial interest in modifying fatty acid composition of vegetable oils with improved health benefits, better performance in cooking applications, and oils that may be used in a variety of different industries. Globally, consumers prefer vegetable oils because of lower content of saturated fatty acid. The oleochemical industries require higher concentration of polyunsaturated fatty acids while hydroxylated oils are required for use in lubricants, paints and polymers. Likewise, lower PUFAs are also desired to reduce the level of *trans*-isomer in hydrogenated food ingredients. Oilseeds are also a good source of protein for human and animals but suffer from some essential amino acids and/or some anti-nutritional substances. Thus, there are many reasons which warrant the fatty acid alteration of oils and improvement of protein quality. Efforts are now underway in different countries to produce specialized 'designer' oilseed crops. In this review, we will discuss fatty acid alteration and protein quality improvement in soybean, rapeseed, cottonseed and groundnut (peanut) through conventional as well as genetic engineering approaches.

8. References

Alford BB, Liepa GU, Vanbeber AD (1996). Cottonseed protein: what does the future hold? *Plant Foods for Human Nutrition* 49: 1–11.

Abeywardena MY, Patten GS (2011). Role of n-3 long chain polyunsaturated fatty acids in reducing cardio-metabolic risk factors. Endocrine, Metabolic and Immune Disorders. *Drug Targets* 11: 232–246.

Alt JL, Fehr WR, Welke GA, Sandu D (2005). Phenotypic and molecular analysis of oleate content in the mutant soybean line M23. *Crop Science* 45: 1997–2000.

Altenbach SB, Kuo CC, Staraci LC, Pearson KW, Wainwright C, Georgescu A, Townsend J (1992). Accumulation of a Brazil nut albumin in seeds of transgenic canola results in enhanced levels of seed protein methionine. *Plant Molecular Biology* 18: 235–245.

Ascherio A, Willett WC (1997). Health effects on *trans* fatty acids. *American Journal of Clinical Nutrition* 66: 1006S–1010S.

Auld DL, Heikkinen MK, Erickson DA, Sernyk JL, Romero JE (1992). Rapeseed mutants with reduced levels of polyunsaturated fatty acids and increased levels of oleic acid, *Crop Science* 32: 657–662.

Barret P, Delourme R, Brunel D, Jourdren C, Horvais R, Renard M (1999). Low linolenic acid level in rapeseed can be easily accessed through the detection of two single base substitution in FADF3 genes. In: *Proceeding of the 10th International Rapeseed Congress*, Canberra, Australia, pp. 26–29.

Bates PD, Browse J (2012). The signicance of different diacylgycerol synthesis pathways on plant oil composition and bioengineering. *Frontiers in Plant Science* 3(147): 1-11.

Bates PD, Stymne S, Ohlrogge J (2013). Biochemical pathways in seed oil synthesis. *Current Opinion in Plant Biology* 16: 1 – 7.

Baud S, Mendoza MS, To A, Harscoet E, Lepiniec L, Dubreucq B (2007). WRINKLED1 specifies the regulatory action of LEAFY COTYLEDON2 towards fatty acid metabolism during seed maturation in Arabidopsis. *Plant Journal* 50: 825–838.

Belo A, Zheng P, Luck S, Shen B,MeyerDJ, Li B, Tingey S, Rafalski A (2008). Whole genome scan detects an allelic variant of fad2 associated with increased oleic acid levels in maize. *Molecular Genetics and Genomics* 279: 1–10.

Bhattacharya S, Dey P, Sinha S, Das N, Maiti MK (2012). Production of nutritionally desirable fatty acids in seed oil of Indian mustard (*Brassica juncea* L.). by metabolic engineering. *Phytochemistry Reviews* 11: 197-209.

Booth JR, Broglie RM, Hitz WD, Kinney AJ, Knowlton S, Sebastian SA (2002). Gene combinations that alter the quality and functionality of soybean oil. United States Patent US 6426448.

Booth JR, Cahoon RE, Hitz WD, Kinney AJ, Yadav NS (2006). Nucleotide sequences of a new class of diverged D9 stearoyl-ACP desaturase genes. European Patent Publication EP1311659.

Bourre JM, Piciotti M, Dumont O (1990). Delta 6 desaturase in brain and liver during development and aging. *Lipids* 25: 354–356.

Braddock JC Sims CA O'Keefe SK (1995). Flavour and oxidative stability of roasted high oleic acid peanuts. *J. Food Sci.*, 60: 489-493.

Broun P, Gettner S, Somerville C (1999). Genetic engineering of plant lipids. *Annual Review of Nutrition*, 19: 197–216.

Bubeck DM, Fehr WR, Hammond EG (1989). Inheritance of palmitic and stearic acid mutants of soybean. *Crop Science* 29: 652–656.

Buhr T, Sato S, Ebrahim F, Xing A, Zhou Y, Mathiesen M, Schweiger B, Kinney AJ, Staswick P, Clemente T (2002). Ribozyme termination of RNA transcripts down-regulate seed fatty acid genes in transgenic soybean. *Plant Journal* 30: 155–163.

Burdge GC, Jones AE, Wootton SA (2002). Eicosapentaenoic and docosapentaenoic acids are the principal products of alpha-linolenic acid metabolism in young men. *British Journal of Nutrition* 88: 355–363.

Burgal J, Shockey J, Lu C, Dyer J, Larson T, Graham I, Browse JN (2008). Metabolic engineering of hydroxy fatty acid production in plants: RcDGAT2 drives dramatic increases in ricinoleate levels in seed oil. *Plant Biotechnology Journal* 6: 819-831.

Burks AW, Cockrell G, Stanley JS, Helm RM, Bannon GA (1995). Recombinant peanut allergen Ara h1 expression and IgE binding in patients with peanut hypersensitivity. *Journal of Clinical Investigation* 96: 1715–1721.

Cahoon EB, Shockley JM, Dietrich CR, Gidda SK, Mullen RT, Dyer JM (2007). Engineering oilseeds for sustainable production of industrial and nutritional feedstocks: solving bottlenecks in fatty acid flux. *Current Opinion in Biotechnology* 10: 236–244.

Cardinal AJ, Burton JW (2007). Correlations between palmitate content and agronomic traits in soybean populations segregating for the fap1, fapnc, and fan alleles. *Crop Science* 47: 1804–1812.

Cases S, Smith SJ, Zheng YW, Myers HM, Lear SR, Sande E, Novak S, Collins C, Welch CB, Lusis AJ (1998). Identification of a gene encoding an acyl CoA: diacylglycerol acyltransferase, a key enzyme in triacylglycerol synthesis. *Proceedings of the National Academy of Sciences, USA* 95: 13018–13023.

Cases S, Stone SJ, Zhou P, Yen E, Tow B, Lardizabal KD, Voelker T, Farese RV Jr (2001). Cloning of DGAT2, a second mammalian diacylglycerol acyltransferase, and related family members. *Journal of Biological Chemistry* 276: 38870–38876.

Cernac A, Benning C (2004). WRINKLED1 encodes an AP2/EREB domain protein involved in the control of storage compound biosynthesis in Arabidopsis. *Plant Journal* 40: 575–585.

Cerutti H (2003). RNA interference: traveling in the cell and gaining functions? *Trends in Genetics* 19: 39–46.

Cheesbrough TM (1990). Decreased growth temperature increases soybean stearoyl-acyl carrier protein desaursae activity. *Plant Physiology* 93: 555–559.

Cheng BF, Wu GH, Vrinten P, Falk K, Bauer J, Qiu X (2010). Towards the production of high levels of eicosapentaenoic acid in transgenic plants: the effects of different host species, genes and promoters. *Transgenic Research* 19: 221–229.

Chopra VL, Vageeshbabu HS (1996). Metabolic engineering of plant lipids. *Journal of Plant Biochemistry and Biotechnology* 5: 63–68.

Clemente Tom E, Cahoon Edgar B (2009). Soybean Oil: Genetic Approaches for Modification of Functionality and Total Content. *Plant Physiology* 151: 1030–1040.

Cox C, Mann J, Sutherland W, Chisholm A, Skeaff M (1995). Effects of coconut oil, and safflower oil on lipids and lipoproteins in persons with moderately elevated cholesterol levels. *Journal of Lipid Research* 36: 1787–1795.

Craig W, Lenzi P, Scotti N, De Palma M, Saggese P, Carbone V, McGrath-Curran N, Magee AM, Medgyesy P, Kavanagh TA, Dix PJ, Grillo S, Cardi T (2008). Transplasto mic tobacco plants expressing a fatty acid desatura se gene exhibit alte red fatty acid pr oles and improve d cold tolerance. *Transgenic Research* 17: 769–782.

Damude HG, Kinney AJ (2008). Engineering oilseeds to produce nutritional fatty acids. *Physiologia Planterum* 132(1): 1–10.

Damude HG, Zhang H, Farrall L, Ripp, KG Tomb J-F, Hollerbach D, Yadav NS (2006). Identification of bifunctional D 12/o 3 fatty acid desaturases for improving the ratio of o3 to o 6 fatty acids in microbes and plants. *Proceedings of the National Academy of Sciences, USA* 103: 9446–9451.

Dirienzo MA, Lemke SL, Petersen BJ, Smith KM (2008). Effect of substitution of high stearic low linolenic acid soybean oil for hydrogenated soybean oil on fatty acid intake. *Lipids* 43: 451–456.

Dodo HW (2007). Allevia ting peanut allergy using genetic engineering: the silencing of the immun odominant allergen Ara h 2 leads to its signicant reduction and a decrease in peanut allergenicity. *Plant Biotechnology Journal* 6: 135–145.

Dormann P, Voelker TA, Ohlrogge JB (2000). Accumulation of palmitate in Arabidopsis mediated by the acyl-acyl carrier protein thioesterase FATB1. *Plant Physiology* 123: 637–644.

Duh PD, Yen GC (1997). Antioxidant efficacy of methanolic extracts of peanut hulls in soybean and peanut oils. *Journal of American Oil Chemists' Society* 74: 745-748.

Eckert H, LaVallee B, Schweiger BJ, Kinney AJ, Cahoon EB, Clemente T (2006). Co-expression of the borage D6 desaturase and the Arabidopsis D15 desaturase results in high accumulation of stearidonic acid in the seeds of transgenic soybean. *Planta* 224: 1050–1057.

Falentin C, Brégeon M, Lucas M O, Deschamps M, Leprince F, Fournier M T, Delourme R, Renard M (2007). Identication of fad2 mutations and development of allele-specic markers for high oleic acid content in rapeseed (*Brassica napus* L.), in: *Proceeding of the 12th International Rapeseed Congress*, Wuhan, China, pp. 117–119.

FAO food outlook (2012). available Online http: //www.fao.org/fileadmin/templates/est/COMM_MARKETS_MONITORING/Oilcrops/Documents/Food_outlook_oilseeds/Food_outlook_Nov_12.pdf.

Flores T, Karpova O, Su X, Zeng P, Bilyeu K, Sleper DA, Nguyen HT, Zhang ZJ (2008). Silencing of GmFAD3 gene by siRNA leads to low α-linolenic acids (18: 3). of *fad3*-mutant phenotype in soybean [*Glycine max* (Merr.)]. *Transgenic Research* 17: 839–850.

Gorbet DW, Knauft DA (2000). Registration of 'SunOleic 97R' peanut. *Crop Science* 40: 1190–1191.

Graef GL, Miller LA, Fehr WR, Hammond EG (1985). Fatty acid development in a soybean mutant with high stearic acid. *Journal of American Oil Chemists' Society* 62: 773–775.

Graef G, LaVallee BJ, Tenopir P, Tat ME, Schweiger BJ, Kinney AJ, Van Gerpen J, Clemente TE (2009). A high oleic acid and low palmitic acid soybean: agronomic performance and evaluation as a feedstock for biodiesel. Plant Biotechnol J 7: 411–421.

Hammond BG, Lemen JK, Ahmed G, Miller KD, Kirkpatrick J, Fleeman T (2008). Safety assessment of SDA soybean oil: results of a 28-day gavage study and a 90-day/one generation reproduction feeding study in rats. *Regulatory Toxicology and Pharmacology* 52: 311–323.

Hawkins DJ, Kridl JC (1998). Characterization of acyl-ACP thioesterases of mangosteen (Garcinia mangosta na). seed and high levels of stearate production in transgenic canola. *Plant Journal* 13: 743–752.

Heppard EP, Kinney AJ, Stecca KL, Miao GH (1996). Developmental and growth temperature regulation of two different microsomal o-6 desaturase genes in soybeans. *Plant Physiology* 110: 311–319.

Holowach LP, Madison JT, Thompson JF (1986). Studies on the Mechanism of Regulation of the mRNA Level for Soybean Storage Protein Subunit by Exogenous L-methionine. *Plant Physiology* 80(2): 561–567.

Hong H, Datla N, Reed DW, Covello PS, MacKenzie SL, Qiu X (2002). High-level production of α-linolenic acid in Brassica juncea using a Δ6 desaturase from Pythium irregulare. *Plant Physiology* 129: 354–362.

Horrobin DF (1992). Nutritional and medical importance of gammalinolenic acid. *Progress in Lipid Research* 31: 163–194.

Hudson K (2012). Soybean Oil-Quality Variants Identified by Large-Scale Mutagenesis. International Journal of Agronomy. Article ID 569817, p. 7

Hu X, Sullivan-Gilbert M, Gupta M, Thompson SA (2006). SA Mapping of the loci controlling oleic and linolenic acid contents and development of fad2 and fad3 allele-specic markers in canola (*Brassica napus L.*), *Theoretical and Applied Genetics* 113: 497–507.

Imsande J (2001). Selection of soybean mutants with increased concentrations of seed methionine and cysteine. *Crop Science* 41: 510–515.

Isleib TG, Wilson RF, Novitzky WP (2006). Partial dominance, pleiotropism, and epistasis in the inheritance of the high-oleic trait in peanut. *Crop Science* 46: 1331–1335.

Jadhav SJ, Nimbalkar SS, Kulkarni AD, Madhavi DL (1996). Lipid oxidation in biological and food systems. In: *Food Antioxidants.* (Eds.): D.L. Madhavi, S.S. Deshpande and D.K. Salunkhe. Marcel Dekker, New York, pp. 5-63.

James MJ, Ursin VM, Cleland LG (2003). Metabolism of stearidonic acid in human subjects: comparison with the metabolism of other n-3 fatty acids. *American Journal of Clinical Nutrition* 77: 1140–1145.

Jha JK, Sinha S, Maiti MK, Basu A, Mukhopadhyay UK, Sen SK (2007). Functional expression of an acyl carrier protein (ACP). from Azospirillum brasilense alters fatty acid proles in *Escherichia coli* and *Brassica juncea*. *Plant Physiology and Biochemistry* 45: 490–500.

Jourdren C, Barret P, Brunel D, Delourme R, Renard M (1996). Specic molecular marker of the genes controlling linolenic acid content in rapeseed, *Theoretical and Applied Genetics* 93: 512–518.

Kanrar S, Venkateswari J, Dureja P, Kirti PB, Chopra VL (2006). Modication of erucic acid content in Indian mustard (*Brassica juncea*). by up-regulation and down-regulation of the Brassica juncea FATTY ACID ELONGATION1 (BjFAE1). gene. *Plant Cell Reproduction* 25: 148–155.

Kinney AJ (1996). Development of genetically engineered soybean oils for food applications. *Journal of Food Lipids* 3: 273–292.

Kinney AJ (2004). Production of very long chain polyunsaturated fatty acids in oilseeds. Patent WO 2004/071467 A2.

Kinney AJ, Knowlton S (1997). Designer oils: the high oleic soybean. In S Harander, S Roller, (eds.), *Genetic Engineering for Food Industry: a Strategy for Food Quality Improvement*. Blackie Academic, London, pp. 193–213.

Knauft DA, Gorbet DW, Norden AJ (1995). SunOleic 95R peanut. *Florida Agriculture Experiment Station Circular* No. S-398.

Knauft DA, Moore KM (1993). Gorbet, D. W. Further studies on the inheritance of fatty acid composition in peanut. *Peanut Science* 20: 74-76.

Korver O, Katan MB (2006). The elimination of trans fats from spreads: how science helped to turn an industry around. *Nutrition Reviews* 64(6): 275–279.

Kridl JC (2002). Methods for increasing stearate content in soybean oil. United States Patent US 6365802.

Krishnan HB (2005). Engineering Soybean for Enhanced Sulfur Amino Acid Content. *Crop Science* 45: 454–461.

Krober OA (1956). Methionine content of soybeans as influenced by location and season. *Journal of Agriculture and Food Chemistry* 4: 254–257.

Lane J, Mansel RE, Jiang WG (2003). Expression of human delta-6- desaturase is associated with aggressiveness of human breast cancer. *International Journal of Molecular Medicine* 12: 253–257.

Lardizabal K, Effertz R, Levering C, Mai J, Pedroso MC, Jury T, Aasen E, Gruys K, Bennett K (2008). Expression of Umbelopsis ramanniana DGAT2A in seed increases oil in soybean. *Plant Physiology* 148: 89–96.

Leonard C (1994). Sources and commercial applications of high erucic acid vegetable oils. *Lipid Technology* 4: 79–83.

Li-Beisson Y, Shorrosh B, Beisson F, Andersson M, Aronde l V, Bates P, Baud S, Bird D, DeBono A, Durrett T (2013). Acyl lipid metabolism. *The Arabidopsis Book*. e0161.

Li R, Yu K, Hatanaka T, Hildebrand DF, Vernonia D (2010). GATs increase accumulation of epoxy fatty acids in oil. *Plant Biotechnology Journal* 8: 184-195.

Liu JW, Huang Y-S, DeMichele S, Bergana M, Bobik E, Hastilow C, Chuang LT, Mukerji P, Knutzon D (2001). Evaluation of the seed oils from a canola plant genetically transformed to produce high levels of α-linolenic acid. In: Huang YS, Ziboh VA, eds. *α-Linolenic acid: recent advances in biotechnology and clinical applications*. Champaign, IL: AOCS Press, pp. 61–71.

Liu Q, Singh SP, Green AG (2002). High-stearic and high-oleic cottonseed oils produced by hairpin rna-mediated post-transcriptional gene silencing. *Plant Physiology* 129: 1732–1743.

Lu C, Napier JA, Clemente TE, Cahoon EB (2010). New frontiers in oilseed biotechnology: meeting the global demand for vegetable oils for food, feed, biofuel, and industri al applications. *Current Opinion in Biotechnology* 22: 1–8.

Madavi DL, DK Salunkhe (1995). Toxicological Aspects of Food Antioxidants. In: *Food Antioxidants*. (Eds.): D.L. Madav, S.S. Deshpande and D.K. Salunkhe. Marcel Dekker, New York, p. 267.

Madison JT, Thompson JF (1988). Characterization of soybean culture cell lines resistant to methionine analogs. *Plant Cell Reproduction* 7: 473–476.

Marik PE, Varon J (2009). Omega-3 dietary supplements and the risk of cardiovascular events: a systematic review. *Clinical Cardiology* 32: 365–372.

Mensink RP, Katan MB (1990). Effect of Dietary trans Fatty Acids on High-Density and Low-Density Lipoprotein Cholesterol Levels in Healthy Subjects. *The New England Journal of Medicine* 323: 439–445.

Meyer BJ (2011). Are we consuming enough long chain ω-3 polyunsaturated fatty acids for optimal health? *Prostaglandins, Leukotrienes and Essential Fatty Acids* 85: 275–280.

Miller MR, Nichols PD, Carter CG (2007). Replacement of dietary fish oil for Atlantic salmon parr (*Salmo salar* L.). with a stearidonic acid containing oil has no effect on omega-3 long-chain polyunsaturated fatty acid concentrations. *Comparative Biochemistry and Physiology-Part B: Biochemistry and Molecular Biology* 146: 197–206.

Minihane AM, Harland JI (2007). Impact of oil used by the frying industry on population fat intake. *Criticial Reviews in Food Science and Nutrition* 47: 287-297.

Moore KM, Knauft DA (1989). The inheritance of high oleic acid in peanut. *Journal of Heredity* 80: 252-253.

Mugendi JB Sims CA Gorbet DW O'Keefe SF (1998). Flavour stability of high-oleic peanuts stored at low humidity. *Journal of American Oil Chemists' Society* 75: 21-25.

Myer RO, Johnson DD, Knauft DA, Gorbet DW, Brendemuhl JH, Walker WR (1992). Effect of feeding high-oleic acid peanut to growing finishing swine on resulting fatty acid profile and on carcass meat quality characteristics. *Journal of Animal Science* 70: 3734-3741.

Napier JA (2007). The production of unusual fatty acids in transgenic plants. *Annual Review of Plant Biology* 58: 295–319.

Nesi N, Delourme D, Brégeon M, Falentin C, Renard M (2008). Genetic and molecular approaches to improve nutritional value of Brassica napus L. seed. *Comptes Rendus Biologies* 331: 763–771.

Norden AJ, Gorbet DW, Knauft DA, Young CT (1987). Variability in oil quality among peanut genotypes in the Florida breeding program. *Peanut Science* 14(1): 7-11.

O'Byrne DJ, Shireman RB, Knauft DA (1993). The effects of a low fat/high oleic acid diet on lipoprotein in postmenopausal hypercholesterolemic women. *Proceedings of the 84th AOCS Annual Meeting; American Oil Chemists Society*: Champaign, IL.

O'Keefe SF, Wiley VA, Knauft DA (1993). Comparison of oxidation stability of high- and normal-oleic peanut oils. *Journal of American Oil Chemists Society* 70: 489-492.

Ogawa T, Tayama E, Kitamura K, Kaizuma N (1989). Genetic Improvement of Seed Storage Proteins Using Three Variant Alleles of 7S Globulin Subunits in Soybean (*Glycine max* L.), *Japanese journal of breeding* 39: 137–147.

Ohlrogge JB, Kuhn DN, Stumpf PK (1979). Subcellular localization of acyl carrier protein in leaf protoplasts of *Spinacia oleracea. Proceedings of the National Academy of Sciences, USA* 76: 1194–1198.

Ohlrogge J, Jaworski JG (1997). Regulation of fatty acid synthesis. *Annual Review in Plant Physiology and Plant Molecular Biology* 48: 109–136.

Osorio J, Fernandez-Martinez J, Mancha M, Garces R (1995). Mutant sunflower with high concentration of saturated fatty acids in their oil. *Crop Science* 35: 739–742.

Pantalone VR, Wilson RF, Novitzky WP, Burton JW (2002). Genetic regulation of elevated stearic acid concentration in soybean oil. *Journal of American Oil Chemists Society* 79: 549–553.

Patel M, Jung S, Moore K, Powell G, Ainsworth C, Abbott A (2004). High-oleate peanut mutants result from a MITE insertion into the *FAD2* gene. *Theoretical and Applied Genetics* 108: 1492–1502.

Perez-Vich B, Fernandez-Martýnez JM, Grondona M, Knapp SJ, Berry ST (2002). Stearoyl- ACP and oleoyl-PC desaturase genes cosegregate with quantitative trait loci underlying high stearic and high oleic acid mutant phenotypes in sunflower. *Theoretical and Applied Genetics* 104: 338–349.

Perez-Vich B, Knapp SJ, Leon AJ, Fernandez-Martinez JM, Berry ST (2004). Mapping minor QTL for increased stearic acid content in sunower seed oil. *Molecular Breeding* 13: 313–322.

Pleins S, Friedt W (1989). Genetic control of linoleic acid concentration in seed oil of rape seed (*Brassica napus* L.). *Theoretical and Applied Genetics* 78: 793-797.

Poudyal H, Panchal SK, Diwan V, Brown L (2011). Omega-3 fatty acids and metabolic syndrome: effects and emerging mechanisms of action. *Progress in Lipid Research* 50: 372–387.

Primomo VS, Falk DE, Ablett GR, Tanner JW, Rajcan I (2002). Genotype X environment interactions, stability, and agronomic performance of soybean with altered fatty acid profiles. *Crop Science* 42: 37–44.

Purkrtova Z, Jolivet P, Miquel M, Chardot T (2008). Structure and function of seed oleosins and caleosin, *Comptes Rendus Biologies* 331: 746–754.

Qiu X (2003). Biosynthesis of docosahexaenoic acid (DHA, 22: 6–4, 7,10,13,16,19): two distinct pathways. *Prostaglandins, Leukotrienes and Essential Fatty Acids* 68: 181–86.

Radcliffe JD, King CC, Czajka-Narins DM, Imrhan V (2001). Serum and liver lipids in rats fed diets containing corn oil, cottonseed oil, or a mixture of corn and cottonseed oils. *Plant Foods for Human Nutrition* 56: 51–60.

Rahman SM, Kinoshita T, Anai T, Takagi Y (2001). Combining ability in loci for high oleic and low linolenic acids in soybean. *Crop Science* 41: 26–29.

Risco CA, Chase CC (1997). *Jr in Handbook of Plant and Fungal Toxicants*, Ed D'Mello JPF (CRC Press, Boca Raton, FL), pp. 87–98.

Robbelen G, Thies W (1980). Biosynthesis of seed oil and breeding for improved seed oil quality of rapeseed. In S. Tsunoda *et al.* (ed.). *Brassica crops and wild allies.* Japan Science Society Press, Tokyo, pp. 253-283.

Roesler K, Shintani D, Savage L, Boddupalli S, Ohlrogge J (1997). Targeting of the Arabidopsis homomeric acetyl-coenzyme A carboxylase to plastids of rapeseeds. *Plant Physiology* 113: 75–81.

Ruiz-Lopez N, Sayanova O, Napier J A and. Haslam R P (2012). Metabolic engineering of the omega-3 long chain polyunsaturated fatty acid biosynthetic pathway into transgenic plants. *Journal of Experimental Botany*, 1-14 doi: 10.1093/jxb/err454.

Saalbach I, Pickardt T, Machemehl F, Saalbach G, Schieder O, Muntz K (1994). A Chimeric Gene Encoding the Methionine-Rich 2S Albumin of the Brazil Nut (Bertholletia excelsa H.B.K.). Is Stably Expressed and Inherited in Transgenic Grain Legumes, *Molecular and General Genetics* 242: 226–236.

Sang S, Lapsley K, Jeong WS, Lachance PA, Ho CT, Rosen RT (2002). Antioxidant phenolic compounds isolated from almond skins (Prunus amygdalus Batsch). *Journal of Agricultural and Food Chemistry* 50: 2459-2463.

Sayanova OV, Beaudoin F, Michaelson LV, Shewry PR, Napier JA (2003). Identification of Primula fatty acid D6-desaturases with n-3 substrate preferences. *Federation of European Biochemical Societies Letters* 542: 100–104.

Scarth R, McVetty PBE, Rimmer SR, Stefansson BR (1988). Stellar low linolenic-high linoleic acid summer rape. *Canadian Journal of Plant Science* 68: 509-511.

Schierholt A, Rücker B, Becker HC (2007). Inheritance of high oleic acid mutations in winter oilseed rape (Brassica napus L.), *Crop Science* 41: 1444–1449.

Shahidi F, Amarowicz R (1994). Antioxidant activity of green tea catechins in a β-carotene-linoleate Modeal System. *Journal of Food Lipids* 2: 47-56.

Simopoulos AP (2000). Symposium: role of poultry products in enriching the human diet with n-3 PUFA. *Poultry Science* 79: 961–970.

Singh MB, Bhalla PL (2008). Genetic engineering for removing food allergens from plants. *Trends in Plant Science* 13(6): 257-260.

Sinha S, Jha JK, Maiti MK, Basu A, Mukhopadhyay UK, Sen SK (2007). Metabolic engineering of fatty acid biosynthesis in Indian mustard (Brassica juncea). improves nutritional quality of seed oil. *Plant Biotechnology Report* 1: 185–197.

Soldatov KI (1976). Chemical mutagenesis in sunflower breeding. In: Proceedings of the Seventh International Sunflower Association, Vlaardingen, The Netherlands. *International Sunflower Association*, Toowoomba, Australia, pp: 352–357.

Stefansson BR, Hougen FW, Downey RK (1961). Note on the isolation of rape plants with seed oil free from erucic acid, *Canadian Journal of Plant Science* 41: 218–219.

Stipanovic RD, Bell AA, Benedict CR (1999). in *Biologically Active Natural Products: Ag rochemicals*, eds Cutler HG, Cutler SJ (CRC Press, Boca Raton, FL), pp. 211–220.

Stoutjesdijk PA, Hurlestone C, Singh SP, Green AG (2000). High-oleic acid Australian *Brassica napus* and *B. juncea* varieties produced by co-suppression of endogenous Δ12-desaturases. *Biochemical Society Transactions* 28(6): 938-939.

Takagi Y, Rahman SM (1996). Inheritance of high oleic acid content in the seed oil of soybean mutant M23. *Theoretical and Applied Genetics* 92: 179–182.

Tang GQ, Novitzky WP, Griffin HC, Huber SC, Dewey RE (2005). Oleate desaturase enzymes of soybean: evidence of regulation through differential stability and phosphorylation. *Plant Journal* 44: 433–446.

Thelen JJ, Ohlrogge JB (2002). Metabolic Engineering of Fatty Acid Biosynthesis in Plants. *Metabolic Engineering* 4: 12–21.

Tian B, Wei F, Shu H, Zhang Q, Zang X, Lian Y (2011). Decreasing erucic acid level by RNAi-mediated silencing of fatty acid elongase 1 (BnFAE1.1). in rapeseeds (*Brassica napus* L.). *African Journal of Biotechnology* 61(10): 13194–13201.

Uauy R, Hoffman DR, Mena P, Llanos A, Birch EE (2003). Term infant studies of DHA and ARA supplementation on neurodevelopment: results of randomized controlled trials. *Journal of Paediatrics* 143: S17–S25.

Ursin VM (2003). Modification of plant lipids for human health: development of functional land-based omega-3 fatty acids. *Journal of Nutrition* 133: 4271–4274.

Vandeloo FJ, Broun P, Turner S, Somerville C (1995). An oleate 12hydroxylase from *Ricinus communis* L. is a fatty acyl desaturase homolog. *Proceedings of the National Academy of Sciences, USA* 92: 6743-6747.

Venkateswari J, Kanrar S, Kirti PB, Malathi VG, Chopra VL (1999). Molecular cloning and characterization of *fatty acid elongation 1* (*FAE1*)gene of *Brassica juncea*. *Journal of Plant Biochemistry and Biotechnology* 8: 53–55.

Walsh TA, Metz JG (2013). Producing the omega-3 fatty acids DHA and EPA in oilseed crops. *Lipid Technology* 25(5): 103–105.

Wang ML, Barkley NA, Chen Z, Pittman RN (2011). FAD2 gene mutations significantly alter fatty acid profiles in cultivated peanuts (Arachis hypogaea). *Biochemical Genetics* 49: 748–759.

Wilcox JR, Shibles RM (2001). Interrelationships among seed quality attributes in soybean. *Crop Science* 41: 11–14.

Williams CM, Burdge G (2006). Long-chain n-3 PUFA: plant v. marine sources. *Proceedings of Nutritional Society* 65: 42–50.

Wu G, Truksa M, Datla N, Vrinten P, Bauer J (2005). Stepwise engineering to produce high yields of very long-chain polyunsaturated fatty acids in plants. *Nature Biotechnology* 23: 1013–1017.

Yu J, Ahmedna M, Goktepe I (2005). Effects of processing methods and extraction solvents on concentration and antioxidant activity of peanut skin phenolics. *Food Chemistry* 90: 199-206.

Chapter II

Sugarcane Yield Plateaus and Potentials under Tropical and Subtropical India

R. Snehi Dwivedi* and Radha Jain

Indian Institute of Sugarcane Research, Lucknow – 226 002, India

Contents

* Retired; Residence 150-J, Southcity, Raebareli Road, Luckbow – 226 025, U.P., India.

1. Introduction

Sugarcane is quintessence of a multistar crop, which has tangible potential of providing raw materials to run many industries like white sugar, paper, alcohol potable and biofuel (*i.e.* blending of ethanol with alcohol), plastic, Co-generation (energy), chemicals etc. production simultaneously and fulfill human requirements, in large, at global level. India is totally dependent on sugarcane to meet its sweeteners requirement. In an agricultural oriented economy state like India availability of land is most precious asset and it is shrinking on per capita basis because of its fast burgeoning population. It was estimated that India would need to produce 348.5 million tonnes of sugarcane with a recovery of above 10 per cent in order to meet per capita requirement of 39 kg sweeteners per annum including g 22.2 kg white sugar and 16.8 kg Gur and Khandsari by AD 2010 (Table 11.1) (Anonymous, 1997). However, India could produce 18.96 Million tonnes of white sugar in 2009-10 which was not sufficient yet, was nearly 11.8 per cent of total sugar production of the world (Table 11.1). The population of country might reach 1.5 billion by 2030 and India might require to produce 520 million tonnes of sugarcane with 10.75 per cent sucrose recovery to meet sucrose and bifuel etc. requirement (Table 11.1) (Anonymous, 2011). Though, a record sugar production of 16.45 million tonnes in 1995-96 was attained in India, the fall in sugar production by 22 per cent in 1996-97 had also been witnessed inspite of about 10 per cent rise in sugarcane cultivated area (Anonymous, 1998). This is because, the productivity of sugarcane has not been rising substantially and attained plateau long back (Figure 11.1).

The cane productivity and sugar recovery patterns in tropical area of India has been found to be significantly superior to subtropical zone since long (Figure 11.2). Considering these facts in mind, whatever, rise in total sugarcane and sucrose production during the current years has been noted (Table 11.1) is mainly due to increase in sugarcane cultivated area to a level of 5.02 million ha which was only 4.0 million ha in year 2000.

Table 11.1: Projections of Sweetner Requirement vis-à-vis Sugarcane Production in India upto AD 2030.

Year	Sweetner Requirement (Mt)			Sugarcane Requirement (Mt)	Biofuel-Ethanol	Sugar Production (Mt)
	Sugar	Gur	Total			
1990-91	12.40	9.00	21.40	241.0	–	12.05**
2000	18.00	13.70	31.70	300.00	–	13.90**
2010	22.17	16.81	33.98	348.50	–	18.96**
2020	27.29	20.69	47.98	415.00	–	27.39
2030	33.0	19.0	52.0	442.0	78.0	33.54

** Production recorded; Anonymous, 1997, 2011.

The maximum yield possible with any genotype is unknown and liable to change because the environmental factors alter the genetic potential. This corroborates the findings that sugarcane has higher solar energy harvesting efficiency of 1.75-2.05 per cent PAR basis, in tropics as compared to 1.02-1.65 per cent -PAR basis, in subtropical area of India as against 3.5 per cent under ideal conditions (Dwivedi, 1994; Dwivedi *et al.*, 1994). Similar differences are observed in said zones for cane and sucrose yield too (Singh, 1987). This in turn enables tropical part of country to contribute more than 60 per cent of total sugar production in spite of the fact that less than 40 per cent of total sugarcane cultivated area of India is located in tropics (Anonymous, 1998).

Area

Yield (t/ha)

Production

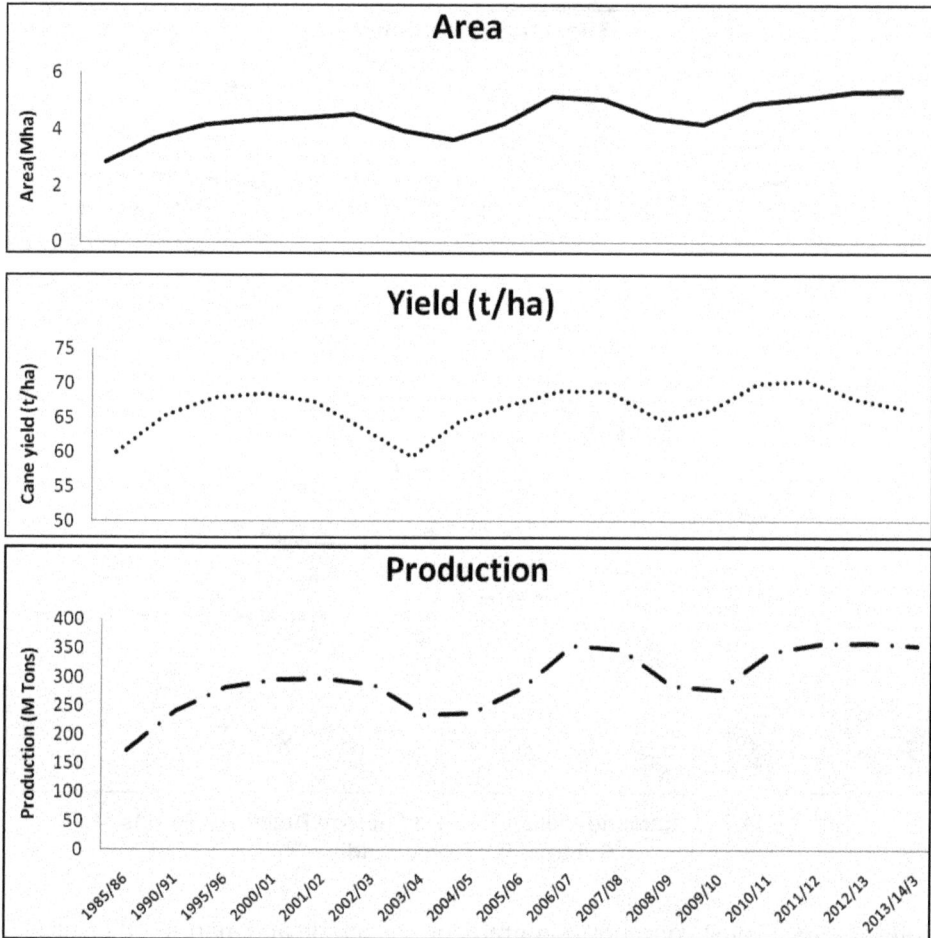

Figure 11.1: Sugarcane Area, Production and Cane Yield Trend of India.

Keeping these aspects in mind, the attempts have been made to present the available pieces of information on genetic potential, cane yield limiting factors, sucrose recovery constraints and specifically a holistic manners in attaining yield maxima. Association of metabolic, physiological, genetic parameters besides molecular basis of yield and sucrose recoveries have also been illustrated.

2. Scope of Raising Cane and Sucrose Yield and Solar Energy Harvest

2.1. Dry Matter, Fresh Cane and Sucrose Productivity

Trapping of solar energy by highly silicate green leaves and translocation and accumulation of economically viable photosynthates in the vertical system of cane column along geotropic forces make sugarcane most endergonic C_4 crop. The exergonic processes including that of uptake, accumulation and transport of inorganic/organic

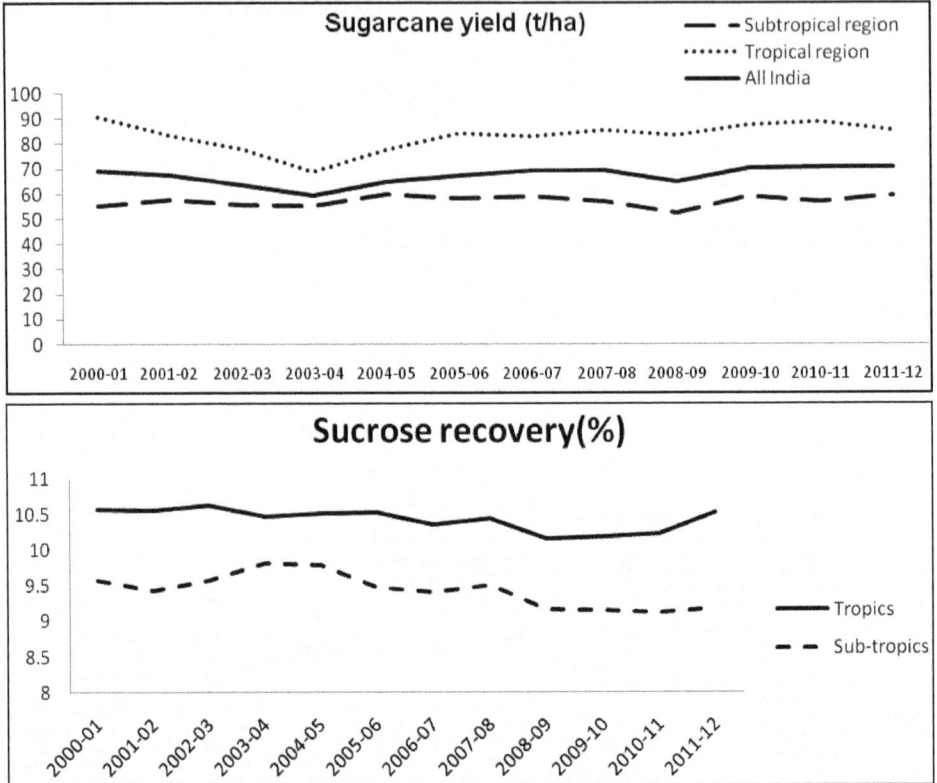

Figure 11.2: Sugarcane Yield (t/ha) and Sucrose Recovery Trends of
Subtropical, Tropical India.

components against concentration gradient are simple and akin to other plants,
hence its efficiency of dry matter production excels most of the crops. Consequently,
the average yield of sugarcane is found to be more than 52 per cent of potential yield
as against that of 20 per cent in cereals and 45 per cent in sugarbeet (Table 11.2).

Table 11.2: Potential Yield, World Record Yield and Percent Yield Achieved in
Various Crops under Normal Conditions (After Dwivedi *et al.*, 1994).

Crops	Potential Yield (t/ha)	World Record Yield (t/ha)	Normal Yield at Optimal Conditions (t/ha)	Percent of Records Yield, Achieved
Maize	52	22.2	4.4-7.0	20-30
Wheat	–	14.5	2.9-4.4	–do–
Rice	20	14.5	2.8-4.5	–do–
Soybean	–	7.1	1.5-2.2	–do–
Sugarbeet	–	121.0	40-60	33-50
Sugarcane	300-805	255-460	125-250	49-54

The synthesis and accumulation of sugar is much more simpler phenomenon than that of proteins, oils/fats and waxes. Hence, under normal conditions, higher productivity and production efficiency are expected in sugar/starch accumulating plants like sugarcane, potato and sugarbeet than others.

Various investigators have worked out the theoretical dry matter production potential of sugarcane (Table 11.3). The theoretical maxima for maize a C_4 plant like sugarcane was calculated by Loomis and Williams (1963). Similar calculations were made by Moore (1987) in sugarcane where respiratory losses were assumed less than 20 per cent as against 33 per cent assigned by Loomis and Williams. In addition, the maximum quantum efficiency was assumed to be 0.125 by Moore as opposed to the theoretical value of 0.100 by Loomis and Williams. Substituting these values in the methodology of Loomis and Williams, a theoretical maximum yield of sugarcane of 129 gm^{-2} d^{-1} equivalent to 1.29 t/ha/day or 470 t/ha/year was obtained. Thus the potential of cane yield (fresh weight) and sucrose yield have been worked as 805 t/ha and 125.5t/ha, respectively by Dwivedi *et al.* (1994) (Table 11.3).

2.2. Energy Harvest, Biofuel (Alcohol) Production, Environmental CO_2 Depletion, O_2 Evolution and Obviation of Greenhouse Effect

Sugarcane being a highly heterozygous complex polyploidy and most endergonic energy crop, has great scope for improvement in solar energy harvest, biofuel (Alcohol) production, environmental CO_2 depletion, O_2 evolution and thereby obviating environmental warming (Dwivedi, 1994). The theoretical yield potential, highest recorded yield and commonly achievable yield are 805.8, 464.03 and 100 t/ha respectively (Table 11.3). These gaps in yield are too much; however, similar variations in energy harvest, CO_2 depletion and O_2 evolution have been recorded in different genotypes of sugarcane which abridge a hope of improvement in sugarcane for aforesaid attributes (Dwivedi, 2000).

2.2.1. Leaf Area Index (LAI) and Solar Energy Harvest

Although LAI is higher at tillering phase, the solar energy, harvest efficiency remains higher at grand growth phase (Figure 11.3) (Dwivedi *et al.*, 1992). An optimum LAI is required for higher solar energy harvest and dry matter production. In spring planted crop, the intensity of solar radiation is higher at early stages of crop growth. This in turn reduces vegetative growth and results in higher percentage of dry matter accumulation in the cane as compared to autumn planted cane (Dwivedi *et al.*, 1994 and Dwivedi, 2000). Studies on the interception of solar radiation by sugarcane at Coimbatore revealed the existence of varietal difference which ranges from less than 50 per cent to 84 per cent of daily radiation.

2.2.2. Environmental CO_2 Fixation and O_2 Evolution

No significant relationship is observed between leaf diffusion resistance, relative water content and photosynthesis. In response of water stress, the activity of phosphoenol pyruvate (PEP) carbozylase decreased, while that of nicotinamide adenine ocleotide phosphate (NADP+) malic enzyme increased. Malate content also increased along with it. In water stress affected tissues, absorption spectrum of chlorophyll occurs in blue region. Rate of leaf initiation is reduced as evident by

Table 11.3: Theoretical Yield Potential and Potential for Yield Increases (Dwivedi, 1994, 2003).

Potential Efficiency	Biomass Dry Weight (t/ha)	Cane Fresh Weight (t/ha)	Sugar (t/ha)	References
A Theoretical Yield Potential				
1 2.4 x 10^-5g CHO Cal^-1 radiation (200 cal radiation cm^{-2} d^{-1})	180	300-350		Bull & Glasziou (1975)
2. (a) 355 μmol CO_2 cm^{-2} d^{-1}; assuming quantum efficiency of 0.125 mol CO_2 mol quanta-1, Energy efficiency = 8.5 per cent; Net production : 118g d. m.m^-2 d^-1	470	805.8 (Dwivedi et al., 1994)		Moore (1987)
(b) Assuming quantum efficiency of 0.053 mol CO_2 fixed mol quanta-1; 150 μmol CO_2 cm^{-2} d^{-1}; Net production 118g d. m.m^-2 d^-1, Energy efficiency 3.6 per cent; 60 per cent is the cane weight in total sugarcane biomass, 65 per cent moisture in total biomass.		339.42 (Dwivedi et al., 1994)		–do–
3. 1 mol CO_2, fixed to the level of sucrose has 120 k cal energy (1 mol CO_2 assimilation will yield 30 g glucose =15 g sucrose)				
(a) Based on 0.125 quantum efficiency and 8.5 per cent solar energy harvesting efficiency.			78.0	Dwivedi et al., 1994
(b) Based on 0.053 quantum efficiency and 3.6 per cent solar energy harvesting efficiency. Assumptions : 1g glucose =4000 cal energy: UDPG synthetase/sucrose P. synthetase			50.22	-do-
Glucose + Fructose ⇌ Invertase Sucrose				
60 per cent cane weight in total aerial biomass (fresh weight basis); Respiration loss 20 per cent.				
4. Assuming 26 per cent sucrose in juice (Naidu, 1987), cane juice extraction 60 per cent and 805.8 t cane ha^-1			125.5	-do-
B. Record Yield				
I 40 g m^-2 d^-1 D.M.	150	250-300		Hames, 1970
I(i) 40 g m^-2 d^-1 D.M.		125	42.0	Anon,1969
I(ii) Krishi Pandits (Most progressive farmers of India) Maharashtra (1970-71); (11 per cent sucrose recovery) Uttar Pradesh (1969-70); 10 per cent sucrose recovery).		464.03 / 335.42	51.02 / 33.54	Dwivedi et al., 1994
(iv) Agriculture Department, Trinidad – Indonesia (1987-88) with 9 per cent sucrose recovery.		360.00	32.4	–do–

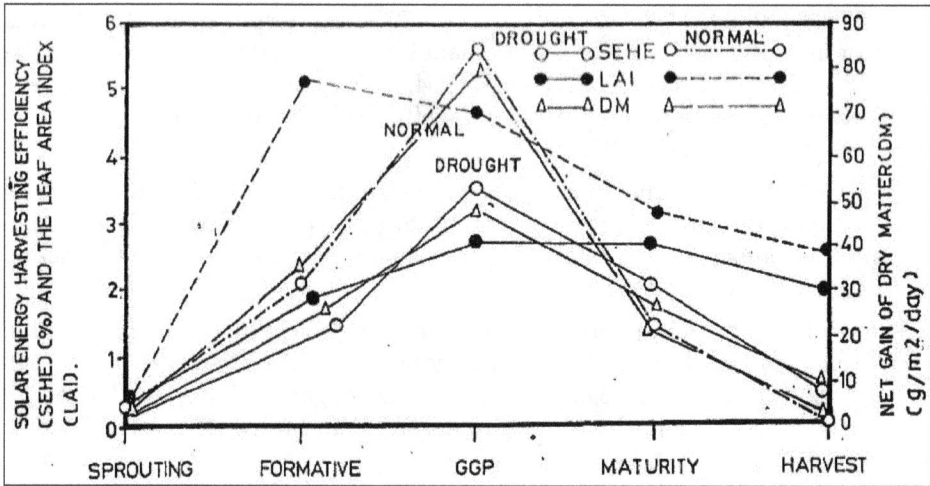

Figure 11.3: Variations in Leaf Area index, Net Dry Matter Production, Solar Energy Harvesting Efficiency at different Growth Stages of Sugarcane.

Source: Dwivedi et al. (1994)

increased plastochron duration. Younger leaves absorb more blue light at early stage and red light at later stages of plant growth. Similarly, RUBP carboxylase was marked at early stage, while that of PEP at later stages of plant growth. Moisture stress (measured as relative water content) depressed $^{14}CO_2$ incorporation and Hill activity (measure of photochemical reaction) in Co 419, Co 1148 and Co 997. These reactions congruently might occur due to overall suppression of photosynthetic process under moisture stress condition. CO_2 fixation and O_2 assimilation rates have been measured under field conditions and presented in Figure 11.4 (Dwivedi and Srivastava, 1993).

2.2.3. Sugarcane is Renewable Source of Self Sustaining Energy Crop, Alcohol, Production, CO_2 Depletion and Obviating Greenhouse Effect

Sugarcane genotypes CoLk 8001, CoLk 7701 and Co 1148 harvest maximum energy. Energy accumulation in biomass range from 84.0×10^6 to 169.8×10^6 kcal/ha/yr in which sugar contributes about 27-36 per cent of total energy produced. The relative ranking of varieties on the basis of biomass energy was observed in the following order : CoLk 8001> Co219> CoLk 7701> Co 1148 > Co 975 > LG 7253 > Co 419 > Co 1336 > Co 764 > Co 997 (Solomon et al., 1988). The total energy output (Solar energy harvest –SEH) of all the test genotypes was greater than the total energy input (SE), tillage, fertilizer, irrigation, sowing and harvesting); the ratio of SEH/SE varied from 6.20 to 12.54. On an average, a 12-month crop of sugarcane produces about 125×10^6 kcal/ha/yr in sub-tropics which comes to about 0.34×10 kcal/ha/day. Studies on energy produced from alcohol (derived from sugarcane) has also been carried out.

Dwivedi and Srivastava (1993) evaluated the energy value of different parts of sugarcane plants and finally the total energy efficiency. The energy values (cal/g dry wt.) in root, stalk, green leaves, dry leaves and leaf sheaths were found to be 2150-

2600, 3750-3900, 3300-3600, 2750-3050 and 2700-3000, respectively. Co 419 was superior in energy accumulation, but ethanol production (l/ha) was found to be 6724, 3800, 3768, and 4770 in Co 1148, CoJ 64, Co997 and Co 419, respectively. A complete picture of aforesaid physiological efficiencies is presented in Figure 11.4 (Dwivedi and Srivastava, 1993).

Visualizing the higher potential of environmental CO_2 depletion and O_2 assimilation sugarcane and great variation in its genotypes, this crop can be used to obviate environmental warming and higher solar energy harvest in CO_2 enriched environment (Dwivedi, 1994).

The total energy output in sugarcane is presented in Figure 11.4 and Table 11.4. Looking at potential energy balance, solar energy harvests to a level of 167×10^6 kcal solar energy equivalent to 739.684GJ/ha/yr, the net gain of 520.80 GJ/ha energy after deducting energy input in crop cultivation proves sugarcane as a most endergonic energy crop (Table 11.4) (Dwivedi, 1994).

Enormous scope thus exists for improving sugarcane productivity but there are several constraints in realizing the potential yield. These constraints are plant constraints and input constraints which do not allow plants to express their genetic yield maxima.

Table 11.4: Energy Balance of Sugarcane Starting from Crop Production to Sucrose Crystallization/Gur Formation (Dwivedi, 1994).

Sl.No.	Particulars	Energy GJ/ha/y
1.	Energy required in crop cultivation	41.88
2.	Fuel required in sugar crystallization	176.88
3.	Total	218.88
4.	Solar energy harvested by crop	739.68
5.	Energy balance (gain)	520.80

3. Constraints in Raising the Productivity of Sugarcane Biomass and Organic Constituents Productivity

3.1. Plant Constraints

Sugarcane is a highly heterozygous complex polyploidy plant. Consequently it has better stability for growth under varied stress environments. However while, considering the improvement in cane yield and its constituents, the chances of improvement become weak because of being under polygenic control. The major plant constraints experienced in the way of improving the cane productivity are as follows:

3.1.1. Poor Germination and Sprouting

(i) In tropical zone when 2-3 bud setts are planted, more than 60-80 per cent germination/sprouting takes place but in subtropics it never exceeds 30-40 per cent (Dwivedi and Srivastava, 1993). This basic process results in

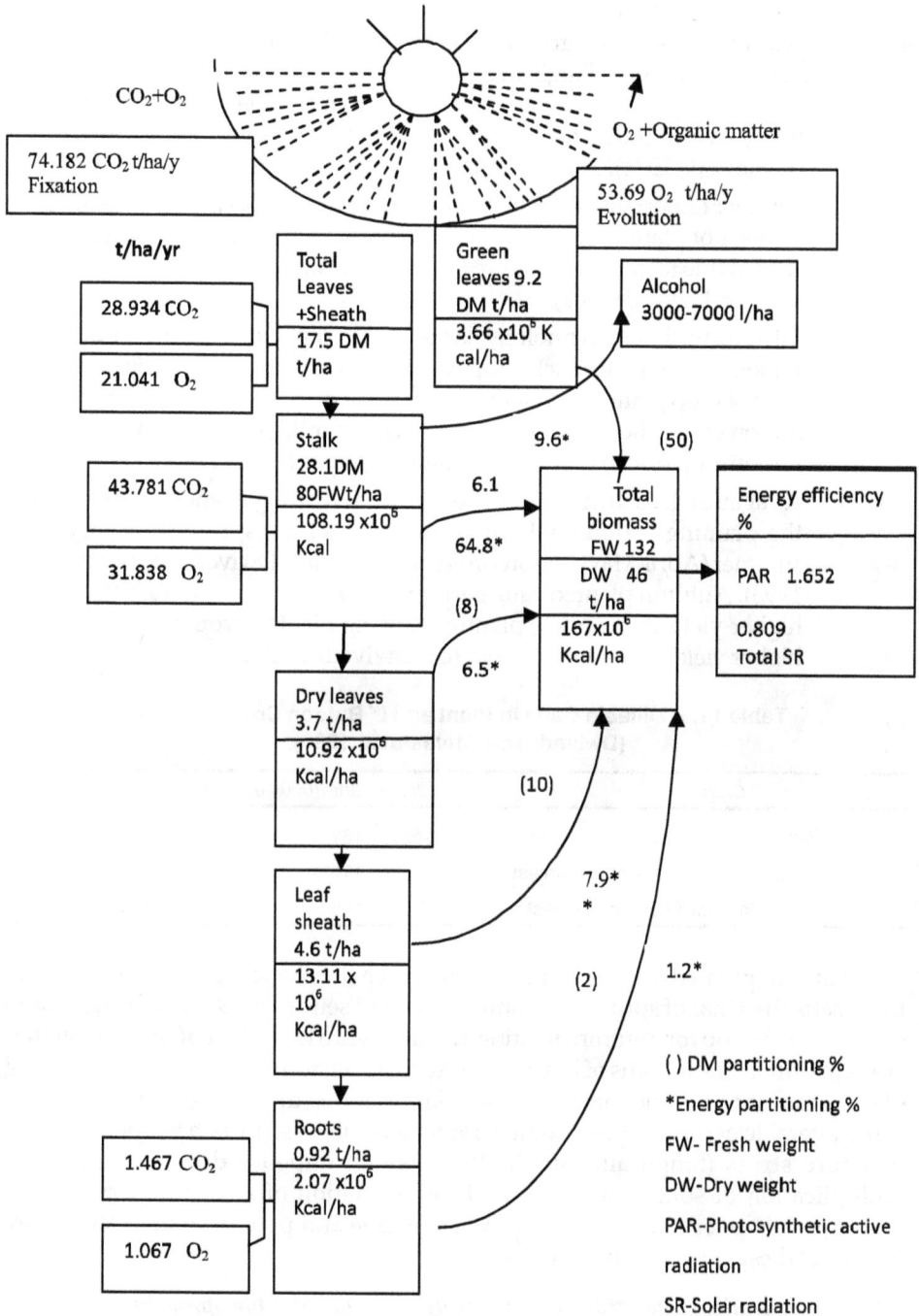

Figure 11.4: Sugarcane as Potential Energy Source of CO_2 Depletion and O_2 Evolution and the Energy (Dwivedi and Srivastava 1993).

poor stand of crop in subtropics. Activation of enzymes, movement of reducing sugars to growing bud(s) and hormonal levels control the sprouting of buds (Singh and Singh, 1964). In subtropics, out of 3 bud in a sett only one sprouts. Proper temperature and higher availability of water, nitrogen and reducing sugars to germinating buds results in sprouting (Panje *et al.*, 1971). This process affects the entire crop productivity.

(ii) Another important problem is very poor sprouting of stubble buds after the harvest of plant crop for raising ratoon crop in subtropical zone. This is caused due to low temperature prevailing in subtropical zone at the harvest time (cf. Dwivedi 2003).

Till date no viable technology has been developed to improve sprouting in ratoon as a result of which poor yield is obtained. The variation in harvesting time of crop plant for raising ratoon is one of major factor for low yield in ratoon crop. The significant difference s in millable cane and the yield of ratoon crop due to this are best exemplified in Table 11.5.

(iii) As against the 8-10 month of planting period of sugarcane in tropical zone, the planting is done only in autumn (October), spring (February) and summer (April-May) seasons in subtropical zone (Dwivedi and Srivasatava, 1993). Autumn planted cane germinates fast and gives at least 10 per cent higher yield than spring planting. Spring planted crop gives significant higher yield than summer ones (c.f. Dwivedi, 2003).

Table 11.5: Millable Cane in Plant and I^st Ratoon Crop and Yield (Dwivedi and Srivastava, 1993).

Sl.No.	Crop	Millable cane (000/ha)	Cane yield (t/ha)
1.	Autumn planted cane	132	85.0
2.	I^st ratoon from December harvest	110	68.0
3.	I^st ratoon from February harvest	125	78.2

Autumn planted setts germinate faster since the reducing sugars are more in those setts than that of spring and summer planted setts (Ali, 1997). A farmer has to sacrifice a *rabi* crop for autumn planting and incur into loss. Most of the *rabi* (winter) crops mature in the months of March and April which result in pushing the planting of sugarcane into summer months. Due to late planting the early growth phase gets short favorable period for proliferation, rather it is subjected to high temperature and moisture stress (Singh and Singh, 1967) which impedes differentiation and multiplication of source and sink and the formation of structural dry matter. Consequently poor tillering *vis-a vis* millable cane and poor growth of shoots are noted. All these congruently reduce yield.

3.1.1.1. Biochemical Changes in Germinating Buds and Stubble Sprouting

(a) ß -Fructo-Furanosidase

Status of ß-fructofuranosidase (acid invertase) in setts and buds of three sugarcane cultivars was determined during germination. A rapid increase in ß -

fructofuranosidase activity was noticed in germinating organs of all the cultivars. It was observed that cultivars Co 453 and Co 1148 showed high activity of ß-fructofuranosidase (Solomon and Kumar, 1987) as compared to the cultivar Co 997. It has been emphasised that ß-fructo-furanosidase, probably is one of the major metabolic factors controlling the germination of sugarcane setts.

(b) Enzyme Activity in Buds Treated with Plant Growth Regulator Triacontanol

Biochemical changes were elucidated in sprouted serial and subterranean buds of sugarcane cultivar Co 1148, pretreated with formulation containing 0.5 per cent triacontanol under laboratory conditions. The treated buds exhibited rapid mobilization of sucrose and high level of reducing sugar. The activity of acid invertase, amylase and starch phosphorylase showed a marked increase in the treated buds whereas, peroxidases and IAA oxidase recorded a marginal change in their specific activity. The triacontanol treated buds showed better sprouting than the water treated control (Solomon *et al.*, 1997).

(c) Poor Stubble Bud Sprouting in Winter Initiated Ratoon

Poor sprouting of stubble buds at low temperatures is associated with a lower level of reducing sugars, reduced activity of acid invertase and higher accumulation of IAA and total phenols. It seems that low temperatures possibly interfere with the metabolism/translocation of certain metabolites in stubble buds essential for their sprouting. On the other hand, higher activities of antioxidant enzymes, such as catalase and peroxidase may protect stubble buds from an oxidative damage, and may perhaps be responsible for maintaining the dormancy of subterranean stubble buds under low temperature stress. The altered activities of these enzymes and certain key metabolites appear to keep the underground buds in a dormant stage under prolonged exposure to low temperature.

3.1.1.2. Effect of Chemical Formulations on Raising Germination and Stubble Sprouting

In resorcinol (0.1 per cent) treated setts, 13 per cent germination was observed in 7DAP (days after planting). After 30 days after planting, the germination was as follows; Control (25 per cent), water primed control (64 per cent) and mixture of phosphates (100 ppm) 68 per cent. By reducing this period, early growth of shoots and leaf development took place which led to attainment of LAI of nearly 1 in 70-75 DAP as compared to 0.3 in control. This also resulted in higher average cane weight.

Out of 36 formulation used, the most effective formulation, containing vitamins (500 mg/l), potassium nitrate (0.1 per cent), Five Phos (100 mg/l) gave 34.5 per cent higher Tmax, 64.7 per cent higher NMC and 73.5 per cent higher cane yield. Calcium, potassium and manganese treatment improved stubble bud sprouting by about 64.3 per cent, 73 per cent, 68.7 per cent, respectively as compared to untreated check (46.0 per cent). The number of tillers per stubble was 157 per cent, 57 per cent and 85.7 per cent more in calcium, potassium and manganese treatment, respectively than control treatment. Sprouted buds showed higher reducing sugars and acid invertase activity and lower contents of IAA and total phenols (Shrivastava *et al.*, 2011).

Table 11.6: Dynamics of Tillering in Autumn, Spring and Summer Planted Cane Ratoon Crop (000/ha)

Planting Months	Autumn (Oct.) Planting	Spring (Feb) Planting	Summer (April) Planting	First Ratoon
March	160	50	–	160
April	400	120	60	380
May	380	320	200	500
June	240	310	258	350
October (NMC)	132	100	95	110
Economically viable tillers (per cent)	33 per cent	31 per cent	37 per cent	22 per cent

3.1.2 Non-Synchronised Tillering (Tiller Dynamics and Mortality)

Tillering is a process of source and sink multiplication. The economically viable tillering phase (millable cane forming tiller production) continue for about five months in autumn planted cane. Out of total tiller formed only 20-35 per cent of tillers result in the formation of millable cane (Table 11.6) (Shrivastava *et al.*, 1985). Tiller formation and tiller mortality takes place simultaneously. This appears to be pure enrgy waste of the plants and planters. Tillers emerged in early period of tillering phase (February-April) form about 70 per cent millable cane. Had there been synchronized tillering, the yield of cane would have risen significantly because of availability of adequate growth period and formation of highest number of millable cane due to uniform competition for inputs and light by limited number of tillers (Dwivedi and Srivastava, 1993). Late tillering phase (after April/May) produces uneconomically viable tillers. Similarly, the water shoots emergence continues till the harvest of crop. Many times it has been found to depress the cane yield and juice quality as it is a plant's stored energy depletion process. However, in April/May ratoon initiation –water shoots add significantly in raising ratoon yield (c.f. Dwivedi, 2003).

3.1.2.1. Raising Synchronization of Tillering trough Physiologically Sound Planting Techniques

(a) Space Transplanting Technique

Based on sound physiological understanding of germination (sprouting), tillering vis-a-vis inter- and intra-plant competition, a scientific crop management procedure, the spaced Transplanting Technique (STP) (raising of one month old settling through single bud sets and transplanting at desired spacing) has been developed which saves the precious cane seed material, ensures higher stalk population (number of millable canes) with a uniform crop stand and higher average cane weight. It has also been a boon for rapid multiplication of seed cane (1:40) *vis-a vis* higher millable cane and higher cane yield i.e. 80-100 t ha^{-1} in subtropics and 120-180 t ha^{-1} in tropics (Dwivedi and Srivastava, 1993). It is labour intensive yet cheaper than conventional planting method. This novel technique is in practice since 1978 all over India for higher and rapid seed multiplication (1:40) and higher yield (Srivastava *et al.*, 1981).

One light irrigation is given immediately after transplanting of settling. All other agronomic practices are followed similar to crops raised through conventional method of planting.

(b) Bud Chip Technique

In India, for conventional system of sugarcane cultivation, about 6 t/ha seed cane is used as planting material, which comprises of about 33,000 stalk pieces having 2-3 buds. This large mass of planting material poses a great problem in transport, handling and storage of seed cane and undergoes rapid deterioration thus reducing the viability of buds and subsequently their poor sprouting. One alternative to reduce the mass and improve the quality of seed cane would be to plant excised auxillary buds of cane stalk, popularly known as bud chips.

With bud chips technology (planting of scooped single bud raised settling or direct planting of encapsulated scooped bud in the field) advantages like a higher seed multiplication rate (1:60) over conventional method (1:10); optimization of shoot population; higher cane height (2.5 m), weight (2.0 kg), number of millable canes per clump(5NMC/clump), and higher cane yield (approx. 120t/ha) as compared to conventional system; higher bud germination (90 per cent) as against 30-35 per cent in conventional system; saving of 80 per cent seed material; and easiness in transportation and seed treatment were obtained. Leftover cane may be processed for juice, sugar/jaggery making. It is labour intensive and requires more care like storage of scooped bud chips at low temperature for maintaining higher duration of viability (7-10 days). Immediate planting of bud chips after scooping does not require care except one light irrigation, yet, cheaper than conventional method of planting as it requires less seed cane and gives higher cane yield (Jain *et al.*, 2010). All other agronomic practices are followed in similar manner that of crops raised through conventional method of planting.

3.1.3. Inadequate Sink

Sink appears to be the limiting factor for sucrose accumulation, since the sucrose synthesized in the leaf is not proportionately mobilized into sink. The total carbohydrate and sucrose content in leaf and leaf sheath are doubled in the month of January and increased further in the month of March (Table 11.7). Had the sink not been the limiting factor, the sucrose and total carbohydrate might have not become double or more in leaves, the site of sucrose synthesis. Thus impeded translocation of sucrose from leaves (source) to stalk (sink) might be another reason for poor sucrose accumulation (Dwivedi and Srivastava, 1993).

The source and sink also differ with genetic variability of crop and the enzymes activities, also differ accordingly and with growth phase (Venkataraman *et al.*, 1992).

In proportion of leaf area, the fresh and dry weight of cane do not vary and so is the case with sucrose accumulation. For example *S. spontaneum* has more leaf area index and photosynthesis rates but fresh and dry weight of cane is found to be lower than *S. officinarum* (Dwivedi *et al.*, 1994) (Table 11.8). Hence sink appears to be a limiting factor.

Table 11.7: Changes in Total Carbohydrate and Sucrose (Per cent dry weight) in Leaf, Leaf Sheath and Stalk at different Growth Phases.

Plant Parts	Formative (June)	Grand Growth (Oct.)	Maturity (Jan.)	Harvest (March)
Leaf	1.603 (2.190)	2.994 (3.385)	3.283 (3.757)	4.203 (6.091)
Leaf sheath	3.177 (4.16)	2.998 (3.245)	2.975 (3.033)	5.826 (7.256)
Stalk	7.395 (11.490)	23.865 (25.585)	31.033 (31.942)	31.666 (32.714)

Values in parenthesis: Total carbohydrates (per cent dry weight).

Source: Dwivedi and Srivastava (1993).

Table 11.8: Phosphoenol Pyruvate Carboxylase (PEP – Carboxylase), Photosynthesis Rate, Leaf Characteristics and Cane Yield of different Species of Sugarcane (c.f. Dwivedi *et al.*, 1994)

Characters	S. officinarum	S. barberi	S. spontaneum
PEP-carboxylase (μ moles NADH oxidized-mg^{-1} protein hr^{-1})	5.299	1.046	2.012
Photosynthesis (mg CO_2 dm^{-2} hr^{-1})	24.17	18.68	28.52
Stomata adaxial (nos. mm $^{-2}$)	84.00	33.00	80.00
SLW (f.w., dm^2 g^{-1})	0.24	0.20	0.26
Yield (g $clump^{-1}$)	810.00	680.00	470.00

3.1.4. Poor Combining Ability for Economic Traits

The commercial constituents of cane crop are regulated genetically. The genetic potential for economically viable constituents e.g. sugar, fibre, alcohol, aconitic acid, waxes and gum in sugarcane is poor and cultivars do not have contrast in these constituents. If the varieties have at least 10t/ha recoverable sugar and 16 per cent fibre, their economic viability in terms of setting different types of industries and co-generation systems and fulfilling human needs would be very well established. The interspecific hybridization gives some hope to solve such problems (Naidu, 1987) (Table 11.9).

Table 11.9: Fibre and Sucrose Content in Hybrids Cerived from CoC 671 and F1 of *Saccharum officinarum* and *Saccharum robustum*, Respectively.

Sl.No.	Clone No	Fibre per cent	Sucrose per cent Juice
1.	G 340	18.06	19.08
2.	G 304	19.41	20.46
3.	G 415	19.62	21.36
4.	G 330	20.16	20.30

Table 11.10: Losses in Sugarcane Yield due to different Constraints in India.

Constraints	Spread of Constraint (Per cent, cultivated area)	Cane Production (Per cent loss)	Cane Productivity (Per cent loss)
Abiotic			
Drought	65	15-20	20-40
Flood	10-30	5-20	5-15
Salt stress	20-25	10-20	20-30
Low Temp.	1-2	2-4	2-5
Mineral Deficiency	20-40	25-45	25-50
Biotic			
Weeds	Every where	10-15	15-20
Diseases	Every where	5-7	5-7
Pests	Every where	5-7	5-7

Source: Dwivedi, 1995.

3.2. Environmental Constraints

The environmental constraints are comprised of biotic and abiotic factors. Among biotic constraints which limit the productivity are diseases, pests, rodents and weeds, The abiotic constraints are drought, flood, salts (salinity and sodicity), frost, low and high temperature, mineral nutrients and winds etc. which congruently cause more damage to crop than biotic one. A summarized form of damage caused by different components on production and productivity of sugarcane are presented in Table 11.10.

3.2.1. Abiotic Stresses

Sugarcane requires about 2000 to 3000 mm water during its life span of 11-18 months. In most of all the sugarcane growing zone of India, the distribution of rainfall is not satisfactory. In subtropical India 5-10 irrigations are required for a good crop of sugarcane whereas in tropical India the irrigation requirement goes as high as 36 in *Suru* (annual crop) and 4-50 in *Adhasali* (16-18 months) in Maharashtra (Kakede,1985). Crop suffers heavily under inadequate irrigation conditions (Singh and Dwivedi, 1995) by disrupting osmoticum and related metabolic activities (Dwivedi *et al.*, 1988).

Heavy irrigation has resulted into higher yield but the soil has become saline. Consequently the chunk of land of Sanghali, Dattatraya, Kolhapur, Pune in Maharashtra state (tropical) are becoming barren and unsuitable for plant growth. In India, about 2000 ha land is becoming saline every year (Dwivedi, 1993) hence, this has become a serious deterrent in limiting sugarcane productivity.

An average crop of sugarcane (100 t/ha) remove 208, 53, 280, 3.4, 1.2, 0.6 and 0.2kg/ha of NPK, Fe Mn, Zn and Cu, respectively from soil (Dwivedi and Singh, 1991). Most of the soils are deficient in available P and K. As a result, the application of 400 kg N, 170 kg P_2O_5 and 186 kg K_2O per hectare has been recommended (Kakede 1985) for sugarcane, depending upon its duration of growth and nutrient status of the soils in Maharashtra (Tropical Zone). In subtropical zone 150kg N, 30 kg P_2O_5 and 40kg K_2O per hectare have been recommended (Anonymous, 1991). In addition, the deficiencies of S and Zn, are emerging due to the continued use of high analysis fertilizers and restricted recycling of organic matter/wastes. These factors, besides limiting cane yield are deterring the juice quality to an extent of 10-40 per cent (Dwivedi and Singh, 1991).

Both the low temperature (< 10°C) and high temperature (40-60°C) are detrimental for sugarcane productivity and specially juice quality (Dwivedi, 1999). Potassium silicate (1.5 per cent) spray strengthen potassium silicate complex poor thermal conductivity layer in tissues and thereby minimize temperature injury as depicted in Figure 11.5 (Dwivedi, 1993).

3.2.2. Biotic Stresses

The damage caused by diseases and insects, in general does not contribute more than 10 per cent loss in productivity. However, the havoc of red rot and wilt and pyrilla, not very common, becomes definitely serious. Weeds are serious problem only before grand growth phase of crop and cause 10-15 per cent loss in cane productivity (Table 11.10).

4. Sucrose Accumulation and Recovery

Sucrose recovery is the consequence of sugar accumulation and juice purity. The following five components and processes controlled by inherent plant attributes and environmental factors regulate sucrose accumulation and recovery, require proper understanding with a view to find out measures to improve sucrose yield.

4.1. Sucrose Synthesis

In India, little work on sucrose synthesis has been done. Sucrose synthesis is the outcome of the interaction of photosynthesis, environmental factors (light, CO_2, water, temperature, mineral nutrition etc.) and the activity of enzymes and related biochemical agents in plant. Investigations are in progress in light of four steps in sucrose synthesis showing in Page 413.

4.2. Sucrose Transport

Considering transport as specific process, little work has been done in India. Application of K in strengthening structural tissues and conducting elements in sugarcane, have been reported (Dwivedi, 1993a). Recent studies conducted by author has indicated that the conducting tissues at the node of different genotypes differ. The standing angle of conducting strand of node has been found to be associated with sucrose accumulation. Figure 11.6 gives an indication that *schlerostachya* sp. (high biomass, low sucrose, 5-8 per cent) has straight conducting tissues at node, whereas CoJ 64 (20-22 per cent sucrose in juice) and Co 1148 (18-20 per cent sucrose in juice) have curved conducting strand at node. In Co J64 and Co 1148 the ratio of standing angle of vascular strand of node: inter-nodal region was found to be 0.61 and 0.91 respectively (Dwivedi, 2003). Further detailed studies are in progress. Chiranjivi Rao (1989) found enhancement in sucrose translocation by using certain mineral elements like K and boron and ripeners.

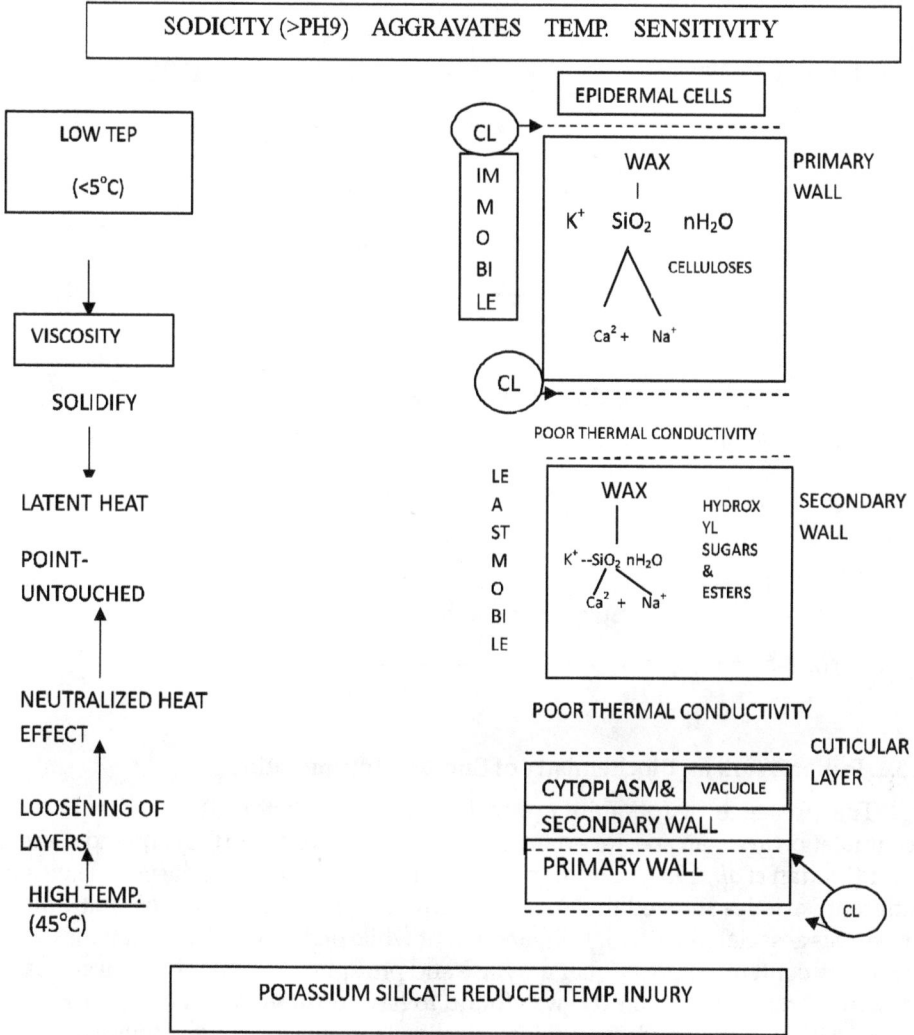

Figure 11.5: Temperature Tolerance in Sugarcane (Dwivedi, 1999).
CL: Cuticular layer.

4.3. Sucrose Accumulation

After unloading of sucrose to sink (stalk), it is accumulated in storage parenchyma and parenchymatous intercellular spaces through various reactions. The unloaded sucrose at apoplast is cleaved by invertases in glucose and fructose which enter in to cytoplasm through electron transport process requiring energy. The glucose and fructose combine with UDPG at tonoplast through various reactions and result in the formation of sucrose in the vacuole in cytoplasm of storage parenchymatous tissues of cane.

UDP transglucosylase

$$UTP + G\text{-}1\text{-}P \longleftarrow \qquad\qquad\qquad \longrightarrow UDPG + PPi$$

Boron

NAD   Mg^{2+}

ATP   ADP + Pi

Sucrose phosphate synthetase

$$UDPG + F\text{-}6\text{-}P \longrightarrow Sucrose\text{-}6\text{-}P + UDP$$

Sucrose synthetase

$$UDPG + Fructose \longrightarrow Sucrose\text{-}6\text{-}P + Pi$$

Sucrose phosphatase

$$Sucrose\text{-}6\text{-}P + H_2O \longrightarrow Sucrose + Pi$$

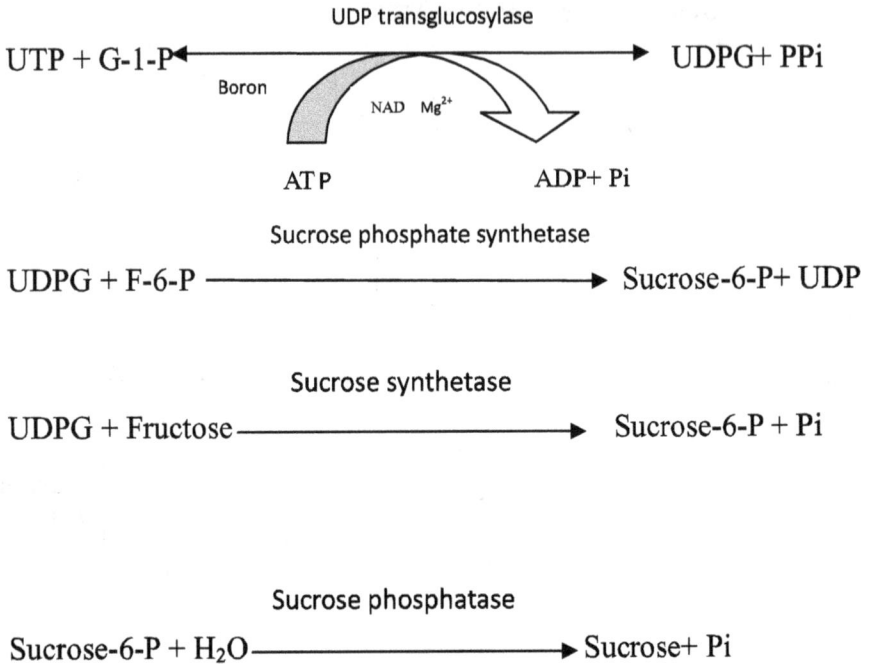

4.3.1. Indian Work on Biochemistry of Sucrose Accumulation

The presence of foliar invertase is one of the factors that control sugar accumulation in sugarcane. Invertase activity is closely related to the cane and sucrose yield (Madan *et al.*, 1980). Comparison of acid and neutral invertases in leaves of different varieties showed that extremely low enzyme activity during maturation phase was associated with high sugar content while higher activity of enzymes with low sugar content. Ratio of foliar amylase and phosphorylase at the initial stages of growth (40 days after planting) was found to have positive correlation with sugar accumulation. Foliar amylase, amylose, amylopectin and starch were found to play a vital role in sugar accumulation (c.f. Dwivedi, 2003). These findings indicate that the enzyme activities in the leaves with special reference to sugar accumulation potential in the cane have a definite trend. After 40 days of planting the acid invertase activity in the leaves was shown to be high in low sugar varieties and low in high sugar ones.

Similarly, amylase activity in the leaves was low in high sugar varieties and vice versa. On the contrary, starch phosphorylase enhanced with the increase in sugar accumulation. It may be concluded that low amylase and high phosphorylase levels in the foliar tissue would permit increased sucrose biosynthesis from starch, while a low acid invertase would check the inversion of sucrose (Dwivedi and Srivastava, 1993).

The sucrose synthetase activity in high sugar (Co 7819 with 20.14 per cent sucrose in juice and 90.24 per cent purity) and low sugar (Co 9304 with 16.59 per cent sucrose

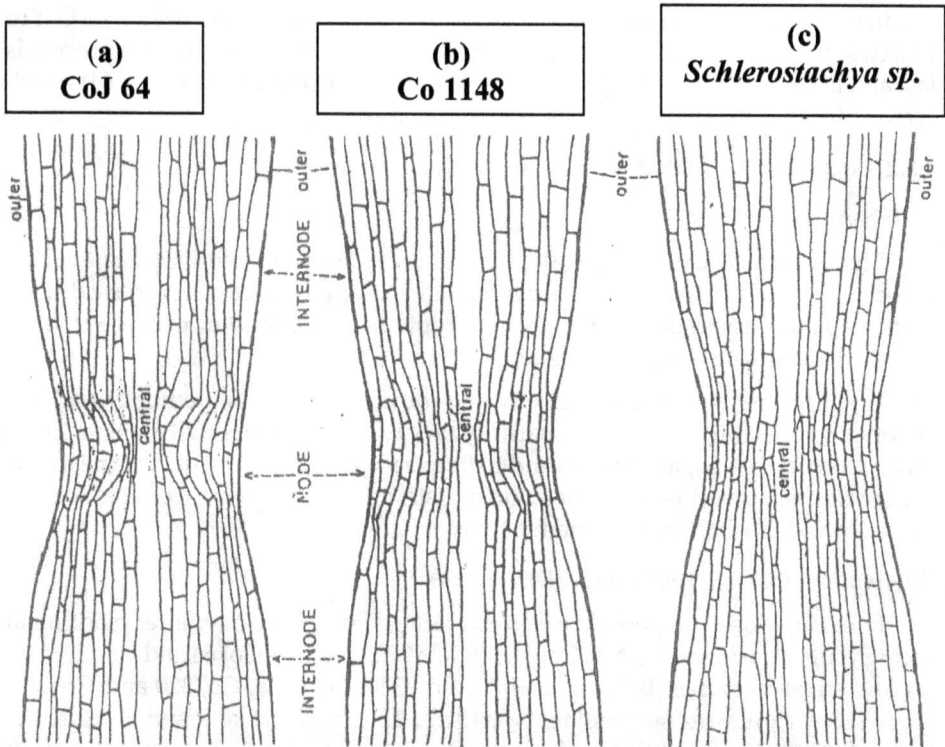

Figure 11.6: Longitudinal View of Sliced Stem; Variation in Conducting Strand at Node and Internode in different Plant Types (Dwivedi, 2003)

(a) CoJ 64- Early maturing high sugar and low biomass, (b) Co 1148- Late maturing, moderate sugar and moderate biomass, (c) *Schlerostachya sp.*- High biomass and very low sugar.

in juice and 86.29 per cent purity) genotype was markedly higher at 240 days of growth but it declined to a level of 11.17 mole glucose/g FWT/hr in former and 10.37 mole glucose/fwt/h in latter genotype at 300 days of growth. On the other hand the acid invertase activity was significantly higher in low sugar genotype as compared to high sugar variety resulting into three times higher acid/neutral invertase ratio in former as compared to latter one (Venkataraman and Ramanujam, 1994).

Studies conducted by Shetiya *et al.* (1991), Dendsay *et al.* (1995) and Dendsay *et al.* (1997) suggested that soluble acid invertase (pH 5.2) of immature internodes was associated with cell expansion/elongation process leading to growth of stalk, while its cessation with sucrose storage in cells. Immature expanding top internodes showed high acid invertase activity but bottom mature internodes show low or negligible activity. Immature inter-node of late maturing (Co 1148) genotype has 2-3 times higher acid invertase activity than early maturing genotype (CoJ 64). (Agrawal *et al.*, 1997) using radio sodium bicarbonate reported the variation in carbon fixation and its translocation with age of crop and maturity of leaf. Naidu et al. (1997) recorded gradual increase in juice sucrose in late maturing varieties (CoA 7602) from Oct to March in tropics, whereas, Dwivedi and Srivastava (1993) found that this process is fast and it starts from last week of August in early maturing varieties (CoJ 64) which is ready for harvest in Nov/Dec in subtropical areas.

4.3.2. Enzymes Associated with Sucrose Accumulation

(a) Invertases

Biochemical analysis of acid and neutral invertases apparently indicated their association with yield and sucrose content and disappearance of invertase during maturation was associated with high sugar content while persistence of the enzymes was associated with low sugar content.

A sharp fall in invertase activity was observed from September in all the genotypes, indicating the onset of maturity phase. The enzyme activity maintained a low pattern in high sugar varieties and it disappeared finally in January while in low sugar varieties invertase activity again registered a rising trend from November onwards and the activity remained high until January.

(b) Amylase, Phosphorylase and UDPG Synthetase

Foliar amylase and phosphorylase were studied at forty days after planting in six high sugar varieties (Co 419, CoLk 7701, Co 771, CoS 510, CoJ 64 and Co 453) and six low sugar varieties (B 41242, CoJ 72, Co 7330, LG 7255, Co7320 and Co 312) grown under uniform management (fertilizer 150 kg, N, 80 kg P and 50 kg K per hectare) (Chandra *et al.*, 2013). Results revealed that high sugar genotypes have much lower amylase activity than the low sugar varieties, while foliar phosphorylase showed an opposite trend with high sugar genotypes having higher phosphorylase activity and low sugar canes having low enzymatic activity.

The findings of these investigations suggest that foliar enzymes such as phosphorylase, amylase and acid invertase activity might be useful biochemical parameters for selection of high sugar genotypes in sugarcane.

UDPG sucrose synthetase, amylase and invertases in relation to sucrose accumulation in some high sugar and low sugar cultivars of sugarcane in subtropical India were studied along with photosynthetic $^{14}CO_2$ incorporation.

Specific activity of foliar UDPG-sucrose synthetase did not differ appreciably in high and low sugar cultivars. The activity of foliar acid invertase was rather high in low sugar cultivars and was negligible in high sugar ones. Amylase activity exhibited the same as that of acid invertase but its magnitude was smaller. The rate of photosynthetic CO_2 incorporation was found to be more or less similar in high and low sugar cultivars.

It appears, therefore, that it is not the photosynthesis and sucrose synthesis rates in high and low sugar cultivars but it is the activity of foliar acid invertase which perhaps determines whether the cultivar will be a high or low sugar type (Chandra *et al.*, 2013).

(c) Effect of Ethephon on Foliar Enzymes in Late Planted Sugarcane

In late planted sugarcane application of Ethephon (500 mg/l) altered the activities of amylase, peroxidase, invertase and nitrate reductase (*in vivo*) in foliar tissues. The *in vivo* NR activity showed a sharp increase in all the cultivars treated with Ethephon.

However, maximum induction in enzyme activity was noted in the cultivar CoJ 64. High nitrate reductase (NR) activity was noted till 120 h in the cultivars CoJ 64 and BO 91 whereas in genotypes Co 1158 the enzyme activity started declining after 48h. This was probably due to genetic differences among cultivars with respect to applied Ethephon.

The foliar amylase activity showed manifold increase in Ethephon treated plants where as acid invertase activity showed a marginal decline (Solomon *et al.*, 1988). Application of Ethephon resulted in increased peroxidase activity in foliar tissues of sugarcane.

4.4. Amalgamation of Impurities in Juice

The term sucrose recovery (crystallization of white sugar from juice) differs from that of sucrose accumulation in the cane. Barring factory level problems, a sugarcane cultivar growing at two distant or close location/fields might accumulate same quantity of sucrose but do not result equal amount of sucrose recovery from juice on unit volume or weight basis. The loss in sucrose recovery and rise in molasses weight and its sucrose content due to prevalence of impurities and disorders in the juice is called as amalgamation effect.

Table 11.11: Rise in Juice Amalgaments and Loss in Sucrose Recovery due to Abiotic Stresses and Juice Quality Associated with Good and Poor Sucrose Recoveries.

Abiotic Stresses	Rise in Amalgaments	Per cent Loss in Sucrose Recovery	Juice Qualities (per cent)	Site I Good	Site II Poor
				Sucrose Recovery (10 per cent)	Sucrose Recovery (9 per cent)
Drought	Amino acid and Minerals	10-15	Fibre	13.84	15.02
Water logging	Starch and Fe	10-15	Brix	20.35	16.57
			Pol	18.85	14.37
			Purity	91.52	86.34
Low temp.	Anthocyanin, wax, Org. acid	10-20	Ash	0.39	0.52
Mineral elements	Amides, Amino acids	8-16	Reducing Sugars N/P	0.40	0.95
				0.55	1.69
			Organic Substances	0.52	1.26

Drought 0.50 MPa; Water logging 1.5 months, Low temperature 10 °C min. and 18-28°C maxi; Salinity Soil pH 8.8 and ESP-25. (Dwivedi, 1993a and 1995).

The undesirable compounds in juice have been reported to be proteins, amino acids, starch, anthocyanin, pigments, pectins (gums), wax, polyphenols, organic acids, mineral elements etc.(Kakede, 1985). Dwivedi (1993a) examined such impurities in the juice of Co J 64 (Table 11.11) grown at two quite close sites. Site I and Site II were at IISR Lucknow and Banthara farm (NBRI) Lucknow, respectively. The latter farm has sodic soil and was subjected to water stagnation during rainy season. Further studies revealed that salt stress results more accumulation of sodium ions; waterlogging cause more Fe and starch accumulation; low temperature raises organic acids, anthocyanin and wax content; severe drought increases amino acids and the minerals and high nitrogen nutrition augments amide and amino acids in juice. How to obviate the accumulation of impurities in juice under stress, needs further studies.

4.5. Staling of Cane Crop

This is very important factor which determines sucrose recovery at factory level. Lot of work on staling of sugarcane has been done in India. The salient points are discussed here to keep up the scope of this article intact.

A well ripen sugarcane crop will lose its sucrose within a few days if not harvested and milled in time or left carelessly in the field. Dwivedi (1993a and 2000) reported that crop standing in the field after maturity/ripening is stale crop. It stales under standing conditions with time and called as "standing crop staling". The "over-stand" is used as misnomer for it. Similarly uncrushed/unmilled cane after crop harvest stales with time and called as "harvested crop staling". The problem of staling of crop is mainly a management problem but scientist cannot be excluded from it because due to staling, the inversion of sugars caused by invertase enzyme activation and microbial attack, the loss in sucrose has been reported to be high as 30 kg/t of crushed cane (Solomon *et al.*, 1997) and still higher in standing staling crop growing under or harvested from adverse environmental conditions (Dwivedi, 2000).

4.5.1. Reasons for Staling of Sugarcane in India

(a) Standing Crop Staling

☆ Inadequate crushing capacity of mills.

☆ Power interruption, sudden break down of plant machinery etc. in mills.

☆ Inadequate availability of labour and machinery for harvesting and transportation of cane to mill.

☆ Lack of planning in planting and harvesting *i.e.* harvesting at maturity stage.

☆ To avoid standing crop staling, February planted crop varieties should be harvested after attaining maturity as mentioned below:

　i) Early maturing varieties: 16 per cent sucrose with 85 per cent purity in juice in Dec. (10 months)

　ii) Mid late maturing varieties: 16 per cent sucrose with 85 per cent purity in juice in Feb. (12 months)

    iii) Late maturing varieties: 15 per cent sucrose with 85 per cent purity in juice in May (13-14 months)

☆ Poor awareness about deterioration caused due to standing crop staling.

(b) Harvested Crop Staling

☆ Absence of a proper varietal balance and scientific harvesting schedule based on cane maturity-

- Early maturing varieties: 16 per cent sucrose with 85 per cent purity in juice in Dec. (10 months)-CoJ 64, CoS 687 etc.
- Mid late maturing varieties: 16 per cent sucrose with 85 per cent purity in juice in Feb. (12 months)- Co 1148, CoLk 8102 etc.
- Late maturing varieties: 15 per cent sucrose with 85 per cent purity in juice in May (13-14 months)- BO 91, Co 8118 etc.

☆ Extension of milling period during summer months when ambient temperature is high (> 40°C)

☆ Practice of harvesting cane 3 to 6 days in advance before its supply to mills, in some areas, this delays is 7-10 days

☆ Limited crushing capacity of the mills resulting into staling of cane at mill yard/cane centres

☆ Inordinate delay in transport of harvested cane from farmers field/cane centers to the mills and lack of an efficient communication network

☆ Complete absence of cleaning system, practice of uprooting, burning and detopping of cane in certain areas

☆ Labor scarcity and power interruption

☆ Lack of understanding regarding cane and mill sanitation program and use of proper biocides during milling

☆ Mechanical harvesting of sugarcane (burnt crop) without proper and timely supply arrangements

4.5.1.1. Changes during Standing Crop Staling and Magnitude of Losses

The cane of standing stale crop results into more pithiness, reduction in tissue moisture, inversion of sucrose, production of side tillers, sprouting of buds on staling cane, drying and fast senescence of left over top leaves. Shriveling and cracking of cane during dry summer in sub-tropics are seen. This provides an excellent opportunity for *Leuconostoc* species to grow which are known to produce dextran at the expense of sucrose resulting to loss in sucrose recovery (Table 11.12).

4.5.1.2. Factors Aggravation Sucrose Deterioration in Standing Crop Staling

Higher temperature (>36°C in tropics and 40-45°C in subtropics), low soil moisture less than 8 per cent), dry atmosphere (RH < 50), salinity (soil pH > 8.8) and ESP > 20), high soil N(>260kgN/ha) and staling sensitive genotypes result more sucrose loss during staling (Tandon *et al.*, 1955, Singh and Behl, 1961, Dwivedi, 2003). These factors besides raising cane temperature and physiological and biochemical disorders, favour the shriveling and cracking of cane and thereby entry of microbes in staling cane. Consequently more loss in sucrose due to rise in sugar inversion process is caused.

Table 11.12: Cane Yield, Juice Quality and CCS Yield of Standing Stale Crop.

Date of Harvest	Cane Yield (t/ha)	Brix Per cent	Sucrose Juice Per cent	Reducing Sugar Per cent	CCS (t/ha)
Feb., 28	79.3	19.4	16.5	0.98	8.8
March, 30	80.7	19.2	16.1	1.52	8.72
April, 30	60.8	19.6	14.8	1.92	5.74
May, 30	38.4	19.6	14.2	2.3	3.08

Source: c.f. Dwivedi, 2003 (Nainital, U.P., Subtropical, Variety Co 1148).

4.5.1.3. Controlling of Standing Crop Staling Deterioration

Selection of varieties such as Co 740, Co 353, Co 7706, Co 62175, CoC 671, Co 419, in tropics (Chiranjivi Rao, 1993) and CoS 92263, CoS 767, Co 1148, CoS 8118, BO 91, in sub-tropics, (Srivastava and Kapoor, 1996), suppression of flowering (Dutt, 1943; Panje *et al.*, 1968) foliar application of chemical lke 2,4-D, 2, 4, 5T, and ripeners *etc.*, Polaris (Singh *et al.*, 1982), and irrigation and trash mulching (Behl and Singh, 1961; Srinivasan, 1987) have been reported to minimize deterioration in standing crop staling.

4.5.2. Changes in Harvested Crop Staling and Magnitude of Losses

The deterioration of harvested cane is primarily a biochemical process followed by bacterial invasion through cut ends or damaged sites of stalk. The time lag between harvests to crushing is crucial importance to achieve maximum sugar recovery. After cane harvesting, endogenous invertases get activated due to lack of any internal physiological and biochemical control mechanism. There are two invertase(s) types based upon pH optima, acid invertase (pH 4.8) and neutral invertase (pH 7.0) in cane stalk and behavior of both the invertases is strongly influenced by variety, pre-harvest burning and storage duration. Glasziou (1962) reported the presence of acid invertase in immature storage tissue of sugarcane, involved intimately in sucrose inversion in the outer space just prior to its movement into storage vacuole. Later on Hatch and Glasziou (1963) reported the presence of neutral invertase (pH, 7.0) in mature internodal tissues. These two invertases are present in mature and immature tissues (Rizk and Normand, 1969). After harvest of crop, due to detrashing and removal of green top from cane the physiological and biochemical controls on cane functioning is lost. Consequently, fast changes in cane metabolism occur as a result of which deterioration in harvested cane is much faster than standing crop staling (Dwivedi 2003).

After 36 hours of harvest the fast inversion of sucrose takes place and it becomes highly significant by lapse of 72 hours. The rapid loss of moisture from the cane, rise in respiration rate and reducing sugars lead such disorders (Solomon *et al.*, 1997, Dwivedi; 1993a). The rise in invertase activity and microbial population further aggravate deterioration in harvested cane (Solomon *et al.*, 1997). Production of dextran due to rise in microbial population like *Leuconostoc mesenteroides* and *L. Dextranicum* is associated with rise in organic acids, gum and reducing sugar in juice and finally decline in sucrose content (Table 11.13) (Gupta and Nigam, 1982).

Table 11.13: Formation of Organic Acids, Gums and Dextran on Storage of Sugarcane for 76 hr. (Var. Co 1148).

Month	Organic (m.equivalent/l)		Gum (mg/l)		Dextran (mg/l)		Sucrose Per cent Juice	
	Fresh	Stale	Fresh	Stale	Fresh	Stale	Fresh	Stale
Nov.	125	200	2790	4590	200	4200	14.58	12.58
Dec.	137	213	2845	4210	200	355	15.70	12.92
Jan.	145	213	2970	3550	–	3200	17.97	16.72
Feb.	250	312	3015	3250	–	3900	18.44	17.47
Mar.	287	300	3330	4250	210	3500	18.62	16.03
Apr.	300	400	1910	6550	1850	6800	17.49	13.20

Source: Modified from Gupta and Nigam (1982).

4.5.3. Biochemical Constituents Affecting Postharvest Sucrose Loss

During staling, enzymatic hydrolysis of starch causes formation of dextran and reducing sugars and as a result specific gravity of the juice increases and sucrose recovery decreases. Das and Prabha (1988) reported the presence of amylase, acid phosphatase, carboxyl-methyl cellulase and fructose 1-6 diphosphatase in the stale cane. Presence of cellulolytic enzymes in cane juice was also reported by Solomon and Kumar (1983). Eggleston and Legendre (2003) emphasized that the enhanced activity of acid invertase could be due to induced invertase and decreased activities of sucrose synthesizing enzymes due to change in pH of cell sap. It has also been noted that the acid invertase activity enhanced the dextran formation (Solomon, 2009), therefore, down regulating this enzyme could minimize sucrose losses after harvest

4.5.4. Microbiological Aspects of Harvested Cane Deterioration

At later stage of staling, besides enzymatic, chemical and respiratory losses, growth of microbes (dextran, alcohol, acid producing) is responsible for huge losses in recoverable sugar. Microorganisms such as yeast (*Saccharomyces, Torula, Pichia*), *Leuconostoc, Xanthomonas, Aerobacter*, acid producing *Streptomyces* are found at the cut ends or damaged sites. Bio-deterioration is caused mainly by *Leuconostoc mesenteroides* or *L. dextranicmu*. These bacteria enter into sugarcane stalk from the soil through the cut ends or damaged sites of stalks and multiply in the mill corners, gutters, pipe lines and the mixed juice tank. These microorganisms convert sucrose into polysaccharides such as dextran catalysed by enzyme dextransucrase or exogenous invertase. The presence of dextran even in very small amount creates problems of filtration, clarification, crystallization and alters the shape of sugar crystals thereby affecting the quality of sugar (Solomon *et al.*, 1997).

4.5.5. Controlling of Harvested Crop Staling

Quick crushing of cane within 24-36 hrs after harvest; proper planting and harvesting with use of CoS 767, Co 1148, BO 91, CoS 8118, CoS 92263 in subtropics (Solomon *et al.*, 1997) and CoC 671, Co 740 Co 6304 in tropics (Balusamy *et al.*, 1990) and Chiranjivi Rao, 1989); avoiding of topping and anticipated delay in transport and crushing; storage of cane in shade covered with trash have been found to minimize loss in sucrose during harvested crop staling.

Efforts have been made to reduce loss in tonnage and sucrose inversion using physico-chemical methods. These include spraying of water, bactericidal solution, use of anti-inversion and anti-bacterial formulations and pre-harvest foliar and soil application of zinc and mangnous compounds. An integrated mill sanitation program and simultaneous use of dextranase could further improve sugar recovery and minimize problems caused by dextran. The possibility of electrolyzed water (EW), fogging to reduce post harvest deterioration in field and mill yard has also been explored (Singh and Solomon, 2011). Some of these methods are useful and present larger options for the industry to minimize after-harvest quality losses in the field and milling tandem (Solomon, 2009).

Controlling the level of invertases at suitable locations is the need of hour which can be initiated utilizing RNAi approach. The reduction of invertase activity soon after harvest of sugarcane crop could be useful in minimizing the post harvest sucrose losses, and it can be initiated by implicating the RNAi approach. The inversion of sucrose into glucose and fructose is the major problem in stale cane which leads to significant loss of sucrose (Chandra *et al.*, 2012).

Chemical formulations containing antibacterial (quaternary ammonium compounds/thiocarbamates), anti inversion chemicals (sodium metasilicate/sodium lauryl sulphate) help in minimizing post harvest sucrose losses in sugarcane. The aqueous formulation(s) are sprayed over freshly harvested cane (whole stalk and billets) followed by covering the treated cane with a thick layer of dried cane leaves (trash). Formulation containing benzalkonium chloride (BKC) + sodium meta silicate (SMS) was found to be most effective and improved sugar recovery by over 0.5 units (Figure 11.7). This method reduces the loss of sucrose from harvested cane up to a period of one week, irrespective of temperature and variety.

Effect of electrolyzed water on post harvest storage quality of sugarcane showed relatively less reduction in quality as compared to untreated cane. The invert sugar formation in Electrolyzed water treated cane was also reduced (Solomon, 2009).

4.5.6. A Common Approach: Use of Metal Ions in Altering Invertase Activity and Improving Sucrose in Cane Juice

Among the different metal ions tested, manganese chloride strongly inhibited the activity of all soluble acid invertase isoforms, and thus may be useful to induce

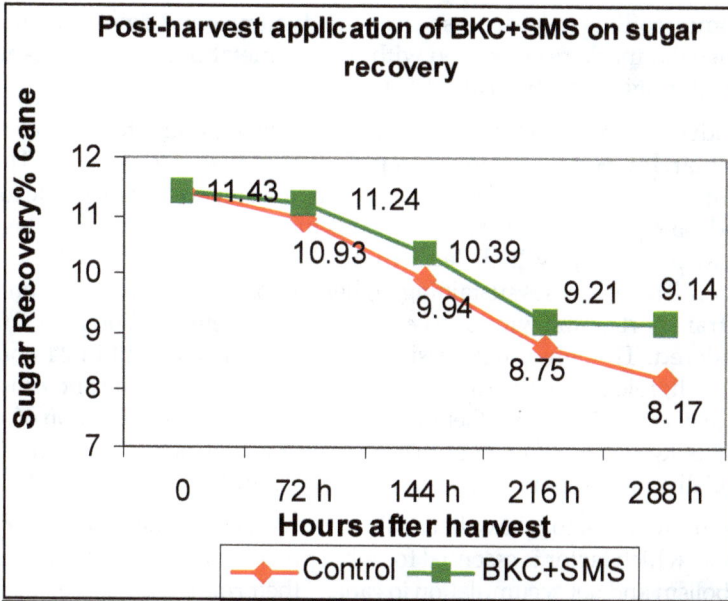

Figure 11.7: Effect of Post-harvest Application of BKC+SMS on Sugar Recovery.
Source: **Solomon (2009).**

early maturity and controlling sucrose inversion in sugarcane, thereby increasing sugar recovery (Sachedeva *et al.*, 2003; Kaur *et al.*, 2002). Vorster and Botha (1998) reported inhibitory effects of Hg^{2+} on the activity of sugarcane neutral invertase. Application of 0.005 M $FeCl_2$, $CuCl_2$, $ZnCl_2$, $CdCl_2$ and $AlCl_3$ reduced invertase activity by 80 per cent, 73 per cent, 32 per cent, 45 per cent and 22 per cent respectively, in sugarcane stalk Mahbubur-Rahman *et al.* (2004). Foliar application of Mg^{2+} and Mn^{2+} ions reduced SAI (soluble acid invertase) expression and increased sucrose per cent juice, S/R ratio and CCS per cent juice in sugarcane stalk of a low sugar genotype BO 91 (Jain *et al.*, 2013).

5. Metabolic Paradigm and Quest for Improving Cane Yield and Sucrose Content in Sugarcane

Since 1950, the Scientist of Plant physiology and Biochemistry Division, IISR, Lucknow have done tremendous effort in improving and advancing germination, biochemical and molecular mechanism for biotic and abiotic stresses, use of chemical ripeners for early and improved sucrose content in cane stalk, physio-biochemical interventions for enhancing stubble bud sprouting of winter initiated ratoons, reducing tiller mortality, development of STP and bud chip technologies to enhance synchronized tillering *vis-a vis* millable cane, seed multiplication rate and higher cane yield., enhancing sucrose content targeting sucrose metabolizing enzymes, bioethanol production, management of post harvest sucrose losses with recent inclusion of metabolic engineering, RNAi technology and gene expression/

transcriptome profiling to address source-sink dynamics and development of gene-tagged molecular markers associated with sucrose metabolizing enzymes and other physiological traits (Chandra *et al.*, 2013).

Dwivedi (1995 and 2000) proposed a metabolic paradigm to raise sucrose and suggest to march towards attaining 26 per cent sucrose in cane juice. They have mentioned a genetically controlled physio-biochemical type of plant mode for higher sucrose and cane productivity. However, this model depicts the physiological and biochemical processes, which lead to sucrose synthesis, translocation and accumulation. The processes sustaining higher sucrose in juice from the time of cane harvest to transport at mill window (*i.e.*, post-harvest deterioration period) has also been considered. Though, the inversion of sucrose in cane within 24 hours after harvest is negligible but such efficient system of transport of cane and crushing do not persist at every mill in India. Generally the post-harvest deterioration of sugars to an extent of 30kg/t cane after 72 hours of harvest in the month from April to June (30°C – 42°C diurnal temperature) automatically occur (Solomon *et al.*, 1997).

In this model 17 factors have been taken into consideration. The main objective is to examine, which factor is essential for what process of sucrose accumulation vis-à-vis metabolism and net accumulation in cane. If their role is not known, the attempts should be made to prove that as essential or unwanted factors. For example, low soil salts, and moderate soil moisture, temperature and light intensity, are essential for all the processes of sucrose accumulation including for least inversion of sucrose in standing crop and harvested staling crop (Dwivedi and Srivastava, 1993). On the other hand, to what extent high K/Na ratio and K/Si ratios are helpful in maintain higher sucrose in juice under delayed harvest and crushing is not known. Similarly, whether, we require low ratio of acid/neutral invertase and low content of non-sucrose (<3 per cent) in the process of sucrose synthesis and thereby attaining higher sucrose is also not known.

In fact, the present knowledge suggest that if the cane is grown on normal and soil (low salts) with better drainage and moderate soil water status (Dwivedi and Srivasatava, 1993) and if the leaf sheath has 2 per cent N, 0.12 per cent P and 2.3 per cent K (4th month age) and have higher sucrose synthetase activity, low acid/neutral invertase and higher phosphorylase/amylase ratio and cane has low fibre (13.5 per cent) the attaining of higher sucrose in juice reaching towards 26 per cent is feasible (Figure 11.8). Secondly, chemical ripeners are also reported to raise sucrose yield but to what extent really these chemicals raise sucrose in juice against the genetic potential is not known. Similarly, low acid/neutral invertase activity is associated with maturity (higher sucrose) irrespective of high and low sugar or early/late maturing sugarcane genotype and to a certain extent, the change in environmental factors at maturity. Hence, based on present knowledge, the use of low ratio of acid/neutral invertase activity as criteria to identify high sucrose genotypes is an illusion and needs further studies. It is therefore, emphasized that to develop a technology to raise sucrose level in the cane, there is a need to probe into depth, the different aspects of sucrose synthesis, transport, accumulation and inversion during standing crop staling and harvested crop staling period as proposed and discussed above.

Metabolic paradigm and quest   for improving cane yield and sucrose content
in sugarcane

| Synthesis | Transport | Accumulation | | Juice sucrose at mill |

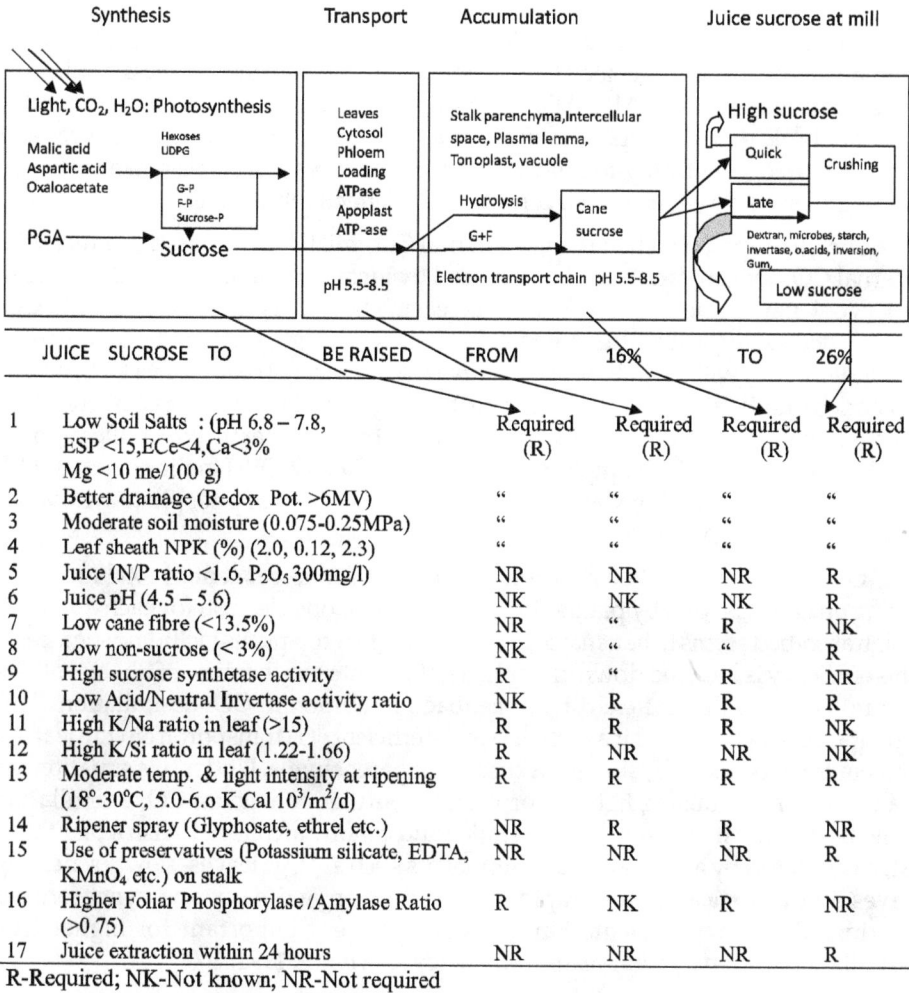

Light, CO_2, H_2O: Photosynthesis

Malic acid — Hexoses
Aspartic acid — UDPG
Oxaloacetate
   G-P
   F-P
   Sucrose-P
PGA ———→ Sucrose

Leaves
Cytosol
Phloem
Loading
ATPase
Apoplast
ATP-ase

pH 5.5-8.5

Stalk parenchyma,Intercellular
space, Plasma lemma,
Ton oplast, vacuole

Hydrolysis   Cane
G+F   sucrose

Electron transport chain pH 5.5-8.5

High sucrose

Quick   Crushing

Late

Dextran, microbes, starch,
invertase, o.acids, inversion,
Gum,

Low sucrose

JUICE SUCROSE   TO   BE RAISED   FROM   16%   TO   26%

		R1	R2	R3	R4
1	Low Soil Salts : (pH 6.8 – 7.8, ESP <15,ECe<4,Ca<3% Mg <10 me/100 g)	Required (R)	Required (R)	Required (R)	Required (R)
2	Better drainage (Redox Pot. >6MV)	"	"	"	"
3	Moderate soil moisture (0.075-0.25MPa)	"	"	"	"
4	Leaf sheath NPK (%) (2.0, 0.12, 2.3)	"	"	"	"
5	Juice (N/P ratio <1.6, P_2O_5 300mg/l)	NR	NR	NR	R
6	Juice pH (4.5 – 5.6)	NK	NK	NK	R
7	Low cane fibre (<13.5%)	NR	"	R	NK
8	Low non-sucrose (< 3%)	NK	"	"	R
9	High sucrose synthetase activity	R	"	R	NR
10	Low Acid/Neutral Invertase activity ratio	NK	R	R	R
11	High K/Na ratio in leaf (>15)	R	R	R	NK
12	High K/Si ratio in leaf (1.22-1.66)	R	NR	NR	NK
13	Moderate temp. & light intensity at ripening ($18°-30°C$, 5.0-6.o K Cal $10^3/m^2/d$)	R	R	R	R
14	Ripener spray (Glyphosate, ethrel etc.)	NR	R	R	NR
15	Use of preservatives (Potassium silicate, EDTA, $KMnO_4$ etc.) on stalk	NR	NR	NR	R
16	Higher Foliar Phosphorylase /Amylase Ratio (>0.75)	R	NK	R	NR
17	Juice extraction within 24 hours	NR	NR	NR	R

R-Required; NK-Not known; NR-Not required

Figure 11.8: Metabolic Paradigm and Quest for Improving Cane Yield and Sucrose Content in Sugarcane (Dwivedi, 2003).

6. A Holistic Plant Model for Higher Productivity

Dwivedi *et al.* (1994) worked on different plant attributes associated with enhancing and stabilizing the following physiological components under normal and stress environments, and proposed a holistic model for efficient light harvest and high cane and sucrose yield (Figure 11.9).

1. Increased light absorption and utilization
2. High CO_2 assimilation rate and improved partitioning

3. Increased sucrose or total fermentable sugar synthesis and accumulation.

4. Efficient use efficiency of inputs (water, nutrient etc.)

The source and sink relationship determine the yield of economically viable components like cane, sugar and alcohol production but for higher biomass production in general, the source (LAI, LAD, plastochron and canopy orientation etc.) play important role than the sink. The effective and balanced functioning of source and sink, of course, genetically regulated but proper expression occur when sufficient inputs and suitable environment conditions are effectively available.

The ideotypes, which maximize internal CO_2 will have increased productivity. Internal CO_2 concentration can be increased by increasing stomatal conductance to CO_2 (Venkataramana *et al.*, 1993) and decreasing boundary layer resistance to CO_2 flow by decreasing leaf width. A genotype, which has high decaboxylation potential in bundle sheath will not only supply pyruvate to mesophyll cells for fast CO_2 fixation but enrich bundle sheath with CO_2 for further augmentation in CO_2 assimilation and sucrose synthesis. Secondly, the high CO_2 in bundle sheath will nullify photo respiration because of very high competition with O_2. NADPH malic enzymes, PEP carboxylase, RUBP carboxylase, sucrose synthetase, sucrose phosphate synthatase therefore, be highly active.

Only the PAR (400-700 nm), which accounts for 45 per cent of total radiation falling on earth, is used by plants. The production of one electron for each quantum of ligh absorbed seems to be a maximum efficiency for crop plants including sugarcane. This efficiency is lowered down under higher light intensity because of photosynthate accumulation in the leaf exerting a feedback inhibition in CO_2 assimilation. This inhibition can be removed if a plant has more efficient : (1) transport from the leaf, (2) phloem translocation, (3) storage of sucrose in parenchyma. Under this situation the manipulation for raising light absorption for further enhancing CO_2 assimilation could be done by improving leaf area index and canopy architect. LAI in a plant type can be increased by aggregation of components such as large leaves, slowly senescing leaves and leaf sheath, more rapid leaf production and early and synchronized tillering. The canopy orientation is, therefore, very important for higher CO_2 assimilation and energy harvesting efficiency (Dwivedi, 1994; 2000).

Figure 11.9 explains the genetic regulation of different components/factors/ processes associated with high cane yield and sugar yield. The plant attributes associated with different frame work cannot make dent on productivity in isolated manner. The congruent effect is the only solution for impact.

The crop productivity is the significant product of multitude of lower level process and their interaction (Figure 11.9). The agronomist is not concerned with finding out plant attributes for raising productivity under a set of environment. However, the plant breeder is concerned with various level of knowledge and plant attributes leading to higher yield. A plant physiologist is/should be concerned with environment plant interaction with special reference to crop productivity. This is holistic approach because different factors *e.g.*, favorable and unfavorable, are interacting with growth and development processes. Hence, yield potential evaluation of genotype suited to a specific environment under normal conditions and vice-versa is not a holistic approach (Naidu, 1987; Singh, 1987; Dwivedi *et al.*, 1994) because it does not last long.

The hierarchical nature of biological processes are exemplified by magnitude of 10^{-27} m^{-3} and 10^{-15} sec. primary biochemical act to 10^5 m^3 and 10^6 sec. for primary productivity of field crop. The gene number controlling characters at different levels thus range from 2^0 to 2^7 (Figure 11.9). In fact, it is easy to work out a significant character related with productivity at lower level where less number of genes are involved but their impact under natural environment cannot be always expressed, achieved and noted. This can be exemplified with the findings of Parthasarathy (1966) and Chiranjivi Rao and Lalitha (1971) who, examined nitrate reductase activity and nitrate nitrogen in the stem tissues of 5-8 months old plants of 4 high yielding and 5 low yielding genotypes; recorded significant correlation between potential nitrate N and yield and suggested the usefulness of this character in selecting poor

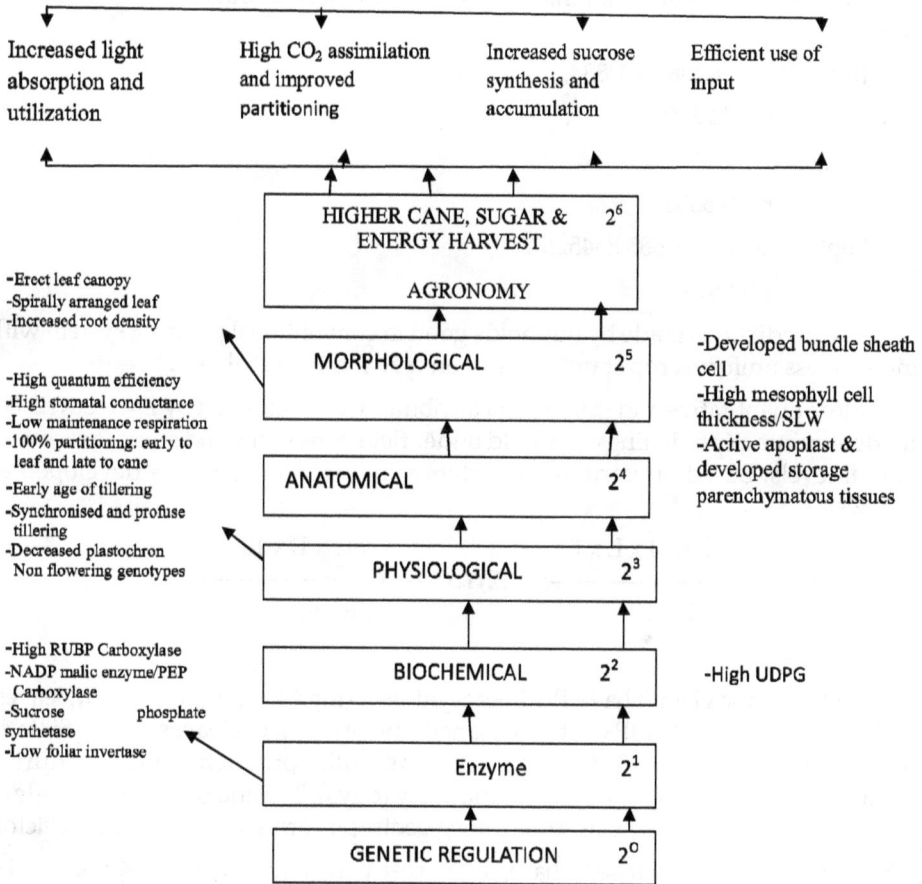

Figure 11.9: Minimum Number of Gene Contributing each Character
(Dwivedi *et al.* 1994, Dwivedi, 2003).

and high yielding genotype. The holistic approach is concerned with development of an understanding of crop yield through analysis of whole process and not single factor. This will supplement the knowledge gained at lower hierarchical level. It is therefore, suggested to selection those yield attributes, which are stable and distinct in expression under natural environment and controlled by preferable less number of genes. The genetic physiology of yield (cane, sugar, alcohol, energy efficiency) regulation characters need special attention for investigation with a view to have stability in wider range of environment and time (Dwivedi, 1999, 2000; Chiranjivi Rao, 1973).

7. Determining Genetic Potential of Sugarcane for Cane Yield at Field Level and Sucrose Recovery at Mill Level

Singh *et al.* (1984) followed by Dwivedi and Srivastava (1993) reported that the fresh weight of 3rd lamina was most suitable attribute to predict crop yield at different growth stages. The correlation coefficient and regression equation (for each sampling occasion) of fresh wt of 3rd lamina (x) in 3 different months and final cane yield (Ye) were as follows:

July – Ye = 0.4708 x 41.834

    r = 0.2209;

August Ye = 1.941 x 54.969

    r = 0.8533;

September Ye = 1.685 x -45.272

    r 0.6818

The predictions made by this holds good in conventional planting system with more or less uniform crop stand. For ratoon crop it is not reliable.

More quantitative and stable plant attributes were examined by Dwivedi (2000) for determining/predicting cane yield under field conditions and sucrose recovery at mill levels. Based on various characters a holistic equation was developed as follows:

$$(1)\ YC\ \frac{P \times E \times Si \times L \times t}{T \times 10}; \qquad (2)YS = \frac{S \times C \times E \times 10}{I \times F \times st \times 100}$$

where,

(1) YC = cane yield (t ha^{-1}), P = Photosynthesis (mg CO_2 dm^{-2} d^{-1}) of LTM leaves, E=PEP Carboxylase/RuBP Carboxylase activity ratio in LTM leaves, Si= K/Si Ratio in LTM leaves, L=Ratio of Plastochron rate m^{-2}/tiller production rate m^{-2} during formative phase, t=crop duration till maturity (days), T- Ratio of standing angle of intermodal and nodal vascular strand along geotropic forces, and 10=constant factor.

(2) YS = Yield of sucrose (t/ha^{-1}), S= Sucrose percentage in juice, C-Cane yield (t ha^{-1}), E= Sucrose Phosphate synthetase/Sucrose synthetase activity ratio in LTM leaves or cane, F=Fiber percentage in cane; I=Neutral invertase/acid invertase activity ratio in LTM leaves, st=Fresh cane/stale cane weight ratio at time of crushing

The data on various aspects based on field performance of sugarcane are presented in Table 11.14. Using aforesaid equations, achievable cane yield at field and sucrose recovery at mill levels could be achieved. It may be mentioned here that, this equation has been developed for considering plant, environment and metabolic processes. Hence, the crop with same rate of photosynthesis cannot give equal cane yield in tropics and sub-tropic. Similarly, in 100 tonnes cane with equal amount of sucrose in juice, or cane in tropics and sub-tropics cannot accrue equal amount of sucrose recovery.

Table 11.14: Example of Plant Attributes Determining Holistic Genetic Potential for Cane Yield and Sucrose Recovery in Tropical and Subtropical Parts of India.

Sl.No.	Attributes	Tropics	Subtropics
	Cane Yield		
1.	Photosynthetic rate (mg CO_2 dm^{-2} hr^{-1})	24-36	22-38
2.	PEP carboxylase/RuBP carboxylase ratio	1.8-3.2	1.4-2.8
3.	K/Si ratio	1.2-2.8	1.2-2.4
4.	Plastochron/tillering ratio	6-12	4-13
5.	Inter nodal/Nodal vascular strand angle ratio	1.12-3.15	1.10-3.20
6.	Crop duration (days)	300-540	300-360
	Sucrose recovery		
1.	Cane yield (t ha^{-1}) presumed	100	100
2.	Sucrose per cent in juice	16-22	15-20
3.	Sucrose Phosphate synthetase/sucrose synthetase ratio	1.6-2.11	1.4-2.22
4.	Fibre per cent in cane	12-18	14-20
5.	Neutral invertase/acid invertase ratio	1.2-2.1	0.8-3.1
6.	Fresh cane/stale cane wt. ratio	1.0-1.5	1.0-1.6

I: Neutral invertase/acid invertase activity ratio in LTM leaves.

T: Fresh cane/stale cane weight ratio at time of crushing.

Source: Dwivedi (2003).

The yield calculated/predicted by this equation has correlation coefficient value of $r = 0.892$ with observed cane yield and that of $r = 0.865$ with observed sucrose recovery.

At present the studies are required to develop a genetically engineered isogenic super plant to excel all previous records of yield.

7.1. Modulating the Expression of Sucrose Metabolizing Enzymes for High Sucrose Accumulation in Sugarcane

Improving sucrose content in sugarcane stalk, foliar application of enzyme effectors; divalent cations (Mn^{++}, Mg^{++} and plant growth regulators (GA and ethrel) was performed before the onset of cane ripening in sugarcane cultivar BO91. Results indicated increased sucrose content, CCS, S/R ratio and reduced content of reducing sugars in low sugar genotype by foliar application of enzyme effectors. Specific activity of SAI decreased while SPS and SS increased by the Application of Mg^{++} and Mn^{++}. Transcript expression levels were studied to verify changes in SAI, SPS and SS enzyme activity due to enzyme effectors. An increasing trend of SS and decreasing trend of SAI was observed due to Mn^{++} and Mg Mg^{++} treatment in cultivar BO 91. SS and SAI transcript levels and enzymes activities were directly correlated with each other. Unlike SPS activity, SPS transcript level was not changed due to chemical treatment (Jain *et al.*, 2013).

7.2. Development and Utility of Gene–Tagged Molecular Markers

Sugarcane is an important cash crop, providing 70 per cent of the global raw sugar as well as raw material for biofuel production. Genetic analysis is hindered in sugarcane because of its large and complex polyploidy genome and lack of sufficiently informative gene-tagged markers. Modern genomics has produced large amount of ESTs, which can be exploited to develop molecular markers based on comparative analysis with EST(Expressed sequence Tag) and about 6,040,000 results (0.24 seconds) datasets of related crops and whole rice genome sequence, and accentuate their cross-technical functionality in complex polyploidy crops and also orphan crops like tropical grasses (Chandra *et al.*, 2013a).

Utilising 246,180 *Saccharum officinarum* EST sequences vis-à-vis its comparative analysis with ESTs of sorghum and barley and the whole rice genome sequence, we have developed 3425 novel gene-tagged markers —namely, conserved-intron scanning primers (CISP) — using the web program GeM prospector. Rice orthologue annotation results indicated homology of 1096 sequences with expressed proteins, 491 with hypothetical proteins. The remaining 1838 were miscellaneous in nature. A total of 367 primer-pairs were tested in diverse panel of samples. The data indicate amplification of 41 per cent polymorphic bands leading to 0.52 PIC (polymorphism information content), and 3.50 MI (marker index) with a set of sugarcane varieties and *Saccharum* species. In addition, a moderate technical functionality of a set of such markers with orphan tropical grasses (22 per cent) and fodder cum cereal oat (33 per cent) is observed. Developed gene-tagged CISP markers exhibited considerable technical functionality with varieties of sugarcane and unexplored species of tropical grasses. These markers would thus be particularly useful in identifying the economical traits in sugarcane and developing conservation strategies for orphan tropical grasses (Chandra *et al.*, 2013a).

Strategies to increase sucrose concentration in sugarcane via transgenic manipulation require a broader understanding of the processes involved in sucrose accumulation, including the possibility that culm sucrose accumulation may be regulated by sink (including storage) demand (Watt *et al.*, 2005). Most efforts to manipulate sugarcane sucrose concentration by transgenesis have targeted single genes encoding putative rate-limiting sucrolytic enzymes in the culm. These include soluble acid invertase (Botha *et al.*, 2001), neutral invertase (Rossouw *et al.*, 2007), pyrophosphate-dependent phosphofructokinase (Groenewald and Botha, 2008) and a yeast invertase carrying various leader sequences for targeting to the apoplast, cytosol or vacuole compartments (Hongmei, *et al.*, 2000). Despite successful transformation, transgenic plants did not result in a significant increase in overall culm sucrose accumulation (Groenewald and Botha, 2008). It is suggested that attempts to increase sucrose content of sugarcane by the transgenic manipulation of sucrose metabolism enzymes, whether singly or in tandem, must take cognizance of regulatory feedback mechanisms known to exist in other plants (Paul 2007). This has, of late, been proposed by McCormick *et al.* (2006, 2009) in sugarcane. To date, attempts to increase sucrose content in the sugarcane culm through the modification of carbon flux and partitioning have ignored this potential regulatory feedback between the culm (sink) and the leaf (source) (Chandra *et al.*, 2011). Papini-Terzi *et al.* (2009) have reported many sugarcane genes associated with sucrose content and also indicated overlap of these genes with drought and cell-wall metabolism processes. To classify the role of these genes as well as to define targets useful for sugarcane improvement, transgenic research is needed. By utilizing sucrose isomerase and proline synthase genes in sugarcane, transgenics for increased sugar concentration and water stress tolerance respectively, have been reported (Wu and Birch, 2007, Molinari *et al.*, 2007). Under these cases increased sugar concentration due to additional accumulation of isomaltulose, a high-value sugar *vis-à-vis* increased photosynthesis, sucrose transport and sink strength has been observed. Similarly, higher biomass yields after 12 days of withholding water was observed along with tolerance to water stress. Controlling the level of invertases at suitable locations is the need of hour, which can be initiated utilizing the RNAi (RNA interference) approach. The reduction of invertase activity soon after harvest of sugarcane crop could be useful in minimizing the post-harvest sucrose losses, and it can be initiated by implicating the RNAi approach. The inversion of sucrose into glucose and fructose is the major problem in stale cane, which leads to significant loss of sucrose. Another problem which comes across after crushing the stale cane is the production of dextran. To minimize this, dextranase is recommended; however, generating transgenic plant using appropriate dextranase gene with appropriate promoter will restrict the production of dextran (Chandra *et al.*, 2012).

8. Future Thrust

(i) Real Time PCR Analysis of Gene(s) of Sucrose Metabolism

Increased accumulation of sugar in sugarcane stalk is a major researchable area. Due to the complexity of genome, major headway has not been made in this direction as far as isolation and characterization of genes associated with sucrose synthesis, transport and accumulation are concerned. Synthesis and cleavage of sucrose before its final storage in vacuole is controlled by many enzymes, transporters and most importantly the signal molecules which direct the source-sink communication. Studies in sugarcane have shown that significant effects are exerted on the sucrose cycle by the enzymes, sucrose phosphate synthase (SPS), sucrose synthase (SS) and invertases which are collectively responsible for the synthesis and breakdown of sucrose in the various cellular compartments.

Although sucrose is synthesized only in the cytosol by SPS or SS enzymes, its distribution to various degrees between the apoplast, the cytosol and the vacuole of the storage parenchyma are monitored and controlled by invertases (âfructofuranosidase) supports the hypothesis that the primary function of invertases is to supply carbohydrates to the sink tissues in general and act as a key regulator for sucrose accumulation in sugarcane stem parenchyma (culm/stalk) in particular. Source-sink perturbation study selectively enhances the photosynthetic activity of leaves (source) depicting availability of active source, hence strength of sink is required to be enhanced to store more sugars in cells.

For better understanding sink activity, the expression behavior analysis of gene(s)/enzymes responsible for the synthesis and breakdown of sucrose is important especially in varieties having ability to accumulate differing levels of sucrose. Associations of gene expression with biological traits have been based on alterations in the timing and intensity of gene expression with various treatments including nature of genotypes and developmental stages. In this direction expression of genes associated with synthesis and cleavage of sucrose with respect to stage in low and high sucrose bearing varieties as well as expression behavior of genes namely CWI, SAI, SPS and SS through real time PCR and their association with low and high sucrose bearing varieties were carried out (Chandra *et al.*, 2013a). Real time PCR gene expression revealed differential expression of SPS, SS, SAI and CWI in two sets of sugarcane genotypes differing in sucrose accumulation and in general expression of sucrose synthesizing genes was higher and for longer time in high sugar accumulating variety than those of low sugar variety. Study also indicated their differential behavior in different portion of cane accumulating the sugar (Chandra *et al.*, 2013a).

(ii) Source-Sink Signal Dynamics to Improve Sucrose Content

In view of complexity of the sucrose metabolism at physiological, biochemical and molecular levels, the most important issues that need to be addressed are : identifying transcription factors regulating sucrose accumulation of sucrose; feed back inhibition studies so as to underpin the modulation of source-sink relationships; bio-physiological changes along with leaf gene expression patterns; in case if sink strength controls the source activity, how the signal is being transported and the potential signal molecules, besides hexoses (so far reported) and how genes of other physio-biochemical responses like drought, ABA signaling and genes belonging to lignin biosynthesis, cell wall metabolism and aquaporins exert regulatory control over sucrose accumulation (Figure 11.10) (Chandra *et al.*, 2011).

(iii) Obviation of Accumulation of Impurities in Cane

How to obviate the accumulation of impurities in cane under field condition is a challenging problem because it contributes significantly higher content in total solid of juice and thereby results significantly higher yield of molasses and low sucrose recovery at mill level (Dwivedi and Srivastava, 1993 and Dwivedi, 2003). Use of gypsum in alkaline soil has raised cane yield but proportional improvement in impurities declined *vis-a vis* sucrose recovery has not been recorded (Dwivedi, 1993 and 1993a). Hence it requires great attention on normal, drought and flood prone and salt affected soil.

(iv) Holistic Approach

Cane yield, sucrose recovery and environment stress tolerance control sucrose

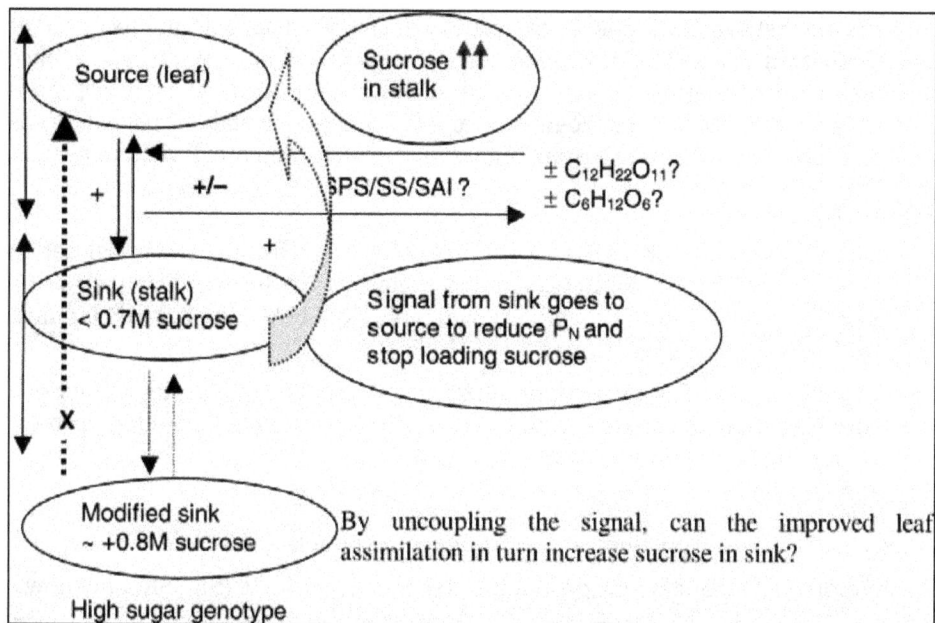

Figure 11.10: Source–Sink Communication Plays an Important Role in Accumulation of Sucrose in Sugarcane Culms.
Source: Chandra *et al.* (2011).

recovery, ethanol production and sucrose yield. Efforts on synchronization of genetic *vis-a vis* molecular approaches to modulate genes, tested markers, enzymes, activators, suppressor etc for improving photosynthesis, translocation and accumulation in sink, reduce photo respiration, sink strength and higher cane yield in combination with abiotic stress tolerance need to be intensified since at present some information are available.

9. Summary

The substantial jump in cane yield to an extent of 50 t ha^{-1} in subtropical and 90 t ha^{-1} in tropical zones of India was recorded in 1950's as against that of earlier corresponding productivity of 15 t ha^{-1} and 35 t ha^{-1} respectively (Devid, 1987). The maximum rise in cane yield in 1960's was found in the range of 40-48 per cent, which has not increased further so far, indicating thereby prevalence of yield plateau since last 3.75 decades. Considering 335 μmol CO$_2$ cm^{-2} d^{-1} assimilation rate and quantum efficiency of 0.125 mol CO$_2$ quanta^{-1}, solar energy (PAR) harvesting efficiency of 8.5 per cent and net dry matter production of 118 g m^{-2} d^{-1} in sugarcane, the potential (theoretical) dry biomass and fresh cane yield have been estimated to be 470t ha^{-1}and 808.8t ha^{-1}, respectively. Similarly, based on 0.125 mol quantum efficiency of crop and assuming 26 per cent sucrose in the juice of 808.8 t ha^{-1} cane, the sucrose yield of 78 t ha^{-1} and 125.5 t ha^{-1}, respectively has been estimated. An enormous scope to raise the productivity of sugarcane therefore, prevails.

The highest yield of 335-464 t ha^{-1} cane and 33.5-50.6 t ha^{-1} sucrose recorded at best Indian farmers field has not been achieved even by progressive farmers. Hence the essentialities for abiotic stress and staling deterioration tolerance in crop have been suggested. A holistic genetic-molecular- physiological- metabolic paradigm has therefore, been modulated to sustain higher productivity and break yield plateaus.

10. References

Agrawal, M., Dhawan, A.K. Dendsay, J.P. S. (1997). Carbon fixation and translocation of vein loaded ^{14}C sucrose in relation to varietal leaf age. In: (eds. Dhawan, et al.). Sucrose synthesis and Recovery in sugarcane, CCS, HAU, RS, Karnal Haryana. pp. 15-19.

Ali, S.A. (1997). Germination and mobilization of carbohydrate in sugarcane setts during germination phase as influenced by different pre-planting sett treatment. Indian Sugar (Sept.) pp. 27-432.

Annonymous (1998). Coop. sugar. 29(9) May; p. 42.

Anonymous (1969). Ann. Rep. Hawaiin Sugarcane Pl. Assoc. Hawaii, U.S.A.

Anonymous (1991). Sugarcane production Technology, AICRP (S), IISR, Lucknow.

Anonymous (1997). Vision 2020, IISR Perspective Plan, IISR (ICAR), Lucknow, p.57.

Anonymous (2011). Vision 2030, IISR Perspective Plan, IISR (ICAR), Lucknow, p. 28.

Balusamy, M., Enyathuyllah Shah, S. and Choklingam. S. (1990). Bhartiya Sugar (July), pp. 36-38.

Batta, S. K. and Singh, R. (1991). Post harvest deterioration in quality of sugarcane. Bharatiya Sugar, 32: 49–51.

Behl, K.L. and H. Singh (1961). Indian Sugar, 11(8): 565-578.

Botha, F. C., Sawyer, B. J. B. and Birch, R. G. (2001). Sucrose metabolism in the culm of transgenic sugarcane with reduced soluble acid invertase activity. *Proc. Int. Soc. Sugar Cane Technol.*, 24: 588–591.

Bull, TA and Glasziou KT (1975). Sugarcane.In (ed. Evans LT) Crop Physiology: Some Case Histories Cambridge Univ.Press London pp. 51-72.

Chandra A., Jain Radha, and Solomon S. (2012). Complexity of invertases controlling sucrose accumulation and its retention: a way forward. *Current Science*, 102(5): 857-866.

Chandra A., Jain Radha, Rai RK and Solomon S. (2011). Revisiting the source –sink paradigm in sugarcane. *Current Science*, 100: 978-980.

Chandra A., R. Bannerjee, Jain Radha, Rai RK, P.Singh, AK Shrivastava, Priyanka Singh and Solomon S. (2013). Six decades of research on physiology and post harvest management of sugarcane. In Souvenir, Diamond Jubilee Celebrations, National Sugar Fest 2013, IISR, Lucknow, pp. 41-51.

Chandra, A, Radha Jain, S. Solomon, S. Shrivastava and Ajoy K Roy Exploiting EST databases for the development and characterisation of 3425 gene-tagged CISP markers in biofuel crop sugarcane and their transferability in cereals and orphan tropical grasses (2013a). BMC Research Notes 2013, 6: 47 doi:10.1186/1756-0500-6-47.

Chiranjivi Rao, K. (1973). Role of Biochemistry in sugarcane research. Kheti, 26(7): 13-15.

Chiranjivi Rao, K. (1989). Bharatiya Sugar. 15(2): 55-67.

Chiranjivi Rao, K. (1993). Management of Post Harvest Losses and over-stand in sugarcane. Proc. National Symposium on improvement sugarcane quality for increasing Sugar Productivity (Abstract), IISR, Lucknow, pp. A-9.

Chiranjivi Rao, K. and Lalitha (1971). Proc. Natl. Symp. On Soil Fert. Evalu. New Delhi, 1: 657-663.

Das, G. and Prabha, K. A. (1988). Intern. Sug. J. 90(1092): 169-71.

Dendsay, J.P.S. Sehtiya, H.L. and Dhawan A.K. (1997). Internodal invertase, stalk growth and maturity in sugarcane. In (eds. Dhawan et al.) Sucrose Synthesis and Recovery in Sugarcane CCS, HAU, RS, Karnal, Haryana, p. 6-14.

Dendsay, J.P.S., Singh, P., Dhawan, A.K. and Sehtiya, H.L. (1995). Sugarcane invertase, Sugarcane, pp. 17-19.

Devid, H. (1987). Research achievement of SBI (1912-1987) SBI, Coimbatore, p. 15.

Dutt,NL. (1943). Control of flowering in sugarcane. Indian Farming, 4(1). 11-13.

Dwivedi, R. (1994). Sugarcane- a hardy crop and an ideal energy crop to obviate greenhouse effect. Agriculture Instrument International (World Crop), 45(11 & 12), 102-106.

Dwivedi, R.S. (1993). Sugarcane Management on Salt Affected Soils (DST. TIFAC), IISR, Lucknow, pp. 82.

Dwivedi, R.S. (1993a). Stress management for higher sucrose recovery in sugarcane. Proc. Nat. Symp. Improvement in sugarcane quality for increasing sugar production. IISR, Lucknow, p. A7-A8.

Dwivedi, R.S. (1995). Constraints in cane productivity in: Proc. Nat. Sem. on I. Sugarcane Production constraints & II. Strategies for Research and Management of Red Rot (eds. G.B. Singh and O.K. Sinha) IISR, Lucknow, p. 88.

Dwivedi, R.S. (1999). Role of potassium as an organometallo- osmoticum in raising abiotic stress tolerance and crop yield. In (eds. Tiwari KN and Modgal SC) Use of K in UP Agriculture, UPCAR, Lucknow & PPIC, Gurgaon India, pp. 120-129.

Dwivedi, R.S. (2000). Physiology of Sugarcane In: (eds Shahi HN, OK Sinha and Shrivastava AK) 50 years of Sugarcane Research in India, IIS, Lucknow, pp. 73-110.

Dwivedi, R.S. (2003). Advances in Plant Physiology (ed. Hemantarajan) Scientific Publishers India, Jodhpur (PB 91), 5: 435-459.

Dwivedi, R.S. and GB Singh (1991).). Zinc nutrition of sugarcane. VP Zinc Sulphate manufacturers association, Lucknow, pp. 63.

Dwivedi, R.S. and K.K. Srivastava (1993). A Scenario of Research on Physiology and Biochemistry of Sugarcane. In: Sugarcane Research and Development in Subtropics, (eds. G.B. Singh and O.K. Sinha) IISR, Lucknow, pp. 143-190.

Dwivedi, R.S., Srivastava, K.K., Solomon, S. and Singh, K. (1988). Rapid test for drought resistance. Sugarcane (Suppl.), pp. 31-32.

Dwivedi, R.S. K.K. Srivastava, Meena Nigam and M. Ram (1992). *Proc. Intern. Soc. Sugar Cane Technol.* Bangkok, CCCXV-XXVI.

Dwivedi, R.S., S. Solomon and K.K. Srivastava (1994). Harnessing solar energy for enhancing biomass and sucrose productivity in sugarcane. J. Agro-tech. and Bioenergy, 1: 76-82.

Eggleston, G. and Legendre, B. L. (2003). Mannital and oligosaccharides as new criteria for determining cold tolerance in sugarcane varieties. Food Chem., 80: 451–461.

Glasziou, K.T. (1962). Accumulation and Transformation of Sugars in Sugar Cane Stalks : Mechanism of Inversion of Sucrose in the Inner Space. *Nature* 193, 1100 (17 March 1962); doi:10.1038/1931100a0.

Groenewald, J. H. and Botha, F. C. (2008). Down-regulation of pyrophosphate: fructose 6-phosphate 1-phosphotransferase (PFP) activity in sugarcane enhances sucrose accumulation in immature internodes. *Transgenic Res.*, 17: 85–92.

Gupta, A.P. and N. Nigam, (1982). Maharashtra Sugar, 7(3): 51-64.

Hatch, M. D. and Glasziou, K. T. (1963). Sugar accumulation cycle in sugarcane. II. Relationship of invertase activity to sugar content and growth rate in storage tissue of plants grown in controlled environments. *Plant Physiol.*, 38: 344–348.

Hongmei, M., Henrik, H. A., Robert, P. and Paul, H. M. (2000).Metabolic engineering of invertase activities in different subcellular compartments affects sucrose accumulation in sugarcane cells. *Aust. J. Plant Physiol.*, 27: 1021–1030.

Jain, R., S. Solomon. A.K. Shrivastava. A. Chandra (2010). Sugarcane Bud Chips: A promising seed material. *Sugar Tech* 12(2): 67-69.

Jain, R., A. Chandra and S. Solomon (2013). Impact of exogenously applied enzymes effectors on sucrose metabolizing enzymes (SPS, SS and SAI) and sucrose content in sugarcane. Sugar Tech, 2013, DOI 10.1007/s12355-013-0211-3.

Kakede, JR (1985). Sugarcane production. Metropolitan Book Co. New Delhi pp. 325.

Kaur, S., Batta, S. K., Sital, J. S., Sharma, P. and Mann, P. S. (2002). Partial purification and properties of soluble invertase isoforms from sugarcane storage tissue. *Phytochemistry*, 16: 443–445.

Loomis R.S. and William, W.A. (1963). Maximum crop productivity: An estimate. *Crop Science*, 3: 67-72.

Madan, V.K., Singh, K., Shivapuri, S., Pande, H.P. and Y.R. Saxena (1980). *Inter. Sugar J.*, 82: 55.

Mahbubur-Rahman, S. M. M., Sen, P. K., Hasan, M. F., Mian, M. A. S. and Habibur-Rehman, M. (2004). Purification and characterization of invertase enzyme from sugarcane. *Pak. J. Biol. Sci.*, 7: 340–345.

McCormick, A. J., Cramer, M. D. and Watt, D. A. (2006). Sink strength regulates photosynthesis in sugarcane. *New Phytol.*, 171: 759–770.

McCormick, A. J., Watt, D. A. and Cramer, M. D. (2009). Supply and demand: sink regulation of sugar accumulation in sugarcane. *J. Exp. Bot.*, 60: 357–364.

Molinari, H. B. C. *et al.*(2007). Evaluation of the stress-inducible production of proline in transgenic sugarcane (*Saccharum* spp.): osmotic adjustment, chlorophyll fluorescence and oxidative stress. *Physiol. Plant*, 130: 218–229.

Moore HP (1987). Physiological basis for varietal improvement in sugarcane.In: Sugarcane Varietal Improvement (Eds. Naidu, KH, Srinivasan, TV and Premchandran MN) SBI, Coimabatore, pp. 19-56.

Naidu KM (1987). Potential yield in sugarcane and its realization through varietal improvement pp. 1-18.

Naidu MV, Rajabapa, Rao and Dora KB (1997). Sucrose synthesis pattern in different maturity group of sugarcane. In: (ed. Dhawan et al) Sucrose synthesis and recovery in Sugarcane, CCS, HAN, Karnal, Haryana.

Panje RR, Gill PS and Singh B. (1971). Germination behavior of sugarcane setts. Ind. J. Agri. Sci., 41: 431-440.

Panje, R.R. T.R. Rao and K.K. Srivastava. (1968). Proc. Intern. Soc. Sugar Cane Technol. (Taiwan): pp. 476-483.

Papini-Terzi, F. S. *et al.* (2009). Sugarcane genes associated with sucrose content. *BMC Genomics*, 10: 120.

Parthasarathy K (1966). Contribution to a biochemical study of sugarcane. III. Nitrate reductase activity and potential nitrate and their use in determining the relation yield capacity of cane varieties. Proc. Ind. Academy Sci., 64B(2): 91-95.

Paul, M. J.(2007). Trehalose 6-phosphate. *Curr. Opin. Plant Biol.*, 10: 303–309.

Rizk, T. Y. and Normand, C. W. (1969). Effect of burning and storage of cane deterioration (part I, II, III). Int. Sugar J., 71: 7–9.

Rossouw, D., Bosch, S., Kossmann, J., Botha, F. C. and Groenewald, J. H. (2007). Down-regulation of neutral invertase activity in transgenic sugarcane cell suspension cultures leads to a reduction in respiration and growth and an increase in sucrose accumulation. *Funct. Plant Biol.*, 34: 490–498.

Sachdeva, M., Mann, A. P. S. and Batta, S. K. (2003). Multiple forms of soluble invertases in sugarcane juice: kinetic and thermodynamic analysis. *Sugar Tech.*, 5: 31–35.

Sehtiya, H.L. Dendsay, J.P.S. and A.K. Dhawan (1991). *J. Agric. Sci.* (Cambridge), 116: 239-243.

Shrivastava, A.K., Saxena Y.R. and Singh, K. (1985). Production physiology of sugarcane cultivar Co1148. Tech. Bull. No.15. Indian Institute of Sugarcane Research, Lucknow, pp. 128.

Shrivastava, AK., S. Solomon, A. Chandra, R. Jain, SP Shukla and Anita Sawnani (2011). Reducing the time for the establishment of the crop stand in sugarcane in subtropical India. Proc. 4th IAPSIT Inter. Nat. Sugar Con. IS-2011, Nov, 21-25, New Delhi, pp. 120-122.

Singh and Singh (1964). Hormones and sugarcane. II. effect of IAA on growth and yield and quality of sugarcane. In: Proc. Fifth All India Conf. on Sugarcane Research and Development Workers held at IARI, New Delhi.

Singh H and Behl, KL (1961). Some studies on late harvesting of cane vs. harvesting of left over crop during the next crushing season. Indian Sugar, 11(4). 271-274; 287.

Singh U.S. and Singh L. (1967). Enhancing cane yield under late planting. Indian Sugar, 17: 475-81.

Singh, G.B. and R.S. Dwivedi (1995). Constraints in increasing sucrose productivity and strategies for improvement. Lead paper Proc. National Seminar on Sugarcane Production Constraints. I.I.S.R., Lucknow, pp. 59-77.

Singh, K., K.K. Srivastava and R.K. Shukla (1984). Sugarcane leaf weight as predetermining index of yield. *India J. Agric. Sci.*, 54: 390-394.

Singh, P. and Solomon, S. (2011). Effect of spraying electrolysed water on post harvest quality of sugarcane at high ambient temperature. In: Proceedings of the 10th Joint Convention of STAI and DSTA, 2011, pp. 43–49.

Singh, R.G., Khan, H.A. Lal, K and A. Bhargara, (1982). Effect of cane ripeners on post maturity deterioration of sugarcane. 46th Proc. Annl. Conv. S.T.A.I., 46: 4-5.

Singh, S. (1987). Physiological basis for varietal improvement under stress environment in sugarcane. In Sugarcane Breeding Under Stress Environment (eds. K.K. Naidu *et al.*) SBI, Coimbatore, India, pp. 57-82.

Solomon, S. (2009). Post harvest deterioration in sugarcane. Sugar Tech, 11: 109–123.

Solomon, S., K.K. Srivastava and K. Singh (1988). Indian Sugar, 36: 583-587.

Solomon, S. and A. Kumar. (1983). Proc. of 47th Conf. of Sugar Technol. Association, India, pp. 117-120.

Solomon, S. and A. Kumar. (1987). Indian J. Sugarcane Technol., 4: 11-16.

Solomon, S., A.K. Shrivastava, B.L. Srivastava and V.K. Madan (1997). Pre-milling Sugar Losses and Their Management in Sugarcane, Tech. Bull No. 37 IISR, Lucknow India, p. 227.

Srinivasan TR (1987). Varietal response to climate –population dynamics, nutrition and other inputs. In: (eds. Naidu KM *et al.*) Sugarcane Varietal Improvement. SBI Coimbatore, pp. 195-220.

Srivastava BL and Kapur R. (1996). Late-harvesting of sugarcane varieties and sugar recovery. Sugar Crops Newsl., 6(2): 14-16.

Srivastava KK and Dwivedi RS (1993). Source and sink relationship of sucrose accumulation and solar energy efficiency in sugarcane. Annual Report, IISR, Lucknow pp. 46.

Srivastava, K.K., Narsimhnan and R.K. Shaklee (1981). Indian Farming, 31: 15-17.

Tandon, R.K. Kapoor GP and Misar GN (1955). Indian J. Agric. Sci.., 25: 31-40.

Venkataramana, S. and Ramanujam, T. (1994). Enzymology of sucrose accumulation. Ann. Rep. SBI Coimbatore, pp. 39.

Venkataramana S, Naidu, KM and Ramanujam, T. (1992). Drought resistance in sugarcane. Ann. Rep. SBI Coimbatore, pp. 36-37.

Venkataramana, S, Naidu, KM and Ramanujam, T. (1993). Source and sink balance in relation to photosynthate production and sucrose accumulation. Ann. Rep. SBI Coimbatore, pp. 38-39.

Vorster, D. J. and Botha, F. C. (1998). Partial purification and characterization of sugarcane neutral invertase. *Phytochemistry*, 49: 651–655.

Watt, D. A., McCormick, A. J., Govender, C., Carson, D. L., Cramer, M. D., Huckett, B. I. and Botha, F. C. (2005). Increasing the utility of genomics in unravelling sucrose accumulation. *Field Crops Res.*, 92: 149–158.

Wu, L. and Birch, R. G. (2007). Doubled sugar content in sugarcane plants modified to produce a sucrose isomer. *Plant Biotechnol. J.*, 5: 109–117.

Index

www.ingramcontent.com/pod-product-compliance
Lightning Source LLC
Chambersburg PA
CBHW050506190326
41458CB00005B/1449